Heating Services in Buildings

Heating Services in Buildings

Design, Installation, Commissioning & Maintenance

David E. Watkins

I Eng, FCIPHE, FSoPHE, MASHRAE, AffCIBSE, MIfL RP

A John Wiley & Sons, Ltd., Publication

This edition first published 2011 © 2011 by John Wiley & Sons.

Wiley-Blackwell is an imprint of John Wiley & Sons, formed by the merger of Wiley's global Scientific, Technical and Medical business with Blackwell Publishing.

Registered office: John Wiley & Sons, Ltd, The Atrium, Southern Gate, Chichester, West Sussex, PO19 8SQ, UK

Editorial offices: 9600 Garsington Road, Oxford, OX4 2DQ, UK
The Atrium, Southern Gate, Chichester, West Sussex, PO19 8SQ, UK
2121 State Avenue, Ames, Iowa 50014-8300, USA

For details of our global editorial offices, for customer services and for information about how to apply for permission to reuse the copyright material in this book please see our website at www.wiley.com/wiley-blackwell.

The right of the author to be identified as the author of this work has been asserted in accordance with the UK Copyright, Designs and Patents Act 1988.

Library of Congress Cataloging-in-Publication Data

Watkins, David E.
Heating services in buildings : design, installation, commissioning & maintenance / by David E Watkins.
     p.   cm.
Includes bibliographical references and index.
ISBN 978-0-4706-5603-7 (pbk. : alk. paper)   1. Heating.   I. Title.
     TH7121.W37 2011
     697–dc22
                          2010051098

A catalogue record for this book is available from the British Library.

This book is published in the following electronic formats: ePDF 9781119971658; ePub 9781119971665; Mobi 9781119971672

1   2011

Contents

Preface

There have been a number of books written on the subject of heating over the years, which would fill a sizable section of any notable library if collected together.

On examining the more recent of these books, that is those published over the last twenty years, it was found that they could be categorised as belonging to one of three groups. These are books written for the DIY market which are of little use to any student who is serious about studying to become a qualified heating professional. Alternatively, there are a number of books aimed at the craft level student concentrating only on the practical aspects of the subject. The third category of technical books, of which there are fewer available, has been written for the qualified professional engineer that assumes the student has previously obtained the basic engineering knowledge that is required to advance to a higher level of their education.

This observation becomes apparent when looking for a suitable technical book to support the NVQ Level 4 Higher Professional Diploma in Building Services Engineering and other design based engineering courses.

The search found that no single book was available to support these courses and the student would have to purchase a large number of publications to cover the subject to the extent required. This would also result in the student incurring a high financial cost to obtain copies of these publications.

The answer to this situation was to produce a number of supporting handout papers that expanded upon the course lectures that eventually developed over the years into a sizable set of notes when bound together.

During the course of developing these supporting notes, the subject of heating buildings, both for domestic residential properties and commercial buildings, has changed enormously, particularly with regard to the need to conserve energy, develop alternative forms of energy and provide controls that are suitable for the system's needs.

This requirement has manifested itself in the form of increased mandatory regulations and improved technology that has been developed to meet these compulsory regulations and conservation targets.

It was that necessity to incorporate explanations and detailed information on these changes that led to the set of supporting notes being developed into the basis of this book.

The aim of this work is to provide in a text and illustrative form a complete guide from basic principles to an advanced level to all the elements that combine to impart the engineering knowledge required on the subject of hydronic heating systems.

The book has been arranged to present the subject matter in a logical order that builds on each preceding chapter and culminates to provide the complete informative material. The book also demonstrates that there is little difference between domestic and commercial heating systems in the approach to the engineering and design of the systems, but makes mention where there is a difference.

This book has been developed over many years from the collection of handout notes to its present volume, where it originally supported a City & Guilds supplementary heating course, which further developed to support the heating design and installation course accredited by the European Registration Scheme (ERS) and other similar academic courses presently run today.

It is also intended that this volume will support Unit 11, 'Space heating technology and design', which is a module contained in the NVQ Level 4 Higher Professional Diploma in Building Services Engineering.

The book is aimed at both craft level plumbing students qualified to NVQ Level 3 standard aspiring to bridge the educational gap to an engineering career, plus school leavers with the necessary academic 'A' level qualifications and employed in a building services engineering consultancy.

Although this volume has been produced to support the NVQ Level 4 course and similar design/engineering courses, it is hoped that it will be of equal interest and use to anyone concerned with the design and installation of hydronic heating systems.

This book has resisted the inclusion of over explaining or illustrating elements in order to provide the information in an affordable manner to all those concerned. This gives the lecturer the opportunity to expand upon each subject and provide further examples in the classroom.

It is also correct to acknowledge that a work of this type has only been possible due to the encouragement and assistance of many other people, most notably Mr David Bantock, whose original set of notes I inherited when I started as a part-time lecturer delivering the course, and who has been instrumental in his encouragement during its development. Also my wife, Jenny Watkins, for proofreading and endless patience, and the many students who encouraged its eventual publication.

Special acknowledgement should also be mentioned for permission to reproduce Figure 5.23, Room Height Temperature Gradients, from Elsevier Publishing, which is based on a similar illustration in their book entitled Faber & Kell's Heating & Air-conditioning of Buildings. Also, for permission granted by Baxi Heating to reproduce Figure 15.8, Illustration of a Micro-Combined Heat and Power Generating Unit and M H Mear Co. Ltd for permission to reproduce Figure 7.3, of a Mear's Slide Rule Heating Calculator.

David E. Watkins

1

Introduction to Heating Services

The broad term 'central heating' is used to describe many types and forms of heating, and some usage is totally misleading and inaccurate, through ignorance of the subject. This chapter is a basic introduction to the mechanics of central heating, which is discussed in greater detail in the following chapters.

If we examine the term, it implies a system where heat is produced from a central source and distributed around the whole building. The method of heat generation and distribution may vary with the type of heating system employed.

Central heating is sometimes referred to as space heating. To be understood fully, this must be described by its type or system arrangement, and may be categorised as being either full, part or background heating.

Full central heating may be defined as being a system of heating from a central source where all the normally habitable or used rooms/spaces are heated to achieve guaranteed temperatures under certain conditions. By today's standards, all heating systems installed in residential dwellings and most commercial buildings should conform to this category, unless there are acceptable reasons for not doing so.

Partial central heating is the term applied where only part of the building is to be heated, but even then the rooms or spaces that are heated should still have guaranteed temperatures under stated conditions. This form of central heating would be a rare occurrence for a residential dwelling but not so uncommon for some commercial buildings, especially where part of the building complex is not normally occupied.

The term 'background heating' is used to describe a form of central heating whereby lower than normal or standard recommended temperatures are aimed at for the type of building involved. The term is sometimes used to refer to heating systems installed in buildings where the room temperatures are not guaranteed. This form of heating is unacceptable by today's standards on both environmental and efficiency grounds.

It should be noted that, unless otherwise specified, full central heating should normally be designed to current regulations and standards and installed in a professional manner. In some instances, usually due to a specific use or financial reasons, the client may only require or specify partial heating to be installed, sometimes with the request that safeguards are included to allow the system to be extended at a later date to achieve full central heating.

Heating Services in Buildings: Design, Installation, Commissioning & Maintenance, First Edition. David E. Watkins.
© 2011 John Wiley & Sons, Ltd. Published 2011 by John Wiley & Sons, Ltd.
As an aid to lecturers and students, full colour versions of the figures in this chapter may be found at www.wiley.com/go/watkins
These figures are © 2011 by John Wiley & Sons, Ltd.

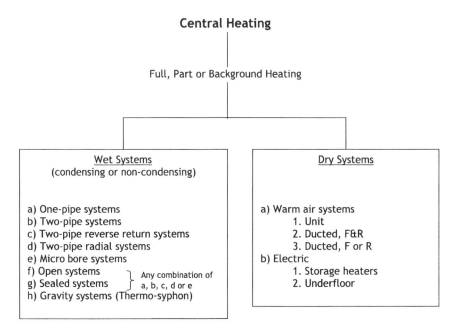

Figure 1.1   Heating system categories

Background heating, where lower than normal or recommended temperatures are aimed at, should only be used when specifically requested by the client for some reason. Even then, agreed temperatures should be incorporated into the design and guaranteed before any installation work commences. Under no circumstances should any heating system be installed without first agreeing specific room temperatures to be achieved when certain conditions exist. These conditions are discussed in Chapter 2.

Having understood the extent of the heating system and its classification, be it full, part or background heating, heating systems may be further divided under the headings of 'wet' or 'dry' systems. The terms wet or dry refer to the medium used to convey the heat from its source of generation to its point of use. Wet systems may be further classified by the piping circulation arrangement, with dry systems being divided into warm air and electric heating.

Figure 1.1 indicates the broad classifications of heating systems.

Heating systems can be sub-divided even further, but this will be explained in Chapter 21.

Wet heating systems

All wet types of heating systems employ a liquid as a medium to convey the heat from its source of generation. It is then distributed around the system to each heat emitter, where it transfers part of that heat through the heating surface of the heat emitters. Finally, the liquid is returned to the source of generation for the process to cycle continuously. The source of heat is commonly referred to as a boiler.

In all domestic heating systems, and most heating systems for other types of buildings, water is chosen as the medium for conveying the heat due to its low cost and being readily available. However, water does have the disadvantages of a low boiling point and high freezing point; it can also be corrosive to metallic materials and has a limited heat carrying capacity. The corrosive nature of the water can be reduced by water treatment, which is discussed later in this volume.

The temperature limitations and heat carrying capacity of water will have to be accepted unless we change the atmospheric conditions of the system, or we can change the liquid. Liquids known as 'thermal fluids' are available and have been used successfully on larger commercial type heating installations. They possess different properties to water, such as being less aggressive to common materials, having higher boiling points and lower freezing points, a greater heat carrying capacity than water and, in some cases, a lower viscosity. The merits of thermal fluids are much superior to those of water but are generally discounted for all domestic heating systems owing to their higher capital cost and not being readily available. They are also rarely used on larger commercial systems for the same reasons, but when conditions are right they can be considered attractive. The difficulty of availability can cause problems when replacement fluid is required immediately, following any emergency maintenance work. Thermal fluids have been used for domestic applications on limited occasions in countries that experience much lower temperatures than in the UK, as the lower freezing point of the fluid can be an important advantage when sub-zero ambient temperatures are experienced for prolonged periods with the heating system in a non-operating mode. They have also been employed as the heat carrying medium for some solar heating systems.

The purpose of the water used in heating systems differs from that used in domestic hot and cold water installations. In those systems, water is the end product or consumable item and after it has been used, it is discharged to waste. The water employed in a heating system is a non-consumable substance. It is the medium used to carry the heat required and, after it has transferred some of the heat, it is returned to the boiler to be re-used over and over again.

Dry heating systems (warm air)

Warm-air dry-type heating systems differ from wet-type heating systems insofar as the fluid employed is not only the medium used to convey the heat, but is also the end product. As the name implies, air is the fluid used to carry the heat from its source of generation, a warm air heater. It is then distributed, usually through a network of ducting, where it is arranged to enter directly into the room under controlled conditions to displace the cooler air. Finally, a mixture of the two is partly returned to the warm air heater for the process to be repeated.

Warm-air heating systems are generally disliked by many occupants of dwellings that have such systems installed, but this is usually because the systems are either not designed correctly, not installed correctly or are, in many cases, incomplete. This is mainly down to ignorance of the fundamental principles of warm air heating, which, if given the respect deserved, can be a very good form of heating. This work exclusively concentrates on wet-type heating systems since it is aimed at students and engineers in the plumbing industry.

Dry heating systems (electricity)

Electrical heating systems may technically be classified as dry systems, but they do not employ a medium as they generate their heat at the point of use. For this reason, electrical heating systems are not included in this book, with the exception of heating systems that use electricity as the source of power to heat the water. Here they are classified as being wet or hydronic heating systems.

Supplementary heating

This is a term applied to describe heating appliances, either fixed or portable, that are used to supplement the central heating system – either during extreme cold spells when the outside air temperature falls well below the base design temperature, or during the heating-off season in spring or autumn, when the outside temperature drops to below that considered comfortable.

Examples of such heating appliances include:

- Radiant electric fires, portable and fixed
- Oil filled radiators
- Oil room heaters
- LPG room heaters
- Gas fires
- Open solid fuel fires.

The list is not intended to be exhaustive, but meant to serve as a general representative selection of supplementary heating appliances.

2

Wet Heating Systems

Wet heating systems, commonly referred to as hydronic heating systems because they use a liquid as a medium, nearly always employ water as the medium to convey the heat from its source of generation, a boiler. This is rather a misnomer, as a boiler must be designed to avoid boiling the water, but is probably a leftover term from the days of raising steam. The heated water is circulated around the system, transferring part of its heat, and returns back to the boiler for the process to be repeated.

The water is fed into the heating system via a fixed piped connection to either a feed and expansion cistern, or a direct connection, as in the case of a sealed heating system. The water is allowed to enter the heating system slowly, thus avoiding creating turbulence, to fill it with all air expelled through the open vent, or by releasing it using manually operated air vents or automatic air release vents.

Water has many advantages as a heat carrying medium when used in hydronic heating systems; not least its plentiful availability. For this reason water is almost exclusively used for domestic heating systems.

Hydronic heating systems are classified by the following basic principles:

- Temperature of medium
- Pressure of system
- Circulation method of medium
- Piping arrangement for distribution.

The classifications are to a certain extent inter-related, as the selection of one of the basic operating principles has an influence on the selection of the others, which is explained in the following discussion.

Heating Services in Buildings: Design, Installation, Commissioning & Maintenance, First Edition. David E. Watkins.
© 2011 John Wiley & Sons, Ltd. Published 2011 by John Wiley & Sons, Ltd.
As an aid to lecturers and students, full colour versions of the figures in this chapter may be found at www.wiley.com/go/watkins
These figures are © 2011 by John Wiley & Sons, Ltd.

TEMPERATURE AND PRESSURE

The classification of hydronic heating systems by the temperature of the circulating water exiting the boiler is closely related to the operating pressure of the system, and the two must be considered together. This is because pressure is required to maintain the water in a liquid form at high temperatures: as water will boil and convert to steam at 100°C at atmospheric pressure when measured at sea level, any increase in that pressure will have a corresponding increase in the boiling temperature of water. Likewise, any decrease in pressure below atmospheric pressure will have the effect of allowing water to boil at temperatures lower than 100°C.

Table 2.1 gives the temperature/pressure classification commonly used in the UK. The minimum pressures listed are those required to prevent the water from evaporating but should not be confused with their vapour saturation pressures, which are lower.

It can be seen from Table 2.1 that water may be retained in liquid form when the operating temperature is above 100°C by pressurising it, giving all the advantages of a liquid and none of the disadvantages of a vapour such as steam. The method of pressurising the heating system is explained later in this chapter.

In contrast to the UK practice of temperature/pressure classification, in the United States of America the classification of heating systems differs slightly, outlined in Table 2.2.

It can be seen from Table 2.2 that the US has higher temperature and pressure classifications than the UK. However, in practice there is very little difference in the operating principles of hydronic heating system either side of the Atlantic.

Almost without exception, all domestic residential heating systems are classified as being low pressure and temperature (LPHW). It is considered safer to install heating systems using materials suitable for working pressures and temperatures below 100°C, therefore avoiding the potential hazard of flash steam occurring in the event of a pipe fracture or valve gland leak.

It has traditionally been the custom to design LPHW systems with a water flow temperature of 82°C and a Δt (temperature difference) of 11–12°C, giving a return water temperature of 71°C. More recently, the Δt has been increased in certain circumstances to take into account the requirements of condensing boilers that are influenced more by lower return temperatures than flow temperatures to function efficiently. This

Table 2.1   Hydronic design operating water temperatures and pressures (UK practice)

Classification	System temperature (°C)	Operating static pressure (bar absolute)
Low pressure hot water (LPHW)	<100	1 to 3
Medium pressure hot water (MPHW)	100 to 120	3 to 5
High pressure hot water (HPHW)	>120	>5*

*Account must be allowed for varying static pressures that would exist in a tall building.

Table 2.2   Hydronic design operating water temperatures and pressures (US practice)

Classification	System temperature (°C)	Operating static pressure (bar gauge)
Low temperature hot water (LTHW)	<120	2
Medium temperature hot water (MTHW)	120 to 175 Normally below 160	<11
High temperature hot water (HTHW)	>175 Normally about 200	<20

has a secondary effect on the increased sizing of the heat emitters, which is discussed in more detail in Chapter 8. Another situation where one should question the return water temperature and the flow water temperature is in heating systems employing underfloor heating sections that require the floor temperature to be limited to an acceptable level.

Low temperature heating systems may be further categorised as being either 'open' systems – where the heating system incorporates an open feed and expansion cistern and operates at atmospheric pressure, plus the static pressure created by the feed and expansion cistern at the traditional flow temperature of not exceeding 82°C – or sealed systems.

With a sealed heating system, the feed and expansion cistern is replaced by a sealed expansion vessel that allows the heating system to operate at a slightly higher pressure above atmospheric pressure and also permits the flow water leaving the boiler to have fractionally higher operating temperatures, in the region of 85–95°C.

If operating water temperatures higher than 82°C are selected for the heating system, then greater consideration must be given to the choice of heat emitters to be used, and all contactable heating surfaces such as traditional panel or column type radiators should be avoided so as to reduce the risk, scalding anyone who comes into physical contact with them.

Low water temperature heating systems are the most commonly used category of operating temperatures and pressures, suitable for all buildings ranging from small domestic residential through to very large and complex developments.

Medium temperature (MPHW) heating systems are favoured where a high heat output is desired so that smaller heat emitters and corresponding smaller pipe sizes can be used. The heat emitters must be of the non-contactable type, such as convectors, low surface temperature radiators and fan coil units. These systems are more suitable to commercial type buildings where the materials used are more robust than domestic low pressure type materials, and the system is more likely to be regularly serviced and maintained. This type of system in a domestic situation would be considered unsafe.

The use of high temperature and pressure systems (HPHW) is normally considered for use in industrial applications as some industrial processes require higher temperatures for manufacturing, or for developments that have a main central plant room that distributes the primary heat at high pressure and temperature to local plant rooms, which then circulate the secondary heat at a lower temperature. This arrangement is ideal for developments that are spread out over a large geographical area, and makes full use of more economical pipe sizes and equipment. As with the medium temperature systems, material selection and maintenance are critical factors.

CIRCULATION

Heating systems can also be classified by the method of circulation employed, i.e. either by gravity (thermosiphon), or forced circulation by a pump, or a combination of both.

Full gravity heating systems have not been installed since the development of the glandless circulating pump. The practice of having a gravity circulation to the domestic hot water cylinder whilst the heating system has a forced circulation, which can have some merit when suitable conditions exist, is no longer permitted by the Building Regulations for residential dwellings, which unfortunately limits the design engineer in the options available. Even where the situation exists that the domestic hot water cylinder is located directly above the boiler at the optimum height, and the occupant's needs are such that heating part of the system is not required for a great deal of the time but domestic hot water is, we are no longer permitted to use this method.

A fully forced method of water circulation for both heating and domestic hot water primaries is by far the most efficient arrangement in the majority of applications and gives freedom in the choice of plant equipment location, but this is not always the best option.

PIPING DISTRIBUTION ARRANGEMENT

Having discussed the temperature, pressure and method of circulating the water, the piping arrangement can be established. The different arrangements listed in Figure 1.1 form the basic systems for which there are numerous variations or modifications, but each may be categorised as belonging to one of the basic forms.

These arrangements each have their own advantages and disadvantages and the final selection should be made on the most efficient and economical method suited to each individual application. Also, a combination of any of the piping arrangements described may be used if it is considered by the design engineer to best meet the needs of the system.

The various piping arrangements depicted on the following pages have been produced to explain the operating principles of each system and are not supposed to be complete. For this reason most control elements and components have been omitted for the sake of clarity as these are dealt with in detail in Chapter 11. Also, the provision to include the means of producing domestic hot water has been included in each case, minus the controls element, to complete the piping arrangement: this may be by gravity primary circulation or by forced circulation. In most cases, either method may be used, unless noted otherwise. It is not the intention here to give the impression that either a gravity primary circulation or a forced primary circulation is the preferred option for satisfying the domestic hot water requirements, but just to show the different options.

ONE OR SINGLE PIPE SYSTEM

Of all the piping arrangements used for heating distribution, the single pipe system is the simplest. It consists of a single pipe main that extends from the boiler around the building as a circuit, or number of circuits, and returns to it with all heat emitters connected to the pipe by their own branch pipe flow and return connections.

Figure 2.1 illustrates the operating principles of the single pipe system and its limitations: a progressive temperature drop around the heating pipe circuit caused by each heat emitter returning its water back into the common circuit pipe. This has the effect of cooling the flow water available to other heat emitters being served by this circuit, which in turn results in subsequent heat emitters having to be oversized to compensate for a lower mean water temperature across the heat emitter.

To avoid heat emitters at the end of each pipe circuit having to be excessively large due to the decreasing mean water temperature available, heating pipe circuits should be limited to supplying water to a few heat emitters each, to restrict the mean water temperature across the heat emitter to no less than 70°C for non-condensing systems, or lower for condensing.

Another effect of pipe circuits suffering from excessive temperature drop is that the piping system on each circuit would also have to be oversized to compensate for the lower circulating water temperature.

The piping arrangement depicted in Figure 2.2 demonstrates that this need not be the case: if the branch circuits supply a minimum number of heat emitters, then the single pipe arrangement is just as suitable for larger domestic residential properties or commercial building applications, as the small domestic heating system.

The object in the design of this system is to limit the temperature drop across each pipe circuit so as to avoid having to significantly increase pipe sizes or heat emitter sizes to compensate, and so lose the lower cost advantage claimed by this system.

From the schematic layout depicted in Figure 2.2, it can be seen that if each piping circuit is limited to a reasonable temperature drop across it, and if each piping circuit is similar in its heat carrying load to each other, then the single pipe heating arrangement is suitable for heating system compositions in larger buildings. It can also be seen that the piping system is fairly evenly balanced in its heat distribution

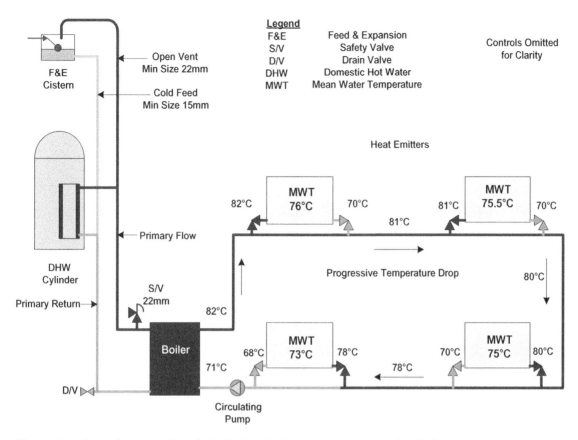

Figure 2.1 Operating principles of single pipe heating system (non-condensing)

in order to achieve the temperature drop required from each heating circuit. This is accomplished by balancing the circulating piping system with the use of regulating valves when the heating system is being commissioned.

The primary flow and return to the domestic hot water cylinder in this illustration is in fact a two pipe arrangement. Lower temperatures may be selected for condensing heating systems.

The single pipe heating system benefits from the employment of special tees, known as 'diverting' or 'inducing' tees. These special fittings are designed to encourage a degree of flow into the heat emitter by creating a resistance to the flow between the flow and return branch connections, in the form of a pressure drop on the single pipe circulating main. This creates the conditions for circulation to occur through the heat emitter, as the resistance of this passage is less than that of the heating main.

The isometric layout illustrated in Figure 2.3 demonstrates the use of these diverting tees, whereby the up feed risers only require one diverting tee to be fitted on the return connection as the thermal head will assist the circulation, but the down feed pipes should be fitted with diverting tees on both the flow and return branch connections because no thermal head exists in this situation.

Diverting tees may be obtained in a copper alloy or malleable iron and are constructed with a venturi shaped restriction inside as shown in Figure 2.4. These tees are similar in design to 'tongued' tees that were commonly used on gravity heating systems.

Figure 2.4 shows how the flow of water through the venturi of the diverting tee induces the flow from the return connection of the heat emitter.

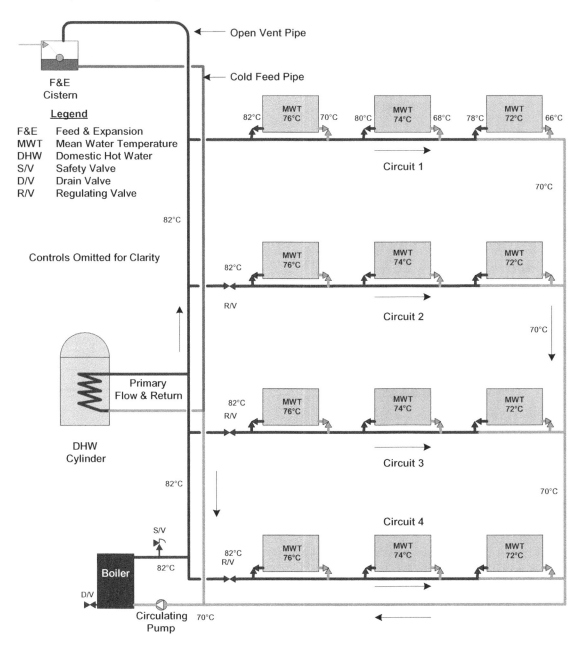

Figure 2.2   Single pipe system for larger building with limited circuit Δt (non-condensing)

Figures 2.5 and 2.6 show how the diverting tees are arranged and how they function for both upward connections to the heat emitters using one diverting tee on the return, and downward connections where diverting tees are employed on both the flow and return connections to the heat emitters.

Diverting tees have been successfully fabricated on site from standard capillary copper pipe fittings, either the end feed type, or integral solder ring type, using a standard tee, a spigot and socket straight reducing fitting, with the larger spigot end cut short but square, which is placed inside one socket end of the tee so that the reduced socket end protrudes past the branch of the tee. The cut spigot end of the reducer must

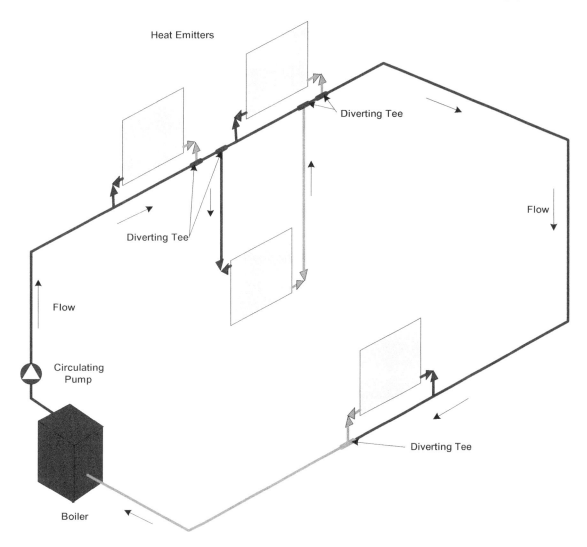

Figure 2.3   Application of diverting tees on single pipe heating system

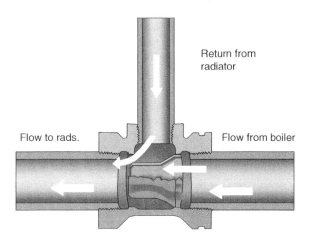

Figure 2.4   Section through a diverting tee fitted on the return connection

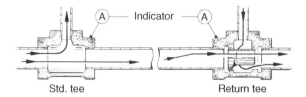

Figure 2.5   Upward branch connections, standard tee on flow and diverting tee on return

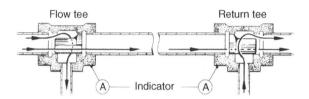

Figure 2.6   Downward connections, diverting tees on both flow and return

be inserted into the tee so that the cut end has cleared the integral ring of solder if this type of capillary fitting has been used. The copper pipe is then inserted into the fitting and the capillary joint made in the normal way.

A 15 mm equal tee would use a 15 mm spigot end × 12 mm socket end reducer, and a 22 mm × 15 mm tee would use a 22 mm spigot end × 15 mm socket end reducer.

A variation of the single pipe system is depicted in Figure 2.8. If compared to the traditional one pipe system illustrated in Figure 2.1, it will be seen that the only difference is the method of connecting the circulating pipework to each heat emitter. The standard pair of radiator valves has been replaced by a single centrally located valve in the bottom of a specially manufactured pressed steel panel radiator that has a single female threaded centrally located tapping incorporated, together with a division baffle plate in the bottom main horizontal water passageway separating the two halves of the radiator. Figure 2.7 shows a cut-through section of this special three-way radiator valve.

The central three-way radiator valve is connected to the radiator via a union radiator tail connection that has a central dividing wall along its length that lines up with the dividing baffle plate incorporated in the radiator, as shown in Figure 2.9. The valve is designed so that when it is fully open it will direct 100% of the flow through the radiator, and when it is closed, it will allow 100% of the flow to bypass the radiator through the valve. Any intermediate position of the valve control will direct an equivalent portion of the flow through the radiator with the remaining portion of the flow being allowed to travel through the valve bypass.

This system heating arrangement was developed in Scandinavia, where it has found most use. Its main merits over the traditional single pipe system are that the three-way valve directs a definite flow of water through the radiator without the aid of diverting tees, and permits individual radiators to be isolated without affecting the flow to the remainder of the circuit. This arrangement, when it has been used with the heating piping installed either below or within the floor, forms a neater appearance with a minimum of pipework on show.

The disadvantages of this system are that the heat emitters are restricted to panel radiators made specifically for use with this valve, which requires confidence that they will continue to be available in the future for replacements, system extensions or alterations. Any thermostatically operated version of this valve yet to be developed would be required to conform to current requirements.

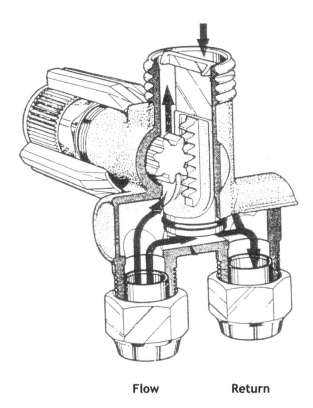

Flow   **Return**

Figure 2.7   Detail of central three-way radiator valve

To summarise, the single pipe heating arrangement continues to be used for heating systems either in part, or in full in commercial buildings when the application is considered suitable, but is rarely installed in domestic residential properties. The reasons for this situation can only be guessed, but one theory is that heating systems installed in commercial properties are normally designed by engineers who approach each building on an individual basis, and consider the client's brief, the architect's design requirements, the structural constraints of the building, the budget available and the energy efficiency targets to be achieved. Only after having considered all of these important design aspects can a decision be made on the piping arrangement to be used, and, quite often the one pipe system meets all of the above requirements.

Unfortunately, this is not normally the situation regarding domestic residential dwellings. A high percentage of installers are also the designers of the heating system and it is quite common for them to install their favourite heating system arrangement in every property because they are familiar with it, regardless of the size or configuration of the building, or the client's own personal requirements.

This text has demonstrated that the single pipe heating system, if selected correctly and after careful design, such as limiting the temperature drop across each piping circuit, will achieve a simpler, less expensive heating system and is suitable for many applications in domestic residential dwellings where normally the installer would not consider it.

It should be pointed out that not all installers of domestic heating systems have this blinkered approach and with ingenuity some very successful one pipe heating systems have been achieved, but for the majority, those who only use one type of heating arrangement regardless of the building layout, more training is required.

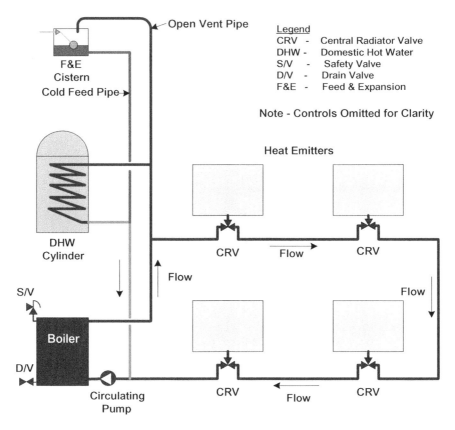

Figure 2.8   Single pipe central valve radiator connection system

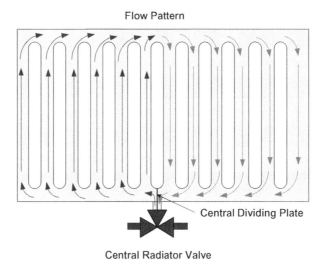

Figure 2.9   Arrangement of central three-way radiator valve and radiator

TWO PIPE SYSTEM

This is by far the most commonly used piping arrangement for wet heating systems, ranging from installations for very large commercial type buildings down to small domestic residential dwellings, due to the versatility of this arrangement.

As the name implies, the two pipe arrangement comprises two main heating distribution pipes in parallel as opposed to one on the single pipe system, see Figure 2.10. One of the pipes is a dedicated flow conveying hot water at boiler temperature to each heat emitter, whilst the other is a completely separate dedicated return that returns the water from each heat emitter, after it has transferred part of its heat, back to the boiler to be reheated.

The main advantage of the two pipe heating system over the single pipe heating system is that water is delivered to each heat emitter at the same temperature as it leaves the boiler, minus any piping heat losses. In turn, the water temperature that exits the heat emitters is the same temperature that is returned back to the boiler, again minus any pipework heat losses, whereas the single piping arrangement suffers from a

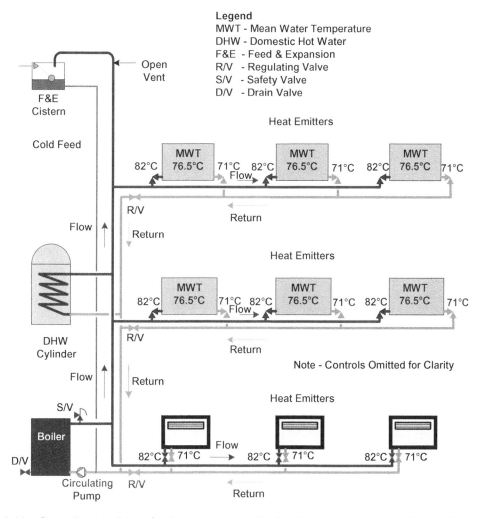

Figure 2.10   Operating principles of a direct return two pipe heating system – non-condensing (condensing systems would employ different flow and return temperatures)

progressive temperature drop across each piping circuit. The main disadvantage is that it is more costly and, with the direct return system illustrated in Figure 2.10, the supply and return lengths are unequal, which results in unbalanced flow and return flow rates.

This imbalance in the system flow rates is due to the tendency for the water to circulate through the heat emitters closest to the boiler, where the resistance to flow is at its lowest, at the expense of those furthest from the boiler where the resistance is at its greatest, resulting in the index circuit heat emitter becoming starved of water. This situation can be rectified by balancing the heating system using balancing devices, such as regulating valves and the lockshield return valves, on the heat emitters during the commissioning exercise on completion of the heating installation: this matter is discussed further later on in Chapter 25.

The direct return method of installing the two pipe heating system, where the flow and return pipes are in parallel to each other is usually simpler, as it is easier to install these two pipes side by side than it is to find alternative routes for them separately.

A variation of the two pipe system is the reverse return two pipe heating system as portrayed in Figure 2.11, which, through its flow path direction of both the flow and return circulating pipes, overcomes the

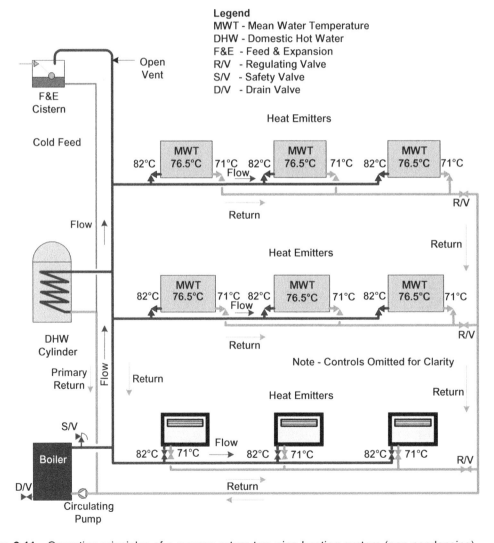

Figure 2.11   Operating principles of a reverse return two pipe heating system (non-condensing)

problem of system flow disparity. It can be seen that the heating system depicted in Figure 2.11 is the same as that shown in Figure 2.10, the only difference being the arrangement of the main return pipe back to the boiler being routed in the opposite direction to that of the main flow pipe. This method of returning the water to the boiler creates an almost equal distance, and also an equal resistance to the flow of water to each heat emitter, thus making the balancing of the system flow much easier.

The reverse return two pipe heating system is more desirable than the direct return system if capital costs can be justified and the building layout does not present complications to the pipe routes, as the heating system will remain in a balanced condition regardless of any tampering with the return regulating valves.

To illustrate the advantage of the reverse return piping system over the direct return arrangement of piping, two identical heating systems are shown in the isometric layout projection in Figure 2.12. Part (a) depicts the heating system with the piping arranged in a standard two pipe direct return layout where it can be seen that the resistance to the flow of heating water from the boiler and back again from heat emitter number 1 is much less than it is to any other of the heat emitters. The flow has to be encouraged by balancing the pipework system by making all the heat emitters have an equal resistance, so forcing the heating water to the index heat emitter and thus preventing the water short circuiting through the nearest heat emitter.

Part (b) shows the same heating system, but the pipework has been arranged as a two pipe reverse return configuration. In this example the resistance of the pipework encountered by the flow is the same to heat emitter number 1 as it is to heat emitter number 4, and all the other heat emitters in between. Therefore, providing that the piping system has been correctly sized, the circulation of the heating water is much simpler.

It can be argued that the additional capital cost of the two pipe reverse return heating system can be justified by the saving of the reduced time spent on balancing the piping network at the commissioning stage and, in turn, offsetting this initial capital cost.

MICRO BORE PIPING SYSTEM

All of the piping arrangements discussed so far have been equally suitable for large commercial heating systems and small domestic residential heating systems, but the micro bore heating system was developed in the 1960s specifically for domestic residential dwellings and small commercial properties, although it has been used successfully on larger building developments where it forms part of the overall larger heating scheme.

There have been a number of myths regarding the terms 'micro bore' and 'mini bore', some of them quite plausible, such as one of them being an open vented heating system and the other one being a sealed heating system, or that one uses imperial sized tubes while the other employs metric diameter tubes, this latter explanation being closest to being correct. The truth, however, is less romantic as they are in fact both proprietary names which have become acceptable to use to describe a heating system that utilises distribution pipes having diameters smaller than the 15 mm (½ inch), which are normally used on small bore heating systems. Mini bore was the first of these names to be used to describe water distribution pipes having diameters of ¼ inch and ⅜ inch, but now micro bore has become generally adopted to describe heating systems employing tubes having diameters of 6 mm, 8 mm, 10 mm and 12 mm diameter.

The micro bore heating system differs fundamentally from any of the heating systems previously described. It employs components that are peculiar to the micro bore system and are not found on small bore heating systems that use conventional pipework and fittings common throughout the plumbing industry.

Figure 2.13 illustrates the basic operating principles of the micro bore heating system, which is still fundamentally a two pipe heating arrangement where the means of circulation is the same as any other form of piping arrangement, but the method of achieving that circulation differs from other piping layouts. The heated water is circulated from the boiler through conventional size pipes arranged as a pair of heating mains to strategically placed manifolds located fairly centrally between a group of heat

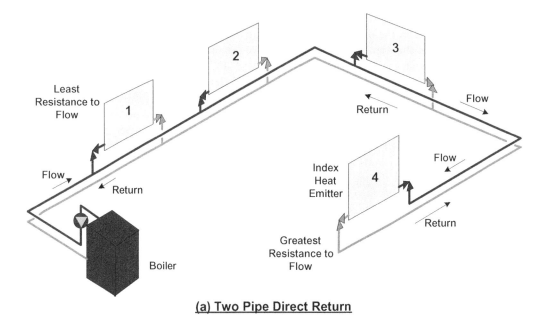

(a) Two Pipe Direct Return

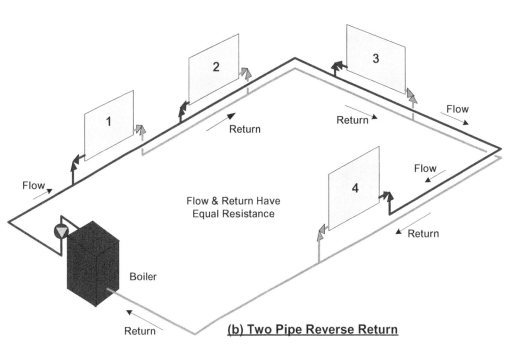

(b) Two Pipe Reverse Return

Figure 2.12   Isometric comparison of two pipe direct return and two pipe reverse return

emitters, this location being chosen to try to achieve reasonably equal micro bore branch pipe runs to each heat emitter that it serves. The manifold may be one of a variety of types, such as an inline multiple tee arrangement that incorporates a central blanking plate, or an end of line multiple reducer, or a number of other forms that are commercially available. The difference between this system and the other piping arrangements is that from the manifold, a separate dedicated micro bore flow and return pipe is

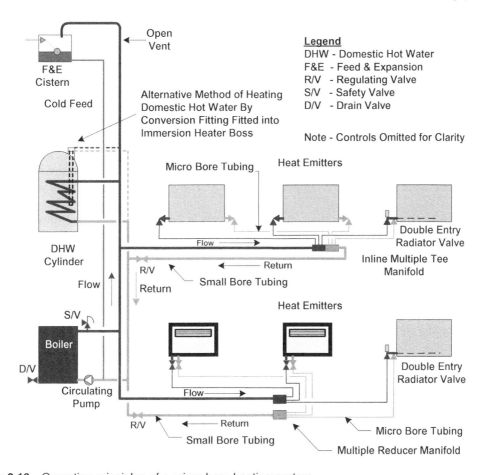

Figure 2.13   Operating principles of a micro bore heating system

extended radially and connected to each heat emitter individually, as shown. There is no limit to the number of manifolds employed, but each pair of micro bore flow and return pipes must be connected to the same manifold.

Heating systems employing conventional pipe sizes, i.e. 15 mm diameter and larger, are designed to have a flow velocity of approximately $1.0\,\mathrm{m\,s^{-1}}$: this has been found to be the most economic regarding the optimum velocity that will not cause noise to be generated or erosion to occur at sharp changes in direction. Micro bore piping arrangements do not conform to this convention and are designed to have flow velocities of $1.5\,\mathrm{m\,s^{-1}}$, which means that as the speed of the water being delivered to each heat emitter is one and a half times faster than that used on traditional piping arrangements, smaller diameter tubes may be used. This higher flow velocity through micro bore tubing is possible as there are no sharp changes in direction such as elbows, etc – and as the tubing is installed in one continuous length, using flexible coiled pipe of either soft fully annealed copper, or a barrier type plastic material, the same volume of water may be supplied to each heat emitter without any noise or erosion problems.

The following advantages are claimed for micro bore heating systems:

1. In existing buildings it is claimed to be easier to install, as fewer floorboards have to be removed when using micro bore tubing. Also, on new build properties the micro bore tubing can be installed by threading the flexible tube through holes pre-drilled in the floor joists from the ceiling below,

allowing the floor to be laid earlier. This is possible as the micro bore tube – be it soft copper, a barrier thermoplastic material such as PEX, or polybutylene – is supplied in soft coil form that allows the tube to be threaded through in a similar way to electric cable. It should be noted that when working with soft fully annealed copper tube, although it is flexible when supplied in coil form, it quickly hardens if it is threaded through joists or below floorboards involving too many turns in direction and may need to be reheated to recover its original soft condition. Special tools are available to form neat labour-pulled bends and to straighten pipework that will be on show.

2. Double entry radiator valves are available and may be used, resulting in a cost saving compared to a pair of traditional single entry radiator valves that would normally be used. However, the type incorporating the provision to isolate the return connection as well as the flow connection should be selected to aid maintenance and allow the radiator to be removed without draining down the entire heating system.

3. A neater pipework installation is claimed as the piping may be hidden more easily in exposed situations, although, conversely, because the micro bore tubing is soft it may be more easily damaged, either accidentally or wilfully; therefore some form of protection should be considered if this potential exists. The pipework may also be passed through smaller openings or holes in the structure than conventional small bore tubing, which sometimes may be a deciding factor in specifying piping systems in existing buildings.

4. As the micro bore tubing is supplied and installed in one continuous length from the manifold to each heat emitter, the piping system requires fewer joints, which in turn means less possibility of leaks.

Manifolds

Figure 2.14 illustrates a typical inline multiple tee manifold that comprises a section of small bore tube, usually 22 mm or 28 mm diameter, which incorporates a centrally placed blank plate that divides the manifold into a separate flow zone and separate return zone, each with an equal number of compression type tees that convert the small bore tube into micro bore tube.

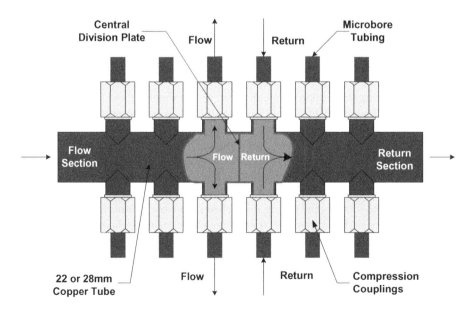

Figure 2.14   Inline multiple tee manifold

Figure 2.15 Linear multiple reducing manifold

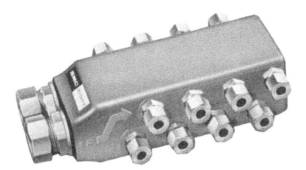

Figure 2.16 Alternative form of micro bore manifold

The inline multiple tee type manifold is the most commonly used manifold of all the different variations commercially available and may be obtained with a varying number of connection tees to suit the application required, with any unused tees being blanked off. It is also where the flow velocity transforms from $1\,\mathrm{m\,s^{-1}}$ in the small bore tube to $1.5\,\mathrm{m\,s^{-1}}$ in the micro bore tube.

Figure 2.15 illustrates an alternative type of manifold that is in common use. It comprises either a capillary or compression linear fitting that is arranged to fit on the end of a small bore pipe that serves to divert the flow into a number of reduced size micro bore tubes. A separate linear type multiple reducing manifold is required for both the flow and return connections.

This type of manifold is restricted in physical size to a limited number of branch connections that are available to connect the micro bore tube into, where the inline multiple tee is not hindered.

Another form of micro bore manifold is shown in Figure 2.16; this type was available commercially for a period of time in a few sizes and was manufactured using cast iron for the body with non-ferrous compression pipe connections. There was an internal division plate arranged horizontally that formed two chambers, one for the flow and the other for the return, which meant that both the flow and return micro bore pipes serving each heat emitter were arranged almost one above the other on the same side of the manifold, thus avoiding having to cross over other micro bore pipes serving different heat emitters.

The higher capital cost of this type of manifold limited its use to pipework that would be on show, as a very neat and professional appearance could be achieved when the manifold was fixed to a wall surface in a cupboard or service duct and the micro bore tubing dressed into the manifold pipe connections without any pipe crossovers.

Double entry radiator valve

Figure 2.17 shows a cut through section of a double entry radiator valve illustrating the flow and return micro bore pipe connections and the application of a short cut piece of micro bore tubing used as a rigid insert to prevent the flow of water short circulating through the radiator. The flexible copper insert supplied with these radiator valves should be discarded as it has a tendency to lay flat on the bottom of the steel panel radiator welded seam, which could promote an electrolytic action between the two metals.

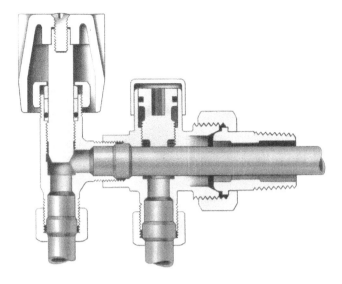

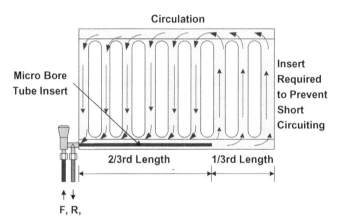

Figure 2.17   Detail and application of double entry radiator valve

The flow enters the radiator through the valve flow connection and into the rigid copper insert where it circulates through the radiator and exits through the body of the valve on the outer side of the insert, as illustrated.

The double entry radiator valve is less costly than a pair of conventional radiator valves but can only be used on single panel radiators, or column type radiators. The construction of double panel radiators and radiators manufactured with back inlet connections prohibits the use of these double entry radiator valves.

TWO PIPE RADIAL SYSTEM

This is the most recent of piping arrangements forming a system of heat distribution that has been derived from the operating principles of the micro bore piping system, but it is not restricted to micro bore tubing or small bore tubing and incorporates components common to underfloor heating (see Chapter 6, Underfloor Heating).

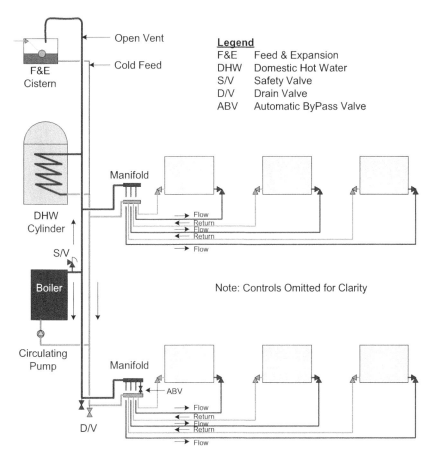

Figure 2.18 Operating principles of two pipe radial system

Figure 2.18 illustrates the piping arrangement of the two pipe radial system. It can be seen that the distribution is very similar to that of the micro bore piping system except that the branch distribution piping may be 15 mm diameter or larger. The configuration of the manifolds used is very similar to those employed for underfloor heating, including the incorporation of individual isolating valves on both the separate flow and return manifolds.

The method of delivering the heat is the same as for the micro bore system. The water is circulated by the pump through the common main flow and return pipes to each manifold located in a service cabinet or service duct, from where it is circulated through individual dedicated branch flow and returns to each heat emitter.

This piping arrangement has been more popular in Northern Europe than in the UK, but has been installed in a number of new housing developments in Britain over recent years. It is more suitable for installing in new build properties that have solid concrete floor constructions, but can be used in existing dwellings and buildings that have raised or ventilated timber floor construction if the property is undergoing a major refurbishment that can help to justify accommodating the building alteration costs.

Figure 2.19 shows a simple floor layout where the heating flow and return pipes are laid in the floor from the manifold and take a direct route below door openings to the heat emitters without any pipe joints, using large radius labour pulled bends.

The piping material chosen for this application is normally either cross-linked polyethylene (PEX), polybutylene or polyethylene, all with an oxygen barrier incorporated within the pipe walls. These pipe

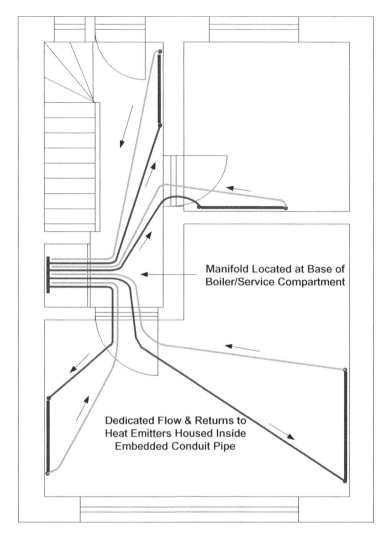

Figure 2.19   Floor plan layout showing radial piping arrangement

materials have the qualities required for this particular application, such as flexibility, and are commercially available in long coiled lengths that allow the pipe to be laid in one continuous length without any intermediate joints. It is good practice to install these pipes inside a second conduit pipe that is directly embedded in the floor screed; see Figure 2.20. This has the advantage of permitting the carrier pipes to remain free to move through thermal movement inside the embedded conduit pipes, and facilitate the removal and replacement of the carrier pipes if necessary for maintenance.

This form of pipework installation makes for a neat and easy to install system, but makes it very difficult for any future alterations such as repositioning of heat emitters and property extensions.

Manufacturers of these systems produce a variety of ingenious components that form a transition from the embedded underfloor pipe to connect up to the heat emitters. Figure 2.21 shows an arrangement that utilises a junction box fitted into the floor screed that facilitates both the transition from a plastic pipe to a metallic pipe riser up to the heat emitter in the normal manner, and provides the access required to withdraw the carrier pipes from the conduit and replace with new if required.

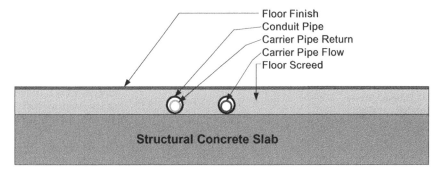

Figure 2.20   Section through solid floor construction showing heating flow and return installed within conduit pipes

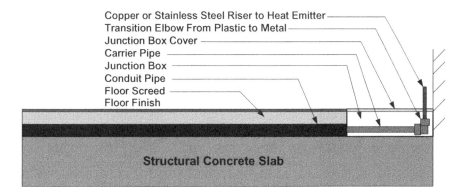

Figure 2.21   Detail of junction box transition for heat emitter connection

The conduit pipe may also be used to contain the flow and return carrier pipes embedded in wall chases or inside the cavity of dry-lined walls; see Figure 2.22. With this arrangement, a commercially available termination junction box, similar in appearance to an electrical power socket outlet cover plate, is fitted flush with the wall behind where the radiator is to be fixed, and the conduit pipes are fixed inside the depth of the walls with the carrier flow and return pipes installed inside them. The wall finish is then made good, and the flexible thermoplastic branch pipes connected to the conduit housed carrier pipes inside the termination junction box before the radiator is fixed to the wall. The flexible branch pipes can then be connected to the radiator valves, making for an aesthetically pleasing installation.

Thermoplastic pipes are extremely suitable for this application, their flexible nature enabling them to be connected from the termination junction box to the radiator.

In newly constructed buildings the radial piping arrangement has a number of advantages, primarily reduced on-site installation time, meaning the time for which other trades are prevented from working whilst the piping system is being installed is reduced. This reduction in programme time results in a lower cost of construction to the developer. The radial piping arrangement also results in a neat, unobtrusive installation. However, the price to pay for these construction phase advantages comes later when the building owner wishes to extend or alter the building, as the resulting pipework alterations involve considerably more work for the builders.

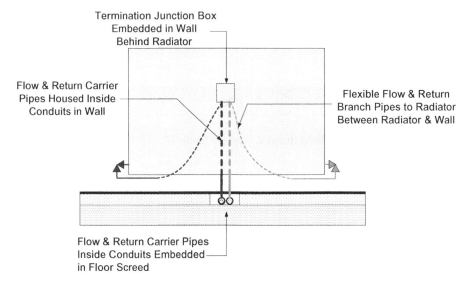

Figure 2.22   Flexible thermoplastic radiator connections from flush mounted termination box

HYBRID (MIXED) SYSTEMS

Figure 2.23 illustrates a somewhat exaggerated example of a mixed piping system arrangement demonstrating the possibilities that are available if the building layout, together with the client's needs, dictate it. It is equally suitable for large or small buildings alike.

Hybrid systems need not just be a mixture of different piping arrangements; they can also be a combination of temperature/pressure configurations.

Figure 2.24 illustrates a hybrid system of sealed heating where there is a mixture of piping arrangements and a mixture of operating temperatures. This form of heating is more common in other parts of Europe than it is in the UK. The system is designed to operate as a medium pressure heating scheme with the water flow temperature leaving the boiler at a temperature of 100–120°C, making it unsuitable for domestic residential dwellings. The high temperature flow water first travels via a single pipe system to pass through a series of low surface temperature radiators or convectors where the high temperature of the water can be exploited to the full, permitting physically smaller heat emitters to be selected. When the high temperature of the flow has been exhausted and the flow temperature falls below 100°C, the water then passes on to a conventional two pipe system, still operating at medium pressure, but permitting standard panel or column type radiators to be employed that have been selected and sized in the same way as for a conventional low pressure heating system. The water is then returned to the boiler at a normal return temperature of 71°C.

This hybrid arrangement has obvious advantages if the building criterion favours it.

OPEN VENTED HEATING SYSTEMS

All the hydronic heating systems depicted so far in this text, with the exception of the sealed heating system illustrated in Figure 2.24, are described as 'open vented heating systems', which is the term used to describe hydronic systems of heating subjected to hydrostatic pressure only. This hydrostatic pressure is determined by the physical height that the water in the feed and expansion cistern is exerting on the rest of the heating system. Heating systems in the past in the UK have traditionally – but not exclusively – been of the open

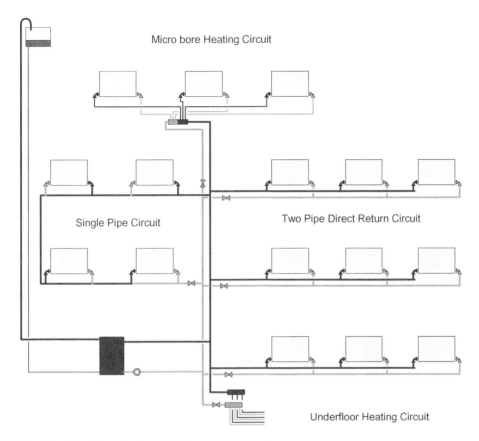

Micro bore Heating Circuit

Single Pipe Circuit

Two Pipe Direct Return Circuit

Underfloor Heating Circuit

Figure 2.23   Hybrid system of mixed piping arrangements

vented type, and the reason why the drawings used in this chapter of hydronic heating systems are of the open vented type is to show the need for, and method of, accommodating the expansion of the water caused by the application of heat.

One of the main features of an open vented system, and the foremost component that makes it an open vented system of hydronic heating, is the inclusion of the feed and expansion cistern.

The function of the feed and expansion cistern may be explained in three ways:

1. The initial purpose of the feed and expansion cistern is to serve as a header tank to enable the heating system to be filled up with water and then to continue to perform as a reservoir, where any water losses due to evaporation may be automatically replenished. The feed and expansion cistern is the automatic point of introduction of water into the heating system, both on the initial fill and any subsequent fills that may be required during maintenance activities. This is the feed part of the feed and expansion cistern that functions before the heating system becomes operational and is only required when water is to be added to the system.
2. The second function of the feed and expansion cistern is to serve as a means of catering for the volumetric increase in water caused by the expansion of the water that occurs when the heating system boiler is operating at its design temperature, without causing any pressurisation of the heating system. This is the expansion part of the feed and expansion cistern and is performed continually during its normal operation.

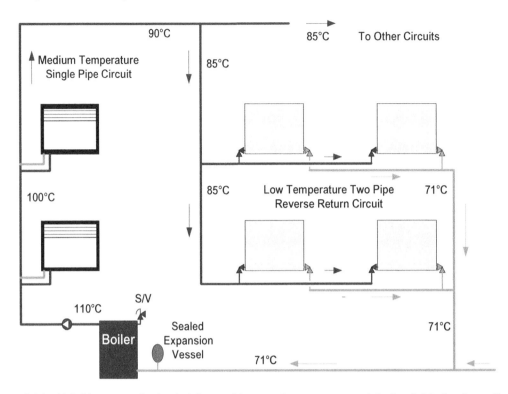

Figure 2.24   Hybrid system of mixed piping and temperature arrangement (not suitable for domestic residential dwellings)

3. The feed and expansion cistern also has a third function, which is equally important but not immediately obvious as the first two points. This third function is one of safety; it serves as a cooling tank in the event of the boiler overheating and high temperature water/steam being discharged into it from the open vent pipe terminating over the top of the feed and expansion cistern. This high temperature water/steam mixes with the water already in the feed and expansion cistern, which is also at a high temperature, but also mixes with any incoming cooler make-up water to replace evaporation losses before returning it back into the heating system to keep the system and boiler wet and thus prevents the boiler becoming dangerous. Because of this safety aspect, it is imperative that the feed and expansion cistern is correctly sized and that it is manufactured from materials capable of withstanding the high temperatures that could exist when this condition occurs. This third function is not always fully appreciated as its importance does not become apparent until something goes wrong with the heating system. If it has not been designed correctly a failure to the feed and expansion cistern could occur compounding a dangerous situation and creating a potentially catastrophic condition.

FEED AND EXPANSION (F&E) CISTERN

When water is heated, as it is in a hydronic heating system, the rising temperature will cause the water to expand. This increase in volume will have to be catered for, otherwise the heating system will become highly pressurised, to a point where it is in imminent danger of fracturing at its weakest part. In the open vented heating system this increase in water volume is accommodated by the feed and expansion cistern.

Table 2.3 Recommended minimum capacities for F&E cisterns and connections

Boiler Rating kW	F&E Cistern Nominal Capacity Litres	Float Valve Size mm	Cold Feed Diameter mm	Open Vent Pipe Diameter mm	Warning Pipe Diameter mm
<25	45	15	15	22	22
<45	70	15	22	28	28
<60	90	15	22	28	28
<75	150	15	22	28	35
<150	225	15	28	35	35
<225	310	22	28	35	35
<300	400	22	35	42	42

Water is at its maximum density at a temperature of 4.4°C; any change in temperature from this point, be it higher or lower, will result in an increase in its volume. If the temperature of water is raised from 4.4°C to 100°C, it will increase in volume by 4.2%, or $\frac{1}{24}$ of its original volume. Therefore the feed and expansion cistern must be sized to accommodate not less than $\frac{1}{24}$ of the heating system volume of cold water, plus the water in the lower part of the cistern required to allow the float valve to operate, a minimum depth of 100 mm and a safety factor.

These requirements are satisfied by allowing a volumetric increase of 5%, or $\frac{1}{20}$ of the heating system's water content when cold and is the figure recommended by BS 5449, and other authoritative engineering guides. This increase in volume should be allowed for above the normal cold water level and below the warning pipe/overflow level as the water should expand into the feed and expansion cistern via the cold feed, but should not overflow through the warning pipe. When the heating system is turned off, the water will cool down and contract back into the heating system by the way it entered, the cold feed.

Table 2.3 gives the minimum recommended nominal capacities for feed and expansion cisterns, together with their service connections, but it should be emphasised that these should only be used as a guide and the final selected size should be calculated, see Box 2.1.

The feed and expansion cistern should conform to the requirements of the Water Regulations exactly as for the main cold water storage cistern, as it is supplied with cold water from the incoming rising main. Figure 2.25 illustrates the general arrangement of the feed and expansion cistern detailing the main points. The construction materials should be suitable for their uses; particularly for the high water temperature that could exist if nearby boiling water is discharged into it during a boiler/system malfunction. If the cistern is not self-supporting, a solid base should be placed under the cistern that extends to a minimum of 50 mm on all four sides and should be of a material that will retain its integrity if wet: marine grade ply is often selected for this use. The float valve that regulates the flow of make-up water into the cistern should conform to BS 1212, either Part 1 or Part 2. Normally a Part 2 copper alloy bodied diaphragm float valve is selected, but a Part 1 Portsmouth pattern float valve may be used provided that a back flow device, such as a double check valve, is installed on the incoming water main, as depicted in Figure 2.25. Care should be taken when cutting any penetration in the wall of the cistern, such as the outlet for the cold feed to the system, to ensure that the flanges of the tank connector do not impose any stress on the base corner of the cistern when the connector back nut is tightened. The cistern cover and warning pipe should be equipped with an insect filter to prevent ingress into the cistern and any penetration in the cover for the open vent should have a flexible seal to permit movement of the pipe through thermal expansion.

The feed and expansion cistern and all pipework in the roof void space, including the warning pipe, must be insulated against freezing, except the space beneath the feed and expansion cistern, which should be left clear so that heat that has passed through the ceiling below will prevent the water temperature dropping below 0°C.

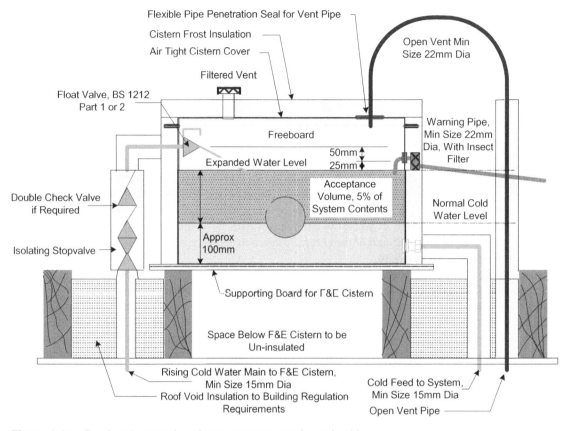

Figure 2.25 Feed and expansion cistern arrangement in roof void

OPEN VENT PIPE AND COLD FEED AND EXPANSION PIPE

The open vent and cold feed and expansion pipe perform an equally important function as the feed and expansion cistern and combine together for the following purposes:

1. When the heating system is being charged with water, either initially, or during any subsequent refilling after any maintenance activities have been completed, the cold water is allowed to enter the heating system under a controlled steady non-turbulent flow via the cold feed pipe from the feed and expansion cistern, whilst air that it displaces is permitted to partially escape through the open vent pipe. Any trapped air is expelled via the manual air vents on the heat emitters, or via automatic air vents incorporated in any non-self-venting pipework on the heating system.
2. When the heating system has been charged with water and the boiler commissioned and fired up, the water as a heating medium in the system will be constantly heated up and cooled down under the dictates of the heating systems control scheme and boiler thermostat. The result of this constant temperature change will cause the water to increase in volume when heated and expand back up the cold feed into the feed and expansion cistern occupying the space calculated for the acceptance volume. When the water in the heating system cools down, the water that entered the feed and expansion cistern will

BOX 2.1　Example calculation of F&E cistern size

Extent of system pipework (Obtained from Engineering Data)

56 m of 15 mm pipe × 0.1452 litres per metre	= 8.14 litres
38 m of 22 mm pipe × 0.3211 litres per metre	= 12.20 litres
18 m of 28 mm pipe × 0.5399 litres per metre	= 9.72 litres

Heat emitters (Obtained from Manufacturer's Data)

Rad 1	= 7.2 litres
Rad 2	= 12.4 litres
Rad 3	= 7.2 litres
Rad 4	= 3.6 litres
Rad 5	= 4.0 litres
Rad 6	= 8.6 litres
Rad 7	= 4.0 litres
Convector 1	= 2.0 litres

Boiler water capacity (Obtained from Manufacturer's Data)

System boiler	= 12 litres

Domestic hot water cylinder (Obtained from Manufacturer's Data)

DHW cylinder heat exchanger only	= 2.0 litres

Total system water capacity　　　　　　　　　　　　　　= **93.03 litres**

Acceptance volume required　　　5% × 93.03 litres　　　= **4.65 litres**

Assuming that the F&E cistern measures 500 × 300 mm and that it contains water up to a height of 100 mm, this would equal 15 litres of water when the system is cold.

15 + 4.65 litres = 19.65 litres of water actual capacity.

Assuming an equal volume of 15 litres above the acceptance volume water line then:

15 + 19.65 litres = 34.65 litres of water nominal capacity

Therefore, capacity of the feed and expansion cistern required =

Nominal capacity	**35 litres**
Having an actual capacity of	**20 litres**
Having an acceptance volume of	**5 litres**

contract slowly back into the heating system and return to its original volume. At the same time water will expand up the open vent pipe, raising the level of water in this pipe equal to that of the acceptance volume water level in the feed and expansion cistern, but not overflowing into the feed and expansion cistern, as shown in Figure 2.25. When the water cools down it will return to its original level in the open vent pipe.

3. The two preceding points describe the function of these two pipes together with the feed and expansion cistern during the normal operation of the heating system, but a third and equally important purpose of the open vent pipe, cold feed pipe and feed and expansion cistern transpires in the event of the heating

system overheating due to some malfunction, such as the boiler thermostat failing. In this situation, the open vent, cold feed and feed and expansion cistern combine to form a safety relief and cold fill system that protects the building occupants by preventing the heating system and boiler from drying out and rupturing, or even – as in the case of cast metallic heat exchangers – exploding. Careful consideration should therefore be given at the design stage to deciding on their location and method of connecting into the heating system. All too often this third point is forgotten, especially in domestic heating systems, in order to reduce the amount of pipework being installed and this compromises the safety of the heating system. The design principles to be applied are quite simple if the operating conditions of this safety aspect are understood. If the burner of the boiler fails to switch off for some reason, the water inside the heat exchanger will continue to rise above the design operating temperature, 82°C for low pressure heating systems, until it reaches boiling point and starts to generate steam. This steam should be allowed to escape up the open vent pipe and discharge this high temperature water and steam mixture into the feed and expansion cistern, thus relieving any build up of pressure. Inside the cistern it mixes with the lower temperature water together with any incoming cold make-up water supplied via the float valve to replace the water lost through steam being generated and escaping into the roof space. This cools the water down slightly before it is returned to the boiler by way of the cold feed pipe, thereby keeping the boiler heat exchanger wet and preventing it from fracturing through stress caused to the metal.

To relieve the boiler and prevent the heat exchanger drying out, the open vent should be arranged to rise vertically from the boiler, either as a separate pipe, or forming part of the heating circuit providing that there are no valves, manual or motorised, that could be shut off, with the cold feed extended down to enter the heating system as close to the boiler as possible. The cold feed can also utilise a heating return pipe if there are no valves on the circuit that could isolate the cold feed from the boiler.

Figure 2.26 (a) shows a good arrangement whereby the open vent rises unhindered to allow the high temperature water/steam to be discharged safely into the feed and expansion cistern; the cold feed also has a direct route back to the boiler to supply into the heat exchanger.

Figure 2.26 (b) shows the dangerous practice of combining the cold feed and open vent into a single common pipe. In this arrangement, the cold feed back to the boiler is prevented by the pressure of the high temperature water/steam pushing up the vent and forcing the cold water with it, thus preventing any water re-entering the system with the result that the boiler will eventually become dry and be in a dangerous condition.

Figure 2.27 illustrates two similar alternative methods of connecting the open vent pipe and cold feed pipe into the heating system. These methods are fairly common for domestic heating systems employing high resistance low water content boilers.

Arrangement (a) shows a method known as a 'close coupled system', where the open vent pipe is taken off the heating circuit 150 mm in front of where the cold feed pipe connects into the heating circuit. This provides a clear passage for both the open vent to relieve pressure and the cold feed to supply water back to the boiler to keep the boiler safe, provided there are no valves between the entry of the cold feed and the boiler.

Arrangement (b) is very similar, except that a proprietary fitting known as an 'air separator' is used to connect the two pipes into the heating circuit as shown.

The air separator is designed to allow any trapped air or free oxygen to escape through the open vent as the velocity of the flow of water through this component is much slower than through the enlarged section of the fitting; however, the use of a short section of enlarged pipe, usually one commercial pipe size bigger than the circuit pipe as shown in arrangement (a), has the same effect.

One of the main reasons why the arrangement of the cold feed and open vent pipes, as depicted in Figure 2.27, has become popular, is that it makes the relationship between these two pipes and the

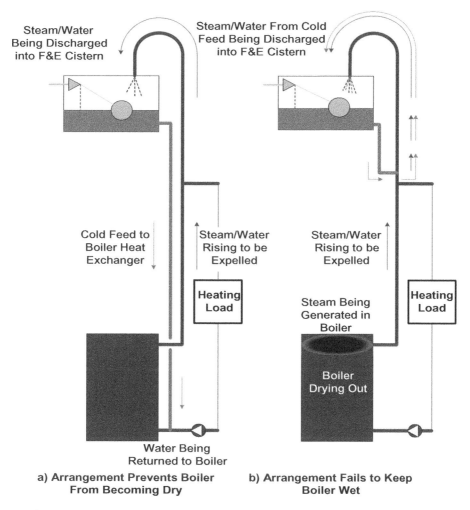

Steam/Water Being Discharged into F&E Cistern

Steam/Water From Cold Feed Being Discharged into F&E Cistern

Cold Feed to Boiler Heat Exchanger

Steam/Water Rising to be Expelled

Steam/Water Rising to be Expelled

Heating Load

Heating Load

Steam Being Generated in Boiler

Boiler Drying Out

Water Being Returned to Boiler

a) Arrangement Prevents Boiler From Becoming Dry

b) Arrangement Fails to Keep Boiler Wet

Figure 2.26   Combination of cold feed, open vent and F&E cistern during boiler malfunction

circulating pump much simpler as there is less risk of pumping over the open vent or drawing air into the heating system, and the domestic heating installer does not have to calculate the pump effects.

This section has concentrated on the relationship between the feed and expansion cistern, cold feed pipe and the open vent pipe during their normal operation and how they combine in the event of a boiler malfunction. There is also an equally important relationship between these two pipes and the circulating pump, which is explored further in Chapter 18, Circulating Pumps.

It should be noted that the arrangements illustrated in Figure 2.27, of connecting the cold feed and open vent pipes into the heating circuit, are only acceptable if there is a clear route for them to function correctly: if there are any manual valves, motorised zone valves or diverting valves which could cause them to close the path from or to the boiler, then an alternative route and method of connecting them should be sought. This is more likely to occur on the return path of the cold feed to the boiler.

The simple principle that should be applied in all cases, whether large commercial heating installations or small domestic heating systems for residential dwellings, is that a clear and unobstructed direct path

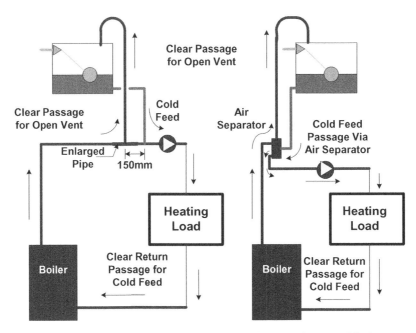

Note - Open Vent & Cold Feed Must Have a Clear Unobstructed Path

a) Arrangement of Open Vent &   b) Arrangement of Open Vent &
Cold Feed Via Enlarged Pipe   Cold Feed Via an Air Separator

Figure 2.27   Alternative arrangement for open vent and cold feed

should exist for both the open vent pipe and the cold feed pipe from the boiler to the feed and expansion cistern.

INTRODUCTION TO AND HISTORY OF SEALED HEATING SYSTEMS

The subject of this chapter is hydronic heating systems and the preceding text has covered the importance of pressure and temperature within these systems. It has also covered the fundamental principles for the different piping arrangements and included many examples in illustrative form of the various pipework layouts. With one exception, all of the piping illustrations have been depicted as being low pressure open vented heating systems. The purpose of these pipework illustrations is to demonstrate the methods of distributing the heat carrying medium from the point of heat generation to the various heat emitters incorporated within the system and back again. These drawings of the heating systems could just as easily have been depicted as low pressure sealed heating systems as the piping arrangements would be exactly the same. Also, the piping arrangements would be the same if the heating system was classified as being either medium pressure or high pressure, but the system would have to be sealed to ensure that the operating pressure was above the distribution temperature to prevent the water from boiling.

The preceding text has discussed the merits and principles of the open vented arrangement. A sealed system of heating may be employed to achieve the same objectives, the only differences being the method of filling the heating system with water and the means of catering for the expansion of the water when heated.

Sealed heating systems are considered by some to be fairly new to the United Kingdom heating market. However, they were first introduced into Britain by an American, named Jacob Perkins (1766–1849), and

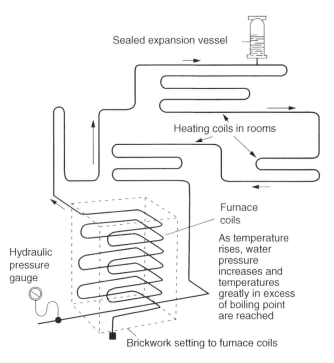

Sealed expansion vessel

Heating coils in rooms

Furnace
coils

As temperature
rises, water
pressure
increases and
temperatures
greatly in excess
of boiling point
are reached

Hydraulic
pressure
gauge

Brickwork setting to furnace coils

Figure 2.28 Perkins system of heating

his son, Angier March Perkins (1799–1881), who, together with both of their familes had emigrated to England from Massachusetts USA. The exact date of their arrival into this country is believed to be somewhere between 1815–1827, following the end of the Napoleonic Wars. The Perkins family were engineers and inventors and had previously experimented with sealed and pressurised heating systems employing high temperature hot water before they left America to settle in the Midlands of England.

The system that they introduced into Britain was quite new at the time. It comprised a number of continuous single pipe loops that distributed heat around the building. Part of it was coiled around inside a brick furnace and the whole system sealed by means of an upturned enlarged pipe arranged to trap air to act as a cushion, which served as an expansion vessel, see Figure 2.28.

The system, which was known as the 'Perkins system of heating', is believed to have had a working pressure approaching 20 bar, with operating temperatures of nearly 150°C. The system operated by gravity circulation with each pipe loop being limited to about 150 m in length, and 15% of each circuit being coiled inside the furnace as the heat exchanger. If one pipe circuit was insufficient to heat the building, then additional pipe loops were incorporated, each with their own expansion vessel and pipe section coiled inside the common furnace.

One of the more interesting features of the Perkins heating system was the choice of piping materials used to convey the high temperature and high pressure water as the heating medium, and in particular, the method of jointing the piping material. Perkins chose steel as the piping material for his heating system, which by today's standards would seem the obvious choice for the high pressure and temperature involved. It must be appreciated however, that at that time, Henry Bessemer had not yet invented the 'Bessemer converter' to enable mass production of cheap steel for pipes and tubes. Therefore, as steel was in short supply, and what steel was manufactured was mainly aimed at military use, Perkins opted to use steel tubes originally manufactured as gun barrels, hence the term 'barrel' when referring to steel tube. Since gun barrels have not been used for piping materials since the mid 1800s, this term should not be used today.

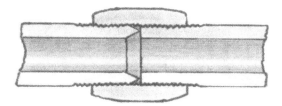

Figure 2.29   Section through a Perkins left- and right-handed thread joint

The steel tube selected had an outside diameter of approximately 1 inch (25 mm), and was made in short lengths of approximately 1–1.5 m long that could be easily bent by the application of heat when placed inside a forge. The only fittings that were required were a straight socket also made from steel for coupling the steel tube together. The steel tube was threaded at both ends, one thread being a right-handed thread and the other being a left-handed thread; one end of the tube was chamfered as shown in Figure 2.29, which when tightened up, and without the aid of any jointing compound or material, would be pulled tight up against the square cut flat face of the mating pipe section forcing the chamfered ended tube to dig in to the flat-faced end.

This type of pipe coupling – having a left-handed thread and a right-handed thread that enabled the pipes to be drawn up together and, if required, to be broken apart without having to undo long lengths of pipeline – may be described as an early form of pipe union.

This joint would have required a high degree of skill on the part of the pipe fitter to ensure that the tube ends were cut perfectly square to make a watertight joint without the aid of any jointing compound when high pressure was exerted. It is believed by many today that this high degree of accuracy in pipe jointing could not be achieved by the majority of today's pipe fitters.

There are still a few remnants of this early form of sealed heating systems in existence in churches and chapels throughout the country.

SEALED HEATING SYSTEMS

Although the sealed system of heating had been first introduced to Britain as far back as the early 1800s, its development took place in North America, Scandinavia and other parts of Northern Europe, where it was considered more important to remove the feed and expansion cistern and its associated pipework from the roof void to reduce the risk of these services freezing in winter. In Britain, designers favoured the open vented low pressure system of heating. The sealed system of heating was still installed occasionally in commercial buildings, but almost never in domestic residential buildings.

As previously established, the only difference between the sealed system of heating and the open vented heating system is the method of catering for the expansion of water due to temperature rise when the boiler is operating. This is achieved by eliminating the feed and expansion cistern, open vent pipe and the cold feed and expansion pipe and replacing them with a sealed expansion vessel. This means that as the feed and expansion cistern no longer exists, an alternative method of filling the heating system with water has to be incorporated.

The earliest form of expansion vessel used was not dissimilar to the method used in the Perkins heating system, which comprised an enlarged pipe with both ends capped off as shown in Figure 2.30, and was located at the highest point of the heating system in an upturned manner to trap air inside when the system was being filled with water. When the water became hot it would expand into the expansion vessel and compress the air at the top of the vessel. It also incorporated a valve at the top to allow the system and vessel to be recharged with air, together with a drain-off valve.

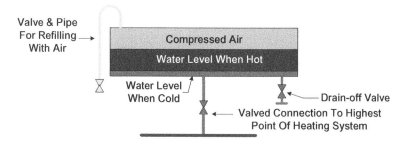

Figure 2.30   Early form of sealed expansion vessel

The common failing of this type of expansion vessel was that the air's oxygen content – approximately 21% – would gradually be absorbed into the heating water as there was no method of separating the air and the water, which eventually led to the air content becoming insufficient to accommodate the expanded volume of water. This absorption of air into the heating water also contributed to corrosion of the heating systems pipework and heat emitters.

There were variations of this type of expansion vessel, including a vertically mounted version and an adaptation that created an air-lock at the top of the boiler heat exchanger by inserting a dip-tube on the flow connection and using the opposite flow connection as the air refilling point. This had the effect of creating an area to trap air at the top of the boiler heat exchanger which would act as a cushion in the same way as the enlarged pipe version of the sealed expansion vessel.

Because of the difficulties in air cushion retention and the potential corrosion problem, coupled with the less severe winters experienced by the United Kingdom compared with countries such as Canada, the USA, Scandinavia and parts of Northern Europe, where it is imperative to remove all services at risk of freezing, it is not difficult to understand why the UK preferred to develop the less troublesome open vented system of heating at that time.

Interest in the sealed heating system re-emerged in Britain during the 1960s, following the resolution of the corrosion problem caused by retention of the air cushion and the elimination of the dissolved oxygen by the development of the sealed expansion vessel, complete with flexible separation diaphragm membrane, which made this system more attractive to heating engineers as an alternative method of heating.

The operation of the sealed heating system is exactly the same as the open vented heating system shown in Figure 2.31. This is a sealed heating system version of the two pipe open vented heating system depicted in Figure 2.10, the only difference being that the increased volume of water caused by expansion is catered for by the sealed expansion vessel in the absence of the feed and expansion cistern, cold feed and open vent pipe. Any of the other open vented heating systems, such as one pipe, two pipe, micro bore or radial piping arrangements previously shown could be arranged as sealed heating systems.

Employing a sealed expansion vessel to cater for the increased volume of water will also result in the heating system becoming pressurised above the normal hydrostatic head, although for a domestic heating system this pressure increase will be only slight.

The advantages claimed for sealed heating systems include:

1. Little if any make-up water is required once the system has been charged and all air vented, because none is lost through evaporation as is the case with the feed and expansion cistern of the open vented system. A feed and expansion cistern with a correctly fitted cover in accordance with the Water Regulations will experience very little evaporation.
2. The absence of the feed and expansion cistern eliminates the possibility of a fresh supply of oxygen being absorbed by the heating water, thus reducing the risk of corrosion occurring in the heating system. Also,

the feed and expansion cistern is an ideal breeding place for bacterial activity, organic impurities and fungi, and the omission of this component eliminates these potential problems.

3. A cost saving is sometimes claimed by the exclusion of the feed and expansion cistern, cold feed and open vent pipe over the installed cost of the sealed expansion vessel, although the cost difference is very small.

4. The removal of the feed and expansion cistern, cold feed and open vent pipe from the unheated roof space eliminates the possibility of these components freezing up in winter. Although they should be insulated in accordance with the Water Regulations to prevent this occurring, if they don't exist, then they cannot freeze.

5. One of the main merits of the sealed heating system, although not normally recommended for domestic heating installations, is its ability to take advantage of the higher operating pressure that occurs and to operate at flow temperatures above the normal 82°C that is used for open vented heating systems. It is possible to operate at temperatures above 100°C without causing the water to boil, or flash to steam. This is possible because the boiling point of water is directly related to pressure: water will boil and vaporise at 100°C, at atmospheric pressure and at sea level. If the pressure is reduced then water will boil at a temperature lower than 100°C and likewise, if the pressure is increased, the boiling temperature will be in excess of 100°C. Therefore, because the sealed heating system is pressurised above atmospheric

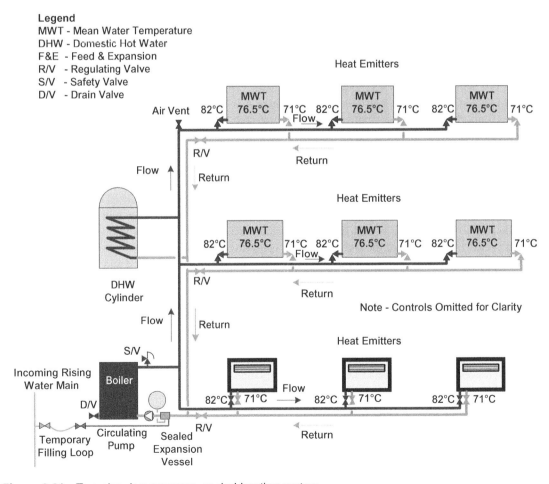

Figure 2.31   Two pipe, low pressure, sealed heating system

pressure, it is possible to operate at temperatures near to or above 100°C without causing the water to vaporise. Operating temperatures higher than 82°C permit the selection of smaller heat emitters and smaller circulating pipe sizes as less water is required to convey the same amount of heat; there is therefore a cost saving compared to an open vented heating system operating at a flow temperature of 82°C and a return temperature of 71°C. Figure 2.32 shows the same heating system depicted in Figure 2.31, but taking advantage of a higher operating temperature.

The version of the sealed heating system shown in Figure 2.32 is still classified as a low temperature and low pressure heating system as the operating temperature is below 100°C and the operating pressure is below 3 bar.

If the operating temperature selected is higher than 82°C flow, which means that the mean water temperature of the heat emitter will be higher than 76.5°C when a return temperature of 71°C is used, then careful consideration should be given to the choice of heat emitters to be used. Standard pressed steel radiators or column type radiators should be avoided as the surface temperature would be high enough to cause skin burns to anybody who came into physical contact with them. The correct choice would be to select convector type heat emitters, or low surface temperature type radiators, explained and detailed in

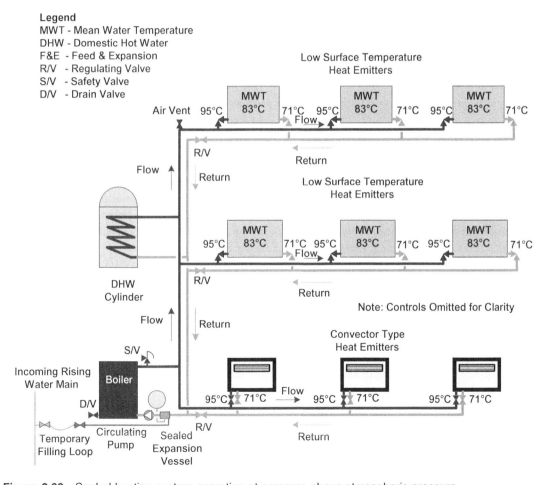

Figure 2.32   Sealed heating system operating at pressure above atmospheric pressure

Chapter 5, Heat Emitters, and illustrated in Figure 2.32, where the mean water temperature in the heat emitters is 83°C.

It also follows that all piping should be installed in a manner that, under normal circumstances, prevents people coming into contact with the surface of the pipe, as the result will be the same as that for standard radiators. Therefore, for safety reasons, the pipework should be installed in service ducts, or in floor or ceiling voids.

It should also be appreciated that the piping material and jointing method specified should be suitable for the higher operating pressure expected and that the standard of workmanship is of the highest quality. If we are operating at temperatures above 100°C, and a fracture of the pipe occurs, or a pipe joint fails, then the release of pressure from the system would cause the water to flash to steam in a violent and dangerous manner. These are the reasons why higher operating temperatures and pressures are not recommended for domestic dwellings as the control of future maintenance to each domestic application cannot be guaranteed.

EXPANSION VESSELS

The function of the expansion vessel or compression tank, as it is sometimes known, is to provide a volumetric space for the increased volume of water caused by expansion when heated, and also to maintain the heating system at a positive working pressure. It will also serve to force the water back into the heating system when the temperature of the water cools down.

Expansion vessels are available in a variety of shapes and sizes, such as direct pipe mounted, floor or wall mounted, circular, oval or cylindrical, but whatever the capacity or shape of the expansion vessel it may be categorised as being one of two basic styles. Both types take the form of a sealed steel outer vessel divided internally into two compartments by either a centrally permanently fixed diaphragm or separate replaceable membrane, each manufactured from a flexible synthetic rubber material. One side of the flexible diaphragm or membrane is sealed from the heating system and is charged with nitrogen or air; the other side is open and is connected directly to the heating system and arranged to accept entry of the expanded water into the vessel. The advantage of having a membrane within the expansion vessel is to separate the air/nitrogen from the heating circulating water, preventing any of the charged gas being absorbed into the water. This reduces the corrosion potential and permits the vessel to be installed in any plane or position, not just in an inverted position above the pipe, as previously required with expansion vessels prior to the development of the flexible membrane expansion vessel, as the air or nitrogen charge would be lost within the circulating water.

Nitrogen is considered to be the most suitable gas to provide the expansion cushion because it is less soluble in water than air in the event of the diaphragm failing and the charge being partially lost. This is due to the fact that uncontaminated air contains approximately 78% nitrogen and 21% oxygen and it is the oxygen content that is absorbed into the water more readily if air is employed, as pure water consists of two parts hydrogen to one part oxygen, H_2O, so that oxygen is easily absorbed. Nitrogen, in simplistic terms, is said to be an inert gas, but Henry's Law has demonstrated that nitrogen will migrate into the water over a period of time, at a quantity and rate dependent upon the temperature and pressure of the water, but any failure in the membrane of the expansion vessel would have been discovered and rectified long before a migration of nitrogen occurred.

A high percentage of expansion vessels used in the domestic heating market employ air as the expansion cushion as it is readily available and the sealed expansion vessels can be recharged or topped-up using a car foot pump connected to the air valve on the vessel, which is similar to that used for a car tyre. Figure 2.33 illustrates the working principles of the sealed expansion vessel showing the difference between the two types.

(a) Fixed Membrane Type

(b) Renewable Membrane Type

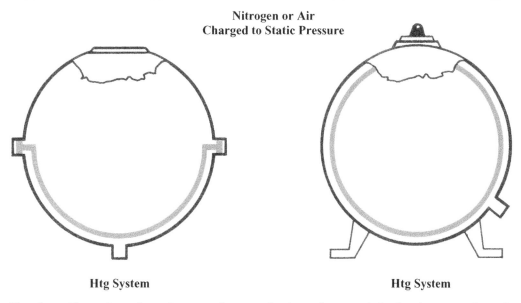

The above illustrations show the expansion vessels charged to equal the heating system's static pressure, which for low pressure heating systems should be the cold fill pressure preventing any of the heating systems water entering the vessel when cold.

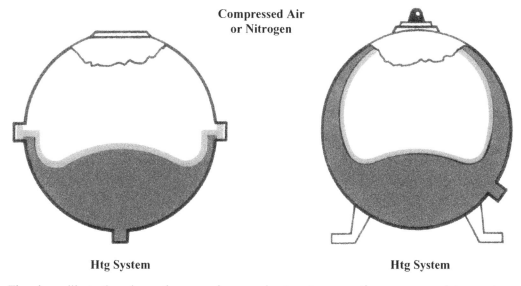

The above illustration shows the expansion vessels at system operating pressure and temperature whereby the heated expanded water has entered the vessel causing the pre-charged nitrogen or air to compress and so create the increase in pressure.

Figure 2.33 Operating principles of the sealed expansion vessel

AIR SEPARATION

The entry and associated problems of air in the heating system that have been touched upon briefly above and are dealt with in more detail in Chapter 21, Water Treatment. The following paragraphs are confined to the elimination of air in the sealed heating system so that the claimed advantage of reducing corrosion in the heating system can be proved.

Air is a cocktail of gases; the exact composition of unpolluted air is given in Chapter 16, Combustion, Flues and Chimneys, but for these purposes and as established earlier, air may be taken as comprising approximately 78% nitrogen, 21% oxygen and 1% other inert gases. What is often referred to as 'air in the system', is in fact oxygen or nitrogen and, if corrosion is occurring, it could be caused by other gases.

Air will exist in the heating system when the installation has been completed and prior to the initial filling up with water; during this process the tedious task of releasing this air from the system high points and heat emitters should be undertaken, but the release of 100% of the air is not always achieved. Also, additional air is being introduced by the water being used to fill the system in the form of absorption, all of which can lead to potential problems if the air is not completely eliminated.

When the heating system is operating, the circulating water is continually heated up and cooled down, and air is released from the water or re-absorbed back into the water, in direct proportion to the temperature and pressure change of the heating system water. As water is made up of two parts hydrogen and one part oxygen (H_2O), it is normally the free oxygen content that comes out of solution or is re-absorbed back into the water rather than air.

The graph shown in Figure 2.34 demonstrates this temperature/pressure relationship of the solubility of air in water, whereby if we have water at a temperature of 20°C at a pressure of 0.7 bar it will hold a maximum of 3% air by volume in solution, but if we raise the temperature of the water to 95°C at the same

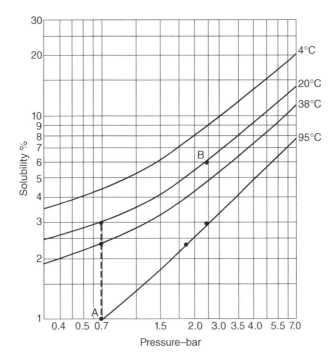

Figure 2.34   Solubility of air in water pressure/temperature graph

pressure it will only hold a maximum of 1% air in solution, having given up 2% of the air that has come out of solution in the form of air bubbles travelling around the heating system. This shows that the colder the water temperature, the greater the capacity to hold air in solution.

Using the same example of water at 20°C and at a pressure of 0.7 bar, if we maintain the temperature of 20°C, but raise the pressure to 2.5 bar the water can now hold a maximum of 6% air in solution. This demonstrates that the greater the pressure of the water, the greater the ability to hold air in solution.

In an open vented heating system this air released from solution by the rise in temperature would be allowed to escape up the open vent pipe to the atmosphere, where it would later be partially re-absorbed when the water cooled back down by surface contact in the feed and expansion cistern. In a sealed heating system if this air is not vented off then it will be re-absorbed into solution when the water cools down where it will remain, only to appear again later when the temperature increases again. If this cycle is permitted to continue unchecked, then the free oxygen content of this air that has come out of solution will contribute to the promotion of corrosion in the heating system, plus the air circulating around the heating system could produce either partial or full air-locking in the piping system and reduce the performance of the heating system.

In a sealed heating system, it is essential that this air is separated out of the circulating water and vented off to the atmosphere. There are two basic components commercially available to perform this separation process, both equally effective and similar to each other.

Air purger

This component consists of a cast iron chamber constructed of two passages. The lower passage permits the flow of water being circulated to pass through the air purger unhindered; the upper passage is designed to channel any air that is being dragged along the pipework in the form of bubbles, to be collected in the domed section at the top where it can then be released via an automatic air vent fitted to the tapping provided.

Figure 2.35 illustrates the application of the air purger whereby the main function is to combine with an automatic air valve to permit any free air to enter the enlarged upper chamber where the velocity will be slower, allowing the air bubbles to escape. It also has additional tappings to allow for a combined high and low pressure switch to be fitted, arranged to cut the boiler out if abnormal pressures occur, plus provision for fitting a pressure gauge so that the pressure can be visually monitored. The air purger can also accommodate the fitting of the sealed expansion vessel in the pendant position, as shown in Figure 2.35, or by extending the connection by piping, as illustrated in Figures 2.31 and 2.32 to an adjacent inverted position. This multifunctional component in addition incorporates the provision to connect the water fill point to the heating system as also shown in Figures 2.31 and 2.32.

The cast iron multi-purpose air purger tee is commercially available for both domestic heating applications and larger commercial sealed heating systems.

Air separator

As shown in Figure 2.27 (b), the air separator arranged for an open vented heating application is a similar device to the air purger and is equally suitable when applied to the sealed heating system to perform the function of separating out the air and eliminating it, but on a sealed heating arrangement the open vent pipe is replaced with an automatic air release valve.

Figure 2.36 shows the operation of an air separator, which consists of an enlarged cylindrical vertically mounted vessel having an inlet connection high up on one side of the cylinder and an outlet located low down on the opposite side of the cylinder. Mounted on the top of the cylinder is a tapping for fitting an automatic air release valve. The circulating water enters the top of the separator where the enlarged area

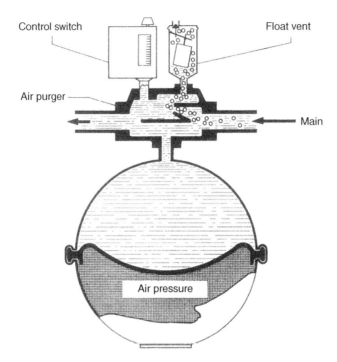

Figure 2.35   Air purger

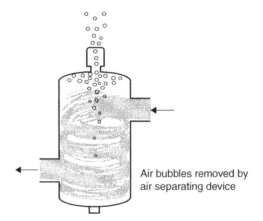

Figure 2.36   Air separator

causes the velocity to slow, allowing the air bubbles to rise and be expelled by the air vent and the water, minus the air that has come out of solution, continues to the outlet to be circulated.

The design of the air separator, or tangential type air separator, is to separate entrained air from flowing system circulating water by the creation of a vortex, allowing free air bubbles to rise vertically in the centre, which is the point of least velocity. Some larger air separators may be constructed incorporating curved vanes, discs or spiral tubes which aid the formation of the vortex and assist the dissolved air to come out of solution. Examples may be seen in Figures 2.37 and 2.38.

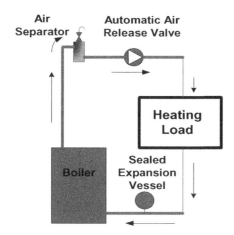

Figure 2.37 Application of spiral air separator/air elimination valve

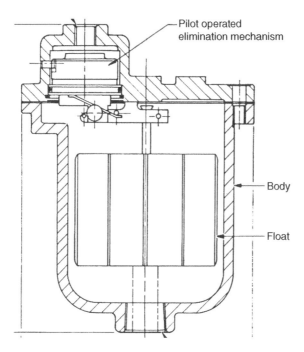

Figure 2.38 Detail of automatic float operated air elimination valve

SYSTEM CHARGING

With an open vented heating system, the feed and expansion cistern is also employed as a reservoir of water to fill the heating system and keep it topped up automatically without the need for any human involvement. For a sealed hydronic heating system, because we have eliminated the feed and expansion cistern in favour of the sealed expansion vessel we have to find an alternative acceptable method of filling the system up with water and keeping it topped up.

The methods used to charge the sealed heating system with water vary from one country to another and here in the United Kingdom the methods permitted have also differed over the years. This is mainly due to the regulations appertaining to water supply in different countries. In England and Wales any connection to the water supply must conform to the 'Water Regulations' and in Scotland to the 'Water Byelaws' as well as the previous water byelaws that were in force at the time of any installation before the current regulations. It should be appreciated that the purpose of these regulations and the previous byelaws is to prevent wastage, misuse and contamination of the water supply and as the water supply in Great Britain is classified as being Fluid Category 1, 'wholesome', which is suitable for drinking water, the quality of this supply should be protected.

The numerous methods, both past and present, of filling the hydronic sealed heating system are reflected in Figure 2.39, but this schematic illustration should not read as a working drawing as it is intended only to demonstrate alternative arrangements: some of the methods depicted are considered no longer acceptable and are only included to help students understand a system if they become exposed to it for the first time.

Box 2.2 summarises three approved filling methods.

Method 'A'

This arrangement indicates a simple form of filling the heating system that was used to a limited degree on small domestic heating systems in the 1960s, but is considered unacceptable by today's standards and is now prohibited. It comprised a clear plastic bodied container with a capacity of 3–5 litres, incorporating a see-through sight contents indicator that was required to be located above the highest part of the heating system. From this container it was connected to the heating system via a flexible hose that included a non-return valve and automatic air release valve. The heating system was filled up from a hose via this container and during normal working operation the system became pressurised by the action of the non-return valve preventing any expanded water returning to the container. Any water loss would be replenished from this container when the heating system cooled down.

This method was dependent upon the occupier of the dwelling frequently checking the water level in the container and topping it up if necessary.

The unit was also described as an automatic top-up container because it would replace any water lost from the system when it cooled down, but as it required manual top-up this was not an accurate description.

Method 'B'

This is another defunct and prohibited method that made use of the main cold water storage cistern serving the domestic hot and cold water for the property where a separate dedicated cold feed was extended from the cold water storage cistern to the heating system.

It incorporated a non-return valve to prevent water expanding back up the cold feed into the storage cistern and was also used in the 1960s before it was common practice to apply chemical treatment to the heating circulating water.

Any existing heating systems found to have this method of water filling should be converted to an approved arrangement as the risk of contaminating the domestic water in the storage cistern is extremely high, with resulting serious consequences to health.

It should also be pointed out that today's practice is to install an unvented domestic hot water system in association with sealed heating systems, so that there is no longer a need for a cold water storage cistern.

Method 'C'

This illustrates what is commonly referred to as a 'temporary filling loop' and has been the most popular method of filling sealed heating systems in domestic residential dwellings as it is an inexpensive, approved arrangement and satisfies the requirement in 'A' above.

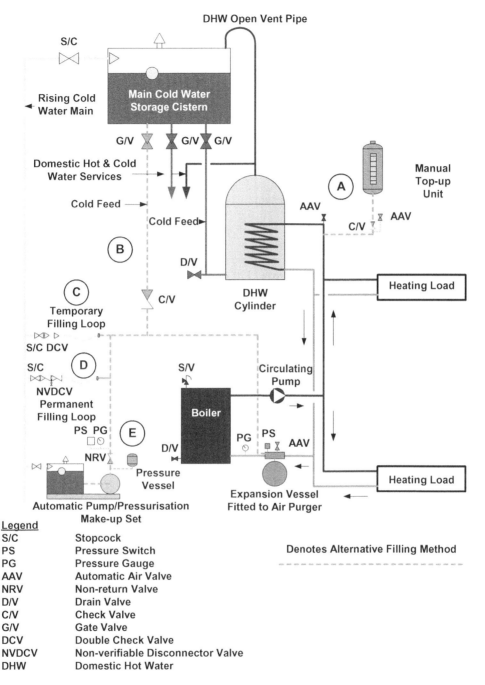

Figure 2.39 Alternative arrangements for filling sealed heating systems

Figure 2.40 illustrates in detail the temporary filling loop arrangement that satisfies the requirements of the Water Regulations. It comprises a braided hose connection made on the pipework with union connections between two resilient seat valves (stopcocks), together with a double check valve. It is permitted on domestic heating systems designated Fluid Category 3 (domestic heating systems are defined as having a heating capacity not exceeding 45 kW). The heating system is filled up by manually opening both of the

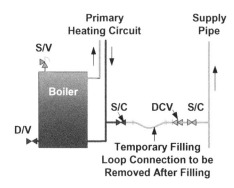

Figure 2.40   Temporary filling loop arrangement

BOX 2.2   Summary of approved filling methods

Approved Filling Methods
The Water Regulations requirements for filling sealed heating systems are:

No supply pipe or secondary circuit shall be permanently connected to a closed circuit for filling a heating system unless it incorporates a backflow prevention device.

A) *A temporary connecting pipe which must be completely disconnected from the outlet of the backflow protection device and the connection to the primary circuit after completion of the filling or replenishing procedure.*
B) *A device which in addition to the backflow prevention device incorporates an air gap or break in the pipeline which cannot be physically closed while the primary circuit is functioning.*
C) *An approved backflow prevention arrangement.*

---

isolating stopcocks to allow water from the supply pipe to charge the heating system. When the pressure in the heating circuit equals the pressure in the water main the isolating stopcocks are shut off and the temporary filling loop removed immediately after the heating system has been filled or replenished.

Method 'D'

This is a more expensive alternative to the temporary filling loop described in method 'C'. It satisfies the requirements of the Water Regulations stated in 'B' on the previous page for a Fluid Category 3 connection, and is illustrated in detail in Figure 2.41.

This filling method employs a permanently fixed backflow prevention device designated by the Water Regulations as a type CA non-verifiable disconnector with different pressure zones. If this valve fails during charging it will prevent backflow occurring by opening the relief port between the two check valves and discharging the water to drain over a tundish installed with a type AA air gap of not less than 20 mm. This arrangement has the advantage of being permanently fixed and although the two isolating stopcocks should normally be in the closed position and only opened to charge or replenish the heating circuit, it is far more convenient to use.

For filling heating systems in buildings other than houses, or where the heating system capacity exceeds 45 kW, a backflow prevention device suitable for a Fluid Category 4 risk is required, as noted previously in

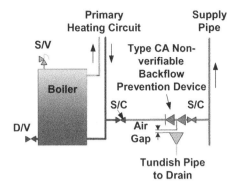

Figure 2.41   Permanently fixed filling pipe arrangement utilising a type CA valve

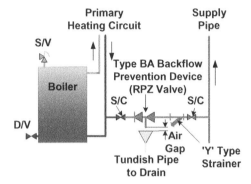

Figure 2.42   Permanently fixed filling pipe arrangement utilising a type BA valve

method 'C'. This backflow prevention device is designated under the Water Regulations as a type BA verifiable backflow preventer with reduced pressure zone, commonly referred to as an RPZ valve assembly, and accompanied by an inline 'Y' type strainer on the inlet for protection, which takes the place of the type CA non-verifiable disconnector shown in Figure 2.41. The arrangement is similar to that described for the type CA valve, including the type AA air gap between the outlet relief port of the RPZ assembly and the tundish drain. This arrangement should also normally be closed and the isolating stopcocks only opened when the system is to be filled, or replenished with make-up water. Figure 2.42 details the arrangement.

It should be noted that if a BA verifiable reduced pressure zone valve is used, to comply with the Water Regulations it must be tested annually by a trained and certified person to ensure that it is in optimum working condition.

Method 'E'

For non-domestic larger heating installations designated Fluid Category 4, an alternative arrangement to the reduced pressure zone valve assembly for filling the heating system is a packaged pressurisation pump set that comprises a small break cistern directly connected to the water supply main controlled by a float valve, a centrifugal pump, pressure vessel, pressure control switches and safety relief valve. The pump draws water from the break cistern to fill up the heating system. The pump is automatically controlled by a low pressure switch to start the pump and a high pressure switch to stop it when the selected pressure setting has been reached. The pressure vessel contains a membrane similar to that in the sealed expansion vessel and is supplied to automatically replenish small water losses and to prevent the pump from 'hunting'.

All the other approved methods of filling the heating systems can only achieve a pressure equal to that of the water supply mains pressure, and consequently can only be used on low pressure heating systems. The pressurisation pump set has the advantage of being able to pressurise the system beyond that of the water supply mains pressure and can therefore be used to provide medium and high pressure heating arrangements.

Although this arrangement is the most costly for the initial installation, it is totally automatic in its operation and does not require any human input to top up the system. It can be arranged to incorporate high and low pressure alarms to alert in the event of a failure.

SIZING THE EXPANSION VESSEL

When designing a sealed heating system it is critical that the size of the expansion vessel is calculated correctly as rule of thumb methods generally lead to either under- or over-sizing.

As previously established, when water is raised in temperature its volume increases by an amount dependent upon the initial and final temperature of the water. When calculating the size for an open vented feed and expansion cistern, an allowance of 5% increase in the heating system's water content is acceptable for arriving at a physical size for the feed and expansion cistern. However, when considering the size required for a sealed expansion vessel, a more accurate approach is necessary to enable it to function correctly.

The graph shown in Figure 2.43 may be used to determine the approximate percentage volume increase due to the expansion of water at different temperatures.

The exact amount that water will expand is dependent upon the total volume of water in the heating system and the mean water temperature (MWT). The MWT is the average water temperature and is found by the following:

For example – flow temperature 82°C, return temperature 70°C.

$$MWT = \frac{82°C + 70°C}{2} = 76°C$$

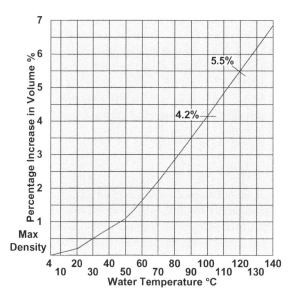

Figure 2.43 Expansion of water above 4.4°C at different temperatures

Assuming a heating system has a total water content of 1200 litres, Figure 2.43 shows that the expansion for MWT at 76°C equals a 2.6% increase in volume, therefore:

$$1200 \text{ litres} \times 0.026 = 31.2 \text{ litres of water increase}$$

The static head or fill pressure of the system must be established to enable the initial gas charge of the expansion vessel to be applied, as the diaphragm within the vessel must be able to support the cold fill water pressure acting upon it. The static head is measured vertically from the expansion vessel to the highest point of the heating system, or equal to the cold fill pressure if from the water supply mains or pressure set. A gas charge pressure of 1 bar will support a column of water 10.2 m high. If the expansion vessel charge pressure is not able to support the cold fill static head, water will be allowed to partially enter the vessel whilst still at its cold pre-expanded temperature and decrease the vessel's holding capacity when hot and expanded.

If the cold water supply is to be from the water mains, then the mains pressure should be monitored over a period of time, using a test pressure gauge incorporating a maximum and minimum slave pointer needle to establish the pressure variation over this period and the vessel gas charge pressure adjusted as necessary at time of charging.

Only a portion of the total expansion vessel's volume is used to accommodate the expanded water. The initial gas charge occupies the total volume of the vessel when cold, but as the system's water heats up and expands, the water begins to be forced into the vessel pushing against the diaphragm and compressing the gas on the other side. This is illustrated in a series of conditions in Figure 2.33.

There is a pressure limitation for the expansion vessel and the maximum acceptance efficiency is the relationship between the volume accepted at working pressure and total vessel volume.

$$\text{Vessel Acceptance Efficiency} = \frac{\text{Acceptance Volume}}{\text{Total Vessel Volume}}$$

The vessel efficiency may be expressed using Boyle's Law, which states that 'the product of the pressure and the volume of a perfect gas will remain constant as long as the temperature is constant' (isothermal relationship). If the pressure increases, the volume decreases.

$$\text{Vessel Efficiency} = \frac{\text{Final Pressure} - \text{Initial Pressure}}{\text{Final Pressure}}$$

Absolute pressures should be used, i.e. gauge pressure + 1 bar.

e.g.   Vessel charge pressure 1.5 bar   (2.5 bar absolute)
       Working pressure 4.0 bar   (5 bar absolute)

Therefore:

$$\frac{5 - 2.5}{5} = 0.5 \text{ Acceptance Factor, or 50\% efficient.}$$

With the above understanding, the following formula may be used for assessing the size of the expansion vessel for heating systems having a design MWT of between 70°C to 140°C with a cold fill temperature of 4°C. Its degree of accuracy is from +0.7% to −10.4%.

Formula:

$$V_t = \frac{(0.000738t - 0.03348)\, V_s}{1 - \dfrac{P_f}{P_o}}$$

Where V_t = Minimum volume of expansion vessel in litres

   t   = Mean Water Temperature in °C

   V_s = Total system water content in litres

   P_f = Initial or minimum operating pressure at vessel expressed in kPa absolute

   P_o = Final or maximum operating pressure at vessel expressed in kPa absolute

Example:

System water content (V_s)	1000 l	
MWT (Flow 82°C, Return 70°C) (t)	76°C	
Minimum operating pressure (P_f) 0.5 bar	50 kPa Gauge	
Maximum operating pressure (P_o) 2 bar	200 kPa Gauge	

Therefore:

1. $V_t = \dfrac{(0.000738 \times 76 - 0.03348)\, 1000}{1 - \dfrac{50 + 101.3}{200 + 101.3}}$ Absolute Pressure

2. $\dfrac{22.608}{0.498}$ litres of expanded water

   Acceptance factor

3. V_t = 45.4 litres minimum volume, plus 10% factor for possible error in formula:
4. V_t = 45.4 + 4.54 = 49.94 (say 50 litres)

Select the standard expansion vessel having a total volume capacity nearest and above this figure.

It should be appreciated that the higher the operating temperature and the resulting continuous flexing of the internal diaphragm caused by the expansion and contraction of the heating water, the shorter the life expectancy of the membrane. To prolong the life expectancy of the membrane the vessel can be oversized by applying an additional 10–15% to the capacity.

By using the gas laws previously mentioned, it is possible to calculate the theoretical pressure developed when a vessel has accepted a given volume of water. The actual pressure developed in the heating system is less than the theoretical pressure due to the expansion of the pipework, etc, which takes up a little of the expansion water.

$$P_1\, V_1 = P_2\, V_2$$

Where P_1 = Initial vessel gas charge (absolute)

   V_1 = Vessel volume

   P_2 = Final gas pressure (absolute)

   V_2 = Final gas volume (Gas volume – Water Acceptance)

Example P_1 = Initial vessel charge of 0.5 bar (1.5 bar absolute)

   V_1 = Vessel volume of 50 l

   P_2 = To be found?

   V_2 = Volume 50 l minus water acceptance of 22 l = 28 l

Therefore

$$P_2 = \frac{P_1\,V_1}{V_2} = \frac{1.5 \times 50}{28} = \begin{array}{c} 2.68 \text{ bar absolute} \\ \text{or} \\ 1.68 \text{ bar gauge pressure.} \end{array}$$

Therefore an expansion vessel having a volume of 50 litres and a pre-charge pressure of 0.5 bar designed to accept the entry of 22 litres of expanded water will have a final operating compression pressure of 1.68 bar, whereas the maximum operating pressure allowed for in the example calculation on the previous page was 2 bar.

GRAVITY CIRCULATION HEATING SYSTEMS

The practice of designing a heating system that creates the conditions for a gravity circulation or thermo-siphon to occur, ceased following the development of the glandless circulating pump in the 1950s in favour of the more versatile forced circulating system with its many advantages. However, this text would be incomplete without some mention and description of this mainly obsolete system, as some heating systems, employing a partial gravity circulation for the domestic hot water primary flow and returns, continued to be installed in conjunction with a forced circulation for the heating system up until recently: see Figure 2.1 for an example. The custom of designing and installing domestic hot water gravity circulating primary flow and returns is no longer permitted by the Building Regulations, even in situations where it would be the best option. This constraint is discussed further in Chapter 24, Regulations, Standards, Codes and Guides.

EXPLANATION OF GRAVITY CIRCULATION

If a tank designed to hold an exact capacity of $1\,m^3$ is filled with water at a temperature of 4.4°C and is then weighed, we would find that the water contents would have a weight of one tonne, or 1000 kg. If we then heated the water to a temperature of 82°C, the water – as previously established – will expand and spill over the top of the tank but still retain its original volume after losing some of the water. If we weigh the tank contents again we would find that it now weighs only 970.6 kg at the same volume; Figure 2.44 illustrates this.

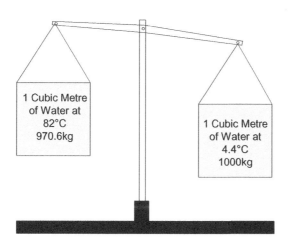

Figure 2.44   Water temperature/weight relationship

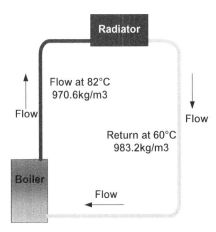

Figure 2.45   Example of gravity circulation (thermo-siphon)

Therefore, we can establish that the higher the temperature of the water the lower the density for a given volume and the lighter the resulting weight. The density of water, together with other properties at differing temperatures, is given in Appendix 14.

If we apply this principle to a gravity heating system, demonstrated in Figure 2.45, the water is heated in the boiler, expands and becomes less dense or lighter in weight, and is pushed up the primary flow pipe by the heavier, more dense cooler return water, which is also heated for the process to be repeated and a circulation to occur.

This demonstrates that all that is required is to heat the water at a central point to create a temperature difference, and the water will then circulate around the system with any heat emitters transferring part of the heat to produce the temperature difference required.

Further explanation is given in Chapter 4, Heat and Heat Transfer.

3

Materials

Today, the choice of different piping materials that are commercially available for the conveyance of fluids is far greater than it has ever been since the discovery of copper pipes that were hand-made approximately 5000 years ago in Egypt. Although the piping materials available today require less skill in the form of craftsmanship than the traditional piping materials of a few years ago, the plumbing and heating engineer/ technician is expected to be fully conversant with this vast choice with their characteristics, properties and associated jointing techniques and therefore requires a greater knowledge of piping materials than ever before in the long history of the Public Health Engineering/Plumbing Industry.

It would be inappropriate and impractical to include fully detailed information on the properties and characteristics of all of these materials as only a select few are considered suitable for heating applications. To cover them all in the detail required would require a separate volume to do it justice.

It has also been considered appropriate only to include limited information on selected piping materials commonly employed for conveying heated fluids, as it is expected that the reader has already obtained the basic knowledge and skills required with these materials at this stage in their education. If further knowledge of these piping materials is desired by the reader then they are directed to an alternative technical volume.

The importance of selecting the correct piping specification for each heating application, as with all plumbing applications, should be emphasised, as each individual piping material has its own attributes and disadvantages making them either more suitable, or less suitable for given applications.

PIPING MATERIALS FOR HEATING APPLICATIONS

The piping materials suitable for heating applications may be classified as belonging to either the metallic group, sometimes inappropriately referred to as traditional materials, or the thermoplastic group, sometimes incorrectly referred to as new materials. It is true that the metallic materials discussed in this section have been known and used for hundreds of years and in one case, thousands of years and therefore are often referred to as traditional materials, although the refining of these metallic materials together with their manufacturing process, has improved over this long period.

Heating Services in Buildings: Design, Installation, Commissioning & Maintenance, First Edition. David E. Watkins.
© 2011 John Wiley & Sons, Ltd. Published 2011 by John Wiley & Sons, Ltd.
As an aid to lecturers and students, full colour versions of the figures in this chapter may be found at www.wiley.com/go/watkins
These figures are © 2011 by John Wiley & Sons, Ltd.

The thermoplastic piping materials included in this section are comparatively new when compared to their metallic counterparts, but they have also been around for many years, in some cases over a hundred years, but it is only more recently that they have been commercially available in piping form.

As well as the individual technical or practical attributes and limitations of each piping material discussed in this chapter, commercial organisations as well as individuals are becoming increasingly aware of the environmental impact on the Earth's resources made by the use of these materials, as well as their ability to be re-cycled, or the energy costs associated with their manufacture or life cycle redundancy. It is no longer considered acceptable to select materials just for technical or commercial reasons: the environmental impact should also be considered when specifying the piping material.

Metallic piping materials

All metallic materials have the common disadvantage of being susceptible to attack by corrosion, some more than others, and the metals used to produce piping components are no exception to this rule and may require some degree of protection. However, all metallic piping materials have the advantage of being easily salvaged by recycling.

Metallic piping materials in common use today for heating systems include:

1. Light gauge copper tube
2. Light gauge stainless steel tube
3. Low carbon mild steel tube

Other metallic piping materials have been used in the past and may well still be used in limited or specialist heating system circumstances, but as they are not in common use they are not included in this work.

Thermoplastic piping materials

As with metallic materials, all thermoplastic materials also have a common disadvantage of being susceptible to temperature and pressure limitations, again, some more than others.

Thermoplastic piping materials in common use for heating systems include:

1. Cross-linked polyethylene tube (PEX)
2. Polybutylene tube (PB)

Other thermoplastic piping materials that have been used to convey heated water, some with limited success, such as Chlorinated Polyvinylchloride, (cPVC), while others such as Polypropylene (PP), or Polyvinylidene Fluoride (PVDF) at present are precluded from common use because of either jointing difficulties or uncompetitive costs, but are used for less common specialist heating applications where their properties are considered more important than their cost implications.

LIGHT GAUGE COPPER TUBE

Copper, over a period of many years, has easily become the most popular piping material for general mechanical services for the conveyance of water in the plumbing and heating industry, used until recently almost exclusively for domestic heating applications, but less so for commercial buildings. This is mainly due to the fact that copper possesses all the qualities that are required for a piping material used in the domestic part of the plumbing industry. It is also a material that the plumber has become very familiar with, having obtained the skills and experience over a long period of time. However, its dominance in the domestic market is being challenged by other materials and it remains to be seen how much of the domestic market share will be lost to these materials.

Copper is an excellent piping material, being neat in appearance, light to handle, bends easily and can be worked, formed and joined together in a number of ways. Scrap offcuts and redundant services can easily be recycled back into base copper ready for re-use.

Copper also possesses good corrosion resistance properties at operating temperatures below 200°C, but can be attacked if no protective precautions are taken when in direct contact with acidic soils or cement mortars.

The British Standard for copper tube has been revised many times since its introduction, but since 1996, light gauge copper tube should be manufactured conforming to the European agreed standard, BS EN1057, which replaces the former standard, BS 2871, Part 1, which had been in existence for nearly 30 years. As far as the installer is concerned, there is virtually no difference between the two standards, as the outside diameters are compatible and the pipe wall thicknesses are identical. The range of sizes remains unchanged, the tubes' temper has been retained, and it handles and works exactly as before.

Although the current British Standard for copper tube is BS EN1057, a description of copper tube conforming to the former withdrawn standard, BS 2871, Part 1 has been included in this text to help those familiar with it to understand the current standard, to help explain the development of BS EN1057, as well as to assist those who are new to the industry to identifying existing copper tube services, which will no doubt remain in existence for a great many years due to the high degree of durability of the material.

Tube conforming to BS 2871, Part 1 – entitled 'Copper tubes for water, gas and sanitation' (Part 2, 'Tubes for general purposes', and Part 3, 'Tubes for heat exchangers' remain unchanged at the present time') was available manufactured in four grades, which have been retained in the current BS EN1057, but expressed differently, as given in Box 3.1.

Table W (*R220*)

Manufactured with outside diameters of 6, 8 and 10 mm and used mainly in micro bore heating systems, this tube was produced in a soft annealed condition supplied in coil form similar to Table Y, but with a thinner wall thickness. This grade of tube enabled long continuous lengths to be laid below timber floors without the need for any intermediate pipe joints.

Now expressed as R220 copper at the same diameters and in the same coil forms.

Table X (*R250*)

Manufactured with outside diameters ranging from 6–159 mm, and supplied in a half hard annealed condition in 6 m rigid straight lengths. Table X tube was the most commonly used grade of copper tube manufactured to the former BS 2871, Part 1, as it could be easily worked by forming bends with the aid of bending springs or bending machines, and it could also be jointed together by all the approved methods available. Table X copper tube was suitable for use in all plumbing and heating services.

Now expressed as R250 copper tube at the same diameters in straight 6 m lengths.

BOX 3.1 Comparison of copper tube classification

Former BS 2871 Designation	Current BS EN1057 Designation
Table W	R220
Table X	R250
Table Y	R220 and/or R250
Table Z	R290

Table Y (*R220* and *R250*)

Manufactured in both a half hard condition in straight lengths, and in soft fully annealed form supplied in coils or straight lengths.

The soft condition copper tube was primarily used for underground water services where a flexible type of pipe is required; it was also suitable for overcoming obstacles such as curved walls where a slow, easy curve could be formed in the pipe so as to follow the walls' contours.

The half hard condition tube had a thicker wall structure than Table X tube, and was therefore suitable for withstanding higher working pressures. The tube was available in both conditions in sizes from 6 mm up to 108 mm in straight 6 m lengths, or in coil form in diameters up to 54 mm.

Now expressed as R220 and R250 copper tube at the same diameters in coil form or straight 6 m lengths.

Table Z (*R290*)

Manufactured in a hard drawn, thin wall condition with outside diameters up to 159 mm in straight 6 m lengths. Introduced during the copper shortage in 1965 as one of the means of reducing the cost of copper tube during that period, its thin wall hard condition meant it was not suitable for forming site-made labour bends and had a slightly lower working pressure than Table X copper tube. This grade of copper tube was less expensive than the other grades, so cost savings could be made when straight lengths of tubing – such as vertical risers in tall buildings, or long straight runs of pipe – were required. Careful site management of the stores was required to ensure that the different grades of copper tube did not get mixed up and wrongly used for the applications required.

Now expressed as R290 copper tube at the same diameters and in 6 m lengths.

CURRENT EUROPEAN BRITISH STANDARD – COPPER TUBE

Copper tube conforming to European Standard BS EN1057 is manufactured to the same outside diameters and wall thicknesses as the previous standard for copper tube, but is now specified by its temper and expressed as an 'R' number, which relates to its tensile strength.

The tube is available in three tempers, namely R220, R250 and R290, the number in each case referring to its tensile strength measured in MPa; therefore R250 copper tube has a tensile strength of 250 MPa.

Relating these tempers to those previously supplied under BS 2871:

- R220 copper tube is a soft annealed tube supplied in coils or straight lengths. It has similar characteristics to Table W and Table Y soft condition tube, which had a slightly lower tensile strength of 210 MPa and was therefore fractionally softer, but in all other respects is the same.
- R250 copper tube is a half hard condition rigid tube supplied in straight lengths, suitable for forming labour bends with the aid of bending machines or springs with wall thicknesses and characteristics, including exactly the same tensile strength as the Table X and Table Y half hard condition. This is now the standard copper tube type for domestic building services.
- R290 copper tube is a fully hard condition tube having similar characteristics to Table Z thin wall tube, although Table Z had a tensile strength of 310 MPa. The thin wall tube thicknesses that Table Z was supplied to are not recommended in BS EN1057.

It should be noted that within BS EN1057, the higher the 'R' number of the copper tube's temper, the harder the condition of the copper.

Copper tube conforming to this standard should be marked on the side of the tube at set intervals with the following information: 'Standard Number' followed by 'Temper' and then 'Outside Diameter' and 'Wall Thickness'.

Therefore, EN1057 − 250 − 15 × 0·7, refers to a 15 mm diameter copper tube with a wall thickness of 0·7 mm, being in a half hard condition with a temper of R250 and conforming to BS EN1057. Copper tube should be specified and ordered using this method of identification.

COATED COPPER TUBE

Copper tube is also commercially available as a standard in sizes up to and including 54 mm diameter, complete with a factory applied protective polyethylene coating or sheaving in a variety of colours, the colour allowing the pipe to be identified by its service; i.e. yellow for gas, blue for water etc.

There are two forms of this protective plastic coating. The simplest comprises a plastic coating – usually PVC or polyethylene – applied as a tight-fitting sleeve to the external surface of the copper tube. Its sole purpose is to create a physical barrier between the copper material and any surrounding buried soil or cement based covering. This enables the tube to be buried underground or cast in a cement screed without fear of attack from aggressive soils or cement compounds.

The polyethylene covering provides an ideal tube for underfloor heating application and has an upper temperature limit of 95°C, but in practice it is better to restrict any continuous exposure to this temperature by limiting the operating temperature to 85°C.

FORMING BENDS IN COPPER TUBE

Light gauge copper tube is one of the metal pipe materials that can be easily bent on site using a number of methods, for example with the aid of a bending machine, a bending spring or by compact pipe loading coupled with heat application, although this last method is normally conducted in a workshop.

The advantage of being able to pull and form labour bends on any pipe is that it achieves a less costly, neater appearance – particularly if more than one pipe is being run side by side, as a bank of different size pipes all with the same bending radius at changes of direction looks far more impressive than a collection of different fittings. Labour pulled bends will also result in the use of fewer fittings and in turn fewer pipe joints, which means a reduced possibility of leaks.

JOINTING COPPER TUBE

Copper as a pipe material has a distinct advantage over other pipe materials in so far as it can be successfully joined together by a number of approved and acceptable methods, each having its own merits. The jointing methods may be categorised as being either proprietary made pipe fittings or some form of fusion bonding or welding; however, these two main categories may be subdivided into a number of related categories, as shown in Figure 3.1.

PROPRIETARY MADE PIPE FITTINGS

Pipe fittings suitable for jointing light gauge copper tube together are classified as belonging to one of the following three groups:

- Mechanical compression type
- Fusion bonded capillary type
- Seal ring flame-free type

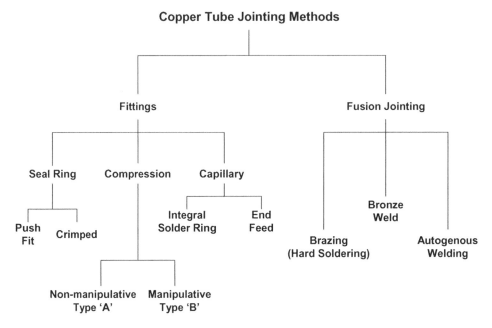

Figure 3.1 Copper tube jointing methods

Regardless of which group the fittings are classified as, they should always conform to their respective standard BS 864. They are manufactured from either copper, copper alloys belonging to either the brass or bronze family of alloys, or – for some of the smaller pipe diameters – from plastic materials, although most of these are mainly aimed at the DIY market. Not all of these fitting types are used for heating applications.

MECHANICAL COMPRESSION TYPE FITTINGS

The British Standard for copper tube fittings, BS 864, categorises compression fittings as being either Type 'A', which is a standard non-manipulative type compression fitting, or Type 'B', which is a manipulative form of compression fitting not commonly used for low pressure heating applications. Both of these types of compression fittings rely on the principle of compressing or squeezing together the two sections that are to be joined.

Figure 3.2 illustrates a Type 'A' compression joint. This is a typical non-manipulative type compression joint, and although the shape of the compression ring and style of the compression nut and fitting may vary from one manufacturer to another, the basic concept and method of assembly are the same. It can be seen that this form of joint works on the principle of compressing the compression ring into the copper tube by the action of tightening the compression nut; the compression achieved can be observed by the formation of the indentation that the ring has created in the wall of the copper tube, thereby accomplishing a water-/gas-tight joint. It should be noted that if the compression nut is not tightened sufficiently, the pressure of the water/gas could force the tube out of the fitting and cause the joint to fail; also, if the compression nut is over-tightened, the ring will distort the tube and the joint will again fail. Therefore it is important that the correct pressure or torque is applied when tightening the compression nut when making this form of joint.

The Type 'A' non-manipulative form of compression joint may be defined as not requiring any further work to it, other than that described above, to achieve compression. They are not normally recommended

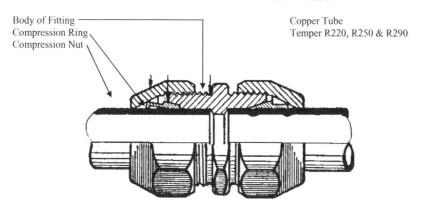

JOINT HAND TIGHT   JOINT TIGHTENED & RING
                                    COMPRESSED

Body of Fitting   Copper Tube
Compression Ring   Temper R220, R250 & R290
Compression Nut

Type 'A' compression joint (6 mm to 54 mm diameter)

ASSEMBLY INSTRUCTIONS

1. Ensure that both the tube and fitting are of the same size, and that the end of the tube has been cut square, de-burred and is perfectly round. Use a re-rounding tool if necessary.

2. Dismantle fitting and place the compression nut and ring over the end of the tube, ensuring that they are facing the correct way round. Also, ensure that the end of the tube and the components of the fitting, including the threads, are free of dirt or any form of contamination.

3. Smear a small amount of an approved jointing compound on the end of the tube and insert the tube end fully into the fitting until it makes firm contact with the tube stop end.

4. Tighten the compression nut by hand and then using a spanner until the compression ring grips the tube, whereby the tube and fitting can no longer rotate. Tighten the nut further by turning it a further half to two thirds of a full turn. Use a correct size spanner or adjustable wrench for this activity.

Note – Do not over-tighten the compression nut and ring as this can weaken the joint.

Figure 3.2 Non-manipulative Type 'A' compression fitting

for applications on pipework installed below fully ventilated timber floors as the joint could pull out without giving any warning of having done so. If a compression joint is required to be installed in an enclosed area such as below timber floors, then a Type 'B' manipulative type joint (not illustrated) should be used because it is less likely to pull out.

CAPILLARY TYPE FITTINGS

As in the case of compression fittings, capillary type fittings are also categorised in two basic forms, namely one which incorporates an integral ring of leadless soft solder arranged inside a bead formed in the socket of the fitting, and the other, which has to have soft solder fed into it manually and is known as an end feed fitting. Both categories work on the principle of creating a soldered joint between the copper tube and the copper or copper alloy fitting by capillary attraction and are ideal for heating pipework installations.

An integral solder ring joint is shown in Figure 3.3. This is a typical capillary joint whereby the fitting is manufactured complete with an integral ring of leadless soft solder contained within a bead formed in the socket of the fitting. This solder normally consists of tin and copper, or tin and silver. Soft solders containing lead should no longer be used, or be available in integral ring type capillary fittings.

The socket formed on the fitting is manufactured to achieve a fine engineered tolerance between the copper tube and the fitting, so that when heat is applied to the tube as indicated in Figure 3.3, it conducts

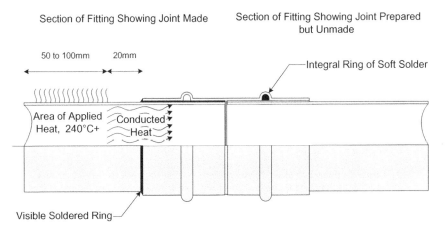

ASSEMBLY INSTRUCTIONS

1. Ensure that both the tube and fitting are of the same size and that the end of the tube has been cut square, de-burred and is perfectly round. Use a re-rounding tool if necessary.

2. Using steel wire wool, clean both the inside of the socket of the fitting and the end of the tube to a distance equal to the socket depth, plus 15–25 mm. The cleaned copper should be bright all over when finished.

3. Smear a small amount of an approved flux on the end of the tube all over to a distance equal to the depth of the fitting socket.

4. Insert the end of the tube fully into the socket of the fitting and rotate the tube at the same time to ensure that the flux has been evenly spread.

5. Apply heat from a gas torch evenly on the tube at a distance of approximately 20 mm from the fitting, not the fitting itself. Heat is conducted into the socket of the fitting causing the solder to melt, tinning the surface and filling the space between the socket and the tube by capillary attraction.

6. Remove the source of heat when the solder appears fully around the circumference at the end of the socket, and allow the joint to cool before handling. Disturbance of the joint before it has cooled could result in the joint failing.

7. Clean joint to remove all residue or traces of flux.

Figure 3.3   Capillary joint with integral solder ring

along the pipe into the socket until the melting point of the solder is reached, causing the solder to melt, tin or wet the copper surface and be spread evenly around the joint by capillary attraction.

Heat should not be applied directly to the fitting, as this will cause the solder to melt before the copper is at the correct temperature to accept the solder and become tinned, with the result that the solder will not attach itself to the copper correctly and the joint will fail.

The manufacturers of these fittings provide sufficient solder in the integral ring for a complete and successful joint to be made, so it is not necessary to add any further solder during the joint making process as this can be detrimental, causing solder to form a bead on the bottom, or even enter inside the pipe joint, creating an obstruction to the flow of water or gas. Also, if the solder that has been added to the joint is of a different composition to that contained inside the ring, it could result in a weakened joint being made, or even an illegal joint, if the added solder contains lead.

The procedure for making this type of capillary joint is described in Figure 3.3. After the tube and fitting have been prepared by cleaning with steel wire wool to remove all dirt and the oxidised coating that would have formed on the surface of them, to produce a bright finish all over, they should be coated with a suitable flux by smearing a thin layer all over the surface. The flux should be of an approved formulation such as a zinc and/or aluminium chloride flux, or made from a mixture of amine hydrohalides in an organic base.

The purpose of the flux is to prevent the cleaned bright surface of the copper from re-oxidising when heat is applied from a flame and allowing the tinning process to occur. If the surface of the copper is allowed to become oxidised, this oxidised coating would act as a barrier and prevent the solder wetting or tinning the surface of the copper, and the joint would fail.

Fluxes, by nature of their purpose, are aggressive in order to realise their function of preventing oxidation, but this also means that they are corrosive to the metal if any excess or residues are not removed after the joint has been completed. The completed joint should be permitted to cool so that the solder has solidified and cured to the point that it is no longer soft before disturbing it by washing off the excess flux or residue with warm water. The completed joint should then be cleaned with steel wire wool to restore its bright appearance and remove any final trace elements of residue flux. When the piping system has been completed, it should be flushed through with warm water to remove any residue flux from the internal surfaces of the pipe walls.

There are fluxes available known as 'self-cleaning fluxes', which contain hydrochloric acid and are extremely aggressive – as would be expected from a flux that eliminates the cleaning process. Care should be taken to remove any trace of this type of flux once the joint has been completed as it could continue to etch into the copper, resulting in pinhole failure.

Fluxes, as in the case of jointing compounds, are required by the Water Authorities not to be able to support microbiological growth, and must be approved by the Water Fittings Directory.

As already established, fluxes are aggressive, and as they are all acid compounds, they pose a health and safety risk, so great care should be taken when handling them and the maker's instructions should strictly followed.

The type of capillary joint depicted in Figure 3.4 is known as an end feed capillary joint. This differs from the integral solder ring form of joint as the solder has to be added while making the joint. Therefore, to achieve this, the heat should be applied on this occasion directly on to the fitting itself, in complete contrast to the procedure recommended for the integral solder ring joint. This raises the temperature of the copper to a level that will allow the solder to melt, wet or tin the copper, and flow freely into the socket of the joint by capillary attraction. Only sufficient solder should be applied to the joint to fill the gap all round the tube; excess solder will just form a bead of solder on the bottom of the joint, or enter inside the pipe and create an obstruction to fluid flow.

As a general guide, when using solder wire in a coil form, a length of solder equal to the diameter of the tube in question should be sufficient to complete the joint.

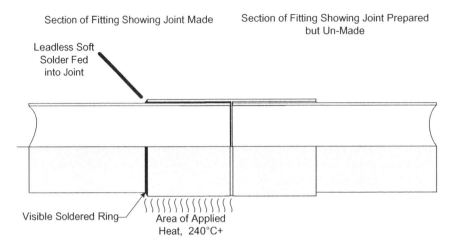

ASSEMBLY INSTRUCTIONS

1. Follow instructions 1 to 4 in Figure 3.3 for preparation of joint.

2. The procedure for making the end feed capillary joint differs only from that of the integral solder ring type, by applying the heat directly onto the fitting and testing the temperature of the copper by touching the end of the joint with the end of a stick of leadless soft solder, until the solder tins the copper and is drawn into the joint to fill the gap by capillary attraction.

3. Complete the joint as per instructions 6 and 7 in Figure 3.3.

Figure 3.4   End feed capillary joint

End feed type capillary fittings are less costly than the type with the integral ring of solder incorporated in them, but they take slightly longer to make, are less reliable in inexperienced hands in ensuring a successful joint each time, and the cost of the solder has to be added to the total cost of the joint to give a true cost comparison.

The Water Regulations require solder to be lead free when used on domestic hot or cold water services; this is not a requirement for heating services pipework, however, it is good practice to only use lead-free solders to prevent getting them mixed up.

Various tools are available that permit the copper tube to be worked, so as to form close fitting sockets that are suitable for making end feed capillary joints without the need to purchase certain fittings.

Capillary joints are suitable for copper tube with tempers of R220, R250 and R290; socket forming tools should not be used on R290 thin wall tube.

One of the advantages that compression fittings have over capillary fittings is that they can be easily disassembled and remade many times over, but capillary fittings can be said to be a permanent joint, unaffected by vibration or movement. With careful handling, they can be taken apart and remade, but not as easily as compression joints.

FLAME-FREE JOINTS

Capillary joints have been tried and tested over many decades and found to be a very cost effective, reliable and permanent joint for copper tube, especially for applications where it was considered an essential requirement that the jointing method of the piping system could be relied upon not to fail by pulling apart or vibrating loose. Unfortunately, capillary fittings require a naked flame – usually from either a paraffin lamp or most likely from a liquefied petroleum gas cylinder and torch – to obtain the heat required to raise the temperature of the copper and solder in order to make a successful joint, thereby creating the conditions for a fire to start and therefore are a potential risk. However, if a properly conducted risk assessment is undertaken to mitigate the risks and with all actions formulated into a comprehensive work method statement and complied with in full, then this potential hazard can be virtually eliminated.

Unfortunately, this has not always been the case and due to either complacency or carelessness, and quite often, as the result of the ignorance of unqualified operatives, fires have occurred.

As a result of a small number of fires occurring, two of them involving historic buildings and causing major damage and the loss of irreplaceable artefacts, developers and the agents of building owners have lost confidence, or become frightened of using capillary fittings for piping systems. This reluctance to undertake tasks that require a naked flame has had the general effect of alternative methods being sought for making pipe joints, which means capillary type fittings are not permitted on certain projects. This applies particularly to the refurbishment of historic buildings or where the construction involves refurbishing buildings that contain a large amount of existing dry timber.

It is regrettable that some authorities have taken this approach to capillary fittings rather than address the real problem of untrained or unqualified operatives being the main risk, but it has given the industry the opportunity to develop and consider alternative materials or methods of jointing pipes that do not require a flame.

Compression fittings are a flame-free jointing method that has also been tried and tested over many decades and found to be a good means of pipe jointing, but the Type 'A' non-manipulative fitting is not always as reliable as the capillary joints, especially if movement or vibration can be expected, and the Type 'B' manipulative joint is financially more expensive when compared to the other forms of pipe joints. This means some other methods and techniques must be sought to overcome this matter.

PUSH FIT RING SEAL JOINTS

The push fit fitting is a fairly recent introduction to the industry, although it has been available for some time now. It makes use of the seal ring technology to form the joint. Originally this type of joint found favour with the DIY market due to its simplicity and requiring little real skill to make, which is why it is only available in the small tube sizes for domestic applications. However, due to the necessity to find a suitable flame-free fitting that could be financially competitive with capillary fittings, it is now aimed at the professional installer as an alternative form of joint.

Unlike the various types of compression and capillary joints previously described, which are all very similar regardless of make, there are many different variations with this form of joint, although they all work on the principle of forming a water tight joint by using an 'O'-ring to form the seal.

Figure 3.5 illustrates a typical copper bodied push fit fitting, although they are also available with brass or non-metallic bodies made from various plastic materials. The early types of these fittings could not be disassembled after they had been made, but now most incorporate a facility that permits them to be taken apart and reassembled again. They nearly all permit the tube to be rotated in the fitting, which means that they must be adequately supported to prevent movement, a feature that has not endeared itself to most professional installers. As with all of these type of joints, they are dependent upon the integrity and reliability of the EPDM (ethylene propylene diene monomer) 'O'-ring seal, which if it does not seal for any reason has to be discarded: it cannot be repaired unlike traditional joints for compression or capillary fittings.

Although the joint requires little skill to make, emphasis should be placed on the importance of bevelling the end of the copper tube with a fine cut file to remove all burrs before pushing into the joint: experience has shown that if there are any slight sharp edges to the end of the tube it can damage the seal ring by putting fine score marks on the ring. These faults can result in latent failures of the joint; these failures

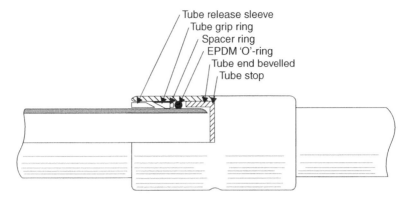

ASSEMBLY INSTRUCTIONS

1. Ensure that the tube and the fitting are the same size and that the tube has been cut square. Also ensure that the tube is clean and free of any grease – do not use any lubricants.

2. The tube end should be bevelled round using a fine tooth file and free of any burrs.

3. Using a pencil, mark the depth required to make the joint on the tube. Do not score the pipe. This will give a visual indication that the tube has been fully inserted.

4. Place the end of the tube into the socket of the fitting keeping it square and push it up to the 'O'-ring. Drive the tube fully home up to the tube stop; a slight twisting action of the tube will help facilitate this.

5. Attempt to pull the tube back out to allow the tube grip ring to dig into the tube and fully engage.

6. To disassemble, use the manufacturer's special tool to push the sleeve in so as to release the tube grip ring and allow the tube to be removed from the fitting.

Figure 3.5   Push fit joint with 'O'-ring seal

usually manifests themselves some months or even years after the faults have occurred. The failures are usually in the form of water weeping past the 'O'-ring seal at the point of the score and dripping out of the end of the fitting; if left for any length of time the leak will gradually become worse.

Push fit joints can be used on all temper grades of copper tube, but caution should be adopted with soft copper tube R220.

CRIMPED OR PRESSED COPPER JOINTS

Another recent jointing method introduced to provide an alternative as a flame-free copper pipe joint is the crimped joint, which has been adapted from an identical jointing method that has been successfully used for many years, for making joints on light gauge stainless steel.

On first appearance the joint resembles a copper bodied integral ring capillary fitting, but on closer inspection it can be seen that it incorporates an 'O'-ring seal instead of an integral solder ring. The joint is made by using a similar procedure to that of the push fit joint, except that the final operation of making the joint is to crimp the socket of the fitting onto the pipe to distort the fitting socket into a hexagonal shape: this is to prevent the pipe from being able to pull out of the joint and to prevent the fittings from being able to rotate on the pipe, which is a common criticism of most types of push fit joints.

Figure 3.6   Crimping tool in operation pressing fitting onto copper tube

The crimping operation is conducted by the use of a special electrically operated crimping tool (illustrated in Figure 3.6) that has interchangeable jaws for tube sizes ranging from 15 mm to 108 mm; it is applied over the socket of the fitting and then activated and presses the fitting onto the pipe to complete the joint. This tool is very expensive, and coupled with the fact that it is recommended that the tool should be overhauled every year by an approved agent of the tool manufacturer, it is only an attractive jointing method for companies that can justify their purchase by the continuous use of these tools. This is unlikely to be considered as a method that will be adopted by small companies that would only have limited demand for the tool and is therefore rarely used for domestic heating applications, unless undertaken on a large-scale development.

The finished crimped appearance of these joints is not conducive to a neat installation, as they can give the impression that the joints have been damaged in some way; therefore, pipes that will be installed on show may have to use joints with a more acceptable finish.

Figure 3.7 shows the procedure.

STAINLESS STEEL TUBE

Stainless steel is the name given to a group of ferrous alloy metals containing a minimum of 9% chromium with varying additions of nickel, molybdenum, titanium, niobium and minor amounts of other elements.

Stainless steel is a very young metal having not been discovered until 1913, whereas the other metals that pipes are manufactured from are thousands of years old. It was discovered by accident by a technician, Harold Brearley, in a Sheffield Steel works. He was working on steel experiments for armaments and noticed that one failed experiment of steel containing 13% chromium had not gone rusty. He then patented it under the name 'stainless' which then gave birth to a world-wide stainless steel industry.

Since that day many forms of stainless steel have been developed, although they can all be classified as belonging to one of three families of stainless steel, namely 'martensitic', 'ferritic' and 'austenitic', with the first two classes being plain chromium steels generally having not less than 11% chromium; both classes are magnetic.

Austenitic stainless steels contain nickel as well as chromium and are used for many applications, including the manufacture of pipes and tubes suitable for conveying many liquids and gases. Austenitic stainless

Joint crimped   Joint prepared for crimping

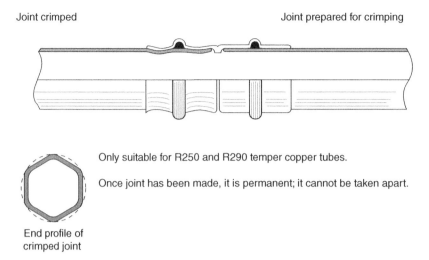

End profile of
crimped joint

Only suitable for R250 and R290 temper copper tubes.

Once joint has been made, it is permanent; it cannot be taken apart.

ASSEMBLY INSTRUCTIONS

1. Prepare and assemble joint using the same procedure as described for the push fit type joints.

2. Assemble the crimping tool, select the correct sized jaws and ensure that the battery is fully charged.

3. Before proceeding, check that the pipe has not moved out of alignment or pulled out of the joint by observing the pencil mark.

4. Fit the jaws of the crimping tool over the ring of the fitting and whilst keeping the tool square with the pipe, depress the tool's start button activating the crimping procedure until complete. Remove tool on completion.

If several joints are pre-assembled first, then a fast installation time can be achieved by crimping all the fittings as a final exercise. Care must be taken by checking the pencil marks to ensure that none of the joints have pulled out and that they are all still at the correct entry depth.

Figure 3.7   Crimped or pressed joint

steels may be differentiated from the other two classes of steels as they are non-magnetic. Typical compositions are given in Table 3.1.

There is a fourth classification of stainless steel, namely 'Duplex', which is made up of approximately equal parts of austenite and ferrite.

STAINLESS STEEL IDENTIFICATION

The many forms of stainless steel can be identified by a number of different methods, such as:

1. Metallurgical structure
2. AISI numbering system (American Iron & Steel Institute)
3. UNS system (Unified Numbering System)
4. EN-standard steel name

Table 3.1   Selective austenitic stainless steels

AISI Number	Chemical Composition %									UNS Number	EN-Standard Number
	C	Mn	P	S	Si	Cr	Ni	Mo	Other		
304	0.08	2	0.045	0.03	1	18	10			S30400	X5CrNi18-10
304L	0.03	2	0.045	0.03	1	18	10			S30403	X2CrNi18-10
316	0.08	2	0.045	0.03	1	17	12	3		S31600	X5CrNiMo17-12
316L	0.03	2	0.05	0.03	1	17	12	3		S31603	X2CrNiMo17-12
321	0.08	2	0.05	0.03	1	18	10		Ti = 5	S32100	X6CrNiTi18-10

C-Carbon, Mn-Manganese, P-Phosphorus, S-Sulphur, Si-Silicon, Cr-Chromium, Ni-Nickel, Mo-Molybdenum, Ti-Titanium.

Stainless steels are commonly identified by either the AISI system, or by the UNS system which was developed by the American Society for Testing Materials (ASTM) and the Society of Automotive Engineers (SAE) and which has been unofficially adopted internationally.

LIGHT GAUGE STAINLESS STEEL TUBE

Following the discovery of stainless steel in 1913 it quickly found use as a material for the manufacture of many products, including the production of pipes, although these were confined mainly to industrial applications and processes at that time.

It was not until the Rhodesian crisis in the mid 1960s that a thin walled light gauge form of stainless steel pipe for domestic use was introduced as an alternative material to copper, as raw copper was in short supply to the UK following the crisis. As light gauge stainless steel tube was developed conforming to BS 4127 as an alternative to copper tube, it was manufactured with the same outside diameters as copper tube so that it was compatible and could use the same compression and capillary fittings as copper tube.

Since then light gauge stainless steel, now conforming to BS EN10312, has established itself as a suitable piping material in its own right, used for the conveyance of wholesome and potable water, domestic hot and cold water including hydronic heating services, although it is a more popular piping material in some other European countries than it is in the UK.

Light gauge stainless steel tube is manufactured from type 304 (sometimes referred to as 18/10 steel) and 316 grade of stainless steel, and has the same external dimensions as copper tube but has a higher tensile strength and is fractionally heavier than copper tube. It also has similar attributes to those of copper and is worked in the same way.

Bending stainless steel tubing

Light gauge stainless steel tubing can be successfully bent to form acceptable bends using the same tooling that is used for copper tubing, although greater strength is required to form these bends.

Jointing stainless steel tubing

Light gauge stainless steel tubing can also be successfully joined together using similar fittings and methods to those of copper tubing.

Figure 3.8 illustrates acceptable methods of forming joints between stainless steel pipes and components. The fittings that are used are manufactured from either stainless steel, brass or copper, although brass and copper detracts from the overall appearance of the stainless steel piping system.

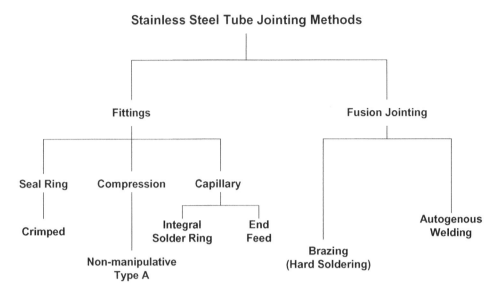

Figure 3.8   Stainless steel tube jointing methods

Compression fittings

Light gauge stainless steel tube can be successfully joined together using a Type 'A' non-manipulative compression fitting made in the same way as described for copper tube, except that due to the higher tensile strength of the metal, a slightly greater effort is required to compress the compression ring into the tube. Due to the higher tensile strength of the metal, Type 'B' manipulative type compression fittings should not be used as the tube end could be split or weakened if swaged out.

Capillary fittings

Light gauge stainless steel tube can also be successfully joined together using both end feed and integral solder ring capillary fittings. The tube and fittings are prepared in the same way as described for copper tube except that a phosphoric acid based flux is required to prevent the metal from oxidising when a flame is applied. It is imperative that any flux residual remaining after the soldered joint has been made is cleaned off immediately and the internal surfaces are flushed through, or the flux will continue to etch through the tube causing it to fail. Care should also be taken not to burn the tube when applying a flame which is too hot and causes black iron oxide to appear: if this does occur, the metal should be cleaned back using emery cloth.

Seal ring fittings

The use of push fit type DIY fittings is not suitable unless the tube and fitting are fully restrained as the fitting can be pushed back off when pressure is applied – this is due to the integral grab ring not being able to dig in to the high tensile metal as in the case of copper tube.

The use of the push fit crimped type joint as described for copper tube, and illustrated in Figures 3.6 and 3.7, was originally pioneered for use with light gauge stainless steel tube and provides an excellent permanent joint.

Stainless steel may also be successfully welded together or joined by a hard soldering or brazing; with this method it is important that the flux used is formulated for stainless steel.

Light gauge stainless steel tube has, since its introduction as an alternative piping material, proved to be a suitable material for hydronic heating systems, as well as for conveying water for other domestic water services. The reader should not assume that the author considers stainless steel to be a lesser material than copper because the text places less emphasis on stainless steel; this is purely to avoid repetition, as many of the attributes and working methods are the same for both materials.

STEEL PIPE

Steel is a ferrous metal that has been converted from iron by controlling the proportion of its carbon content; it has the advantage of being physically stronger when compared to other pipe materials, both metallic and non-metallic. Steel can resist a high degree of mechanical damage and is an ideal material for use in applications where a robust material is required or vandalism may be expected. It has been used to convey most liquids and gases at high or low pressures, including hot and cold water, heating, steam, soil and waste services, rainwater, goods, oils, compressed air and all forms of gases. Steel pipes may also be worked by forming bends and jointing together in a number of ways. However, steel does have one major disadvantage when compared to other metallic pipe material: it does not have any inbuilt resistance to corrosion and so must be protected if it is to be of use.

LOW CARBON MILD STEEL TUBE

The most common steel tube used for general building services, including hydronic heating systems, is low carbon mild steel tube conforming to BS EN10255 (formally BS 1387), or mild steel tube, as it is generally known due to its versatility. It is manufactured by passing continuous strips of steel plate through a series of angled rollers to form a tube shape over the length of the manufacturing process culminating in the seam being electrically welded. The base tube is then subjected to a further series of rollers to accurately form the desired pipe diameter before proceeding on to the many non-destructive testing and checking processes. Selective samples are taken and tested to destruction as part of the quality control procedures. The final part of this manufacturing process is to apply any protective coatings or finishes on to the tube before despatching it from the factory for use. The tube is cut into random lengths of between 6.4 m and 6.55 m before leaving the tube mill, although different lengths may be cut to an exact size as a special order if required. It can be supplied in plain ended form, or complete with threads cut on both ends of the tube.

Too much carbon content in steel renders the material brittle, therefore low carbon mild steel is an ideal pipe material for building services as it can be worked in a number of ways.

The tube is referred to as 'low carbon', due to the low content of carbon permitted by BS EN10255 in the steel quality.

Low carbon mild steel tube conforming to BS EN10255 is available in five grades expressed in weights which are directly related to the wall thickness of the tube. This standard of steel tube is in common use for commercial and industrial hydronic heating systems due to its physical attributes, but rarely considered for domestic hydronic heating installations.

Grades 'L' and 'L1' are thin walled lightweight tubes that are in common use in many European countries but not in the UK where they are not generally available.

The other three grades, namely lightweight, medium weight and heavyweight, are copied direct from BS 1387, which has been withdrawn and replaced by the harmonised European standard BS EN10255.

Lightweight tube L2

Lightweight grade tube can be identified by a 50 mm broad brown coloured band painted on each end of the tube and is suitable for non-pressure applications. Before the advent of natural gas, when the gas fuel was manufactured, lightweight tube was used extensively as the main piping material for conveying the gas inside each property downstream of the meter, because of the extremely low supply pressure used. Lightweight tube is very rarely used today because a piping material as the smaller outside diameter of the tube will not permit a thread to be cut in complete compliance with BS 21.

Medium weight tube

Medium weight tubes are identified by a 50 mm wide blue coloured band painted on each end of the tube and are used for low pressure applications such as cold water down services, domestic hot water distribution, low pressure heating systems and natural gas services. Medium weight tube is suitable for cutting threads conforming to BS 21.

Heavyweight tube

Heavyweight tube is supplied with a 50 mm wide red coloured band painted on each end of the tube to identify it, and is used for higher pressure applications such as water mains and fire hydrant mains for hose reels and wet and dry risers. It is also employed for higher pressure and temperature hydronic heating systems. Like medium weight tube, heavyweight tube is suitable for cutting threads conforming to BS 21.

On studying the dimensional table of the three different weight grades of tube, namely lightweight L2, medium weight and heavyweight illustrated in Appendix 5 and depicted in the graphical illustration in Figure 3.9, it can be seen that the outside diameter of both medium weight and heavyweight tube are identical, but the lightweight grade tube L2 has a slightly smaller outside diameter. This means that by employing a standard set of pipe dies, a standard pipe thread complying with BS 21 may be cut on both medium weight and heavyweight tube; however, using the same standard pipe dies on the lightweight L2 grade tube will still achieve a pipe thread, but it will not be in total compliance with BS 21 due to the lesser outside diameter of the tube. When used with a standard malleable iron pipe fitting, a complete pressure tight joint between the two threads cannot be achieved, hence lightweight tube is no longer used as a piping material. It can also be observed that the wall thickness of the heavyweight tube is greater than that of the medium weight tube, which in turn is greater than that of the lightweight tube, both of which have a similar although not identical inside diameter.

This results in the medium weight and the lightweight L2 grade of tubes having a larger internal pipe bore than the heavyweight grade of tube.

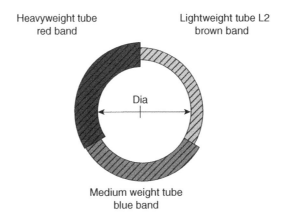

Figure 3.9 Graphical comparison of low carbon mild steel tube weights for lightweight L2, medium weight and heavyweight tube

Figure 3.9 shows in graphical form the relationship between the external and internal diameters for the three commonly used mild steel pipe grades. The exact dimensions for each individual pipe size may be read off the table in Appendix 5, and this illustrates the reason why a fully compliant BS 21 pipe thread using standard malleable iron screwed pipe fittings cannot be cut on the lightweight L2 grade of steel tube.

Low carbon mild steel tube can be supplied with a plain black finish, with or without a lacquered coating, that is suitable for closed circulatory circuits such as hydronic heating systems. It can also be supplied with a protective galvanised coating, both externally and internally, that is suitable for domestic hot and cold water services, including hydraulic hose reel installations together with dry and wet risers.

FORMING BENDS IN STEEL TUBE

Site-made labour pulled bends can also be made on steel tubes, although because of the nature of the material greater strength or power is required to achieve this. The methods available to us on site may be categorised as either:

1. Cold pull methods, limited to pipe sizes of below 100 mm diameter. These include the use of hydraulic bending machines.
2. Heat assisted methods, not suitable for tube having a galvanised finish.
3. Part welded heat assisted labour bends, also not suitable for galvanised finished tube.

JOINTS FOR LOW CARBON MILD STEEL TUBE

Low carbon mild steel tube may be joined together either by using one of a number of mechanical methods available, or by fusion such as welding. The choice of jointing steel tube together is not as great as for copper or stainless steel tube, but there is still a variety of methods to select from.

MECHANICAL TYPE JOINTS FOR STEEL TUBE

Ever since steel tube has been used for the conveyance of liquids and gases, screwed type mechanical jointing methods have been used as they have been tried and tested and found to be reliable for well over two hundred years.

Conversely, they are labour intensive for cutting and making a threaded joint. In its finished form the lack of flexibility is sometimes an advantage, but at other times a distinct disadvantage. More recently other types of mechanical pipe joints have been developed that do not require as much labour time to complete and some incorporate a degree of flexibility; none the less, the threaded screwed pipe joint remains the traditional mechanical method of steel pipe jointing.

Threaded screwed joints

The use of threaded screwed joints on steel tube can be traced back to the Perkins heating system and the threaded steel tube he used to construct the piping circuit in the late 1700s (see Figure 2.29). Many different types and forms of threads have been used for connecting steel tubes together since the introduction of steel as a piping material, but it was not until Sir Joseph Whitworth developed his standard engineering screwed thread in 1841 that a reliable method was found, despite many attempts to produce a sound screwed thread up until this point in time. Prior to this, each manufacturing factory produced its own form of

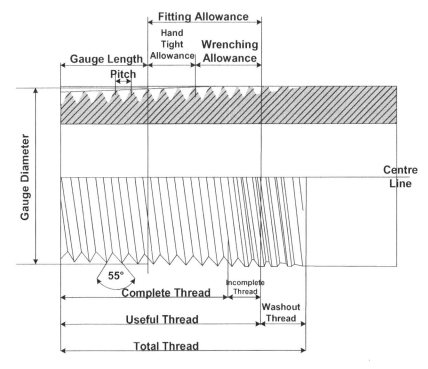

Figure 3.10   Explanation of terms relating to screw threads

thread which was not compatible with those of other factories. Whitworth's thread quickly became the accepted method of producing a threaded screw and was adopted to form the basis of BS 21, British Standard Pipe Thread, which in turn became internationally adopted, with the exception of North America, and was used to become the International Standards Organisation ISO 7 standard pipe thread.

Terms relating to screw thread fixings are shown in Figure 3.10.

Sir Joseph Whitworth devised his standard engineering thread based on a 55° thread angle having rounded roots and crests, and although now superseded by the metric thread, which has a 60° thread angle, it has been retained as the standard pipe thread by both BS 21 and ISO 7 in both the parallel form and taper form. The taper form has a tapered angle of 1 in 16, which means that the diameter over the crest of the thread decreases by 1 mm in every 16 mm of pipe length.

The British Standard Pipe Thread BS 21 is abbreviated to BSP thread, which refers to the parallel form of thread. The tapered version would normally be abbreviated to BSPT or British Standard Pipe Taper thread, and to make a successful pipe joint at least one of the threads needs to be tapered.

International Standards Organisation standard ISO 7 has been adapted from BS 21 which in turn adopted the standard Whitworth thread, but it differs in the way that parallel and tapered threads are specified. ISO 7 signifies a parallel thread by the abbreviation of R Pl behind the specified pipe size, and just the letter R for tapered threads, but to avoid confusion it is safer to write in full either 'taper' or 'parallel' when specifying pipe threads by either BSP 21, or ISO 7. BS21 is now incorporated in BS EN10226.

Table 3.2 indicates the main characteristics of a standard Whitworth or British Standard Pipe thread.

MALLEABLE IRON

Pipe fittings for low carbon mild steel tube should be made from malleable iron conforming to BS 143/1256, partially replaced by BS EN10242.

Table 3.2 British Standard Pipe thread (Whitworth)

BSP WHITWORTH DETAILS (BS 21)									
Pipe Size		Threads per 25.4 mm	Pitch mm	Depth of Thread	Gauge Length mm	Eq No of Turns	Fitting Allowance mm	Eq No of Turns	Wrenching Allowance mm	Eq No of Turns
mm	Ins									
8	¼	19	1.34	0.85	9.7	7.25	3.7	2.75	2	1.5
10	⅜	19	1.34	0.85	10.1	7.25	3.7	2.75	2	1.5
15	½	14	1.82	1.16	15.2	7.25	5	2.75	2.7	1.5
20	¾	14	1.82	1.16	14.5	8	5	2.75	2.7	1.5
25	1	11	2.3	1.48	16.8	8	6.4	2.75	3.5	1.5
32	1¼	11	2.3	1.48	19.1	8.25	6.4	2.75	3.5	1.5
40	1⅓	11	2.3	1.48	19.1	8.25	6.4	2.75	3.5	1.5
50	2	11	2.3	1.48	23.4	10.12	7.5	3.25	4.6	2
65	2½	11	2.3	1.48	26.7	11.5	9.2	4	5.8	2.5
80	3	11	2.3	1.48	29.8	13	9.2	4	5.8	2.5
100	4	11	2.3	1.48	35.8	15.5	10.4	4.5	6.9	3
150	6	11	2.3	1.48	40.1	17.4	11.5	5	8.1	3.5

25.4 mm is a direct conversion of 1 inch.
BS21 Thread originally devised by Sir Joseph Whitworth and used for engineering nuts and bolts.

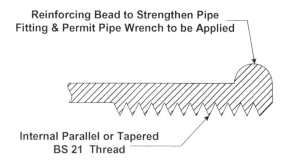

Reinforcing Bead to Strengthen Pipe
Fitting & Permit Pipe Wrench to be Applied

Internal Parallel or Tapered
BS 21 Thread

Figure 3.11 Detail of pipe fitting reinforcing bead

Cast steel or wrought iron are not suitable for manufacturing screwed pipe fittings due to their lack of malleability, which would result in the fitting splitting when being tightened by the application of a standard pipe wrench.

Malleable iron is an iron–carbon alloy which combines the outstanding properties of cast iron and steel to produce a material which can still be cast, but has improved strength and ductility. A controlled heat treatment of annealing is applied to the cast material, which changes the structure and reduces the carbon content. This leaves the material less hard, no longer brittle and possessing good malleable and ductile properties.

The malleable iron pipe fittings have an internal female thread and are reinforced on the outside by the inclusion of a bead on the end of the fitting above the thread that serves to strengthen the fitting, and allows the teeth of the pipe wrench to be applied, as in Figure 3.11.

PLASTICS

With apologies to chemical engineers, in simple terms, plastics are hydrocarbon by-products and may be classified as:

- *Thermosets* – These are plastics that have had a hardener added to the polymer to produce a rigid material that cannot be melted, fused or deformed by products such as electrical light switch plates or small power socket outlets, even when subjected to overload conditions.
- *Thermoplastics* – These are plastics that produce a material with a degree of flexibility that will distort and melt when heated and solidify again and are therefore ideal for extruding into pipes and injection moulding to produce fittings. All plastic pipes are thermoplastics, but only the following are suitable for hydronic heating systems.

CROSS-LINKED POLYETHYLENE (PEX)

Polyethylene (PE) is a thermoplastic that has been used to manufacture tubing since its conception in the early 1930s, first as a low density form (LDPE), which would collapse if subjected to warm temperatures, then as a high density form (HDPE), which can become brittle and crack at higher temperatures, and more recently as a medium density form (MDPE), which has proved to be suitable for waste pipes conveying waste water having intermittent high temperatures. Although all these versions of polyethylene have been successfully used to manufacture tubing appropriate for general plumbing services where high temperatures are not involved, none have been suitable for hydronic heating systems employing the continuous flow where the temperatures of the water exceeds 50°C.

Cross-linked polyethylene, commonly abbreviated to PEX, formerly XLPE, is a form of polyethylene that has been cross-linked in the polymer structure, changing the thermoplastic into an elastomer and preventing the material reverting to a liquid when heated. The cross-linking is normally achieved using high density polyethylene during the extrusion stage of the tubing and produces a piping material that is flexible and suitable for conveying low temperature/pressure water for hydronic heating systems at <85°C. The process also improves its low temperature properties.

European manufactured cross-linked polyethylene is made conforming to any one of three classifications that depict the method of cross-linking employed, but these classifications are not related to any form of rating system.

- *PEX-A* – This method uses peroxide that is mixed with the high density polyethylene and is also referred to as the Engel method. The cross-linked bonds are between carbon atoms and this is a consistent and uniform method.
- *PEX-B* – This method is known as the 'silane method' and the cross-linking is produced in a secondary post-extrusion process. The cross-linked bonds are formed through silanol condensation between two grafted vinyltrimethoxysilane units connecting the polyethylene chains.
- *PEX-C* – This is the oldest method where cross-linking is produced through the electron beam processing method. It is the most environmentally friendly method as it does not involve other chemicals: it uses high energy electrons to split the carbon–hydrogen bonds to facilitate the cross-linking.
- *PEX-Al-PEX* – This is a fourth type of cross-linked polyethylene tubing that is a composite pipe made from a layer of aluminium sandwiched between an external layer and internal layer of cross-linked polyethylene (see Figure 3.12).

Hydronic heating systems employing any form of thermoplastic piping material must use a barrier composite type pipe, as indicated in Figure 3.12, to prevent the ingress of oxygen through the thermoplastic walls of the pipe, causing corrosion to occur to ferrous metals that are contained in the heating system. The permeation of oxygen through the walls of all types of thermoplastic tubes must be prevented or premature failure of steel radiators will occur.

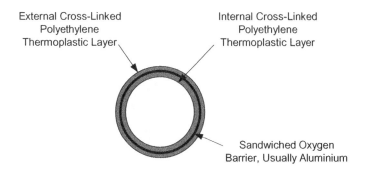

External Cross-Linked Polyethylene Thermoplastic Layer

Internal Cross-Linked Polyethylene Thermoplastic Layer

Sandwiched Oxygen Barrier, Usually Aluminium

Figure 3.12 PEX-Al-PEX composite barrier pipe

POLYBUTYLENE (PB)

Polybutylene, normally abbreviated to PB, is a thermoplastic material used to manufacture tubing suitable for low temperature/pressure hydronic heating systems at temperatures below 85°C. It is a semi-crystalline material with a density in the range of other thermoplastic pipes such as polyethylene.

Polybutylene is a flexible tubing supplied in rolls of varying lengths. Its flexibility at lower temperatures is better than cross-linked polyethylene which may require heating by the application of warm water being passed through it to allow bends to be formed.

Polybutylene is also a lower cost material than cross-linked polyethylene and can be recycled more easily at the end of its life when it can be broken back down to form a raw material ready for using again to produce new products.

Polybutylene tubing is also available as a composite type barrier pipe suitable for use on hydronic heating systems incorporating ferrous materials in the system.

JOINTING PEX AND PB TUBES

The jointing methods and procedures are similar for both polyethylene and polybutylene tubing, whereby electro-fusion and mechanical type crimped compression joints can be successfully used, although cross-linked polyethylene is not suitable for jointing by the electro-fusion method. None of these piping materials can be joined by the solvent welding process that is commonly used on PVCU and ABS tubes.

Electro-fusion joints

Both polybutylene and polyethylene tubing may be successfully joined by employing the spigot and socket type electro-fusion fittings (see Figure 3.13), but only polyethylene may be joined by the butt-welded electro-fusion method.

The preparation of making an electro-fusion joint is critical. The pipe spigot must be shaved to remove the oxidised film that has formed and the fitting socket is supplied in an air sealed bag to prevent oxidation so that the two surfaces can be successfully fused together. If the surfaces remain oxidised they will still melt but will only stick together, but not be electro-welded.

The electro-fusion welding machine is connected to the fittings terminals by the connecting leads and the machine set for the pipe size and ambient temperature. It will then automatically apply electrical heat and weld the joint together plus advise curing time before the joint can be disturbed.

Ideally, thermoplastic pipes should not be joined below floors; where this is unavoidable, the jointing below floor areas should be performed by using the electro-fusion type of fitting.

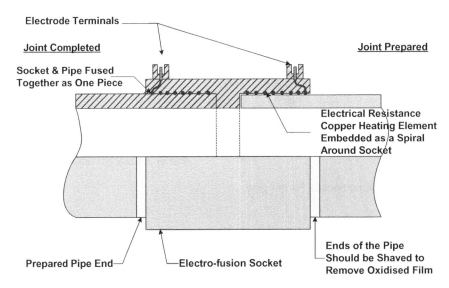

Figure 3.13   Detail of spigot and socket electro-fusion joint for PE and PB pipe

This method of forming pipe joints is very reliable if correctly prepared and conducted, but it is not normally considered financially viable for domestic heating systems. However, as these applications are quite small, the piping can be arranged as continuous runs not requiring any joints to be made below floors or any other inaccessible places.

Mechanical crimped compression joints

There are many variations of these types of joints depending upon the manufacturer, but they all work on the same principles of compressing the pipe ends onto the fitting to form a permanent sealed joint that cannot be unassembled once it has been crimped. They are ideal for domestic heating applications and are suitable for both cross-linked polyethylene and polybutylene tubing where the tubing sizes being used are quite small, thus making the crimping tools affordable and manually operated.

VALVES

There are numerous types of valves used for the control of fluids in the plumbing/public health industry, but this volume is only concerned with those engaged for controlling fluids in hydronic heating systems. Some specialist valve types are explained within appropriate sections of this text, whereas the following valves are common to many systems included throughout this volume.

Most valves may be categorised as being one of the following:

- Isolating – Valves designed to be fully open or fully closed to control the flow of fluid through a pipeline.
- Regulating – Valves designed to vary the flow of fluid through a pipeline, that can be regulated anywhere between fully open and fully closed.
- Automatic – Valves designed to automatically control the flow of fluid in a pipeline to perform its particular function.

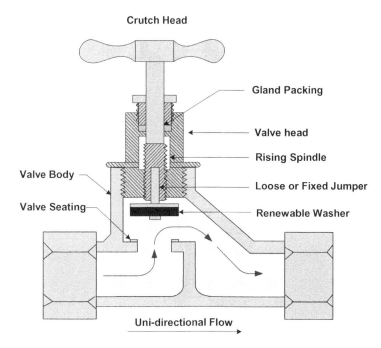

Figure 3.14   BS 1010 stopcock pattern stopvalve

STOPCOCK PATTERN STOPVALVE

This type of valve is a uni-directional flow form of isolating valve, commonly used to control the flow of water into each property. It is manufactured from DZR brass and conforms to BS 1010. It offers a high resistance to the flow and is ideal for higher pressure applications where a 100% tight shut-off is required. It is available in pipe sizes up to and including 50 mm (2 inch) diameter with pipe connections arranged with either a female BSP thread, or Type 'A' or Type 'B' compression coupling or capillary ends suitable for copper/stainless steel or thermoplastic tubing. Although these valves are predominantly employed for isolating flows through pipelines, they do possess reasonable regulating characteristics due to their under-and-over flow path. They have also been manufactured with a loose or fixed jumper, the loose jumper pattern serving as a non-return valve to prevent the reversal of flow through the valve. Fixed pattern jumpers should be used for applications where the washer may become stuck to the seating if the valve is left in the closed position for prolonged periods, such as in hot water systems.

   An example is shown in Figure 3.14.

GATEVALVES

These types of valves are a full-way bi-directional flow form of isolating valve, commonly used to control the flow of water from cold water storage cisterns on cold feeds and down service distribution pipes. They should be manufactured from DZR brass or bronze. They offer the minimum resistance to the flow and are ideal for controlling the flow of water when the minimum of pressure loss is required. They are available in virtually unlimited pipe sizes but rarely used in diameters exceeding 600 mm due to less costly forms of valves being available. Gatevalves can be arranged with female BSP connections, as indicated in Figure 3.15, or compression and capillary couplings suitable for copper tube, stainless steel and thermoplastic

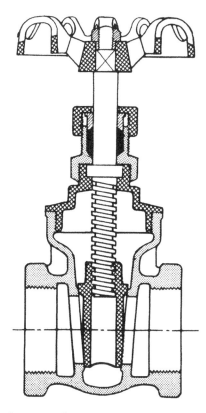

Figure 3.15   Non-rising stem wedge gatevalve

pipes with larger diameter valves which have flange connections and are usually manufactured from ductile iron or forged steel.

Gatevalves are designed for isolating purposes only; they should be either fully open or fully closed and have very poor regulating properties.

There are numerous variations of gatevalves used within the building services industry, but Figure 3.15 shows the non-rising stem wedge type gatevalve, which is the most commonly used valve type for building services. Gatevalves are equipped with a wheel-head top which, when operated, turns the threaded stem which then permits the wedge shaped gate to be drawn up on the stem thread and into the bonnet or upper part of the valve body, hence the stem just turns but does not rise, unlike the spindle of the stopcock. As the name implies, the gate is wedge shaped, which permits the valve gate to be lowered into the shut-off position and come into contact with the valve side seating forming a metal-to-metal seating. Any foreign matter on the seating will be knocked off by the descending gate and drop into the trough at the bottom of the valve body.

It should be noted that in situations where there is a high pressure on the valve upstream side together with a low pressure on the downstream side of the gate, the valve gate can distort and even bend the stem, resulting in the gate becoming jammed and unable to move.

Figure 3.16 portrays a form of gatevalve known as a parallel slide type, which is a two-piece valve used extensively on high pressure/temperature heating systems and is manufactured from bronze, ductile iron or forged steel in sizes of 50 mm and larger.

It differs from the previous gatevalve described by the isolation being effected by means of two sliding discs which are traversed across the flow path of the fluid, the fluid pressure ensuring a tight joint between

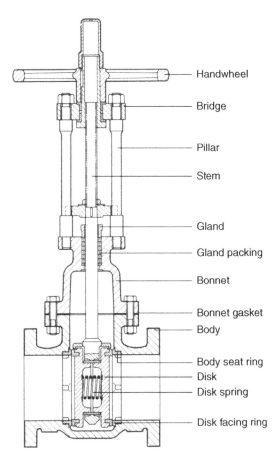

Figure 3.16 Parallel slide rising stem pattern gatevalve

the downstream disc and its seat. The floating action of the discs, which are separated by a spring, reduces the effects of mechanical thermal distortion and together with the sliding action assists the closure of the valve and prevents jamming.

The valve has a rising stem that is threaded along its length, and when operated will cause the stem to rise up into the cover above the wheel head when it is in the open position; this has the advantage of giving an instant indication as to the valve's open or closed status. When the stem is raised it permits the spring to pull the two discs together, thus releasing the pressure between the discs and the side seating.

BUTTERFLY VALVES

This is another form of full-way bi-directional isolating valve that also offers the minimum resistance to fluid flow and is simpler in construction in virtually unlimited pipe sizes above 50 mm diameter.

The valve comprises a circular metal disc of the same diameter as the valve body, which is manufactured from ductile iron, that when operated, just requires a quarter of a turn for the valve to travel from fully open to fully closed. In the fully open position this circular disc adopts a position of being in line with the fluid flow, as shown in Figure 3.17, whereby it will offer the minimum resistance for the fluid to flow past the disc. When the valve is closed the disc adopts a position of being 180° against the fluid flow and comes

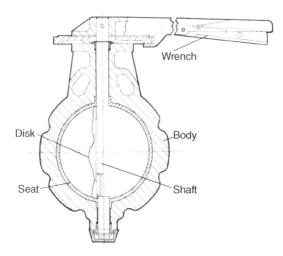

Figure 3.17   Wafer pattern, lever operated butterfly valve

to rest against a resilient seating, forming a reasonable leak free seal. As with the gatevalve, high differential hydraulic pressures across the valve disc can cause distortion to the valve shaft or disc, causing the valve to become jammed.

These valves are now used quite extensively for isolating purposes within the building services industry, but should not be used on high velocity pipelines unless it has been arranged for them to operate slowly by an electric actuator, as the fast shut-off could cause surge conditions (water hammer) to occur.

Butterfly valves have the advantage of having a slim body occupying the minimum of space and are fitted between two mating flanges with extra long bolts clamping the valve body between them. The valve body can have recesses for the flange bolts to pass over, as shown in Figure 3.17, or lugs on the periphery of the valve body. Alternatively, butterfly valves can be supplied with mating flanges cast onto the valve body.

As the butterfly valve body is slim, the valve disc will protrude past the valve body when in the open position; therefore care should be taken to avoid the valve being fitted in direct contact with some other pipeline component or equipment that will foul the operation of the butterfly valve.

Figure 3.17 indicates a butterfly valve with a quarter turn lever operation, which has the advantage of giving instant indication as to the valve's open or closed status. They can also be arranged to have a side turned wheel-head operation through a simple gearbox, or be equipped with an automatic electrically operated actuator.

DIAPHRAGM VALVE

The diaphragm valve is another bi-directional isolating valve originally developed for the control of compressed air but now adopted for the control of many types of fluids (liquids and gases), as it accomplishes a reliable tight 100% shut-off.

The valve comprises a cast body component that is available in a wide variety of both metallic and thermoplastic materials and an upper bonnet section incorporating the non-rising stem assembly, with the diaphragm spreading compressor with a flexible diaphragm bolted between the flanges of these two body components. The flexible diaphragm is obtainable in a vast choice of materials that can be selected to suit the fluid being controlled. This is a valve that not only achieves a reliable shut-off, but also prevents the fluid from entering into the bonnet of the valve, which eliminates the need for leak-proof glands, a

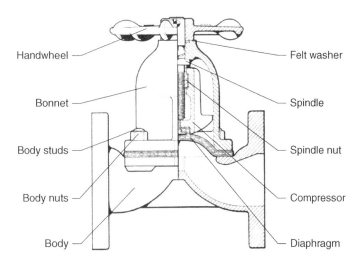

Figure 3.18   Diaphragm isolating valve

particular advantage where very searching liquids or gases are being controlled. This separation of the fluid's entry into both sections of the valve also ensures that the lubricated operating parts of the valve spindle in the bonnet cannot contaminate the pipeline fluid, thus guaranteeing the purity of the line fluid.

Diaphragm type isolating valves are seldom selected to control water distributing or circulating systems in building services, but due to their reliability and effectiveness they are ideal for controlling the flow of higher pressure gases, particularly liquefied petroleum gases.

Maintenance to diaphragm valves is restricted to replacing the diaphragm itself, which is a simple task of just unbolting the valve bonnet, lifting it off the valve body to expose the diaphragm, removing it from the valve compressor and replacing with a new one of the same material. The valve can then be reassembled.

The flow path through diaphragm pattern valves is so shaped that the fluid has to pass over the raised seating section formed in the bottom of the valve body, therefore if the fluid is a liquid, the valve should be installed turned on its side to aid drain-downs by preventing liquid from being trapped in the pipeline.

This valve is illustrated in Figure 3.18.

PLUG VALVES AND COCKS

These are simple quarter turn bi-directional isolating valves offering a straight through low resistance to fluid flow and available in a range of pipe sizes from 10 mm up to 100 mm diameter.

The spherical plug valve, sometimes referred to as a ball valve, comprise a spherical plug with a passageway through the middle connected to a vertical spindle with a lever attached to drive the plug valve 90° from fully open to fully closed.

Spherical plug valves are an inexpensive method of providing a service valve to isolate individual sanitary fixtures and fittings, plus a means of isolating individual heating circuits for maintenance purposes. The orifice through the middle of the spherical plug should be of equal diameter to that of the pipe it is connected to or the flow will be restricted if this passageway is smaller.

Figure 3.19 shows a spherical plug valve in the open position with the valve orifice turned to be in-line with the fluid flow without restricting the stream.

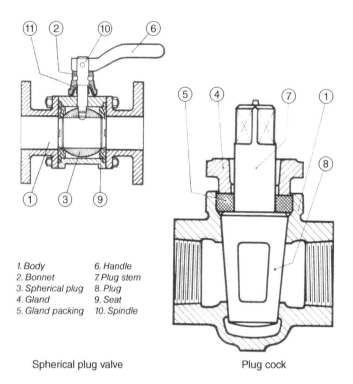

1. Body
2. Bonnet
3. Spherical plug
4. Gland
5. Gland packing

6. Handle
7. Plug stem
8. Plug
9. Seat
10. Spindle

Spherical plug valve   Plug cock

Figure 3.19   Quarter turn plug valves and cocks

Quarter turn plug cocks are similar to the spherical plug valve except that the plug is tapered and has an elongated passageway through the middle of it. Plug cocks are an ideal valve for controlling the flow of low pressure gas to appliances including gas fired boilers.

Lubricated versions of the plug cock are suitable for higher operating pressures and are used in many industries, including the building services industry where higher gas supply pressures are involved.

NON-RETURN AND CHECK VALVES

Non-return and check valves are essentially the same type of valves as they perform the same function, which is to automatically prevent the reversal of fluid flow through a pipeline. They operate entirely by the motion of the fluid passing through the valve. However, a definition that is generally accepted to differentiate between the two names is:

- *Non-return Valve* – A valve that will prevent the reversal of fluid flow when the downstream pressure reverses and becomes higher than the upstream pressure, but if both upstream and downstream pressures are equal it will not prevent a trickle reversal of flow back through the valve.
- *Check Valve* – A non-return type of valve that not only prevents the reversal of fluid flow when the downstream pressure becomes higher than the upstream pressure, but will also remain tightly closed when pressures are equal on both sides of the valve, preventing any trickle back flow occurring.

All non-return and check valves are uni-directional by design and automatic in operation.

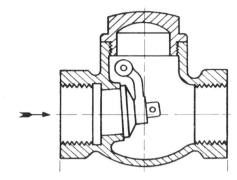

Figure 3.20 Swing gate type non-return valve

NON-RETURN VALVES

Figure 3.20 illustrates a form of non-return valve known as a swing gate type. This is manufactured from a wide variety of materials including DZR brass, bronze, ductile iron and forged steel with pipe connections arranged for copper, steel, flanged connections and adaptors for thermoplastic tube and pipe sizes which are also virtually unlimited.

The valve comprises a gate that is suspended from a spindle that traverses across the valve body to enable the gate to swing from it, as shown. Because of this design where the gate functions by gravity, hanging down from the spindle where it is free to swing, the valve must be installed in the horizontal plane as depicted, or vertically in conjunction with an upward path flow. This enables the fluid flow to push open the gate as it passes through and then swing shut by gravity to cover the valve inflow port when the flow ceases. When this return backflow pressure is greater than the pressure on the other side of the gate, a positive leak free shut-off is achieved. If the valve has been installed in the vertical plane with a normal upward flow, the valve will also achieve a positive shut-off due to the weight of the gate closing by gravity, even when the pressures on either side of the gate are equal. These valves are ideal for protecting pumps, as explained in Chapter 18.

There are a number of variations of this type of non-return valve including a version that resembles the body and flow path of a stopcock, where the flow lifts the valve off its seating which is then allowed to fall back on cessation of flow to prevent reversal.

CHECK VALVES

Figure 3.21 indicates the basic operational concept of a single non-verifiable check valve. It comprises a two-piece body housing a spring assisted stop valve complete with a resilient seating attached to a compressive return spring, which is braced against internal lugs forming part of the valve body.

The fluid flow enters the valve from the left side as shown, and the pressure pushes the valve off its seating against the resistance of the compressive spring to allow the fluid to pass through the valve. When the fluid ceases to flow, the tension of the spring returns the valve back on to its seating to prevent any backflow occurring and any fluid trickling back past the valve seating when the pressures either side of it are equal.

The term 'non-verifiable' means that there is no provision in the valve to test that it is functioning correctly. To accommodate this, a separate test valve would have to be installed on the upstream side of the check valve located between an isolating valve and the check valve, to enable that section of pipe to be

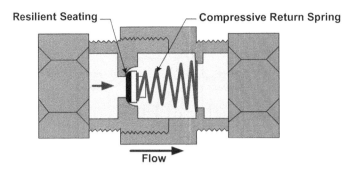

Resilient Seating — Compressive Return Spring

Flow

Figure 3.21　Single non-verifiable check valve (WRAS designation type EB)

drained to prove that the valve is shutting off correctly. This type of valve has been designated as a type EB by the Water Regulations Advisory Scheme (WRAS).

A similar single check valve that incorporates a small test point on the upstream side of the valve seating to enable it to be tested, is termed 'verifiable' and is designated by WRAS as a type EA verifiable single check valve.

These types of valves are used as a less expensive means to prevent backflow occurring in non-critical applications. They are manufactured from DZR brass with either Type 'A' compression couplings for copper tube, or provided with BSP female threads. They are only available for small pipe diameters for domestic applications. It should be noted that check valves generally offer a greater resistance to the fluid flow, due to the inclusion of the compressive return spring, than does the swing gate type non-return valve.

The WRAS valve designation is only applicable if the particular valve has been submitted by a valve manufacturer for testing and found to be suitable for the designated purpose.

Both type EA and EB single check valves are an approved means of preventing backflow and back-siphonage for fluid category 2 as defined by WRAS, which is wholesome water whose aesthetic quality is impaired owing to:

- A change in the water temperature
- The presence of substances or organisms causing a change in its taste, odour or appearance, including water in domestic hot water distribution systems.

DOUBLE CHECK VALVE

Figure 3.22 indicates a verifiable version of a double check valve, WRAS designation EC, which is simply two single check valves incorporated in a single common three part body. The verifiable test points enable this type of valve to be tested to prove that each of the two check valve sections are functioning correctly.

The non-verifiable version of the double check valve is exactly the same in design, except that the verifiable test points have not been incorporated in the valve body and it has been given a WRAS designation of type ED. The application of the non-verifiable version of the double check valve is depicted in Figure 2.40, in conjunction with a temporary filling loop.

These types of valves are used as a less expensive method of preventing backflow than two single check valves and are also manufactured from DZR brass with either Type A compression couplings for copper

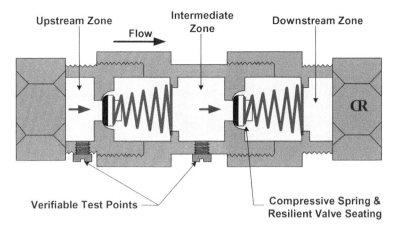

Figure 3.22   Double verifiable type check valve (WRAS designation type EC)

tube, or provided with BSP female threads. They are also only available for small pipe diameters for domestic applications.

Both type EC and ED double check valves are approved means of preventing backflow and back-siphonage for fluid category 3, as defined by WRAS, which is a fluid that represents a slight health hazard because of the concentration of substances of low toxicity, such as found in a sealed hydronic heating system, including any fluid which contains:

• Ethylene glycol, copper sulphate solution or similar chemical additives
• Sodium hypochlorite (chloros and common disinfectants).

NON-VERIFIABLE DISCONNECTOR VALVE

This is a form of non-verifiable double check valve that incorporates an intermediate reduced pressure zone that opens to drain the water contained in this reduced pressure zone, and so establish a break between the upstream and downstream zones, hence the term 'disconnector valve'.

The valve comprises two check valves separated by a spring-assisted sliding valve over a relief port. During normal operation, as depicted in Figure 3.23, the flow of the fluid opens the two check valves as normal, but also has the effect of sliding the intermediate valve over the relief port to close it. When the flow ceases, the two check valves are assisted by their springs to close normally, which allows the action of the intermediate valve to slide back by the decompression of its spring to expose the opening of the relief port and drain the small quantity of fluid from the intermediate reduced pressure zone, creating a void between the upstream zone and the downstream zone. For this reason, the relief port of the valve must be piped to drain via a tundish similar to that of the relief drain associated with unvented domestic hot water systems.

Figure 3.24 illustrates the valves operation during backflow conditions when the downstream check valve has become stuck open, or jammed by the action of foreign matter lodging between the valve and its seating. In this situation the upstream check valve has closed as normal, allowing the intermediate valve to slide and open the relief port allowing the backflow fluid to drain, reducing the risk of contamination to the water main.

WRAS also designates the type CA valve suitable for protecting at fluid category 3, but permits this valve to be used to protect the water supply on permanent supply connections to domestic sealed type hydronic heating systems as shown in Figure 2.41.

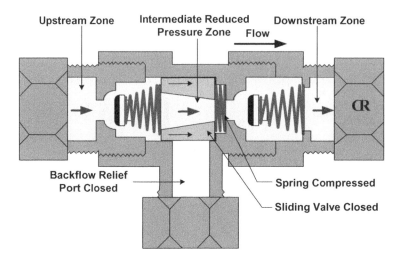

Figure 3.23 Non-verifiable disconnector valve, shown in normal flow condition (WRAS designation type CA valve)

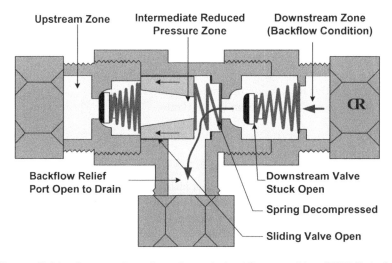

Figure 3.24 Non-verifiable disconnector valve, shown in backflow condition (WRAS designation type CA valve)

REDUCED PRESSURE ZONE (RPZ) VALVE

Backflow preventers are designated as type BA valves by WRAS, but are commonly known as RPZ valves because they incorporate a reduced pressure zone. They are the highest form of mechanical protective devices permitted under the Water Regulations.

RPZ valves are permitted to protect against backflow for fluid category 4, which is defined as a fluid that represents a significant health hazard because of the concentration of toxic substances, including any fluid which contains:

- Chemical, carcinogenic substances or pesticides (including insecticides and herbicides)
- Environmental organisms of potential health significance

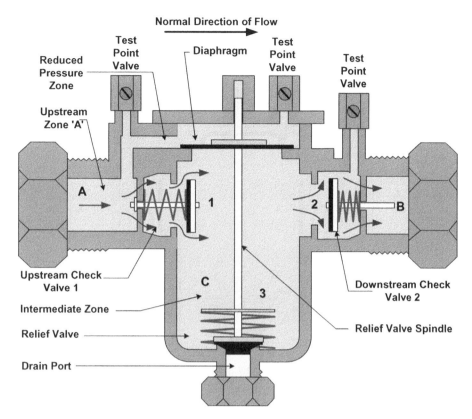

Figure 3.25 Reduced pressure zone (RPZ) valve, shown during normal operation (WRAS designation type BA valve)

Figure 3.25 illustrates a reduced pressure zone valve operating under a normal flow condition. It consists of a double check valve arrangement with an intermediate zone chamber, which itself has a diaphragm operated relief valve fitted at the bottom of the intermediate chamber to serve as a drain. Under normal flow conditions the hydraulic pressures being exerted on both sides of the diaphragm, as well as all other parts of the valve are equal, therefore the valve will perform like any other double check valve. When the flow ceases, both check valves will close assisted by their springs and the hydraulic pressure will remain equal throughout the valve.

RPZ valves have only been permitted for installation in the UK since the introduction of the Water Regulations in 1999; prior to that, only air gaps were permissible to protect systems.

Figure 3.26 illustrates the reduced pressure zone valve during backflow conditions. In situations where the upstream flow fails and backflow conditions are created, the hydraulic pressure in upstream zone 'A', the pressure above the diaphragm, will be reduced, causing the backflow valve 3 to rise and thereby open the drain port to drain the intermediate chamber. If the downstream check valve 2 fails to close fully, then the backflow water from the downstream zone 'B' will enter the intermediate zone 'C' and drain out through the relief valve port.

RPZ valves are approved for installing on permanent water supply connections to sealed hydronic heating systems as depicted in Figure 2.42.

It should also be noted that RPZ valves are required to be installed by operatives who hold a valid competency certificate. The building owner is also required to have the valve tested and certificated annually by a competent operative.

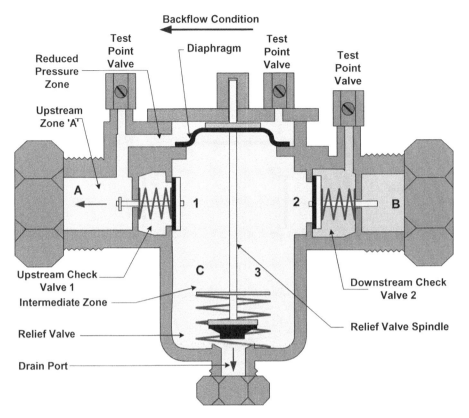

Figure 3.26   Reduced pressure zone (RPZ) valve, shown during backflow condition (WRAS designation type BA valve)

VACUUM BREAKING VALVE

An anti-vacuum valve or vacuum breaker is a mechanical device designed to automatically prevent backflow conditions from siphoning water out of the piping system downstream of the device. During normal system operation, the system water pressure keeps the valve disc closed against its seating, but if upstream backflow conditions occur and the water level starts to fall due to lower atmospheric pressure prevailing, the valve disc will fall to below its seating to open the valve to the atmosphere and allow air to enter into the piping system, and thus prevent siphoning the water out of the system.

An example can be seen in Figure 3.27.

FLOATVALVES

The function of a floatvalve is to automatically control the flow of fluid into a storage cistern, water tank or other water storage receptacle by allowing water to automatically enter it when required, and close it off when full. This is achieved by the action of a piston sliding up against an orifice seating, the motion of this piston being controlled by the movement of the floatvalve lever arm, which is attached to a float at the other end of the arm to raise or lower the float at the dictates of the water level within the storage cistern.

The diameter of this orifice seating will be selected to suit the pressure of the incoming water and is available in four interchangeable sizes, namely high, medium and low pressure, plus full way.

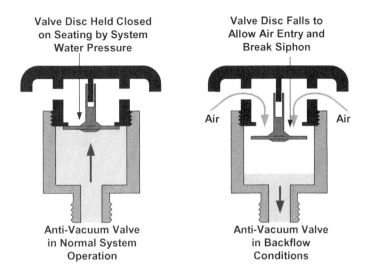

Figure 3.27 Basic operating principles of vacuum breaking valve (WRAS designation type DA valve)

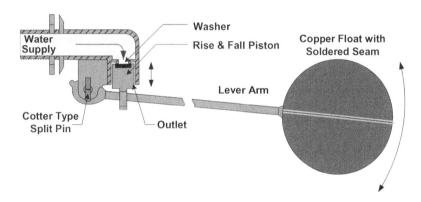

Figure 3.28 Obsolete Croydon pattern type floatvalve

Floatvalves have sometimes been incorrectly termed 'ballvalves', due to the shape of the float, but to prevent confusion with the inline spherical isolating valve commonly referred to as ballvalves and depicted in Figure 3.28, the correct name 'floatvalve' should be used.

There have been many designs for floatvalves over the centuries that include the vertical rise and fall piston type Croydon pattern floatvalve, which has been obsolete since the 1920s, but there are still a few examples in working use today. Figure 3.28 depicts the basic operating arrangement of the Croydon pattern floatvalve, which also had a number of design refinements available at the time.

Floatvalves should now conform to BS 1212, which consists of four parts. Part 1 covers the Portsmouth pattern floatvalve, which in its Ministry of Health (MOH) pattern superseded the Croydon type of floatvalve. Part 2 covers the now preferred design known as the diaphragm type of floatvalve constructed from brass. Part 3 is also a diaphragm type floatvalve but manufactured from plastic materials, and Part 4 covers diaphragm floatvalves designed to fit into more confined spaces such as water waste preventors.

The Portsmouth pattern floatvalve conforming to BS 1212 Part 1 differs because the sliding piston moves horizontally instead of vertically, as with the Croydon pattern floatvalve that it superseded. This gives the

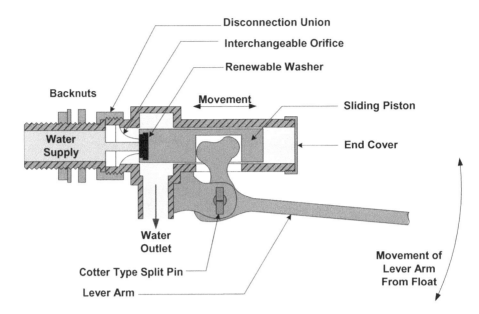

Figure 3.29 Portsmouth pattern floatvalve, BS 1212 Part 1

advantage that the piston will slide along the barrel of the body when moved by the float/lever arrangement, to freely expose the valve outlet for the water to enter into the cistern unhindered, which was not possible with the obsolete Croydon pattern floatvalve.

The Portsmouth pattern floatvalve has changed very little from its original conception as an MOH specification to its present specification of BS 1212 Part 1. The lever arm attached to the float, but not shown in Figure 3.29, moves up and down under the dictates of the water level, which will cause the sliding piston to move horizontally within the barrel of the valve body. When the water level in the cistern is at its highest the washer end of the piston will be held tight up against the orifice thus closing the valve. The orifice comprises a venturi shaped disc with a calibrated hole drilled through the centre: it is interchangeable with four different sizes of orifices where the diameter of the passageway through the middle varies from the smallest, or high pressure orifice, through medium pressure and low pressure to the largest, which is termed full-way.

The final water level in the cold water storage cistern was achieved by manually bending the floatvalve lever arm down by approximately 15° to attain a distance of around 50 mm between the water level and the floatvalve water outlet, a practice that is now discouraged.

Portsmouth pattern floatvalves are manufactured from brass or bronze materials and are available in sizes from 15 mm up to 150 mm diameter, with larger variations of these valves, above 150 mm diameter, being produced from ductile iron.

Although the Portsmouth pattern floatvalve is no longer the preferred type of automatic floatvalve for domestic residential dwellings, it is still used for larger commercial type properties as a means of controlling the entry of water into cold water storage cisterns.

The Water Regulations requires that Portsmouth pattern floatvalves should only be used in conjunction with a backflow protection device, such as a non-verifiable double check valve type ED, to prevent water from being siphoned back into the valve if the valve body has become submerged.

Floatvalves, regardless of the design type, can be affected by higher or fluctuating pressures where the inlet water pressure, acting against the washer, pushes the piston back and forces the float down to allow the valve to open momentarily before the float bounces back up by the buoyancy of the water. This situation can continue, resulting in surging water supply conditions and producing the effect known as 'water hammer'. In severe situations such as these, an equilibrium pattern floatvalve should be selected, as depicted

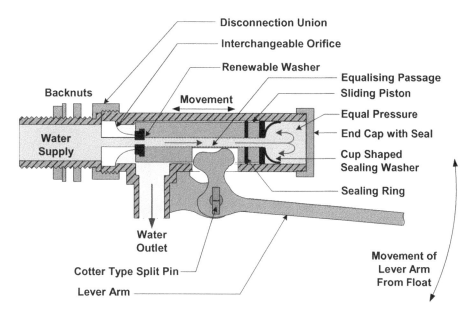

Figure 3.30 Equilibrium pattern Portsmouth floatvalve

in Figure 3.30, which has a hole drilled through the length of the piston to allow the water pressure to pass through to the opposite end of the valve body to equalise the pressures at both ends.

This permits the valve piston to freely move along the valve body without being affected by unequal water pressures. The far end of the piston is equipped with sealing washers and/or rings to prevent leakage and maintain pressure equalisation.

Because of the pressure equalisation arrangement this type of valve will not be affected by high fluctuating pressures pushing the valve piston off its seating.

The most recently introduced type of floatvalve is the diaphragm pattern valve for domestic residential dwellings and other smaller applications, sometimes referred to as the 'Garstan' pattern floatvalve. It was so named, in keeping with naming floatvalves after places as this valve was developed by the Building Research Station located at Garstan in Hertfordshire.

These floatvalves should conform to BS 1212 Part 2 (brass body), or BS 1212 Part 3 (plastic body).

This improved design arranged the water outlet to issue from the top of the valve body, thus reducing the risk of back-siphonage of the cistern water occurring back into the water supply main, as the outlet will still be above the floatvalve body in situations where the valve becomes submerged.

Another design improvement was to change the sealing washer for a flexible diaphragm that served both as a means of providing a water-tight seal when pressed up against the interchangeable orifice seating, and also to provide a physical barrier to prevent water from the storage cistern entering back into the valve body when the floatvalve body becomes submerged.

A further improvement with this form of floatvalve is the incorporation of the provision to be able to adjust the water level without having to physically bend the lever arm. This is achieved by either adjusting the main sliding valve float connected to the turned down section of the lever arm, or by regulating the fine adjustment screw that comes into contact with the diaphragm pin.

Because of these design improvements, the Water Regulations do not require any additional backflow prevention devices, as in the case of the BS 1212 Part 1 Portsmouth pattern floatvalve. Figure 3.31 illustrates the basic operating principles of the diaphragm pattern floatvalve.

Unlike the Portsmouth pattern BS 1212 Part 1 floatvalve, the diaphragm pattern BS 1212 Parts 2 and 3 is only manufactured in the small domestic size of 15 mm diameter.

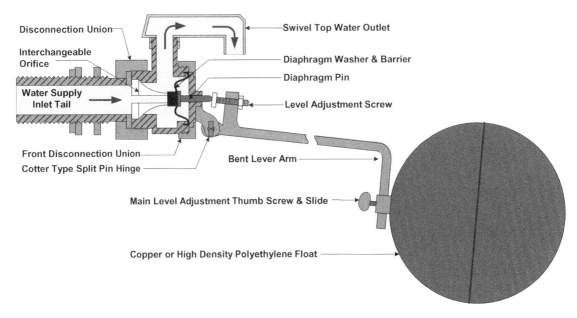

Figure 3.31   Diaphragm pattern floatvalve, BS 1212 Parts 2 and 3

GLOBE VALVES

A globe type valve is a simple uni-directional form of regulating valve that can be employed to balance the flow to zoned circuits on simple hydronic heating systems.

In its body shape, the valve resembles a stopcock but has a gatevalve type wheel head. The flow of fluid through the valve is similar to that of a stopcock where the passageway is an under and over path, but the flat disc type washer is replaced by a domical shaped washer that improves the flow regulating properties over the seating compared to those of a stopcock, and far greater than those of a gatevalve.

Globe valves are manufactured in sizes ranging from 15 mm up to very large diameters and in various materials including brass, bronze, ductile iron and forged steel. They are available having female BSP threads or flange connections on sizes above 50 mm.

The traditional style of globe valve is depicted in Figure 3.32 and can also be used as a stopvalve to control the flow of fluid through the pipeline.

The value of the globe valve as a regulating type of valve can be observed in Figure 3.33, which depicts typical flow curves for various forms of valves used in the plumbing and heating industry. Using the diagonal black straight line shown in Figure 3.33 as the ideal flow characteristic, it shows that when this ideal commissioning valve is only 10% open it has a corresponding flow that also equates to 10% and when it is half open it will permit half the total flow rate possible. A flow characteristic that corresponds to this ideal straight black line, where the flow rate increases in direct proportion to the amount of valve opening, would make the commissioning exercise of a hydronic heating system simpler and quicker than it actually is.

In contrast to this ideal valve flow characteristic, the standard gatevalve which is indicated by the top line has a relatively poor flow characteristic. The flow rate through the gatevalve increases rapidly when the gatevalve has been slightly opened, producing a steep incline on the characteristic flow curve: this means the gatevalve has very poor regulating properties and should only be used as an isolating valve, either fully open or fully closed. If the valve is only 30% open the corresponding flow rate could be as high as 80%, which if accompanied by high flow velocities, could produce a turbulent flow resulting in erosion to the gate of the valve and premature failure.

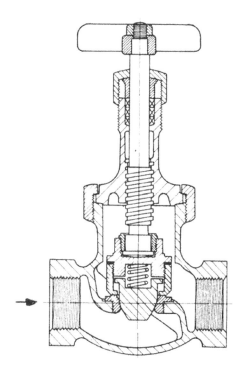

Figure 3.32 Traditional style globe type stopvalve

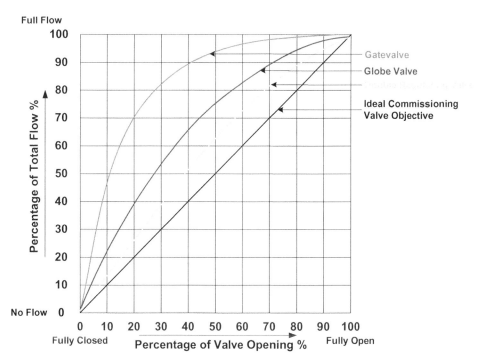

Figure 3.33 Typical valve flow curves (typical values indicated for comparison only – consult manufacturers, flow curves for specific products)

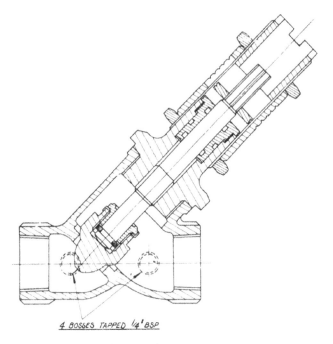

4 BOSSES TAPPED ¼' BSP

Figure 3.34   Oblique pattern double regulating valve

The traditional globe valve is depicted by the second flow characteristic curve marked in Figure 3.33. Although it does not achieve the ideal straight flow line, it does produce a better regulating property curve than the gatevalve, making it a suitable and affordable valve for regulating the flow to individual heating zone piping circuits. However, the globe valve on its own does not have the facility to measure the flow rate passing through it, or the pressure drop encountered: measuring test points would have to be installed separately to accommodate this requirement if desired.

The double regulating valve is also shown in Figure 3.33. It produces an improved flow characteristic curve compared to the simple globe valve.

Figure 3.34 illustrates an oblique pattern double regulating valve, which resembles an angled pattern globe valve; it is also a uni-directional flow valve with the domical shaped disc and seating angled in the direction of the fluid flow to reduce the sharp change in direction for the flow to provide a more linear flow path. This angled design produces a further improvement in the flow characteristics, making the task of balancing the system easier.

The body of the valve incorporates tapped bossed test points located both upstream and downstream of the valve seating to facilitate the use of an inclined manometer, which will indicate the resistance being obtained when the valve is being regulated. When the valve has been set to provide the frictional resistance to the fluid flow required, the valve may be locked to prevent it being altered or tampered with, but may be used as a stopvalve which can then be reopened as far as the pre-set lock will permit.

The valve shown in Figure 3.34 also has the added provision of a lockshield cover to prevent unauthorised interference, and is also available with standard wheel head operation.

Double regulating valves are manufactured in similar pipe sizes and materials as standard globe valves with female BSP connection threads, or flanges.

4

Heat and Heat Transfer

One of the main objectives in the design of a heating system involves the distribution and transfer of heat. In order for us to progress in the design process of the heating system, we need to understand what heat is and how it may be transferred.

HEAT

The amount and intensity of heat that we receive from the sun varies throughout the world and also varies seasonally. Countries such as the UK that normally experience cold climatic conditions during the winter season, and usually part of the spring and autumn seasons too, require heat to be added to living spaces otherwise life as we know and expect would not be the same.

Warmth alone is not the only criterion to achieve comfortable conditions by today's standards: the design engineer must also attain a pleasant atmosphere within the living space.

In warmer climates and many commercial buildings constructed today this would be achieved by an air-conditioning system with summer cooling and winter heating. Confining our attention to the heating aspect alone, the design engineer must achieve an even temperature coupled with the correct form of heat distribution, thermostatic control and where necessary, adequate ventilation that includes a fresh air supply together with humidity control. All these factors, plus conforming to current regulations and energy efficiency, must be considered and duly incorporated into the heating scheme.

We all have a preconceived idea of what heat is until we are required to define it, as heat is often described in simplistic terms as a condition of being hot. The correct definition of heat is that it is a form of energy arising from the random motion of impacting atoms and molecules and requires a temperature imbalance to flow. The unit of heat is the joule, so named after the English physicist James Prescott Joule (1818–1889), whose work first identified this energy form. Heat and temperature are not the same; temperature is the effect that is produced from heat energy.

Heating Services in Buildings: Design, Installation, Commissioning & Maintenance, First Edition. David E. Watkins.
© 2011 John Wiley & Sons, Ltd. Published 2011 by John Wiley & Sons, Ltd.
As an aid to lecturers and students, full colour versions of the figures in this chapter may be found at www.wiley.com/go/watkins
These figures are © 2011 by John Wiley & Sons, Ltd.

TEMPERATURE

This is the measure of the relative warmth of an object or substance that allows it to be compared to other hotter or cooler substances, and is described simply as the thermal state of a substance at any given time. The temperature may be measured by a thermometer using a number of different scales or units, although the Celsius scale has now been internationally adopted and should be used when referring to such matters.

Thermometers

There have been many forms of thermometers over the course of history, but it was not until 1701 that Ole Christensen Romer, a Danish astronomer, created the first practical thermometer. Romer's thermometer used red wine as a temperature indicator contained in a bowl with a glass tube. The tube was sealed at the top and had its open bottom end immersed in the bowl of red wine. The temperature scale that Romer created started at 0, which was the freezing point of a salt water mixture, about −14°C, with the freezing point of water being 7½° and the boiling point of water being recorded as 60° on this scale. This early form of a recognised thermometer was only used for a few years before being superseded by Fahrenheit's thermometers and the first of five scales used to measure temperature.

Fahrenheit scale

Gabriel Daniel Fahrenheit (1686–1736), a German born scientist, is best known for the 'Fahrenheit temperature scale' that has been used for many years. It was Fahrenheit who first invented the alcohol thermometer in 1709, and then the mercury thermometer in 1714, both of which are still in common use today. The scale chosen by Fahrenheit for his thermometers was based on creating a practical temperature scale of both the maximum and minimum climatic temperatures experienced in Western Europe, whereby 0°F was the lowest temperature normally encountered and 100°F was the hottest. Using these parameters, the freezing point of water was recorded as 32°F, and the boiling point of water established as 212°F. Fahrenheit made certain assumptions that temperature varied with atmospheric conditions that were proved wrong, but he later established the effects that pressure has on the boiling point of liquids. The Fahrenheit scale was first adopted by Britain where it remained in use until it was officially replaced by the Celsius scale in 1971, which was in common use in most of Europe. The Fahrenheit scale is now mainly confined to use in North America, and even there the Celsius scale has started to be used by some industries.

Although the Fahrenheit scale is no longer in general use, the important contribution that Fahrenheit made in the form of temperature measurement, as well as other works, should not be diminished.

Reamur scale

René Antoine Ferchault de Reamur (1683–1757), a French physicist, devised in, 1730, a simpler temperature scale than that created by Fahrenheit. Reamur measured the expansion of a mixture of water and alcohol as its temperature increased. The liquid mixture was contained in a glass bulb at the base of the tube as in a thermometer. When it was at freezing point the tube was marked 0°R accordingly. The remainder of the tube was graduated into equal units, each of which was equal to one-thousandth of the volume of the liquid in the bulb and tube when it was at freezing point. When the liquid reached its boiling point it was found that its length in the tube had increased to 1080 units, so it had risen 80 units. Consequently, Reamur's scale ran from 0°R at freezing point to 80°R at boiling point. Reamur's temperature scale was used originally in France and part of Germany, but later replaced by the Celsius scale.

Celsius scale

Anders Celsius (1701–1744), a Swedish born professor, created his temperature scale in 1742 for his meteorological observations, but it was slightly different from the one that we are used to today. The scale that he devised was graduated in the familiar 100 equal units, but it was inverted, whereby the boiling point of water was represented by 0°C and the freezing point of water was 100°C. It was not until Celsius's death in 1744 that the Swedish doctor Carl Linnaeus proposed reversing the temperature scale so that 0°C represented the freezing point of water and 100°C represented the boiling point. The centigrade scale as it was then known – because it was divided into 100 degrees between the freezing and boiling points of water – gradually became popular throughout the world. The centigrade scale officially changed its name to the Celsius temperature scale in honour of its creator in 1948, at the Ninth General Conference on Weights and Measures. Britain officially adopted the Celsius temperature scale in 1971 as part of the metrication programme.

Celsius was a great scholar and like Fahrenheit, his work comprised much more than the creation of the temperature scales that we use on a daily basis.

Kelvin scale

Sir William Thomson (1824–1907), better known by his title Baron Kelvin of Largs, in 1848, proposed a thermodynamic temperature scale for scientific purposes based on the centigrade scale, as it was then known, by assigning 0 to thermodynamic absolute zero. This resulted in the freezing point of water being 273.15°K and the boiling point being 373.15°K. In 1967, at the Thirteenth General Conference on Weights and Measures, it was officially agreed to change the symbol °K, to K, which in the metric system can be somewhat confusing, as K represents 1000 units.

Rankin scale

William John Macquorn Rankin (1820–1872), a Scottish born engineer and physicist, produced in 1859 a version of Kelvin's thermodynamic temperature scale but based on Fahrenheit's units, whereby absolute zero was equal to −459.67°F. This imperial measurement version of the thermodynamic temperature scale was more popular in Britain, as well as North America, until Britain adopted the metric system of measurement in 1971. Degrees Rankin are expressed as °r, so as not to be confused with degrees Reamur, which is written as °R.

The Kelvin scale was officially adopted as the metric thermodynamic temperature scale in 1954 and is universally used in scientific measurement around the world.

Absolute zero

Zero degrees on the Celsius scale is based on the freezing point of water, but we all know that the temperature that naturally occurs can go much lower than 0°C and it is expressed as a minus figure. The thermodynamic temperature scale basis zero degrees is what used to be considered the lowest temperature possible. Absolute zero of a gas is a theoretical consequence of the law of expansion by heat, assuming that it is possible to continue the cooling of a perfect gas until its volume has contracted to a point where it diminishes to zero, termed 'absolute zero'.

The lowest temperature that exists in the universe is said to be 3 K, whereas the lowest temperature that has ever been achieved is recorded at 12 μK, or 12-millionth of a Kelvin, which was achieved in a laboratory at Lancaster University.

The scientific thermodynamic temperature scale has very little to do with the subject matter of this work, but it is important that engineers have a knowledge of the full range of the subject of heat, and the means to measure it.

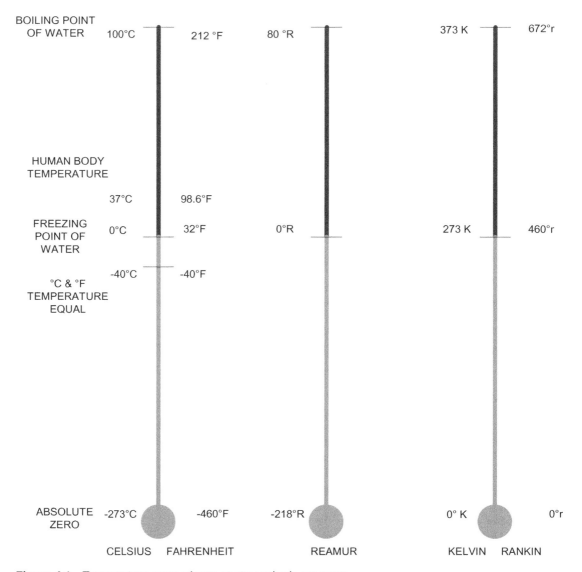

Figure 4.1   Temperature comparisons at atmospheric pressure

Figure 4.1 shows a graphical comparison of the five temperature scales with the boiling point of water given based on atmospheric pressure at sea level.

Box 4.1 shows some temperature scale conversions.

SPECIFIC HEAT CAPACITY

The specific heat is defined as the thermal capacity of a substance and is expressed as the amount of heat required to raise a unit mass through a unit temperature: in the metric system this is the number of calories required to raise one gram of substance through 1°C. In the imperial system of measurement this was related to water as being unity, but in the metric system is taken as kJ kg^{-1}°C^{-1}, and although specific heats vary with temperature, water may be taken as approximately 4.2 kJ kg^{-1}°C^{-1}.

BOX 4.1 Temperature scale conversion

$°C = \frac{5}{9}(°F - 32)$	or	$\frac{5}{4}°R$
$°F = \frac{9}{5}°C + 32$	or	$\frac{9}{4}°R + 32$
$°R = \frac{4}{5}°C$	or	$\frac{4}{9}(°F - 32)$

BOX 4.2 Specific heat of water at various temperatures

Temperature/°C	Specific heat capacity of water/kJ kg^{-1} °C^{-1}
0	4.21
15	4.186
35	4.178
60	4.185
71	4.1916
82	4.1999
100	4.219

The unit 'calorie' is no longer recognised in the SI system of measurement, but one calorie equals 4.2 joules, hence the specific heat of water in SI units is 4.2 kJ kg^{-1} K^{-1}

The examples in Box 4.2 serve to illustrate the effect that temperature has on the specific heat capacity of water. A value of 4.2 kJ kg^{-1} °C^{-1}1 is chosen for the majority of equations found in this work.

LATENT HEAT

When heat is applied to a substance, it is absorbed by that substance when it is converted from a solid to a liquid and from a liquid to a gas, as in the case of water. Latent heat is the quantity of heat absorbed when passing through these phases and it can be observed that as heat is added, the temperature remains the same; i.e. there is no recorded rise in temperature, until the conversion from a solid to a liquid, termed latent heat of fusion, and from a liquid to a gas, termed latent heat of vaporisation, is completed.

As illustrated in Figure 4.2, to raise 1 kg of water from freezing to boiling point, i.e. 0°C to 100°C, or 100 degrees, would require 100 × 4.2 (specific heat of water in kJ) = 420 kJ. However, to convert that same 1 kg of water at boiling point into 1 kg of steam would require some 2300 kJ at the same temperature (100°C). Likewise, to convert 1 kg of ice into 1 kg of water at the same temperature would require 330 kJ.

Latent heat may be described as heat added during a change in state without any corresponding change in temperature.

SENSIBLE HEAT

Unlike latent heat whose temperature remains unchanged when heat is being applied and can therefore be regarded as potential energy, sensible heat differs insofar as a temperature rise can be observed when heat is applied and is therefore called kinetic energy. If water is heated to raise its temperature from 0°C, freezing point, to 100°C, boiling point, the rise in temperature can be sensed and observed by a thermometer; consequently, as it can be sensed and the rise in temperature is seen to be proportional to the amount of energy in the form of heat being applied, it is termed 'sensible heat'.

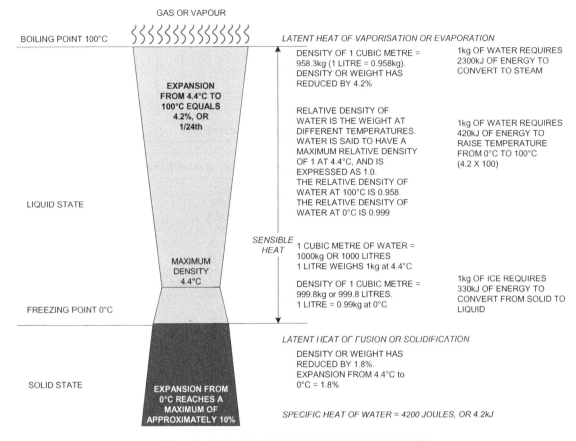

GAS OR VAPOUR

BOILING POINT 100°C

LATENT HEAT OF VAPORISATION OR EVAPORATION

DENSITY OF 1 CUBIC METRE = 958.3kg (1 LITRE = 0.958kg). DENSITY OR WEIGHT HAS REDUCED BY 4.2%

1kg OF WATER REQUIRES 2300kJ OF ENERGY TO CONVERT TO STEAM

EXPANSION FROM 4.4°C TO 100°C EQUALS 4.2%, OR 1/24th

RELATIVE DENSITY OF WATER IS THE WEIGHT AT DIFFERENT TEMPERATURES. WATER IS SAID TO HAVE A MAXIMUM RELATIVE DENSITY OF 1 AT 4.4°C, AND IS EXPRESSED AS 1.0. THE RELATIVE DENSITY OF WATER AT 100°C IS 0.958. THE RELATIVE DENSITY OF WATER AT 0°C IS 0.999

1kg OF WATER REQUIRES 420kJ OF ENERGY TO RAISE TEMPERATURE FROM 0°C TO 100°C (4.2 X 100)

LIQUID STATE

SENSIBLE HEAT

1 CUBIC METRE OF WATER = 1000kg OR 1000 LITRES 1 LITRE WEIGHS 1kg at 4.4°C

MAXIMUM DENSITY 4.4°C

DENSITY OF 1 CUBIC METRE = 999.8kg or 999.8 LITRES. 1 LITRE = 0.99kg at 0°C

1kg OF ICE REQUIRES 330kJ OF ENERGY TO CONVERT FROM SOLID TO LIQUID

FREEZING POINT 0°C

LATENT HEAT OF FUSION OR SOLIDIFICATION

DENSITY OR WEIGHT HAS REDUCED BY 1.8%. EXPANSION FROM 4.4°C to 0°C = 1.8%

SOLID STATE

EXPANSION FROM 0°C REACHES A MAXIMUM OF APPROXIMATELY 10%

SPECIFIC HEAT OF WATER = 4200 JOULES, OR 4.2kJ

CONDITIONS BASED ON ATMOSPHERIC PRESSURE AT SEA LEVEL

Figure 4.2   Physical properties of water

HEAT TRANSFER

The term 'heat transfer' implies a transfer of heat or thermal energy from one region to another. The mode of transfer may be by any one or combination of three ways, namely conduction, convection and radiation. Regardless of which form of heat transfer or combination of forms occurs, each of them requires a temperature difference between the two regions or substances for the motion of heat to flow: if the two substances are at equal temperature, then no transfer of heat will occur.

Each form of heat transfer is important in the design of the heating system and the equipment employed, whereby the temperature difference between the two regions involved in the transfer has a particular effect, especially with regard to conduction and convection.

CONDUCTION

Thermal conduction may be described as a transfer of heat energy from one body to another by contact. Materials that are said to be good conductors of heat are generally good conductors of electricity also. Copper is an excellent conductor of both heat and electricity and if one piece of copper is heated and then

Copper at High Temperature   Copper at Low Temperature

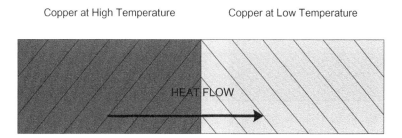

HEAT FLOW

Figure 4.3 Heat being transferred by conduction

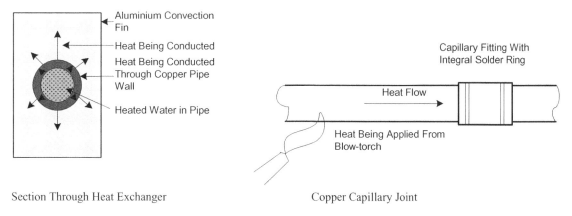

Aluminium Convection Fin

Heat Being Conducted

Heat Being Conducted Through Copper Pipe Wall

Heated Water in Pipe

Capillary Fitting With Integral Solder Ring

Heat Flow

Heat Being Applied From Blow-torch

Section Through Heat Exchanger   Copper Capillary Joint

Figure 4.4 Some common examples of conduction in plumbing services

placed in contact with a second colder piece of copper, the heat will flow by conduction from the heated piece of copper to the colder piece of copper until the two pieces equal out at some intermediate temperature (see Figure 4.3), when heat will cease to flow.

There are many examples of thermal conductivity in the plumbing services industry. A typical example is the making of a copper capillary joint using a fitting containing an integral ring of solder; here the pipe is heated at a point approximately 100 mm from the fitting and the heat travels along the pipe by conduction into the joint where it melts the solder when the temperature is hot enough (see Figure 4.4).

Another example is a heat exchanger comprising a copper tube waterway fitted with aluminium convection fins to increase the heating surface area, which is the form of heat exchanger used in many applications, including both natural and forced convectors. Here the heat being carried by the water is conducted through the walls of the copper tube and continues to conduct into the convector fins in contact with the copper tube whereby it transfers its heat by convection. Both copper and aluminium are excellent conductors.

Other examples include heat exchangers inside domestic hot water cylinders and the transmission of heat through building fabric.

The rate of thermal conduction will depend upon the resistance of the material conducting the heat and the temperature difference of the two bodies involved. This rate may be calculated if the resistance value of the material is known together with the cross-sectional area of the material where the heat will flow, plus the temperature difference between the two bodies.

Conduction plays a major part in the process of heat transfer in the subject of heating.

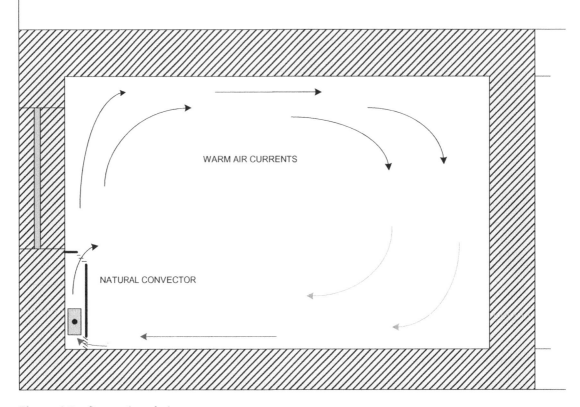

Figure 4.5   Convection of air

CONVECTION

The form of heat transfer known as convection may be described as a transfer of heat involving the motion of a medium such as a liquid or gas, which may be natural due to the difference in density at the disparity of temperatures, or forced as described in later chapters.

There are two classic examples of convection found in heating systems, one involving a gas and the other a liquid. The first example is depicted in Figure 4.5 and involves a natural convector type heat emitter that warms the air that is in contact with the heat exchanger (by conduction), whereby it immediately expands and becomes less dense or lighter than the rest of the air in the room and so rises, producing an upward warm air current. This circulates around the room giving up the heat added and eventually returning to the convector to pass over the heat exchanger again for the process to be repeated.

The other example is shown in Figure 4.6. It is a typical gravity circulation heating system, whereby the water heated in the boiler expands and consequently becomes less dense or lighter and begins to rise, so producing an upward circulation of water. It then gives up part of its heat through the heat emitters before returning to the boiler for re-heating. When the water is heated and becomes less dense, the rise is caused by the heavier, more dense colder water pushing the warmer and lighter water up to create the circulation, in the same way as for air convection.

Convection cannot occur in a vacuum and requires a medium capable of movement for transporting the heat, and use can be made of either air or water, as described in the two examples shown in Figures 4.5 and 4.6.

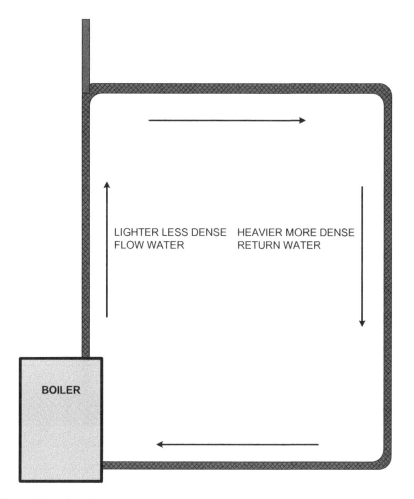

Figure 4.6 Convection of water

The two examples given describe the process of natural convection, be it for air or water, but convection of both these mediums can be accelerated or forced by the application of fans in the case of air, or pumps in the case of water. Both these methods are dealt with in greater detail later in this chapter.

RADIATION

Unlike conduction and convection which, although they are not visible, can be visualised and are easy to understand, radiation is not so straightforward to explain or visualise. The radiation energy transfer process – be it heat, light, X-rays or radio waves – is the consequence of energy carrying electromagnetic waves emitted by atoms and molecules, as the result of changes in their energy content.

Also, unlike conduction and convection, radiation does not depend on the presence of an intermediate material or medium as a carrier of energy, and does not depend upon warming the medium through which it travels, as it will readily occur across a vacuum, as it does across a room filled with air. However, it can be impeded by the presence of any object within its path which serves to either partially deflect or absorb the radiation, at an amount depending upon the material of the object. This is depicted in Figure 4.7.

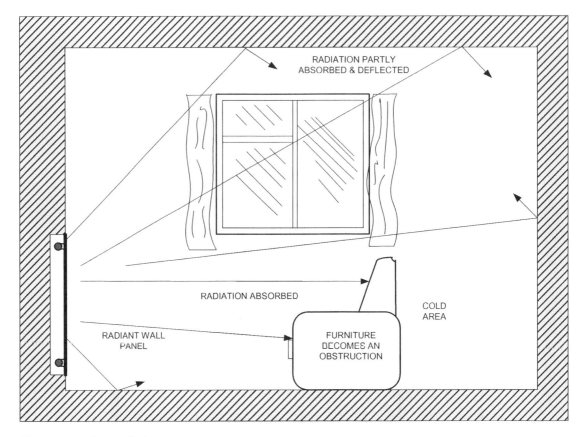

Figure 4.7   Heat radiation waves

The amount and characteristics of radiant energy emitted by a material will depend on the size, colour, texture and general nature of that material. Smooth matt black surfaces are found to make the best radiators, and incidentally the best absorbers of radiation as well, and polished metallic surfaces, such as bronze and gold, the worst. This is why black asphalt type roofs are covered with white stone chippings to reduce the radiation effect on the surface. Conversely, there is no effective reduction in the radiation emission value at operating temperatures below 100°C by painting radiator type heat emitters, except when metallic paints, such as aluminium or bronze are used. These finishes can result in a reduction of the radiation value of up to 20% for panel type radiators and 50% for column radiators: the convection value remains unchanged, but in the case of the column type radiators the radiation value might represent 20% of their total heat emission, which therefore would result in a reduction of 10% of its total combined heat output. This reduction in heat emission may be partially restored by the application of a clear lacquer.

Figure 4.7 illustrates the radiation heat waves pattern from a typical radiant wall panel. Although these panels emit part of their heat by convection, this has been omitted for clarity. The drawing also shows how radiation can be deflected and/or absorbed.

HUMIDITY

Derived from the Latin word *humidus*, meaning moist, humidity refers to the amount of water vapour contained in the air at any given moment. When the degree of water vapour in the air reaches its maximum, it is said to be saturated and this is termed 'dew point', at which point condensation begins to occur, or even rain.

Relative humidity may be defined as the degree of saturation for a given volume of air and is expressed as a percentage; it is defined as the ratio of the actual partial pressure of the water vapour to the partial pressure of the water vapour if the space were saturated at the same temperature. Therefore the maximum amount of moisture which can occur for a given volume of air is dependent only on temperature, and is expressed in terms of 'kg per kg' of dry air.

If the relative humidity is at 100%, no evaporation can occur in the form of moisture, but if the relative humidity is much lower, evaporation could be rapid. Humidity that is too high or low humidity causes discomfort, as at a high relative humidity evaporation, in the form of perspiration from the skin does not occur, and at low relative humidity, sore throats and dry eyes can result. In general it is considered that most people feel comfortable with a relative humidity of 40% to 70% RH, coupled with temperatures in the region of 18°C to 22°C.

It should be noted that condensation occurs when the air volume has a high relative humidity, approaching saturation point, and the air temperature is lowered, having the effect of reaching dew point: the moisture then begins to precipitate out by condensing on colder surfaces, such as walls or glass windows.

It can therefore be said that the higher the temperature of the air, the higher the moisture carrying capacity, and likewise, any lowering of this temperature results in raising the relative humidity percentage until it reaches saturation, at which point it begins to give up part of its moisture content.

Humidity can be measured on a simple hygrometer, but for an indication of the relationship between temperature, moisture content and relative humidity, a sling psychrometer containing both a standard dry bulb thermometer and a wet bulb thermometer, which has a wick kept wet by a reservoir of distilled water, can be used. The sling psychrometer is twirled around vigorously in the air to record both the dry bulb temperature and the effect of evaporation on the wet bulb temperature: at high relative humidity the evaporation will be negligible and record a higher temperature, but at low relative humidity the evaporation will be extensive, and a lower temperature will be observed. The results of this exercise may be plotted on a psychrometric chart where the relative humidity RH percentage, together with the moisture content, will be discovered.

5

Heat Emitters

Heat emitters – or disseminators as they have been termed in some textbooks – may be described as a form of heat exchanger that transfers part of the heat generated and distributed from the heating system to the room or space where they have been located. They are usually classified as being either a form of radiator or convector, the names of which implie the method of heat transfer. This is certainly true in the case of convectors, but the term 'radiator' is a misnomer, as some types of so-called radiators emit a vast percentage of their heat by convection as well as radiation. Even in the case of the best radiant panel type heat emitters, a certain amount of the heat emission is by convection.

It could be argued that heat emitters have changed very little over the years, but recently a wider selection of designs, styles, finishes and materials of manufacture have been available for the heating designer and the client to choose from. This has given greater freedom to produce more individual and exciting designs for the heating system, particularly when the aesthetic appearance is important.

Now there is a greater imagination in the design of heat emitters, it is even more important that the design engineer has the knowledge and a clear understanding of the working principles of each type of heat emitter, so that they may be selected and used to their best advantage. It is imperative that they are chosen correctly to suit both the client's wishes and the intended design operating philosophy; especially the comfort control scheme of the heating system. This subject is discussed further in Chapter 11, Controls, Components and Control Systems, as well as in this chapter.

The larger choice of heat emitters now available also includes a greater selection of materials used to manufacture them, therefore it is also essential that these materials are compatible with the other materials used in the installation of the heating system, or that precautions, by way of chemical treatment are included and applied properly to negate any adverse effects. This subject is also discussed further in Chapter 21, Water Treatment, as well as in this chapter.

It should be pointed out that it is not the intention of this work to include descriptions and details on every form of heat emitter ever made, as this would not serve any meaningful purpose, but to include information on each main type regardless of the numerous variations that are available, so that a full appreciation may be gained.

Heating Services in Buildings: Design, Installation, Commissioning & Maintenance, First Edition. David E. Watkins.
© 2011 John Wiley & Sons, Ltd. Published 2011 by John Wiley & Sons, Ltd.
As an aid to lecturers and students, full colour versions of the figures in this chapter may be found at www.wiley.com/go/watkins
These figures are © 2011 by John Wiley & Sons, Ltd.

PRESSED STEEL PANEL RADIATORS

The introduction of the pressed steel panel radiator coincided with the birth of domestic central heating as we know it, and contributed to its widespread popularity and affordability during the 1950s. Its contribution to the domestic central heating industry was due to its lower cost, mass production and unobtrusive appearance, helping to make domestic central heating systems both possible and affordable to most people, who, up until that time, could not contemplate having a fixed heating system in their own homes. These attributes are as applicable today as they were when they were originally introduced, and although the market share of all heat emitters sold is not quite so dominant, radiators still remain the most common form of heat emitter used in both domestic and commercial applications. This reduction in market share is due to the greater choice of heat emitters now available, and a desire by architects and home owners to create something more individual.

They are manufactured by either a fully automated or semi-automated mass production process line, from either a single sheet or two sheets of light gauge steel. In the case of a single sheet, the steel is pressed into the desired design shape and bent double in the middle of the sheet by rolling it to form the top. The sides and bottom seam are electrically welded to form the vertical and horizontal waterways shape of this type of rounded top heat emitter. The other method is very similar, except it is manufactured from two separate sheets of light gauge steel that are also pressed into shape, placed together to form a front and rear panel and their seams continuously electrically welded around the seams. The finished appearance of this type is one of a continuous weld seam.

There is no real difference in performance or quality between the two styles of the rolled top or welded seam pattern pressed steel radiators, it is purely a matter of personal preference on behalf of the individual. There are now slight variations in the pattern style and method of providing the water connections between each manufacturer which have no detrimental effect on performance.

Plain pressed steel panel radiators emit their heat by part radiation and part convection; the exact percentage will contrast slightly from one make to another, but for single panel versions installed in unrestricted locations, the emission may be taken as approximately 50% convection and 50% radiation.

Pressed steel panel radiators are commercially available in a range of heights and lengths, the exact dimensions varying between manufacturers, but this difference in physical dimensions may help the designer to select one make rather than others, if the dimensions of one particular manufacturer's models are more suitable for a given set of circumstances. They may also be obtained in a single panel version, double panel version, and to special order from some of the bigger manufacturers as treble panel and quadruple panel versions. Although the three and four panel versions can emit large quantities of heat, they are quite bulky, protruding out from the wall by a significant amount, and when fully charged with water, they are also very heavy.

When pressed steel panel radiators were originally introduced they were fixed to the walls by a complex form of clamp bracket arrangement, adapted from the type of bracket fixing used on the heavier cast iron column type radiators that were in common use at the time. Now, because the pressed steel panel radiators are reasonably lightweight compared to the cast iron column type radiators, they are fixed to the wall by a comparatively simple galvanised steel angle bracket so that the radiator is just hung on by the steel straps. The number of brackets used on each radiator is dependent upon the size and weight of the radiator. Common examples are shown in Figure 5.1.

As the heat emission is both by convection and radiation, a double panel radiator will not emit twice the amount of heat of that of a single panel radiator of equal length and height, as the radiation value is lost from any additional rear panels that are blocked by the obstruction of the front panel. Therefore, as previously established, the radiation value amounts to approximately 50% of the heat emission from each panel, and if this quantity of heat is lost from the rear panel a double panel radiator will only emit approximately half as much heat again compared to a single panel radiator. Any subsequent third or fourth panel will also only add an additional 50% of the total value of heat emission of the front panel.

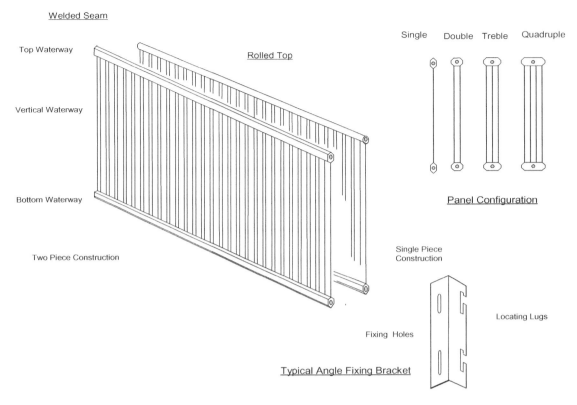

Figure 5.1   Pressed steel panel radiators

Plain light gauge pressed steel panel radiators have the advantage that they may be bent to suit curved wall surfaces by specialist engineering companies, who will undertake this prerequisite without charge when furnished with details of the panel radiator in question together with a paper template of the wall profile and a sample wall support bracket. The engineering companies who take on this type of work will, using the paper template supplied, allow for the wall support bracket stand-off dimension and set their roller press machine to suit. This machine comprises a set of adjustable rollers that pass the panel radiator backwards and forwards and by the operation of the press will begin to form the desired curved shape of the radiator. This process is normally only suitable for single panel radiators, but curved shapes have been successfully achieved on double panel radiators by separating the two panels, forming the curves individually on each panel, adjusting the difference in the length of each panel caused by the bending process and welding them back together again. This is not usually done for multiple panel radiators, only slight curves are normally formed. The same engineering companies will also undertake any requirement to cut and mitre plain light gauge pressed steel panel radiators and weld them back together again to form angles to suit corners formed in the walls, thus making full use of wall space available.

It should be noted that any work that involves altering the original shape of the panel radiator will no doubt render any manufacturer's warranty null and void.

BS EN442 requires light gauge pressed steel panel radiators to have a maximum working pressure of 8 bar, although higher working pressures are available as a special order. It is normal for the panel radiators to have an external factory applied coat of paint/enamel and an internal thin film of an oil based product, to protect the panel radiator from corrosion during transit and storage. However, this oil film must be removed on completion of the heating installation, discussed further in Chapter 21, Water Treatment.

Because pressed steel panel radiators are light in weight and fairly versatile, they have been used successfully in unorthodox situations, such as installing them on end when sufficient wall space is not available.

CONVECTOR RADIATOR

The convector radiator, or high efficiency radiator as it is sometimes called, is a development of the plain light gauge pressed steel panel radiator, whereby additional heating surface in the form of metal convection fins are fixed to the back of the radiator panel. This has the effect of increasing the convection heat emission value, which results in a smaller physical sized radiator compared to that of a standard plain pressed steel panel radiator of the same heat emission output.

The proportion of heat emission will vary from one make to another, but for average values it may be taken as 70–75% convection and 25–30% radiation; it should be noted that any additional panels will only result in the lower radiation heat emission value being lost from the extra panels compared to 50% for the plain pressed steel panel radiator.

This style of heat emitter has become quite popular, especially in the domestic heating market where available wall surface is limited, due to its high efficiency.

Apart from the additional heating surface, the appearance is exactly the same as a standard plain pressed steel panel radiator with four pipe connection tapings available, one in each corner for choice of connection, but because of the convector fins, this type of radiator is not suitable for forming curved sections.

Figure 5.2 depicts one form of convector radiator, showing the convector fins attached to the back of the panel, together with back entry side turned water connections. (These type of water connections are not suitable for micro bore double entry radiator valves.)

The examples shown in Figures 5.2 and 5.3 are typical convector radiators that have been developed from the standard plain light gauge pressed steel panel radiator, whereby additional heating surface in the form of fins has been added to the radiator. These are also available commercially in single, double and treble panel versions, and may also be fitted with side infill panels and top outlet grilles for a more aesthetic appearance.

There are other heat emitters purporting to be convector radiators, but in reality as they emit their heat almost entirely by convection, they are discussed below in the section on natural convectors.

Figure 5.2 Typical convector radiator

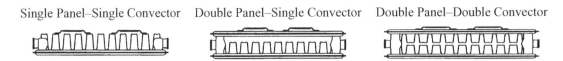

Single Panel–Single Convector   Double Panel–Single Convector   Double Panel–Double Convector

Figure 5.3   Plan view section through a typical convector radiator

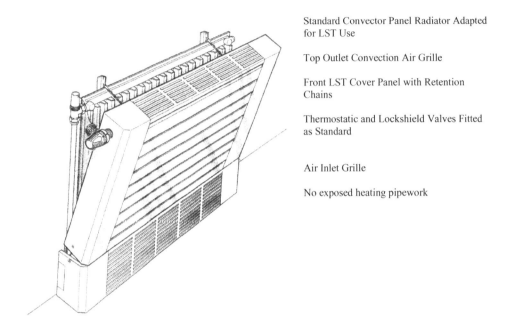

Standard Convector Panel Radiator Adapted for LST Use

Top Outlet Convection Air Grille

Front LST Cover Panel with Retention Chains

Thermostatic and Lockshield Valves Fitted as Standard

Air Inlet Grille

No exposed heating pipework

Figure 5.4   Typical LST radiator

LOW SURFACE TEMPERATURE (LST) RADIATORS

Pressed steel panel radiators have many advantages, hence their popularity since their introduction, but they do have one distinct disadvantage as previously established, and that is the high surface temperature of the radiator panel.

Pressed steel panel radiators are only suitable for low temperature heating applications if there is a possibility of anyone coming into contact with the surface of the heat emitter, which will be the case with most heating systems employing pressed steel panel radiators. With a water flow inlet temperature of 82°C and a return outlet temperature of 71°C or lower in a condensing system, giving a mean water temperature of 76.5°C or greater in the radiator – this is still a high surface temperature that could burn anybody whose exposed skin comes into contact with it for more than a few seconds. Hence, in situations where people do not possess the physical or mental ability to realise that it is hot and to pull away from it – such as handicapped people or infants etc – and who could come into contact with a directly heated surface of a pressed steel panel radiator or its convective variation, even if the heating system is of a low temperature design, then alternative forms of heat emitters should be selected. This normally applies to certain areas within hospitals, care homes, some types of clinics and depending upon the age group catered for, some types of children's nurseries.

The low surface temperature radiator illustrated in Figure 5.4 is a typical example, whereby a standard pressed steel panel radiator, complete with additional convection fins, has been adapted by enclosing it

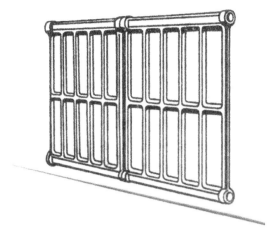

Figure 5.5   Cast iron panel radiator

within a panel casing complete with air inlet grilles and warm air outlet grilles. This has the effect of converting the radiator into a convector heat emitter, as the panel casing obstructs and nullifies any heat that has been emitted by radiation. The total heat emission is now 100% by convection.

The advantage of this form of LST radiator is that it conforms to NHS standards for safety; it limits the surface temperature of the panel casing to below 43°C for flow temperatures of 85°C and below, and it is suitable for heating systems designed to have operating temperatures of 90°C to 95°C and higher, providing that the expected occupancy of the building in which the heating system is to be installed does not fall into any of the above categories.

Detailed explanation of the workings of natural convectors is included later in this chapter, including other variations that appear to be either radiators or low surface temperature radiators.

CAST IRON PANEL RADIATORS

This form of panel radiator is now obsolete, but was perhaps the forerunner to the pressed steel panel radiator. It was much heavier than the steel type and required more substantial wall support fixings.

Figure 5.5 shows a typical cast iron panel radiator; they were manufactured in a number of different heights, but only one length – usually around 400 mm long – with the lengths increased by joining panels together in a similar manner to that of the cast iron column radiator. The cast iron panel radiator depicted in Figure 5.5 comprises two panels joined together. They are very robust with a good resistance to corrosion, hence the number of examples still in working condition.

COLUMN RADIATORS

The column type radiators have been in existence longer than any other form of heat emitter used today, with probably the only exception being the radiant panel heat emitter.

They have been manufactured over the years in a vast variety of forms, including floor standing types, wall hung types, window types, easy clean pattern types, types with decorative ornate castings and columns ranging from two to six or more in a diversity of heights. They are made from a number of different materials including cast iron, cast steel and more recently, polypropylene. However, the choice available today is more limited than previously, particularly during the heat emitter halcyon days from the end of the 1800s through to the mid-1900s, when it was almost exclusively used.

Cast iron was the original choice of material for the manufacture of this form of heat emitter. It is a robust and corrosion-resistant material and easily lends itself to producing interesting and attractive patterns cast on to the sections; hence, the great multiplicity of variations that have been available in this material over the years. Cast steel is slightly lighter in weight compared to its cast iron counterpart and is less expensive. Although cast steel is now the most common form of material from which column type radiators are manufactured, it is more susceptible to corrosion than the cast iron type.

Column type radiators, mainly manufactured from cast iron, were once used almost exclusively as the choice of heat emitter for all heating applications, for commercial and residential buildings alike, due to their many attributes. However they quickly lost popularity for domestic applications in favour of the less expensive, easily cleaned and unobtrusive pressed steel panel radiators when the latter were introduced in the 1950s, resulting in the demise of column type radiators for many years. Conversely, they have experienced a resurgence in recent years, being selected once again for domestic heating applications, due mainly to the demand for greater choice in heat emitters and the character that they can add to certain styles of buildings; the cast steel column type radiators have a narrow selection of styles and offer less choice.

Column type radiators are more compact and occupy less wall surface than pressed steel panel type radiators of equal heat output, although they protrude further off the wall and can be more difficult to keep clean. They are ideal for public buildings and other applications where a strong, more vandal-resistant and enduring heat emitter is called for.

Whether manufactured from cast iron or cast steel, column type radiators are mis-termed, as they emit their highest proportion of heat by convection – approximately 70% to 80% of their total heat emission being by convection, with the remaining 20% to 30% by radiation.

These radiators are manufactured in sections of varying heights and widths, the width being dependent upon the number of columns per section. This permits the radiator manufacturer to assemble the sections to form an almost infinite number of different lengths. The sections may be joined together to produce a radiator of practically any length to achieve the required heat output, but this length is normally governed by the practicality of the combined weight of the sections, rather than anything else.

The individual radiator sections, have in the past, been fastened together by either a steel tapered push-fit friction nipple, or a parallel threaded screwed nipple. The push-fit nipples are rarely used today: they comprised a machine tapered steel nipple that was placed between the two main waterways of each section to be joined, with the whole radiator being secured together by threaded tie rods that passed through the vertical columns of the radiator.

Today, the common method of joining the sections of radiators consists of a parallel threaded nipple having one standard right hand thread, with the other one having a left hand thread. This threaded nipple also has a pair of locating lugs on the inside so that a special assembly bar can be passed through the column sections to permit both the top and bottom nipples to be tightened up simultaneously at an even rate; this has the result of drawing the two adjoining column sections together. It is important that the top and bottom joints are tightened up at the same rate to prevent any of the threads binding or cross threading. A paper washer impregnated with a jointing compound is placed between each joint before commencing the assembly procedure.

The jointing nipples of either type are normally 25 mm (1 inch) in diameter to suit the tapped thread on the end of each section. The tappings on the end sections are then fitted with a 25 × 15 mm (1 × ½ inch) reducing bush to permit radiator valves, air release valves and blank plugs to be fitted.

Both methods described offer the advantage of allowing damaged or faulty sections to be replaced without having to replace the whole radiator, as the broken section can be removed without having to dismantle all the sections in between.

Figure 5.6 illustrates a number of typical column type radiators that are available, but it is not intended to be exhaustive as there are many variations.

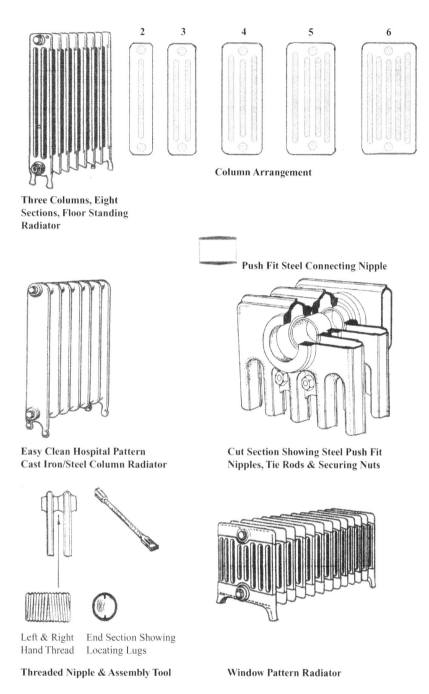

Figure 5.6   Typical column type radiators

Floor standing radiator models are equipped with support legs cast on to the end sections that bear the weight of the radiator, with separate wall ties to secure the whole unit to the wall and prevent any movement.

Wall hung column type radiators do not have support legs cast on to the end sections and are usually much narrower, normally restricted to no more than three columns wide, or the equivalent width of a solid wall easy clean hospital pattern type radiator. They are hung on concealed cantilevered brackets fixed or built into the wall with the brackets incorporating some form of bolted clamp, to prevent them from being moved. It goes without saying that because of the weight of these heat emitters when fully charged with water, the wall should be of a solid construction capable of supporting the weight.

The version known as the window pattern radiator is much shorter in height and may consist of a greater number of columns to compensate for the heat emission reduction due to the lack of height. They are so named as they are designed to fit beneath windows or window seats. They are usually floor standing, complete with end support legs.

The design of the column type radiator, particularly deep radiators of four columns or more with their mass of vertical inaccessible airways, makes cleaning extremely difficult as the dust carried by the convection air currents deposits particles of dirt on the inner columns and heating surface. This, along with the price, probably contributed to the column radiator's decline from its almost total domination of the heat emitter market and the rise in preference for the pressed steel panel radiator.

Where greater need for cleanliness is essential, the so called 'easy clean' or 'hospital' version of the column radiator is used. This form of radiator is manufactured by the same method used for the standard column type, except each section has a solid infill between the columns that reduces the area for dust to collect, and slightly reduces the available heating surface, thus making cleaning easier but resulting in greater weight. Although easier cleaning is still a valid reason for specifying this form of heat emitter, modern vacuum cleaning equipment is more efficient and can overcome the problem of dirt and dust collecting on the internal surfaces of the heating columns. This has, no doubt, contributed to the revival of the column type radiator in recent times, along with its other merits.

The column type of radiator has been manufactured in many different styles in both cast iron and cast steel for over one hundred years, but more recently the hospital easy clean version has been manufactured in limited arrangements and sizes in the non-metallic material polypropylene, which has the advantages of being non-corrosive and lighter in weight. However, its expansion ratio compared to its metallic counterparts is greater, and it should be limited to smaller assembled arrangements.

ALUMINIUM RADIATORS

Die-cast aluminium radiators have been made and used in Europe for many years, with the main manufacturers being in Italy, where these radiators have become very popular and represent a large proportion of the heat emitter market. Although introduced into the UK back in the early 1970s, their higher cost means they have not been able to compete with the pressed steel panel radiator that dominates the heat emitter market. They are generally considered too expensive for the average UK homeowner, costing two to three times more than a pressed steel panel radiator of equal heat emission output. Hence, aluminium radiators are limited to use in high rental, designer style commercial offices and luxury high cost residential properties.

They are similar in many ways to the cast iron/cast steel column radiators as they are manufactured in separate vertical sections, but cast into more interesting and imaginative shapes. The sections are assembled in a similar manner to that described for the cast iron column radiators, and therefore they have the same advantage of versatility in allowing any length within reason to be assembled, but they are not as robust as cast iron: aluminium is more easily damaged and should not be considered for situations where vandalism or accidental physical damage could occur.

Like other forms of column type radiators, aluminium radiators are predominantly convectors, emitting between 70% to 80% of total heat by convection with the remaining 20% to 30% by radiation. They are compact and aesthetically pleasing heat emitters.

Aluminium is also an excellent conductor of heat, and it can be argued that its response time in heating up from cold to emission temperature is very rapid, achieving an efficient transfer of heat.

The attributes of aluminium radiators with regard to their aesthetic value, compact size, design versatility and heat transfer efficiency have been discussed along with their disadvantages, including the higher cost and greater susceptibility to mechanical damage when compared to the cast iron column and pressed steel panel radiators, but the main criticism directed towards aluminium radiators is one of corrosion.

Aluminium as a metal comes low on the electro-motive series table, which is dealt with in more detail in Chapter 21, Water Treatment. This means that in a typical heating system, be it domestic or commercial, the possible mix of metals such as copper, copper alloys, steel, cast iron and stainless steel, could all react with the aluminium in the form of electrolytic action to the detriment of the aluminium. There have been numerous cases of early failure of aluminium radiators, which could have been avoided with knowledge of the chemical properties of the metals involved and the corrosion process. In the worst case scenario, where no thought has been given to the mixture of metals employed in the heating system, it is imperative that a properly formulated corrosion inhibiting chemical, suitable for all the metals used to construct the heating system, is added to the circulating water.

Conscious of this disparagement, some manufacturers of aluminium radiators have attempted to address this problem by creating an artificial coating of aluminium oxide on the inside wetted surface of the radiators, to protect them from dissimilar metals in the heating system. This has enabled the manufacturers to give a conditional guarantee on the longevity of the radiators, providing that the circulating water has been suitably treated.

There can be certain merits in a heating system whereby all ferrous materials have been excluded, with the result of avoiding rust accumulation in the water, which in turn prolongs the life of other main heating components such as the boiler, circulating pump and valves etc, but a careful informed design approach is needed to accomplish this.

There is no doubt that, providing the designer has a full understanding of all the implications involved with the use of aluminium radiators within a heating system, together with their mitigations, and as long as cost is not a crucial factor, then they can provide the homeowner with the opportunity of being more creative in the design layout of the heating system.

For further information on the corrosion process and the chemical treatment of such, please refer to Chapter 21, Water Treatment.

Figure 5.7 illustrates a typical example of an aluminium radiator. There are a number of variations, including a panel version that incorporates a series of aluminium fins and emits its heat almost entirely by

Convection Air Currents

Figure 5.7 Typical eight section aluminium radiator (may be obtained with or without end support legs)

convection. They are available as wall hung models supported on concealed wall brackets that are lighter in weight than their cast iron/cast steel counterparts, or as floor mounted versions that either incorporate support legs on the end sections, or can be fitted with separate bolt-on legs to adapt standard wall hung radiators to floor mounted types.

NATURAL CONVECTORS

Unlike the types of heat emitters so far described, the natural convector emits its heat almost purely by convection. The natural convector type of heat emitter is designed to allow for a convection air current to flow through the unit by natural means. There are many variations in the design of natural convectors, but this section concentrates on unit types; continuous forms are discussed later. The varieties available differ in their design appearance, shape and configuration, but they all comprise a finned tubular heat exchanger element that is mounted near the bottom of the casing, just above the air inlet opening and incorporating a damper regulated heat outlet grille at the top.

The efficiency of the natural convector depends on the temperature difference between the heating water in the heat exchanger and the room air temperature, and the height of the casing of the natural convector, which affects the stack effect within the casing. The greater the temperature difference, the greater the convection air circulation, and the higher the casing, the greater the convection air current. The design of all natural convectors is based upon the creation of this stack effect so as to encourage a rising column of warm air within the casing and produce a flow of warm air that issues from the outlet into the room, whilst cooler, more dense air is drawn into the base of the convector.

Natural convectors may be obtained in a number of designs and models, including floor standing types, wall mounted types, recessed wall types, continuous under window types and floor trench types.

Figure 5.8 depicts a typical unit type natural convector whereby the flow of air may be controlled by an adjustable regulating damper that spans the length of the unit's casing; this serves to restrict the air flow

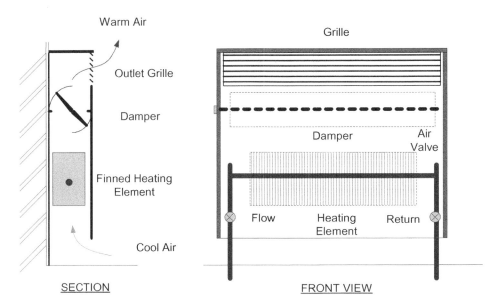

Figure 5.8   Unit type natural convector

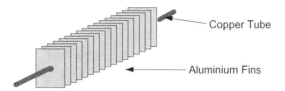

Figure 5.9   Section of aluminium finned heat exchanger heating element

through the unit and is operated either manually or thermostatically, although it is common practice to control the temperature by thermostatic control valves.

The continuous form is particularly suited for commercial buildings with uninterrupted glazing, and is designed as a modular form with units being joined together to produce a curtain of warm convection air below the window sill. The heating element and casing can extend from one end of the building to the other, with each section resembling the single unit convector and having its own means of damper regulation. This method of heat emission is fairly inexpensive for commercial buildings, but because the heat transfer will reduce as the temperature within the flow falls as it travels along the length of the continuous convector, its length should be limited to the maximum heat carrying capacity of the flow pipe connection to the heating element (unless a separate flow and return are installed below the heat exchangers).

The natural convector is ideally suited for heating systems designed to operate at temperatures above 82°C, such as sealed pressurised heating systems, as the heating element cannot be touched and the casing only gets warm. It should be noted that if the damper is in the closed position, with the effect of stopping the convection process, the front of the casing will get warmer, by an amount depending upon the flow temperature, size of the casing and the material from which the casing is manufactured. It is possible that the casing front panel could emit by radiation up to 10% of the total unit's heat emission value. If the unit is to be used as a LST radiator, then provision should be made to conceal all of its piping connections so that they also cannot be touched.

Building maintenance, such as decorating rooms in which natural convectors are installed, is made easier as only the casing needs to be removed to decorate, leaving the heating element and connections undisturbed and the system operational, but the removal of the casing will eliminate the stack effect and so reduce the heat emission to almost zero.

The heating element or heat exchanger normally consists of a single or double looped pipe, usually 15 mm or 22 mm diameter copper tube, with a series of closely placed aluminium fins located along the entire length. The heat output is higher for a double looped element than for that of an equal length single pipe element. An example layout is shown in Figure 5.9.

Convector type heat emitters achieve a faster heat-up response time compared to that of a heating system employing heat emitters that are predominantly radiant.

FLOOR TRENCH CONVECTOR

The modern floor trench convector is similar to the traditional church heating system, which employed steel or cast iron pipes laid in shallow floor trenches below decorative cast iron grilles, and which subsequently filled up with centuries of dirt and dust.

Floor trench convectors are intended to be incorporated in the structural floor, to ideally be located running parallel to any full height windows or glazed doors to offset the cold down-draught from the glass surface (see Figure 5.10).

The components comprise a steel channel set into the floor. This channel incorporates a continuous aluminium finned heating element of either a single or double pipe run, contained below a fairly substantial

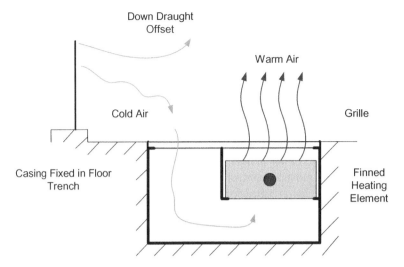

Figure 5.10   Section through trench convector heater

cover grille. The grille, which is normally made from polished brass or has a similar finish, is laid either in sections or continuous 'rolled-up' form. The channel box section is arranged to allow cold air to enter on one side to pass through under the heating element and issue as warm air from the top.

Floor trench convectors still suffer from collecting dust and require frequent maintenance by way of cleaning to keep them functioning efficiently. They are effective at off-setting cold down-draughts from fully glazed areas, but cannot alone provide sufficient heat to satisfy the requirements of the room.

The construction of these units defies the normal design parameters for natural convectors: due to the shallow sunken arrangement, the stack effect is almost lost entirely, hence the low heat emission per metre run compared with continuous wall unit convectors.

The heat emission expected from this type of unit is said to be around 150 to 200 watts per metre run, but a fan-assisted version is also available that can increase the heat output to around 900 watts per metre run.

FAN, OR FORCED CONVECTORS

Unlike natural convectors, fan convectors are not constrained by having to be constructed to create a stack effect through them and this therefore allows manufacturers to produce more compact designs with higher heat outputs. In addition, they free the designer from constraints in the choice of location, as high level models that blow hot air in a downward direction are available, as well as heaters designed to fit below kitchen units, thus saving space where space is limited.

The fan convector is comparable in its construction to the natural convector, consisting of the basic casing having an outlet grille at the top, an inlet opening or grille at the bottom – some models may be equipped with a base section if required, which extends through an external wall and thus allows a proportion of fresh air to enter from the back to mix with the re-circulated room air. Each unit also incorporates a removable and washable air filter located between the air inlet and fan assembly, with the fan being arranged to have a number of speed settings. The high or boost speed is intended to provide a rapid heat-up of the room where the additional noise level generated is considered unimportant. The fan convector should not be selected on its boost speed heat output, but should be designed to operate on its lower speed setting for normal circumstances. Fan convectors may be controlled by their own on – off

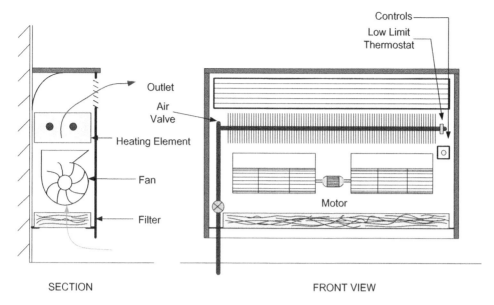

Figure 5.11 Wall unit twin fan convector

and speed selector switch integrated into the unit, or arranged to be controlled independently by a room thermostat.

Forced convectors offer the advantage of being able to achieve a very fast heat-up time compared to the other kinds of heat emitters described and they require less wall space for an equivalent heat output; however, they do require an electrical supply, although their power consumption is very low.

Fan convectors are more expensive than most other forms of unit type heat emitters and generate a certain amount of noise, which can become progressively worse if the unit is not maintained on a periodic basis. Because of this inevitable noise level, fan convectors are unsuitable for living type room areas where peace and quiet are important.

Figure 5.11 illustrates a basic low level wall mounted fan convector, complete with a double looped aluminium finned heating element, twin fan arrangement and constructed in a similar fashion to the natural convector.

As the name 'fan convector' implies, the convection air current is forced through the unit by the action of the centrifugal fan making the creation of a stack effect independent of the height of the casing. Most fan convectors are equipped with a low limit thermostat. This is arranged to prevent the fan from being switched on until it has sensed a rise in temperature within the heat exchanger; this prevents the possibility of cold air being blown out by mistake. This low limit thermostat can normally be overridden by a separate summer time switch to enable it to be operated in the summer season when the heating is not on, thus providing a degree of air movement for cooling purposes.

The versatility of fan convectors is shown in Figure 5.12.

SKIRTING HEATING

Skirting heating (sometimes referred to as 'base board' or 'perimeter' heating) is a form of natural convector heating, although a radiant pattern was made and used but is now obsolete. It is so called because it is installed around the perimeter of the room in place of the skirting board and resembles a continuous mini-ature natural convector. It is neat and unobtrusive and may be installed along one or all of the walls of the

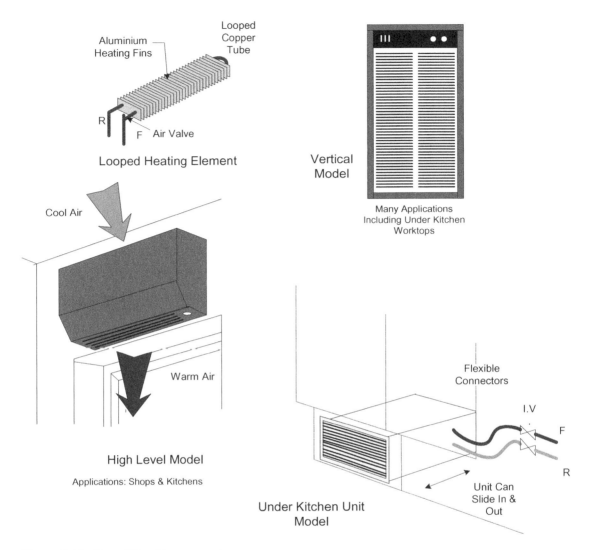

Figure 5.12   Versatility of fan convectors

room, depending upon the heat output required. Skirting heating's main advantage over all other forms of heat emitters is that it provides a uniform distribution of heat evenly around the room and is relatively un-affected by the positioning of furniture etc, whereas all unit forms of heat emitters, such as radiators and unit convectors can, if not designed correctly, create hot and cold spots in the room. Hence, skirting heating achieves one of the main fundamentals in the design of a heating system in accomplishing 'comfort conditions'.

The convective form of skirting heating, similarly to natural and forced convectors, comprises a continu-ous tubular finned heating element consisting of a standard 15 mm diameter copper tube with aluminium fins. The heating element is supported by a series of multi-purpose brackets housed within a light gauge sheet steel casing formed of separate back and front panels, with limited air movement control facilitated by a manually adjusted continuous damper strip. The casing sections are usually finished in a strong enamel paint that provides a certain degree of mechanical protection. Space and provision is usually available, if required, to accommodate the return pipe within the casing when it is not intended to install the skirting heating all the way around the room.

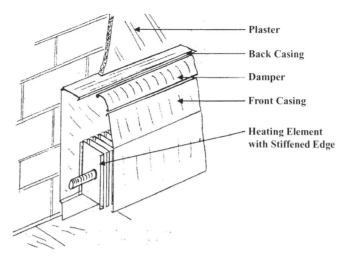

Figure 5.13   Convective form of skirting heating

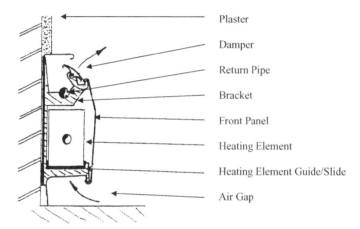

Figure 5.14   Section through convective skirting heating

Example installations can be seen in Figures 5.13 and 5.14.

The convective form of skirting heating is also available with false dummy sections that do not contain any heating element and are intended to complete the perimeter installation when the heat output has been achieved. Corner negotiating units, valve boxes and joining strips are also available, but curved walls – such as bay windows – and angles other than 90° or 45° cannot be accommodated. Convective skirting heating is ideal for square or rectangular shaped rooms.

The installation procedure requires the back corner casing panels to be fixed first, followed by stringing a line between the corner sections to provide a straight edge to which the remaining back panels can be fixed. The heating element and pipework can then be installed with the front casing cover snapped on to finish the installation. When fixing the casing back sections into place, consideration should be given to the installed height so as to maintain a sufficient gap below the casing for convection air to flow unhindered by deep pile carpets, which would reduce the heat emission available.

This form of heat emission is often claimed to be more expensive than other forms of heat emitters, but it has been found that when compared to a heating system employing pressed steel radiators installed on a traditional two pipe system, the overall cost of labour and materials works out about the same.

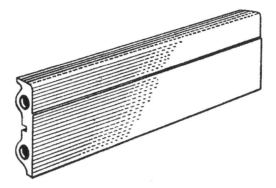

Figure 5.15   Radiant form of skirting heating

The radiant form of skirting heating (shown in Figure 5.15) consisted of much heavier cast iron sections that emitted heat by both radiation and convection. The sections were joined together in a similar manner to that described for cast iron column radiators, using running nipples with cover strips placed over the joins to hide them. This form is now obsolete as it was more expensive, heavier to install and difficult to adapt to angled corners other than 90°; it did, however, have the advantage of being much more robust – able to withstand greater physical damage than its convective counterpart.

The front panelling of the convective form can be damaged by indentation if treated roughly or being knocked into. Also, the front panels should be removed periodically and the entire convective finned heating element cleaned by vacuum to remove the dust collected and so restore the original heat emission.

Another disadvantage of this type of heating is that if small objects find their way into the casing and get lodged between the heating element and the front panel, usually put there by infants, and noise can be generated by the vibration from the circulating pump. This is transmitted through the piping arrangement, amplified through the thin sheet steel front panel and reverberates into the room.

However, the merits of the convective form of skirting heating far outweigh any of the disadvantages. It is a much under-used form of heating, particularly in domestic residential properties, where its main advantage of achieving an even temperature and heat distribution is better than almost any other form of heat emission. It is unobtrusive and not affected by the placement of furniture. The reason for its under-use is probably due to a lack of understanding on the part of both the installer and the homeowner or building developer, as the installation costs compare reasonably with those of a heating system employing any other form of heat emission.

RADIANT PANELS

Plain, flush mounted panels that are predominantly radiant in their method of heat transfer may be installed either in wall recesses or form part of a false ceiling construction. Moreover the panels will be constructed exactly the same as each other, although the heat emission ratio between convection and radiation will vary with application.

No heated surface is able to provide a wholly radiant heat output; the nearest that we are able to get to it is a radiant ceiling panel, which can attain nearly 90% radiation, although the same panel installed in a wall recess will only manage approximately 60% radiation.

The radiant panels comprise a flat smooth panel made usually from sheet steel – or sometimes cast iron, although this material is not normally used today. On the back of the panel is either a ladder system of steel pipework, or a sinuous pipe coil fixed by welding to the panel.

Although the panels are designed for, and intended to be used in recessed locations, providing a flush inconspicuous heat emitter, they have also been used successfully surface mounted directly on walls in a similar way to radiators, or suspended from a ceiling slab in industrial workshop or warehouses where the appearance is not so important. When they are installed within a recess, the piping control valves and air release valve should be housed in an adjacent recess behind an accessible panel. As the radiant panel is designed to fit flush with the wall or ceiling surface, it should be arranged to have a clearance of approximately 10 to 15 mm on all sides to allow the panel to expand without damaging the surface finish. This gap is normally concealed by a cover strip fixed to the wall or ceiling, under which the radiant panel slides during thermal movement.

Examples are shown in Figure 5.16.

Recessed radiant heating panels attain a continuous flush surface to any wall or ceiling finish causing no obstruction. In the case of radiant ceiling panels it is necessary to consider the room height, as this form

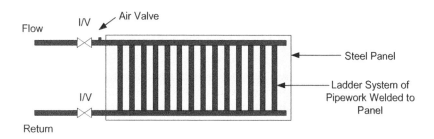

Radiant Panel Construction

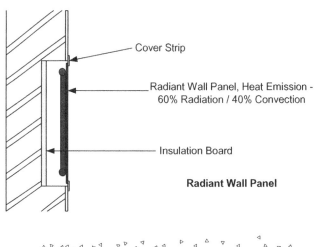

Radiant Wall Panel

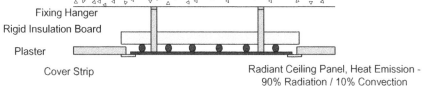

Radiant Ceiling Panel

Figure 5.16 Radiant panels

of heat emission is best suited to rooms with a ceiling height of between 3 m and 5 m. When radiant panels have been installed in ceilings lower than 3 m, the occupants have often complained of feeling uncomfortable, especially if they have been standing up for long periods: this is due to the downward radiation, coupled with high level convection, combining to create unpleasant room temperature gradients and conditions. With ceilings that are higher than 5 m, the effectiveness of the radiant ceiling panels is reduced as the temperature gradient within the room is increased. To offset this, an allowance of an additional 1% to the heating surface of the panel should be made for every 300 mm of height above 5 m.

The space above the radiant ceiling panel should be insulated to reduce the upward heat transmission by as much as possible.

Radiant panels are rarely used for domestic residential buildings, although if the application is right there is no reason for them not being specified, providing that the overall aesthetic value has been considered. Radiant panels are, however, ideally suited to providing an economic method of heat emission to industrial and commercial type premises, particularly when an unobtrusive and robust heat emitter is desirable.

EMBEDDED RADIANT PANELS

The preceding section on radiant heating panels dealt with specific manufactured heat emitting equipment that is obtained and installed and emits its heat in a concentrated fashion, and although these radiant panels require a large amount of builders' involvement to accommodate them, they remain individual components that may be maintained or replaced without affecting the building structure to any major degree.

The history of embedded radiant panel heating can be traced back to examples in ancient Korea and Roman times, with many developments over the centuries to reach where we are today.

Embedded radiant panels differ from the radiant panels previously described as they are formed within the building construction to become a permanent part of the structure. They also have to be formed on site – or in some cases, pre-fabricated or modulised off site then delivered to the site, but this still requires a large degree of civil works involvement during the construction phase.

Embedded radiant panels are created by forming either continuous pipe coils or electric resistance cables within ceilings, walls or floors, whereby the structure becomes the heat emitter. They have the obvious advantage of providing an even and uniform temperature by their method of heat transmission, without any visible heat emitters to cause constraints on the positioning of furniture. However, they do restrict the fixing of things such as pictures and cupboard units to walls, or hanging items from the ceiling.

Both electric embedded radiant heating and mechanical pipe coils embedded into the structure have been fashionable during certain periods over the last 100 years or so, but have suffered from a lack of understanding in both material selection and construction methods, with resulting pipe or cable failures, or cracking of the structure surface: the remedial work has been major and expensive, resulting in this form of heating becoming less popular.

More recently, embedded pipe coils forming underfloor heating have had a resurgence, and although technically classified as a form of heat emitter, the subject is dealt with separately in the following chapter.

RADIATOR FIXING AND SHELVES

Radiators should always be fixed to the walls in accordance with the manufacturer's instructions, ensuring sufficient clearance for the fixing and operating of the radiator valves. The wall selected to hang the radiator on should also be strong enough to support the weight when full of water. The radiator should also be fixed leaving adequate clearance between the top of any skirting board and the bottom of the radiator for the path of convection air to pass up between the back of the radiator and the wall, as any obstruction

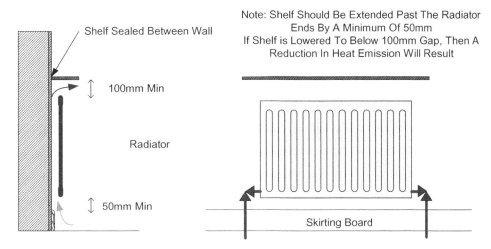

Figure 5.17   Radiator shelves

would restrict the convection circulation and reduce the convection heat emission value of the radiator, making it less efficient.

When radiators are fixed to plain walls that are not directly beneath windows, a phenomenon known as convection or pattern staining can occur directly above the radiator, due to small particles of dust being picked up and carried by the convection air current flowing over the radiator to the wall surface above. To prevent this, a correctly made and fixed radiator shelf offsets the convection air current away from the wall surface.

It is important that the shelf is fixed correctly as illustrated in Figure 5.17, with the ends of the shelf extending past the radiator and with sufficient clearance above so as not to reduce the convection heat transmission value of the radiator. The back of the shelf should be sealed against the wall if it is to be effective.

Shelves that are designed to clip on to the top of the radiator should be avoided as they seriously obstruct the convection air current and reduce the overall heat emission.

EXAMPLES OF CONVECTION STAINING

The phenomenon of convection/pattern staining may also be observed on the internal decorative finish of walls constructed from thermal block, especially if the finish is textured and painted white or a similar light colour.

In Figure 5.18 the lines of the blockwork become clearly visible as the dust particles have impinged on the surface where the cement mortar joints occur. This is due to the thermal resistance of the cement mortar being considerably less than that of the blockwork, and therefore the heat flow through the wall is greater at the joints than it is at the blocks. This form of convection/pattern staining is often mistaken for damp, even though the internal surface of the block wall has been cement rendered and plastered and feels dry to the touch. The only solution is to lower the room temperature, or decorate more frequently.

Another example of convection/pattern staining is where the spaces between timber ceiling joists have been insulated to a point whereby the thermal resistance of the joist itself is a lot lower than the thermal resistance of the insulated space, resulting in dark lines appearing on the ceiling, as in Figure 5.19.

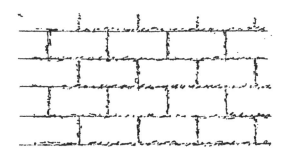

Figure 5.18   Pattern staining on block wall

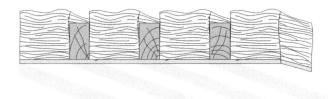

Figure 5.19   Pattern staining on ceilings

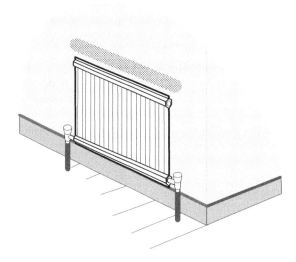

Figure 5.20   Pattern staining above radiators

Convection/pattern staining will occur above radiators fixed on walls without a window above if a shelf is not fitted, as can be seen in Figure 5.20.

Convection/pattern staining is the result of dust particles being deposited on the surface of the wall, which serves to filter them out as the heat flows through. If the thermal resistance of the material is equal, then this phenomenon will not happen, but in situations as above where the thermal resistance is not equal, staining will occur. Pattern staining above radiators differs insofar as the dust is impinged on the textured wall surface as the convection air current passes over it.

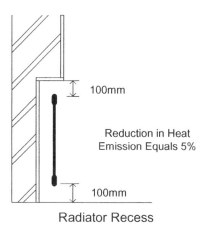

Radiator Recess

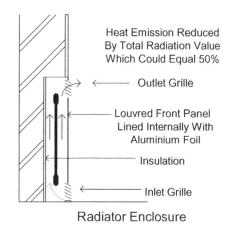

Radiator Enclosure

Figure 5.21 Radiator enclosures

RADIATOR ENCLOSURES

It is sometimes desired for architectural reasons to install radiators within enclosures or recesses in the wall, either to improve appearance or for practical reasons, when any projections from the wall need to be avoided. When radiators are to be located within such features, then allowance for additional radiator heating surface should be made to counteract the reduction in heat emission caused by the enclosure's resistance.

Two basic forms of enclosure are commonly used (see Figure 5.21): one takes the form of an open fronted recess formed within the wall whereby the radiator can be neatly located within it; the other has a similar construction but with a louvred panel fixed to its front. The open fronted recessed version can result in a reduction of heat emission by about 5%, whereas the totally enclosed method of construction has the effect of turning the radiator into a convector, as the total radiation value of heat emission is lost. This could amount to about 50% of the heat emission if the radiator is a plain pressed steel panel type.

If the latter form of enclosure is to be used, then a natural convector would be a more suitable choice, because converting a radiator would require over-sizing by as much as 50% to compensate.

RADIATOR CONNECTIONS

Most forms of radiators, be it panel or column type, are provided during the course of manufacture with four pipe connection tappings, one located at each of the four corners. These pipe connection tappings are normally ½ inch female BSP threads, but on larger radiators may be ¾ inch or 1 inch diameter. Because there are four tappings to choose from, there are a number of variations with regard to how the pipework flow and return connections are made. This gives the installer greater freedom in selecting connections to suit particular piping arrangements, but consideration should be given to the reduction in heat emission that could result from the pipe connection selected and the radiator should be sized to compensate for this reduction in efficiency.

Radiator manufacturers that manufacture in conformance with BS EN442, base their published heat emission data on laboratory tests where the test rig is arranged with pipe connections of *Top, Bottom Same End*, for some unexplained reason. Any other combination can result in a reduction of the stated heat emission value, as shown in Figure 5.22. The exact reduction that may be experienced varies, depending upon the source of the test data, but in practice the variation in emission is rarely taken into account. The

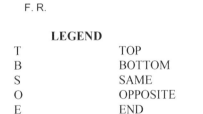

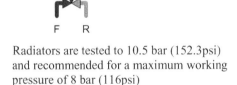

LEGEND

T	TOP
B	BOTTOM
S	SAME
O	OPPOSITE
E	END

Radiators are tested to 10.5 bar (152.3psi) and recommended for a maximum working pressure of 8 bar (116psi)

The spare tapings may be used for the manual air vent and the remaining taping plugged off.

Figure 5.22   Radiator connection and effects

radiator selected normally has additional heat emission capacity and is selected to the nearest heat output above the calculated heat output emission. It is good practice to allow 5% additional heat emission above that required for B.O.E, T.B.O.E and double entry radiator valves; never use B.T.S.E and do not add a percentage for central radiator valves as these are tested using this arrangement.

ROOM HEIGHT EFFECT

When designing a heating system, and in particular when locating and positioning heat emitters within rooms, the chief object is to accomplish a uniform temperature throughout the height of the heated space. Certain modes of heating cause different vertical temperature gradients and it therefore appears reasonable to make an allowance for the height of a room, bearing in mind that warm air will rise towards the ceiling.

Consequently, the heating system must be designed to maintain a comfortable temperature in the lower two metres of the room, as this is the living area occupied by people, and we must accept that a higher temperature will occur in the upper part of the room nearer the ceiling, which in turn will inevitably lead to greater heat losses through the higher parts of the building structure.

Figure 5.23 illustrates the different temperature gradients for various forms of heat emission and it can be seen that the gradients are greater with predominantly convective systems, which rely on warming the air contained within the room for the conveyance of heat, as in the case of convectors and convector type radiators. The worst example is that of warm air systems, as with this type of system the air is supplied into the room already heated, and it can be seen that the temperature gradient for a high level warm air outlet grille arrangement is greater than that for a low level outlet.

The most uniform temperature gradient exists with underfloor heating, followed by ceiling heating.

Careful consideration when locating heat emitters can reduce the temperature gradient, especially under large cooling surfaces such as windows.

Table 5.1 gives the recommended percentage increase in heating surface emission for various room heights. The 'not applicable' entries in the table relate to forms of heating not normally recommended for these ceiling heights.

LOCATION OF HEAT EMITTERS

As briefly mentioned in various parts of this text, thorough consideration should be given to selecting locations for installing heat emitters as a number of factors must be taken into account. The correct sizing of the heat emitters to attain the designed room temperature under specified conditions is just one important aspect, but the location of the heat emitters is just as important if comfort conditions are to be achieved.

The following points should be considered when selecting suitable positions to install heat emitters:

- The physical size and shape of the room, including the height.
- The location and physical size of all the windows including the space beneath the window sill.
- In the case of existing properties, the type of curtains such as full length or sill height. Also, the positioning of furniture, particularly for heat emitters having a high radiation emission value.
- The proposed locations of heat emitter should be discussed with the client or homeowner in order to achieve satisfaction and comfort conditions.

Skirting heating, underfloor heating and radiant ceiling heating provide the most uniform distribution of heat as these modes of heating are less affected by objects within the room, provided that the previously mentioned conditions are applied. Also, these heating methods can be evenly dispersed within the room so that an even temperature throughout is obtained.

Because of their method of construction, these forms of heating have the natural advantage of being able to achieve a uniform temperature, but in the case of unit type heat emitters such as radiators and convectors, careful thought should be applied in the selection for the location of them.

The ideal arrangement would be to evenly disperse the unit type heat emitters around the perimeter of the room, or at least along any external walls, as this would accomplish the same excellent comfort conditions as those attained by skirting heating. However, this would be neither practical nor economical, but it is critical to the success of the heating system that all unit forms of heat emitters are positioned correctly, and in the case of large or irregular shaped rooms, the total heat emission should be provided by two or more heat emitters to avoid uneven temperature gradients across the room.

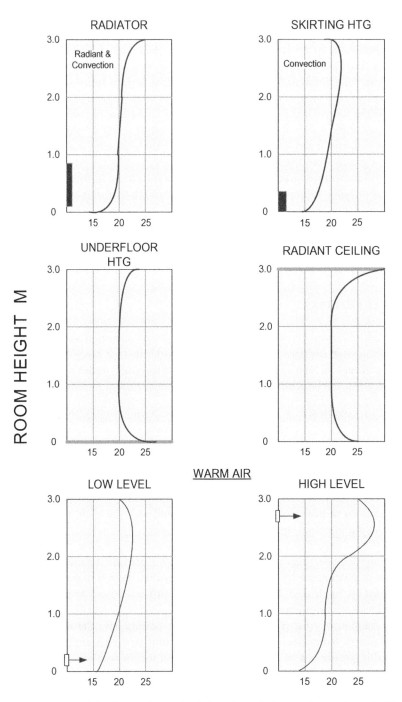

Vertical temperature gradients based on a design room temperature of 21°C

Figure 5.23 Room height temperature gradients. From Faber & Kell's Heating & Air-Conditioning of Buildings, 10th Edition, D. Oughton & S. Hodkinson, figures 4.12 and 4.13, Copyright Elsevier (2008).

Table 5.1 Allowance for height of heated spaces

TYPE OF HEATING	PERCENTAGE ADDITION OF HEATING SURFACE FOR HEIGHT OF HEATED SPACE		
	Below 5m	5 to 10m	Over 10m
Underfloor Heating	0	0	0
Ceiling Heating	0	0–5	Not Applicable
Convective Heating	0	0–5	Not Applicable
Radiant Wall Heating	0	0–5	Not Applicable
Warm Air, Low Level	0–5	5–15	15–30
Warm Air, High Level	0–5	5–10	10–20

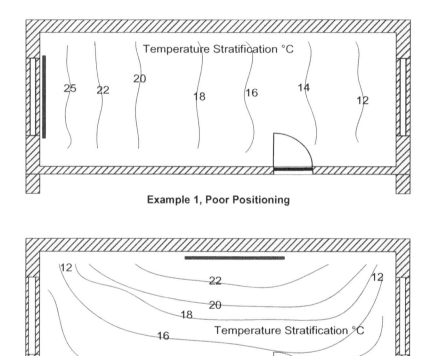

Figure 5.24 Room temperature stratification

The proportion that the heat emission is divided by will depend upon the shape and format of the room: whereas in fairly square or rectangular shaped rooms the heat emitters will be equal in their heat output, with 'L' shaped rooms the heat output of the heat emitters would probably be unequal.

Figure 5.24 illustrates two examples of poor radiator positioning within a room and the temperature stratification that could be expected. Example 1 has a radiator positioned at one end of a long room, resulting in a dimensioning temperature at the opposite end. Example 2 has the radiator positioned centrally in the same room but still does not produce good comfort conditions, because cold spots exist at the ends of the room.

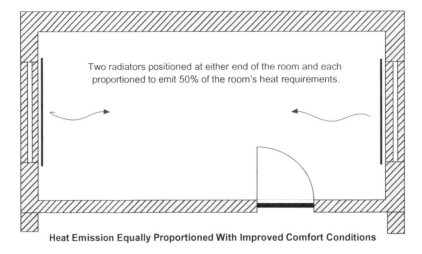

Heat Emission Equally Proportioned With Improved Comfort Conditions

Figure 5.25   Improved temperature stratification (equal proportion)

In both examples, comfort conditions have not been achieved in the room illustrated; this is due to the fact that a single unit heat emitter cannot satisfy the requirement of attaining an even temperature throughout the room and produces uncomfortable hot and cold spots.

Figure 5.24 demonstrates that a single unit heat emitter is inadequate and that a minimum of two radiators would be required to provide an acceptable level of comfort conditions in the room.

Figure 5.25 depicts a similar sized and shaped room, but having two radiators installed one at either end of the room. Here the temperature stratification of the room is much improved and better comfort conditions are obtained.

There are no hard and fast rules regarding the number of heat emitters that should be installed per room, or the ratio of heat emission for each of them; only experience will enable the design engineer to successfully decide this, whilst retaining the principle of providing a cost effective heating system.

As a general guide to achieving an acceptable compromise between providing comfort conditions and an economic heating system installation cost, additional radiators or convectors should be considered for every 3.5 kW to 4.5 kW of total room heat requirements, and the proportion of heat emission for each heat emitter may be split directly in relation to the cubic volume of the room when divided into smaller area zones. It should be noted that this is only a general guide, as there are many circumstances where this would not wholly apply, such as rooms having only one external wall, or where the room has been constructed having square, or nearly square, dimensions whereby a single but larger heat emission capacity heat emitter would suffice. On the other hand, long rectangular rooms, or irregularly shaped rooms, or rooms having a large percentage of external walls would certainly benefit from having more than one heat emitting unit to furnish the rooms heat requirements.

When selecting locations for heat emitters, each room should be considered on its own individual merits in assessing the likely heat flow pattern and the stratification that could exist, with the aim of limiting the temperature stratification to a minimum, whilst still maintaining an economic and cost effective heating system.

The ideal position and first choice for heat emitters in all rooms, is to install them beneath windows, as this serves to offset the cold radiation downdraught from the glass surface which would help to reduce the room temperature gradient (see Figure 5.26). The physical size and dimensional proportions of radiators to be located below windows should be correctly selected, as the radiator should be arranged to span the full length of the window opening if it is to be fully effective in offsetting downdraught. A high narrow

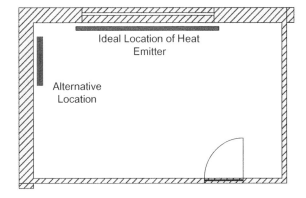

Figure 5.26   Heat emitter locations

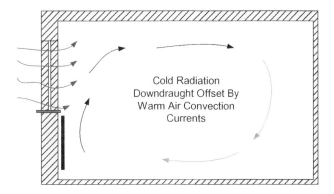

Figure 5.27   Air flow pattern (radiator beneath window)

radiator will create a fountain of warm air moving upwards that would increase heat loss through the glass and produce an equally narrow cascade of colder air falling either side of the radiator.

If it is not possible to locate the heat emitters below the windows because of insufficient wall space or if the windows are full height, then an alternative location will need to be sought. Figure 5.26 shows an alternative option of locating the radiator on the side wall, which is also an external wall – this being preferable to the other opposite internal wall, which locates the heat emitter at right angles to the window. The air flow pattern from this alternative location will not be as good as from below the window, but it is the best alternative available. Heat emitters should be fixed to external walls whenever possible as this will be more successful in reducing the temperature stratification effect.

Figure 5.27 shows the air flow pattern for a radiator fitted beneath a window, where the warm air convection currents can be seen to deflect and offset the cold radiation downdraught from the glazed area of the window. Without this deflection of the cold radiation downdraught, comfort conditions could not be achieved.

Figure 5.28 illustrates the air flow pattern for a radiator fitted at a right angle to the window. Here it can be seen that the cold radiation downdraught from the glazed area of the window is only partially deflected, resulting in incomplete comfort conditions. If it is not feasible to locate the heat emitters beneath window areas, then this arrangement is the next best option, providing that the wall chosen for the radiator is an external wall.

Figures 5.29 and 5.30 show the effect that the physical proportions of the radiator have on deflecting any cold radiation downdraught from glazed areas of windows. A correctly proportioned radiator should extend

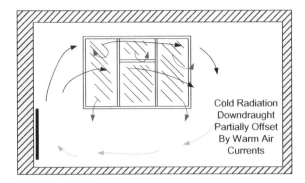

Figure 5.28   Air flow pattern (radiator at right angles to window)

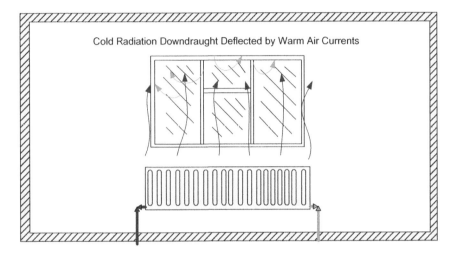

Figure 5.29   Correct proportioning of radiator

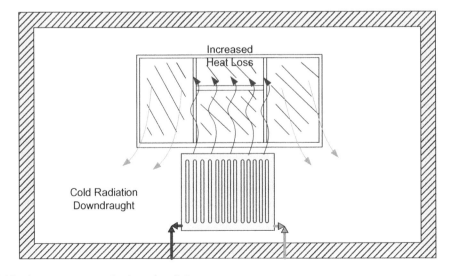

Figure 5.30   Incorrect proportioning of radiator

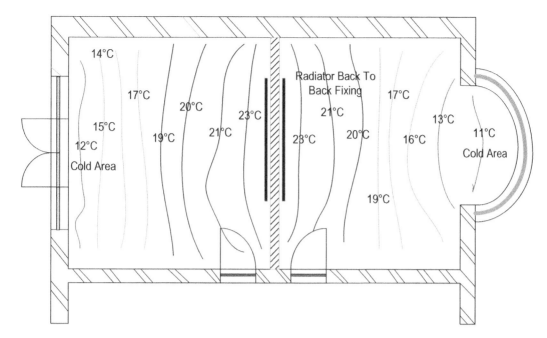

Figure 5.31 Incorrect radiator locations

the full length of the window opening, plus a further 50 mm to 100 mm beyond so as to offset 100% of the downdraught.

If the radiator is incorrectly proportioned, then cold radiation downdraught can penetrate through the window at each end and cascade down where it is not deflected by the warm air currents. Here, only the central part of the downdraught is offset, which in turn can result in a greater heat loss through the upper part of the window. Incorrectly proportioned radiators located beneath windows are only partially effective at achieving comfort conditions in the room and in comparison, are about as effective as the alternative option of locating the radiator at right angles on the side wall.

Figure 5.24 illustrates the expected temperature stratification that occurs when radiators are wrongly located. Equally as bad is the poor practice of fixing radiators in the so-called 'back to back' situation whereby the heat emitters are fixed on the opposite sides of an internal wall (see Figure 5.31). This positioning of radiators has been carried out in the past purely on economic grounds, as it is cheaper due to using less tube, and it completely ignores any attempt to obtain comfort conditions. The temperature gradient that occurs with this poor practice results in a large temperature difference from one end of the room to the other.

RADIATOR ACCESSORIES

Standard radiators are provided with four ½ inch or ¾ inch BSP female threaded connections to choose from, but only two will be used for connecting it to the flow and return piping system (Figure 5.22). The remaining two connections, one of which will be a top connection, will be employed to incorporate the manual air release valve supplied with the radiator, with the final tapping being blanked off using a threaded steel plug.

Figure 5.32 shows a section cut through a typical radiator manual air vent that is supplied with the radiator and fitted on site by the plumber. Its purpose is to allow air that has become entrapped to be

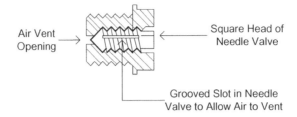

Figure 5.32 Section through a radiator manual air release valve

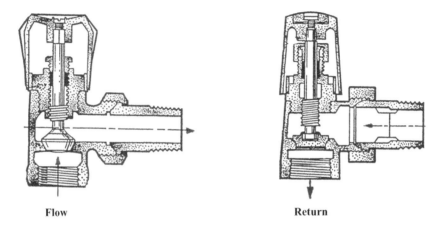

Figure 5.33 Flow and return radiator isolating valves

manually released when filling or refilling the heating system and is operated by a manual radiator air release key. The manual air vent takes the form of a simple needle valve that has a grooved slot along its threaded body to assist air to escape when being operated.

The manual air release vent valve is fitted to one of the top radiator tappings by either a hexagonal Allen key or square headed radiator valve key, depending upon the manufacturer's style. Alternatively the body of the valve may have a hexagonal nut shape formed to accept an open ended spanner, or adjustable wrench. The air release valves are normally manufactured from brass or stainless steel.

Each radiator or any heat emitter should be equipped with a pair of isolating valves to assist in maintenance activities and commissioning the heating system. On radiators, these are normally termed 'radiator valves' and comprise a flow valve and a return valve, as shown in Figure 5.33. The valves are similar in design to each other, but have definite purposes: they are designed to fit either on the flow with the water passing through the valve before entering the radiator, or water exits the radiator through the return valve.

The valve designed to fit on the flow gives the occupant of the room the facility to open or close the valve at will, turning the radiator on or off when the need requires. To enable this to be achieved, the valve is fitted with a manually operated standard wheel-head arrangement that the occupant will easily understand, as it is similar to turning on or off a normal pillar valve or bib-tap found in the bathroom or kitchen.

The valve that is designed to be fitted on the return is known as a 'lockshield' type valve. Its main purpose is to enable that part of the heating system to be balanced during the commissioning stage, this procedure is explained later in this volume. Consequently, lockshield valves are equipped with a plastic cover that may be turned by anyone, but does not turn the valve on or off. By removing this plastic cover, that is normally held on by a single grub screw, the plumber can access the valve's square head. This may be operated by a valve key and can set the valve to regulate the flow of return water from it. The two valves combined can be used to isolate the radiator if it needs to be removed.

The radiator valves illustrated in Figure 5.33 have female BSP threaded connections suitable for low carbon steel tube pipework; they can also be obtained with compression connections, suitable for copper tube normally used in domestic applications.

It can also be seen in Figure 5.33 that the radiator valves both have detachable radiator connection tails that couple to the valve body by screwed union connections. When the valves have been turned off, these union connections permit the radiator to be removed from the wall without draining down the heating system, or disturbing the piping arrangement. The radiator valve tails have male BSP threads that can be screwed into the radiator tappings selected for the flow and return connections: this is achieved by the use of an adjustable wrench on the hexagonal part of the tail's external barrel if provided, or by a radiator Allen key designed to fit the hexagonal socket on the end of the radiator tail. The tails on some radiator valves intended for commercial applications are manufactured with internal lugs that require a special radiator spanner designed to fit inside them, in order to tighten them up.

DESIGNER RADIATORS AND TOWEL RAILS

The text of this chapter has concentrated on standard heat emitters that are readily available, but there has been in recent times an influx in the heating market of so-called designer radiators that offer a more interesting or unusual appearance. These designer radiators are too numerous to discuss here, but all follow the same criteria described in this text. Designer radiators are generally aimed at the luxury end of the heating market, but can be used in certain applications when a conventional heat emitter is not suitable, and therefore should not be dismissed without consideration. If designer radiators are to be used then the design engineer should check to ensure that the heat emission values claimed are based on the same heat output data required by BS EN442; see Chapter 8, Heat Emitter Selection and Sizing.

Towel rails have been available to be connected to the heating system for as many years as there have been heat emitters: some incorporate small radiators within them, which will emit a certain amount of heat; however, they generally only provide enough heat to warm for any towel draped over them and add little heat to the room. If towel rails are desired, then a separate heat emitter may be required to satisfy the heat requirements of the bathroom.

6

Underfloor Heating

The history of underfloor heating (UFH) has been quite chequered: versions of it have been used since Roman times, albeit mainly using warm air. Water as a medium for conveying the heat and providing underfloor heating has been used over the last two centuries, employing steel tube embedded under floors and in effect turning the floor into a heat emitter. Recent history of underfloor heating using materials that are more familiar today, has been no less chequered, with some disastrous results occurring, particularly during the 1970s when underfloor heating became fashionable for a short period of time on new-build projects. If we study the reasons why so many underfloor heating systems failed during that period it is easy to see that there was a lack of understanding of both the system and, more importantly, the behaviour of the materials used together in the heating system and their reaction with the building materials that they came into contact with. There was no excuse for these failures as the knowledge regarding the system materials and their performance in proximity to the building materials was widely available, and on further investigation it was found that in the majority of cases the system had been specified by the developer without consultation with a qualified design engineer.

As previously mentioned, there has been a resurgence in the United Kingdom in the installation of underfloor heating systems. Lessons have been learnt from previous mistakes and an engineering approach has been applied to the subject. Underfloor heating has in the past been classified as a form of heat emitter, as the floor becomes a radiant panel that emits its heat into the room. Although this statement is true, underfloor heating is a lot more than just employing the floor as a heat emitter; it is a system and should be treated as such. Hence the subject is given its own chapter here.

The main advantage claimed for underfloor heating is that it achieves an even distribution of heat emission within the space, a merit that can also be claimed by other forms of heat emission, elaborated on in Chapter 5. It also frees up wall space to allow furniture to be positioned at will. However, it does add constraints to the occupier with regard to the choice of suitable floor coverings.

Heating Services in Buildings: Design, Installation, Commissioning & Maintenance, First Edition. David E. Watkins.
© 2011 John Wiley & Sons, Ltd. Published 2011 by John Wiley & Sons, Ltd.
As an aid to lecturers and students, full colour versions of the figures in this chapter may be found at www.wiley.com/go/watkins
These figures are © 2011 by John Wiley & Sons, Ltd.

OPERATING PRINCIPLES

The general operating principles of an underfloor heating system employing water as the medium to convey the heat, are not too dissimilar to the other forms of piping arrangements described earlier in this volume. The water is heated centrally in a boiler and circulated around the building whereby it gives up part of its heat and returns to be reheated. The difference from other forms of heating is that rooms heated by underfloor heating have a pipe coil in the floor forming an extension to the circulating piping arrangement. This serves to heat the floor structure, and in turn the floor surface becomes the heat emitter and transfers part of that heat to warm the room or space.

Underfloor heating systems also differ from the other forms of heating explained in this work in that they involve close co-ordination with the building contractor, which may also lead to periods of absence from site between the first fix pipework arrangement and the second fix fit-out and commissioning. It is important that the building contractor understands the system components and operating principles, and that the floor layers are closely supervised to ensure that the pipe coils are not damaged or disturbed during the floor laying activities.

The design and specification of the floor structure must also be agreed with the architect and builder at the design stage as the floor construction materials must not only be suitable for the temperatures involved, but must also be non-aggressive to the pipework components and thermally conductive to allow the heat to be transferred into the space to be heated unhindered.

When floors are used as heat emitters, as in the case of underfloor heating systems, heat is emitted first by conduction through the floor structure from the pipe coil to the floor surface, and then into the room, mainly by radiation, the amount being dependent upon the material of construction of the floor and the texture of the floor surface. This radiation value will normally be between 60% and 80%, with the remaining heat emission by convection.

The entire floor surface area of the room becomes the heating surface available to emit the heat into the room, which when compared to a conventional panel or column radiator, is a significantly greater area. This permits a lower floor surface temperature to be acceptable than that produced by the surface temperature of a radiator.

The subject of floor surface temperatures in relation to comfort conditions has been debated for a number of years, without any real conclusion. However the temperatures given in Table 6.1 may be used for guidance providing that the floor construction and surface finish are suitable.

Figure 6.1 illustrates a typical conventional two pipe heating system employing pressed steel panel radiators on the upper floors and incorporating the provision for domestic hot water, all controlled independently by motorised zone valves. The lower floor is shown being heated by underfloor heating complete with its own control arrangement.

Table 6.1   Floor surface guidance temperatures

Room or Area	Minimum Temperature	Maximum Temperature
Rooms where occupants will be standing for long periods	24°C	25°C
Rooms where occupants will normally be seated but active	26°C	27°C
Rooms where occupants will normally be seated but inactive	28°C	29°C
Rooms where people will normally be passing through	29°C	30°C

These temperatures are for guidance only and may vary if circumstances prevail.

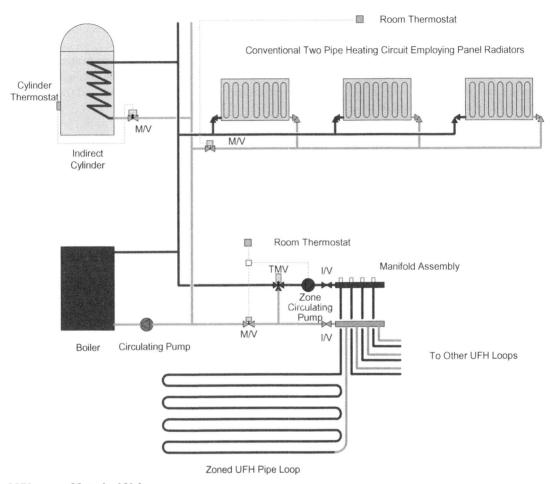

Figure 6.1   Diagrammatic layout of heating system with part UFH

UFH PIPING MATERIALS

Numerous piping materials have been used at one time or another for underfloor heating applications either embedded or buried, but today these are normally just limited to the following few, which have proved to be successful, if used correctly and are commercially available.

Metallic pipes

The only metallic pipe material used successfully today for underfloor heating applications is soft copper tube conforming to BS EN1057, having a temper of R220, which is classified as a soft fully annealed condition supplied in coil form with a minimum coiled length of 25 m. If the tube is to be buried within a floor screed the copper tube should be obtained complete with a plastic coating, usually PVC or polyethylene, to protect the copper from being attacked by the cement screed. There are two types of pre-wrapped plastic coating available; the one that should be preferred has a serrated castellation that forms passageways for air to travel between the copper tube outside wall and the protective plastic coating. This form of coating not only protects the copper tube from corrosion, but also allows the tube to move freely within limits inside the protective sleeve as it expands and contracts when heated. (A common cause of failure in the past has been when copper has been unprotected and firmly embedded and unable to move.) It should be noted that this covering on the copper tube will have certain insulating properties that should be taken into account (see Figure 6.2).

The advantage that copper tube can claim over the plastic materials is that it can be easily bent to a tight radius using a hand held bending machine, and can be laid directly on top of the concrete structural slab without the need for any form of pipe guide/restraining rail, as copper tube will retain the shape of the bend when heated up.

A plastic covered copper pipe can be seen in Figure 6.3.

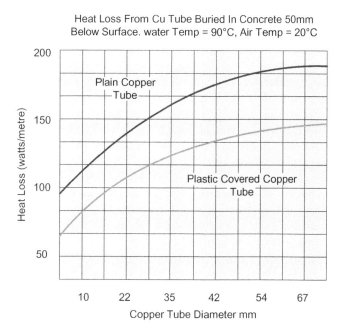

Figure 6.2   Graph showing the insulating effect of coated copper tube

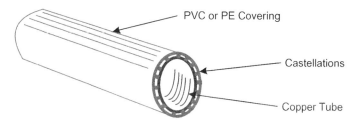

Figure 6.3   Plastic covered copper tube with castellated air voids

Plastic pipes

A number of thermoplastic piping materials have been used with varying degrees of success for conveying hot water in heating systems. For underfloor heating systems these have been mainly limited to three, namely polybutylene (PB), cross-linked polyethylene (PEX), formerly abbreviated to XLPE, and polypropylene (PP). All three of these piping materials have the attributes required for underfloor heating pipework. These include being resistant to corrosion or attack from cement floor screeds; their suitability for the water temperatures involved, providing that the temperature does not exceed 82°C (they should not be connected directly to the boiler); and their commercial availability in long coils that can be laid without any intermediate pipe joints within the floor screed. They can also be bent to a reasonable bending radius of 200 mm, providing that the pipe is being installed at ambient temperatures in excess of 0°C. A bending radius of 5× the pipe diameter can be achieved at an ambient temperature of 20°C, but 8× the pipe diameter can be achieved at an ambient temperature of 0°C. In winter conditions, if there is no room heater available the pipe may need to be filled with water and pre-heated by an electric portable heater to about 40°C. In all cases, thermoplastic pipes must be laid and held in restraining guides or rails as the pipes will have a tendency to return to their original shape when hot water is passed through them.

All thermoplastic pipe materials have a coefficient of expansion far greater than the metallic pipe materials, but whereas metallic pipes will suffer from stress if they are not allowed to move linearly – as in the situation with embedded pipes – thermoplastic pipes, if restrained from moving linearly, can compensate by expanding their wall thickness. However as a note of caution, excessive expansion in long runs of pipework will cause the pipe to ripple and form corrugations, which will eventually result in the pipe failing.

It is important that the characteristics of all pipe materials are fully understood and that they are used correctly. Otherwise the pipe will fail, which in the case of underfloor heating could be extremely expensive to rectify.

Although thermoplastic pipes have certain advantages over metallic pipes, they also have certain disadvantages, which include increased linear expansion, low resistance to mechanical impact and the requirement for almost continual support when conveying hot water. Another characteristic, sometimes referred to as a disadvantage, is the phenomenon of oxygen being able to ingress through the walls of the thermoplastic pipe material. This has the effects of causing frequent air locks in the heating system and accelerating corrosion to any ferrous materials such as steel radiators or cast iron heat exchangers. Therefore, all thermoplastic pipe materials used in heating systems should incorporate a physical barrier to prevent oxygen from entering the circulating water, as depicted in Figure 6.4.

Barrier pipes are sometimes referred to as 'composition pipes', as they incorporate a physical barrier midway between two layers of the same thermoplastic material. Usually this barrier consists of a thin sheet of light gauge soft aluminium that is rolled into a tube shape with the edges overlapping and welded or not sealed. A true composition pipe is manufactured from two or more different materials that are bonded

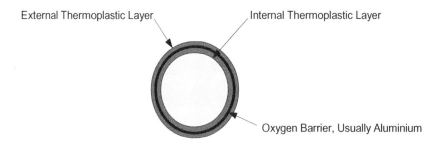

Figure 6.4   Cross-section through a thermoplastic barrier pipe

together in order to obtain particular characteristics from each one, whereas a pipe, such as the one shown in Figure 6.4, should be correctly referred to as a barrier pipe.

Thermoplastic pipes are often specified for applications that they are not best suited for, but they are ideal for underfloor pipe loops, provided that they are laid correctly.

It should be noted that if any joint is unavoidable below the floor, then it should be made and housed in a floor access cover box for future inspection.

PIPING ARRANGEMENTS

There are two basic pipe layout configurations used for underfloor heating arrangements, namely the 'series pattern' and the 'spiral pattern'. The choice of which piping arrangement to use will depend upon the following factors:

- Piping material
- Piping spacing required
- Room configuration
- Extent of any external walls
- Floor temperature fluctuations
- Floor material construction

If the above factors permit, there is no reason why both piping arrangements cannot be used for different areas on the same heating system.

Series pattern

As depicted in Figure 6.5, the series pattern piping arrangement takes the form of a continuous pipe coil laid forwards and backwards across the room, in what is sometimes referred to as a 'meander' or 'serpentine' pattern – serpentine because it resembles a serpent in its movement. The flow pipe should be directed towards the external wall that encompasses any window area as soon as it enters the room, as this will be the area of the room that has the highest heat loss. This arrangement creates a gradual floor temperature reduction across the room as the pipe loop transfers its heat and loses part of its temperature en route. At pipe loop spacings of 250 mm or greater, there are few installation problems as the pipe bend can be formed cold, but for tighter radius bends, thermoplastic pipes may require pre-heating. Figure 6.5(a) shows a piping loop arrangement at constant specified spacing, but when a greater amount of heat is required the spacing can be laid closer, for example, immediately below windows or in door areas, as shown in Figure 6.5(b).

If closer pipe spacing is required than is possible due to the bending radius restriction of the piping material used, then the pipe may be laid with loops as shown in Figure 6.6.

Spiral pattern

Figure 6.7 illustrates the alternative, more popular, spiral pattern piping arrangement whereby the flow pipe is looped around the room, starting from the outer edge and working towards the middle in ever diminishing loops. The pipe is then laid in the opposite direction, looping it back between the flow pipe so as to form the final pattern, where the pipes within the loop alternate between flow and return.

This piping arrangement achieves a more even floor temperature with the highest pipe temperature being located along the outer edge of the room, which is particularly important against any external walls that may incorporate windows.

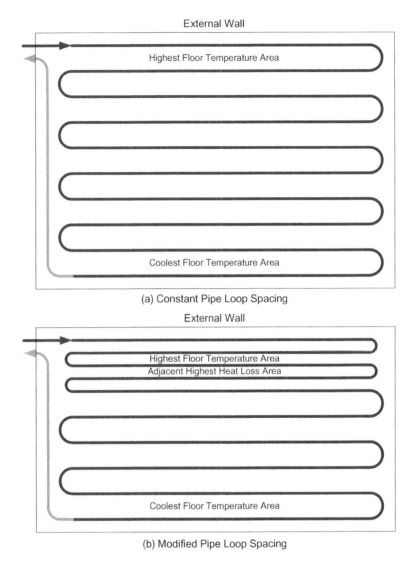

Figure 6.5   Series pattern underfloor heating piping loop arrangement

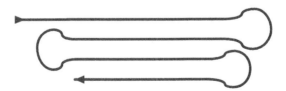

Figure 6.6   Forming closer spacing on thermoplastic pipe loops

This method of laying the pipe loops also eliminates the need to form tight radius bends in the pipes with the possible exception of the central return loop, but this can normally be negotiated with the pipe bends formed to an acceptable radius.

The spiral pattern of laying the pipes also permits closer pipe spacing between the alternating flow and return pipes than that possible with the series pattern.

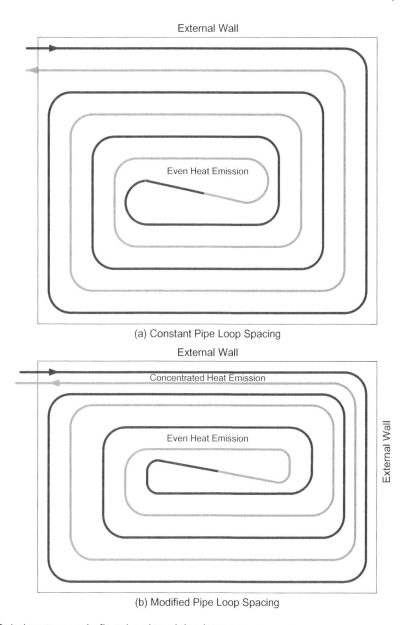

Figure 6.7　Spiral pattern underfloor heating piping loop arrangement

Figure 6.7(a), shows a constant pipe spacing arrangement where the pipes are laid at pre-determined centres, and Figure 6.7(b), shows the modified version with closer pipe spacing along the external walls where the greater heat loss will occur, or when a higher heat output is required.

TRANSITING PIPEWORK

Where large numbers of UFH pipes have been grouped together on transit to other parts of the building, they should be insulated to prevent unduly high floor temperatures occurring, which could have an adverse effect on the floor finish or structure as well as creating uncomfortable conditions in the areas that they have been located in.

FLOOR STRUCTURES AND FINISHES

Floor structures and finishes are builders' work and not normally of interest to plumbing or public health engineers, but in the case of underfloor heating systems, they are important aspects of the design and installation phases of the system. The design engineer should be consulted at the design stage as the floor structure and finish will determine the piping arrangement chosen and affect the heat output emission required from the floor.

UFH systems can be accommodated in most floor structures provided that the following basic criteria are adhered to:

- The temperature limitation of the floor surface and finish that the heating system will have to operate within.
- Suitability and extent of the floor insulation, as large buildings, other than domestic residences having large floor areas, may only be insulated around the external edge of the building.
- The desired floor structure and finish and the ability not to impede the flow of heat through them.

It should be noted that most common floor finishes can be used in-conjunction with underfloor heating systems, but fully carpeted floors should be avoided unless its Tog value (thermal resistance) is known.

Floors may be classified as being either solid construction or timber joists.

Solid floors

Solid floors are normally constructed as a solid concrete structural or over-site concrete slab with a sand/cement screed on top to finish it off flat and level, either as a finished floor surface itself, or as preparation for some other floor finish. Solid concrete constructed floors have an advantage over timber joist constructed floors, especially ground floor level slabs in so far that the compacted ground below the concrete slab serves as a form of insulation to the heat being lost through it. Figure 6.8 illustrates how heat is lost through a solid concrete floor which has no insulation, whereas the compacted ground below the concrete slab provides a certain degree of insulating properties, with the most heat being lost around the outside edge of the building.

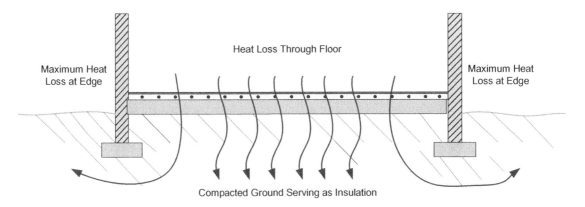

Figure 6.8   Heat loss pattern through uninsulated solid concrete floor

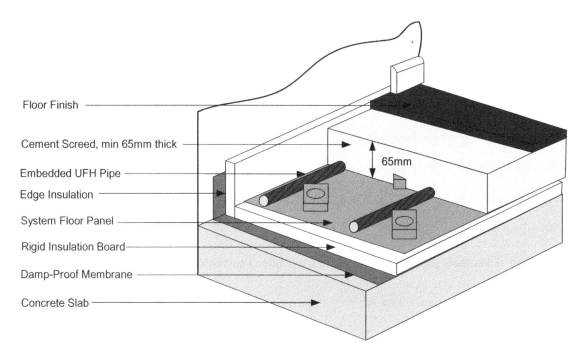

Floor Finish

Cement Screed, min 65mm thick

Embedded UFH Pipe

Edge Insulation

System Floor Panel

Rigid Insulation Board

Damp-Proof Membrane

Concrete Slab

65mm

Figure 6.9   Embedded UFH piping arrangement in solid floor construction

Figure 6.9 illustrates the make-up of a typical underfloor heating pipework arrangement for a solid ground or intermediate floor construction in a new build property. It comprises a damp-proof membrane (ground floors only) laid over the top of the structural concrete slab, or over-site, with a rigid insulation board required to meet the thermal requirements of the Building Regulations. An edge insulation strip is also applied around the outer edge of each room. This edge insulation strip not only provides resistance to the flow of heat through the external walls, but also provides a cushion to absorb any thermal expansion of the floor and reduce the likelihood of cracking; therefore it should be applied to all rooms regardless of whether they are external or internal walls. The insulation used is normally rigid fibreglass boards or polystyrene slabs. Each should be capable of supporting light pedestrian traffic when working on the piping installation.

Figure 6.9 also illustrates a system floor panel, which is a sectional pre-formed panel with raised guide blocks for conveniently locating thermoplastic pipes into position at pre-determined centres, and holding them there until the floor screed is laid. It also serves to prevent the UFH pipes from coming in direct contact with the floor insulation material.

Copper pipes do not require a system plate to be used as the pipe material will remain in position after any labour bends have been formed on the tube, so the floor insulation can be laid before the over-site concrete slab is constructed. The copper pipes can therefore to be laid directly onto the concrete slab, although it is good practice to locate the pipes in position with dabs of cement mortar to reduce the risk of them being displaced when the floor is being laid. If the floor is to be constructed with the insulation laid as shown in Figure 6.9 – on top of the concrete slab and waterproof membrane – then either timber or aluminium guide strips should be used in place of the dabs of cement mortar to raise the copper pipes above the floor insulation and prevent direct contact between the two.

Figure 6.10 illustrates an underfloor heating piping spiral arrangement laid on a 'system plate' prior to the cement screed being constructed.

It should be noted that when underfloor heating pipes pass over a structural floor expansion joint or pass through a wall (edge insulation strip), the pipes should be sleeved as shown in Figure 6.11. This is to

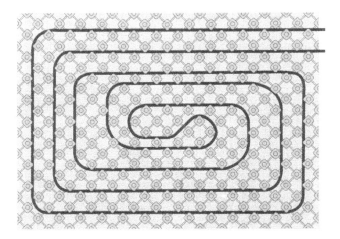

Figure 6.10   UFH spiral pattern piping laid onto a pre-formed sectional system plate

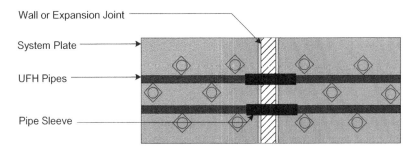

Figure 6.11   Detail of buried UFH pipe sleeve through walls/expansion joints

allow movement across the expansion joint or between rooms without creating stress in the pipes that would risk fracturing the tubes. The sleeves used for copper pipes usually take the form of one commercial size of copper tube larger than the pipe passing through; hence, 15mm diameter tube will be equipped with a 22 mm diameter sleeve.

Thermoplastic underfloor heating pipes are sleeved using what is known as flexible conduit pipe for the same size tube. This is piping that is manufactured in a flexible corrugated form, intended to be buried directly into concrete or cement floor screed to allow the pipe passing through to be removed without cutting up the floor slab, although this is easier said than done.

All UFH pipes should be pressure tested before the cement screed is laid to prove that they are sound and that there are no leaks. The test pressure should be 1½ times the working pressure – providing that the piping material used is capable of withstanding that pressure without weakening or de-rating the tube. Manufacturers recommend that thermoplastic pipes are kept pressurised with water to a pressure varying between 4–6 bar – depending upon the material and the manufacturer's recommendations – whilst the cement floor screed is being laid. This reduces the risk of the tube wall being crushed or collapsing under the weight. The piping system should remain pressurised until the cement floor screed has cured.

Almost all types of hard material floor finishes are suitable for applying to solid constructed floors – including marble, stone terrazzo, slate, brick and granolithic finishes – but if they are to be fixed using a cement mortar, this should be applied by spreading rather than using small cement dabs, which would create cavities and reduce the heat transmission value being conducted through the floor structure. Softer material finishes can also be applied – such as seasoned and kiln dried timber planks or blocks, cork and

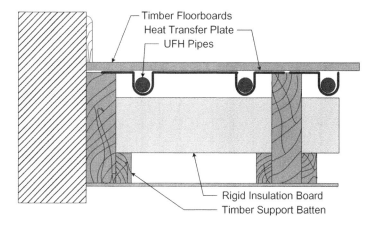

Figure 6.12   UFH piping arrangement for suspended timber floor

vinyl tiles and carpets – providing that the thermal Tog value has been allowed for at the design stage. Mastic compounds used to fix these softer materials must be suitable for the temperatures involved and not soften when heated.

Timber floors

The construction of timber floors is common in the UK as they are somewhat lighter than solid concrete floors and can be erected more quickly on site. They comprise structural timber joists that support either timber floorboards fixed across the joists or timber composite flooring sheets.

There are two basic methods of installing underfloor heating in buildings with timber constructed floors, one used for new-build properties and the other for existing properties, although both are similar in their operation.

Figure 6.12 shows an arrangement that is equally suitable for existing buildings and new-build properties. The timber joists are installed in the normal manner spaced at 400 mm to 450 mm centres, which is a standard spacing for timber joists, but supporting timber battens are fixed to the bottom of the joists to allow for rigid insulation boards to be placed in between the joists. The underfloor heating pipes are laid into a heat transfer spreader plate at 200 mm to 225 mm centres in a series pattern. The transfer plate is manufactured from aluminium and designed to support the pipes and spread the heat evenly, by conduction across the floor, it also permits an air gap to be maintained between the pipes and the tongue and grooved timber floorboards that have been fixed across the joists as the floor finish. This air gap is required to prevent uneven temperatures on the timber boards.

Figure 6.13 illustrates a plan view of a suspended timber floor with the heat transfer spreader plates placed across the joists and the UFH pipes laid into them in a series pattern. It should be noted that where the pipes have to cross the joists they should be notched in accordance with the requirements contained in the Building Regulations. These floors are normally finished by fixing floorboards or floor-grade sheeting across the joists, taking care not to puncture the pipes with nails or screws where they cross the joists.

An alternative method, shown in Figure 6.14, can be employed for installing underfloor heating in buildings with suspended timber floors where the joists are spaced wider apart than 450 mm or closer together than 300 mm, or where it would be undesirable to notch the timber joists, providing that the floor can be raised by 25 mm.

In this arrangement, 75 × 25 mm timber battens are laid and fixed across the main timber joists at centres to suit the pipe spacing with the aluminium heat transfer plate fixed to them. The battens should be cut short at alternate ends, as shown in Figure 6.15, to allow the underfloor heating pipe laid in the heat transfer

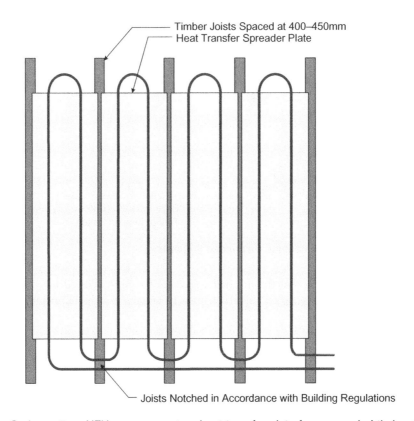

Figure 6.13   Series pattern UFH arrangement on heat transfer plate for suspended timber floor

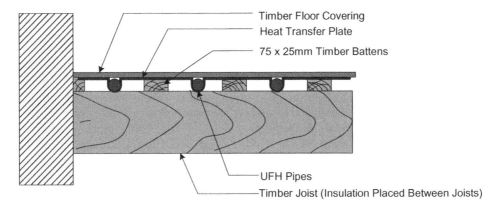

Figure 6.14   Section through raised timber floor UFH arrangement

spreader plate to be looped back on itself to form the series pattern arrangement. The floor can be finished off by laying timber floorboards or flooring-grade timber boards.

Fibreglass or some other approved type of insulation should be placed between the timber joists, similar to the arrangement shown in Figure 6.12.

This arrangement is not usually installed in existing properties unless major refurbishment is being undertaken, as it is not usually practical to raise the level of the floor to accommodate the underfloor heating pipes.

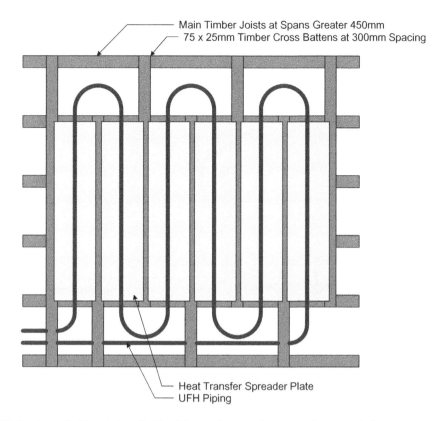

If it is not practical to cut short the 75 × 25mm timber cross battens at alternate ends for reasons of floorboard support, then they may be fixed spanning the full length with the UFH pipe arranged to pass under them.

Figure 6.15   Series pattern UFH arrangement on heat transfer spreader plate for raised timber floor

This arrangement can also be used in a modified form in buildings with solid concrete floors, whereby timber battens are fixed across the floor in one direction with further timber battens fixed on top of them, but at 90° to the first set of timber battens. It must be taken into account that this would cause the floor level to be raised by about 50–60 mm.

Floating floors

The practice of constructing a floating floor enables underfloor heating to be installed on top of an existing solid sub-base or timber floor. However, if installed on top of an existing timber floor, maintenance to any services below that floor would involve taking it all up again, which would be expensive as well as disruptive.

Figure 6.16 shows a detailed section of an underfloor heating piping system incorporated into a floating floor construction. Here, the sub-base floor is prepared by removing all floor imperfections and vacuuming it clean. A pre-formed polystyrene insulation sheet is then laid onto the floor. This insulation sheet has a grooved pipe pattern formed in it that includes bends and straight pipe run. A heat transfer diffusion plate is then laid across the insulation sheet and pushed into the grooves, but with the ends left clear to allow the pipe return bends to be formed. The underfloor heating pipe can then be rolled out into the pipe recesses in the heat transfer plate in a series pattern. After successfully hydraulically pressure testing the

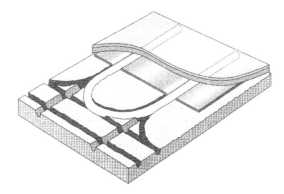

Figure 6.16 Detail of UFH floating floor arrangement

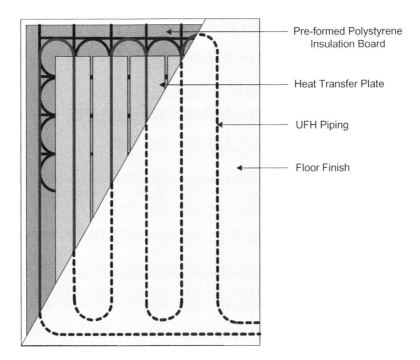

Figure 6.17 Plan layout of UFH arrangement in floating floor

underfloor heating piping system, the floor finish can then be fitted. This normally takes the form of sheet flooring or laminated tongued and grooved flooring sheets.

This method of floor construction is said to be structural, meaning that it can be walked on, but it will not support heavy loadings or withstand physical punishment.

Figure 6.17 depicts the constructional sequence layout of a floating floor series pattern underfloor heating arrangement. This shows pre-formed and grooved polystyrene floor insulation board with pipe tracks, heat transfer diffusion or spreader plate, underfloor heating tube installed in a series pattern and with flooring grade laminate boarding as the floor finish.

FLOOR FINISHES

All floor coverings and finishes on top of an underfloor heating system will create a resistance to the transfer of heat into the room, which in turn will increase the time-lag for the room to heat up. Therefore careful consideration should be given to this important aspect of building design and construction at the design stage. Also, advice should be given to the building owner or occupier regarding any limitations in the choice of suitable and non-suitable floor coverings that can be used. This is particularly relevant to soft floor coverings such as carpets, vinyl tiles and floor laminates, all of which are easily changed when rooms are decorated by DIY enthusiasts.

Floor coverings such as granolithic floor screeds, terrazzo, marble, stone and ceramic tiles are more permanent. They are probably fitted during the construction of the building and are therefore less likely to be changed so readily. These floor finishes can be accommodated during the design of the heating system and the resistance to the flow of heat catered for.

The effect that floor coverings have on rooms heated by underfloor heating is to increase the time-lag between when the heating system is turned on and when the room reaches its design temperature. As underfloor heating has a greater time-lag than most other forms of conventional heating, thick and dense floor coverings could increase this time-lag to an unacceptable level. The occupier of the building should be made aware of this time-lag situation so that the heating system can be turned on earlier than normal. In this way, the rooms will achieve their respective design temperatures in good time.

The thermal resistance added to a floor with underfloor heating can be seen in Table 6.2, expressed in $W\,m^{-2}\,{}^{\circ}C^{-1}$ and in the textile industry's Tog value measure of thermal resistance.

Tog value of floor coverings should not exceed 2.5.

ASSESSMENT OF FLOOR HEAT EMISSION

Underfloor heating systems rely on the floor surface to emit heat into the room, but the heat emission rate will depend on a number of factors that impact on its efficiency.

These factors include:

- Selected design room temperature (variable, dependent upon client's choice).
- Mean water temperature in UFH pipe (will be determined by floor construction).
- Designed floor temperature (will be selected by room usage).
- Floor construction and finish (selected by architect following consultations at design stage).
- Heat emission requirement (obtained by heat loss calculation).

Table 6.2 Thermal resistance of floor coverings

Floor Covering	Thermal Resistance/ $W\,m^{-2}\,{}^{\circ}C^{-1}$	Tog Value
Vinyl Tiles, Cork Tiles, Ceramic Tiles, Epoxy Coating, Granolithic Cement Skim	0.00	0.0
Marble Tiles, Terrazzo, Cushioned Vinyl & Linoleum	0.05	0.5
Medium Pile Carpet, 12.5mm Hardwood Strips	0.10	1.0
Deep Pile Carpet, Wood Blocks, 20mm Laminate Sheets	0.15	1.5

The heat loss from the room or rooms intended to be heated by underfloor heating would be arrived at in the same manner as for the rest of the building during the heat loss calculation exercise. However, the heat loss calculated through the floor element of the rooms, although included for assessing the boiler capacity, would be discarded when assessing the heat emission requirements for the rooms to be heated by the underfloor heating pipe coils.

The procedure for calculating the extent of the UFH piping would be as given in Box 6.1.

BOX 6.1 Converting heat loss to heat requirements

Having completed the heat loss calculations for each room or area to be heated in the building, the heat emission rate required from the floor surface can be calculated by applying the following formula:

$$q = \frac{\text{Heat Loss}}{\text{Floor Area}}$$

Whereby:

q = heat emission quantity expressed in $W\,m^{-2}$.

Heat loss calculated for individual room in watts with floor heat loss omitted.

Floor area available in the room measured in metres.

In most situations, the floor area available will be the same as the total floor area for the room, but in the case of certain rooms such as kitchens or bathrooms, part of the floor area would certainly be covered by kitchen units or bathroom fixtures, meaning that the available floor area would be less.

Example 1 – Calculate the extent of the underfloor heating piping spacing having a flow temperature of 50°C, for a room with a floor area measuring 5 m × 5 m having a heat loss of 1820 watts after the floor loss has been deducted and a room design temperature of 21°C. (Building built to current Building Regulations.)

$$\frac{1820 \text{ watts}}{25 \text{ m}^2} = 72.8 \text{ W m}^{-2}$$

Using Table 6.3, a 16 mm PEX pipe spaced at 100 mm centres will emit 74 W m^{-2} with a floor surface temperature of 29°C, or a 18 mm PEX pipe spaced at 150 mm centres will emit 74.5 W m^{-2} with a floor surface temperature of 28°C.

Example 2 – Repeat Example 1, but for a building built in the 1920s to the standards appertaining for that period whereby the heat loss amounted to 3050 watts.

$$\frac{3050 \text{ watts}}{25 \text{ m}^2} = 122 \text{ W m}^{-2}$$

From Table 6.3 it can be seen that even if the flow temperature is increased to 60°C, a room temperature of 21°C cannot be achieved.

In this situation, a lower heat emission would have to be accepted from the floor, such as selecting an 18 mm PEX pipe spaced at 150 mm centres which will emit 74.5 W m^{-2} giving a total heat emission of 1862 watts and a floor temperature of 28°C. The remaining 1188 watts required by the room would have to be supplemented by a conventional form of heat emitter.

Table 6.3 Typical UFH emission values

Solid Floor Heat Output, Floor Resistance 0.00 W/m²/°C – 12 mm cu/16 mm PEX Pipe

Flow Temperature °C	Room Temperature °C	100 mm Centres		150 mm Centres		200 mm Centres		250 mm Centres		300 mm Centres	
		Emission W/m²	MFT °C	Emission W/m²	MFT °C	Emission W/m²	MFT °C	Emission W/m²	MFT °C	Emission W/m²	MFT °C
40	18	53	23.2	44	23	39	23	38	22.5	36	22
	20	40	24.8	38	24.2	36	24	34	24	31	23
	21	35	25.6	33	25.5	32	24.5	29.5	24.5	29	23.5
	22	30	26	29	26	27	25	26.5	25	27	24
45	18	60	25	58	25	55	24	44	23	40	23
	20	56	26	53	26	42	25.5	38	25	36	23.5
	21	51	27	40	26.5	36	26	34	25.5	32	25.5
	22	39	27.5	35	27	33	26	30	26	29	26
50	18	81	27	79	26.5	77	25	75	24.5	73	24
	20	76	28	74	27	72	26	70	25.5	68	25
	21	74	29	72	28	70	27	68	26.5	66	26
	22	73	29.5	71	29	69	28	67	27	65	26.5
55	18	94	29	92	28.5	90	27.5	88	26	86	25.5
	20	92	29.5	90	29	88	28	86	27	84	26
	21	91	30	89	29.5	87	28.5	85	27.5	83	26.5
	22			87	30	85.5	29	84	28	82	27
60	18					112	28	110	27.5	107	26.5
	20					109	29	107	28	105	27
	21					107	30	106	29	103	28
	22					105	30.5	104	29.5	101	29

Table 6.3   (*Continued*)

Solid Floor Heat Output, Floor Resistance 0.00 W/m²/°C – 15 mm cu/18 mm PEX Pipe

Flow Temperature °C	Room Temperature °C	100 mm Centres		150 mm Centres		200 mm Centres		250 mm Centres		300 mm Centres	
		Emission W/m²	MFT °C	Emission W/m²	MFT °C	Emission W/m²	MFT °C	Emission W/m²	MFT °C	Emission W/m²	MFT °C
40	18	59	24	54	23.5	48	23	44	22.5	39	22
	20	52	25	48	25	42	24	38	24	34	23
	21	48	26	44	25.5	40	25	36	24.5	32	24
	22	45	26	41	26	37	26	33	25.5	30	25
45	18	75	25	68.5	24.5	62	24	56	23.5	50	23
	20	68	26	63	25.5	56	25	52	24.5	46	24
	21	65	27	59	26.5	54	26	49	25.5	44	25
	22	62	28	56	27.5	51	27	46	26.5	41	26
50	18	92	27	84	26.5	76	26	69	25	62	24
	20	85	28	77	27	70	26.5	63	25.5	57	25
	21	82	29	74.5	28	67	27	61	26.5	55	26
	22	79	29	72	28.5	66	28	59	27.5	53	27
55	18	110	29	100	28	90	27	82	26.5	74	26
	20	104	29.5	94	28.5	83	27.5	76	27	68	26
	21	100	30	91	29	82	28	74	27.5	66	27
	22	96	32	87	31	78	30	71	29	64	28
60	18	130	31	120	30	106	30	98	29	88	28
	20	122	31.5	114	31	98	30.5	92	30	82	28.5
	21	118	32	110	31.5	96	31	88	30.5	80	29
	22	114	33	108	32.5	94	32	86	31	78	30

Not Recommended

The two examples emphasise the suitability of UFH installed in a modern building complying with current thermal resistances required by the Building Regulations, but for the older property with a lower thermal resistance, resulting in a greater heat loss, UFH would require supplementing with some form of conventional heat emitter in order to attain the desired room temperature. This would also be the case for most conservatories.

Table 6.3 can be used as a typical guide for establishing underfloor heating emission values, but in practice, when assessing underfloor heat emissions, specific manufacturer's data should be used.

UNDERFLOOR HEATING COMPONENTS

Underfloor heating systems employ materials and components that are common to all wet type heating systems, with the exception of the central distribution manifold. Manifolds are normally used to collate the flow and return piping connections to each underfloor heating circuit at a centrally convenient location, the only exceptions being in situations where only a single room is being heated by underfloor heating or where the total floor area is less than $25\,m^2$.

Figure 6.1 at the beginning of this chapter, illustrates the use of the manifold arrangement where a number of pipe circuits can be taken from this point. The exact form that the manifold takes varies slightly between manufacturers, but all manifolds comprise a flow header and return header, which is basically a common barrel or multiple tee arrangement. The manifold incorporates main and individual isolating valves, lockshield regulating valves on the return piping loops, drain valves, manual air release valves and a common wall support bracket. The flow branches from the manifold normally include a flow indication meter to aid commissioning of each circuit. A pressure and temperature gauge should also be installed on the flow manifold for the same reason. The manifold should be limited to about four or five circuits to avoid large numbers of pipes causing congestion in the floor near the manifold; although they are available with up to 12 tees on each header if required, this could have the effect of creating a high floor temperature locally around the manifold location.

Figure 6.18 shows a typical manifold arrangement complete with wall support bracket, where the return header is set further back close to the wall to allow the return pipes from each circuit to pass behind the flow header.

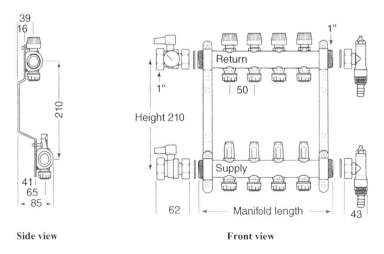

Figure 6.18   Typical UFH flow and return manifold arrangement

The manifold arrangement is normally housed inside a cupboard or purpose-made cabinet, located somewhere that is both convenient for the heating system layout, and acceptable to the building occupants.

UFH WATER TEMPERATURE CONTROL

The main control for a heating system incorporating underfloor heating either in part or in full, does not differ from any other heating system, except that whereas all other heating systems operate with a water flow temperature of 82°C, underfloor heating operates with a water flow temperature of 60°C or less, as temperatures above 60°C could result in high floor temperatures, or even damage to the floor structure or finish. Therefore, if the boiler is operating to provide a water flow temperature of 82°C, additional controls need to be furnished to maintain a lower water flow temperature for the underfloor heating portion of the overall heating system.

There are a number of methods for accomplishing this. All involve mixing the flow and return waters to attain a suitable flow temperature for the floor structure or finish employed.

Figure 6.1 illustrates how the underfloor heating element of the system is incorporated into the heating scheme, and the primary heating circuit is controlled by a motorised zone valve, but the control of the water flow temperature to heating circuits can differ.

Figure 6.19 shows a simple less expensive arrangement to control the temperature of the flow water to an underfloor heating system. This arrangement is ideal for a small UFH section of a domestic property. It comprises a common manifold that serves to collate a small number of flow and return pipes from individual underfloor heating loops, which in turn is connected to the main heating system via a primary heating circuit. The water is circulated through the underfloor heating loops by its own dedicated UFH circulating pump, which draws water through a manually operated, non-electric temperature sensing thermostatic mixing valve. This mixes a proportion of lower temperature return water with a proportion of higher temperature flow water to maintain a constant pre-selected flow temperature through the UFH loops.

The overall room temperature control is by a single room thermostat located in one of the rooms heated by the UFH system. The thermostat switches the circulating pump either on or off depending upon temperature dictates. This method has its limitations and its only advantage is its lower cost.

Figure 6.20 illustrates a variation of the temperature sensing valve arrangement which – due to its lower installed cost – is also a suitable and popular method of controlling the circulating water temperature in domestic residential properties incorporating underfloor heating.

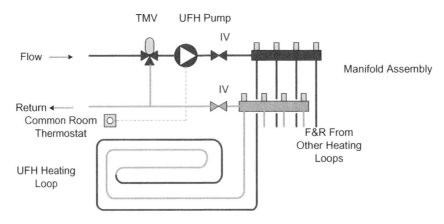

Figure 6.19 Temperature sensing thermostatic mixing valve arrangement

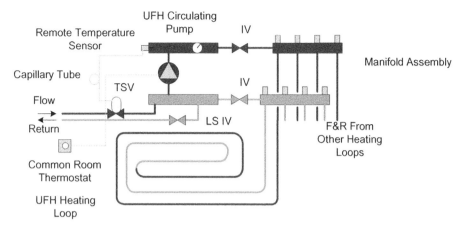

Figure 6.20   Temperature sensing injection valve arrangement

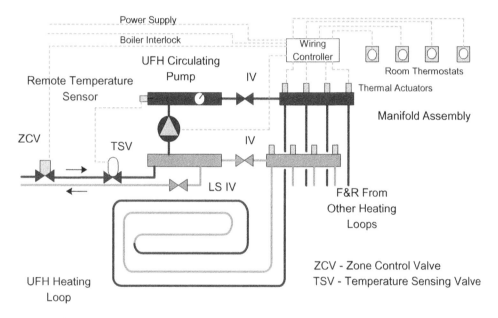

Figure 6.21   Individual room temperature control arrangement coupled to water flow temperature control

With this system, the three-way temperature sensing mixing valve has been replaced by a two-port temperature sensing valve (TSV), referred to as an injection valve, where the temperature sensing element has been removed from the valve and located remotely by clipping it onto the side of the extended flow manifold to sense the temperature of the water flowing through the flow manifold. This is also a non-electrically operated valve and is connected to the injection valve by a capillary tube. It allows boiler water to mix with the UFH return water when required to raise the water temperature.

Figure 6.21 shows a further variation of the water temperature control system that incorporates additional control components. The two port temperature sensing injection valve has been replaced with a more sensitive electrically operated version, although the non-electric capillary valve could still be used.

The temperature in each room is individually controlled by its own room thermostat that is arranged to operate a thermal actuator incorporated in the flow manifold. These electrically operated thermal actuators replace the manual isolating valves that are normally provided. They control the flow to each underfloor heating pipe loop, which in turn controls the motorised zone control valve, together with a boiler interlock that both closes the zone control valve and shuts down the boiler and main circulating pump when all room thermostats are satisfied.

This arrangement is more expensive than the previous control systems described but is equally suitable for both domestic residential heating applications and commercial type properties.

A constant water temperature control system, as shown in Figure 6.22 is more commonly used in larger domestic residential or commercial properties, due to the higher cost of the system. Constant water tem-

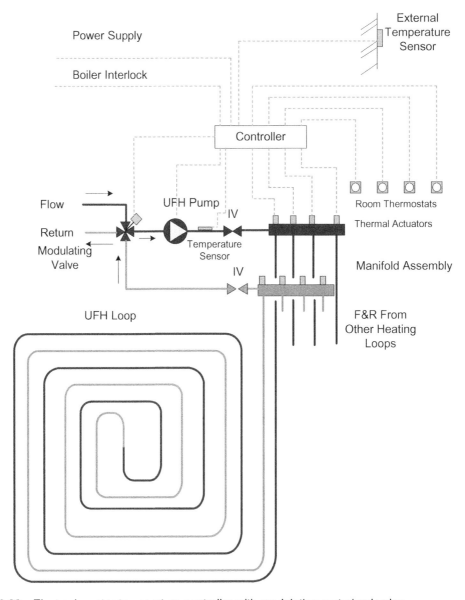

Figure 6.22   Electronic water temperature controller with modulating motorised valve

perature controllers operate by modulating a three or four port motorised valve to maintain the mixed flow water temperature to the underfloor heating loops connected to the manifold. A pipe mounted temperature sensor is fixed to the flow pipe of the flow section of the manifold, to sense the water temperature and relay the information back to the electronic controller.

These types of control systems often incorporate a variable temperature control by the addition of a weather compensator, as illustrated in Figure 6.22. An externally mounted temperature detector senses the outside temperature and any temperature change and relays this information back to the electronic controller. The system functions by combining the information received from the external temperature detector, together with the information received from the internal temperature detector and pipe mounted temperature sensor, and causes the modulating motorised valve to mix the boiler flow water with the return water from the underfloor heating pipe loops to create the optimum flow temperature.

This method of control provides a constant temperature to each heating pipe loop by operating the modulating motorised valve to maintain the set room temperatures, regardless of any variation in either the internal or external temperatures.

SYSTEM TESTING AND COMMISSIONING

The main subject matter of testing and commissioning heating systems is dealt with in detail in Chapter 25, Testing and System Commissioning; however, it is relevant to include here a few words on system testing and commissioning applicable to underfloor heating.

Testing

The soundness of the piping system should be established by hydraulically testing the heating scheme in the normal way, by subjecting the pipes and all the components to a pressure equivalent to a hydraulic pressure of 1½ times the working pressure. The only deviation from this procedure would be when a plastic material pipe has been used for the underfloor heating loops. BS EN1264 requires a test pressure of this section of the system to be hydraulically pressure tested to twice the working pressure, with a minimum pressure to be applied of 6 bar prior to the floor screed being laid.

It is certainly important to prove that any embedded section of piping is sound before being covered over by a floor screed. The installer must be satisfied that the piping material is capable of withstanding a minimum test pressure of 6 bar that would not result in a permanent pressure de-ration of the piping material that would be less than 1½ times the working pressure at the operating temperature. For low pressure domestic heating schemes a hydraulic testing pressure of 4 bar would be adequate to prove the soundness of the system.

After a successful hydraulic pressure test has been completed, the pressure within the plastic material pipe loops should be maintained at 4 bar, or 6 bar if following BS EN1264, during the laying of the floor screed. This pressure should be held until the floor screed has cured sufficiently so as not to be at risk of crushing or deforming the pipes under its weight.

This practice of pressurising the underfloor piping loops also serves to give an indication – by way of a pressure drop – of any damage that might have occurred to the embedded pipework during the laying of the floor screed.

Commissioning

As with the modus operandi of applying a soundness test, the commissioning activity also follows the normal procedure that would be applied to all heating systems; the only additional activity is that associated with the embedded floor loops.

A frequent criticism made by plumbers is that floor layers do not take adequate precautions to protect the underfloor pipe loops when laying the floor screed. Such precautions include laying raised boards above the pipes when moving wheelbarrows over them, and using kneeling pads that span the pipes when spreading and levelling the cement screed. It can equally be said that plumbers do not understand the finer points about the cement screed and if adequate care is not taken, the screed could shrink and crack.

Therefore to prevent damage occurring to the floor screed, the following precautions should be adhered to:

- The cement floor screed should not be laid if the ambient temperature is expected to fall below 5°C. The ambient temperature should be maintained at 5°C or above for at least three days after the floor screed laying has been completed. Some floor screed cement compositions may contain anti-frost additives to protect them during freezing conditions.
- The floor screed should be allowed to cure for at least 28 days before the initial heating-up operation commences – or, in the case of anhydrite screeds, only 7 days. In all cases the cement additive manufacturer's instructions should be followed.
- The water flow temperature during the initial heating-up period should be set to supply water at a temperature of between 20°C and 25°C, which should be maintained for at least three days. Following this, the maximum design temperature of the flow water can then be set and it too maintained for a period of a further four days.

7

Heat Requirements of Buildings

The principal objective when designing a heating system – regardless of whether the system is hydronic or dry, and regardless of the type of heating system – should be to enable the occupants of that space to pursue their normal activities in comfort. Only where human occupation is of a secondary requirement to a particular process, or when designing a heating system for a space used to store certain temperature-sensitive materials, should other criteria be considered.

To enable this primary objective to be achieved and arrive at an efficient and economical size for the proposed heating system, it is necessary to establish the amount of heat that will be lost from the building during the winter months, and this requires heat loss calculations to be undertaken for each and every part of the building to be heated.

Heat loss calculations are required to:

- Assess the correct boiler or warm air heater output rating and therefore permit the heating system to operate efficiently and consume the minimum amount of fuel.
- Select heat emitters with the correct heat emission or warm air terminal points, and locate them to achieve effective comfort conditions.
- Calculate economical pipe sizes or warm air duct sizes and subsequently arrive at any forced circulation duties.
- Estimate fuel consumption and determine annual operating costs.

The amount of heat that will be lost from the building structure has to be calculated so that an equal amount of heat can be put back and in the correct proportion, and thus maintain the desired room temperatures selected. This can only be achieved accurately by undertaking a recognised engineering heat loss calculation procedure using current engineering data available; anything else could lead to an inefficient and ineffective heating system.

Having established the necessity of calculating the heat losses from the building, to facilitate this procedure, it is essential to gather the data and information required to commence the calculations and to ensure they are conducted correctly. Without this information it would not be possible to carry out the heat loss calculation exercise.

Heating Services in Buildings: Design, Installation, Commissioning & Maintenance, First Edition. David E. Watkins.
© 2011 John Wiley & Sons, Ltd. Published 2011 by John Wiley & Sons, Ltd.
As an aid to lecturers and students, full colour versions of the figures in this chapter may be found at www.wiley.com/go/watkins
These figures are © 2011 by John Wiley & Sons, Ltd.

This information can be obtained from a number of sources depending upon whether the building is a proposed new development or an existing property or structure. For new buildings, much of the information can be obtained from the architect's drawings and specification; for existing buildings, the information can be obtained from as-built record drawings if they exist and/or a site survey. Other sources of information are consultations with the client to obtain their requirements, recognised engineering reference data and general observations.

The information required to permit the heat loss calculations to commence includes:

- The materials used, or to be used, in the construction of the building – including the location and exact physical size of such items as windows and doors. Also, the type and configuration of the roof and floor construction, together with details of free air openings such as air bricks, room ventilation openings and open fireplaces.
- The physical size, dimensions and configuration of each room including ceiling heights and window heights. Establish the thickness of walls: if they are existing walls, this can be measured at door or window openings. Determine the overall external dimensions of the building.
- The desired room temperatures and the external base design temperature: these may be obtained by consulting recognised published engineering reference sources and data, or by discussions with the client.
- The air infiltration rate expressed as the number of air changes per hour for each room to be heated: this may be ascertained by noting the size of any free openings such as fireplaces, air bricks and any mechanical ventilation that has been, or will be, installed. Also, for existing buildings, note the conditions of all the windows and doors regarding their air-tightness, if in poor condition it may be more economical to recommend that they are replaced, the replacement costs being offset by savings on the operating cost of the heating system.
- Location of the building with regard to its exposure to weather conditions – such as coastal, hilltop, valley, town or suburban sheltered by surrounding buildings. Also, by use of a compass or architect's drawing, establish the north aspect of the building and the prevailing wind. Obtain local knowledge of weather conditions from published meteorological data or reliable local sources.
- A Standard Assessment Procedure (SAP) rating may need to be undertaken at this stage: further explanation into SAP ratings is given later in this volume when discussing Building Regulations, the importance of energy efficiency and their effects on heating systems. At this stage we are only concerned with the assessment and calculation modus operandi to establish heat losses from buildings; the other important related subjects will be better understood when the heat loss calculation procedure has been mastered.

INTERNAL DESIGN TEMPERATURES

In order to accomplish comfort conditions in the heated space and enable heat losses to be calculated, it is necessary to select the internal design room temperatures that need to be achieved and conditionally guaranteed when the heating system has been commissioned and operating.

The temperature of a heated space in winter forms just one element of achieving comfort conditions for the occupants, along with humidity and ventilation rate. The temperature at which people will feel comfortable in any given situation will fall within a band, with some disagreement between individuals when subjected to the higher and lower ends of the range. Generally, for sedentary occupation and general rest/living type areas, a temperature band of 19°C to 23°C is considered by most people to be comfortable whilst wearing normal clothing, whereas for leisure activities or active occupations the temperature band would be between 13°C to 20°C, with the exact temperature depending upon the degree of physical activity involved.

Because of the variance in opinion regarding comfortable temperatures, the exact temperatures to be selected at the design stage should be discussed and agreed with the client – although if the client is also the permanent occupant of the dwelling as in the case of domestic properties, some form of guidance may

Table 7.1   Temperature effects on comfort and health for the elderly and infirm

Temperature	Effects
24°C	Top range of comfort conditions
21°C	Recommended living room temperature
20°C	Below 20°C – risk of death begins
18°C	Recommended bedroom temperature
16°C	Resistance to respiratory diseases weakened
12°C	More than two hours' exposure to this temperature raises blood pressure and increases risk of heart attack and strokes
5°C	Significant risk of hypothermia

Source – West Midlands Public Health Observatory, United Kingdom.

be required. If the client is a property developer or architect, design room temperatures may have been specified as part of the design brief or enquiry documents.

Table 7.1 lists the effects that certain temperatures can have on the elderly and infirm, illustrating the risk that this group of the community faces during the winter months.

When no specified design temperatures are available, Table 7.2 may be used as guidance for the recommended design room temperatures for different applications. It should be noted that the temperatures listed are for occupants in normal living or working environments. If the occupants are resident in a rest home – such as the elderly or infirm – then the design room temperatures should be increased by one or two degrees from those listed in the table for each given application.

It can also be observed in Table 7.2 that the temperatures recommended for dwellings vary for different types of rooms, depending upon their intended purpose: this is because human body temperature and comfort levels vary for diverse living such as light physical activities, sedentary activities, resting and sleeping – hence the variation in the recommended design room temperatures.

It has become the practice by some involved in the design and installation of heating systems in domestic residential dwellings to design the system to heat each room to the same temperature. This single temperature approach reduces the amount of calculations required as it uses the 'whole house' assessment method, only requiring the heat loss through the external elements of the structure to be calculated and ignoring any internal partition walls or floors, as there is no flow of heat from room to room. However, this practice can lead to over-sizing the heating system, resulting in uncomfortably high bedroom temperatures, and even though the control system could reduce these temperatures, this is not a very energy-efficient method of design.

EXTERNAL TEMPERATURE (BASE DESIGN TEMPERATURE)

The recommended room space temperatures previously discussed and selected for the situation required, are those that will have to be guaranteed and maintained by the heating installation during the operating periods of the system during the winter months. Any room temperature specified can only be guaranteed when the external temperature is known as this dictates the rate of heat flow from internal to external: the greater the temperature difference, the greater the temperature gradient, and if the external temperature is lower than that designed for then the internal room, temperatures will not be able to be achieved.

Therefore, to design a heating system that can maintain the specified internal room temperatures during the winter months, the calculations must be based on a realistic external temperature that can be expected during the winter months for the locality of the building.

The study of weather conditions, including high and low temperatures, shows that they can vary considerably throughout the world; indeed, they can vary from one part of the country to another. Restricting our interest to winter temperatures in the United Kingdom, records show that temperatures vary with the

Table 7.2   Recommended selected room temperatures and air infiltration rates

Building Type	Room	Temperature °C	Air Change Per Hour
Dwellings	Living/Dining Areas	21	1.5
	Bedrooms	18	1.0
	Kitchens	20	2.0
	Bathrooms	23	3.0
	Cloakrooms	18	2.0
	Halls/Landings	18	2.0
	Studies	21	1.5
	Bed Sitting Rooms	21	1.5
	Garages/Workshops	12	2.0
Hotels	Bedrooms	21	1.0
	Bathrooms	24	3.0
	Function Rooms	21	2.0
	Foyers	20	2.0
Offices	Private/General/ Open Plan	21	1.0
Public Houses/Bars		20–22	3.0–5.0
Churches		19–21	1.5
Schools	Day	19–21	1.5
	Boarding Living Areas	20–22	1.5
Factories	Sedentary Work	21	Air change will
	Light Work	16–19	depend upon type
	Heavy Work	11–14	of work
Hospitals	Wards	22–24	2.0
	Treatment Rooms	22–23	1.5
	Operation Theatres	17–19	As required
	Waiting Rooms	20–21	1.5
Libraries	Lending/Reference	19	1.0
	Reading Areas	21	1.5
Museums/Art Galleries	Display	20	Dictated by artefacts
	Storage	18	
Shops	Shopping Malls	19–21	2.0
	Retail Shops	20	2.0–3.0
Restaurants		20–23	2.0
Leisure Centres	Games Halls	12–20	6.0–10.0
	Swimming	23–26	10.0
	Changing Rooms	22–24	6.0–10.0

Note – Where mechanical ventilation is involved, then the air change rate used for the ventilation should be adopted as the air change rate used for the heat loss calculation.

Air change rates listed assume that smoking is not permitted.

If large open fireplaces exist, or are to be incorporated, then the air change rate to be used in any heat loss calculations should be increased accordingly.

Air change rates are dependent upon occupancy and usually expressed in litres per second of air per person which should take precedence.

Table 7.3 Recommended UK base design temperatures

Type of Building	Exposure	Base Design Temperature
House & multi-storey buildings with solid intermediate floors up to and including 4th floor: England & Wales	Normal, sheltered in town & city centres surrounded by other buildings	−1°C
House & multi-storey buildings with solid intermediate floors up to and including 4th floor: Scotland, Northern England & Northern Ireland	Normal, sheltered in town & city centres surrounded by other buildings	−3°C
Single storey buildings	Normal	−3°C
Houses in coastal or high altitude areas including exposed rural areas	Exposed	−4°C
Multi-storey buildings above 4th floor and single storey buildings in coastal or exposed and high altitude areas	Exposed	−5°C

north of Scotland experiencing the coldest and more prolonged cold temperatures, and the south-west of England being the mildest, but it is not unusual for any location to experience extremely cold or extremely mild temperatures during the winter months. For further information and details of weather conditions in the UK, or any other part of the world, refer to Meteorological Office records and data, or CIBSE Guides 'A' and 'J'.

In the United Kingdom it is normal to encounter temperatures of below minus degrees Celsius in all parts of the country during the winter months, with −2°C to −6°C being usual, however, during severe winters it is not unusual for temperatures to fall to −15°C or lower.

It would be uneconomical to design a heating system capable of maintaining the desired room temperatures during these extreme weather conditions as they occur infrequently, plus the heating system would be over-sized, which would cause the boiler to 'hunt' on its thermostat and thus shorten its working life. Therefore a more acceptable base design temperature must be used, based on external temperatures that are normal for the building's locality.

Table 7.3 lists the recommended base design temperatures for use in the United Kingdom, but local knowledge should, in some situations, be applied as exposed type conditions are not always obvious.

Table 7.3 gives different base design temperatures for sheltered locations and severe exposed locations. Generally, normal sheltered exposures are for low rise buildings surrounded in close proximity by other buildings, such as in towns and cities including the inner suburbs of cities. Exposed conditions are for high or low rise buildings located in coastal areas, at a high altitude including hill top locations and general rural locations.

The British Isles enjoys a milder winter climate than that experienced by other Northern European countries, where the base design temperature will be lower than that used in the UK, and although the heat loss assessment process will be the same, the base design temperature used will be different. For further information on weather and climatic conditions for other countries, refer to the CIBSE Guide Book 'A' or the ASHRAE Handbook of Fundamentals.

THERMAL PROPERTIES OF BUILDINGS

To enable the heat loss from a building or structure to be calculated, it is necessary to know what the thermal properties of the building are. New buildings today have to comply with Part 'L' of the Building Regulations, which gives minimum thermal requirements for each element of the building and stipulates maximum 'U' values that are permitted, therefore information on the materials to be used in constructing new buildings

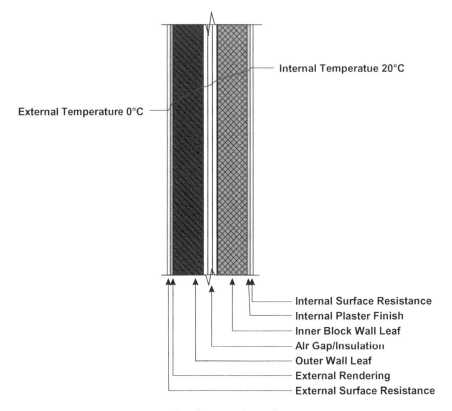

External Temperature 0°C

Internal Temperatue 20°C

Internal Surface Resistance
Internal Plaster Finish
Inner Block Wall Leaf
Air Gap/Insulation
Outer Wall Leaf
External Rendering
External Surface Resistance

Figure 7.1   Example of 'U' value composition for a cavity wall

is easily obtained. Further information on this together with an explanation on SAP ratings is included in Chapter 24, Regulations, Standards, Codes and Guidelines.

When designing a heating system for an existing building, the materials that were used in the construction of that building have to be established by surveying the building so that the 'U' values for these materials can be ascertained. In the case of either a new building or existing structure, the 'U' values for the construction materials and their combined construction elements must be accurately established.

'U' VALUES (THERMAL TRANSMITTANCE COEFFICIENT)

The thermal transmittance coefficient, or 'U' value, is a measurement of the thermal resistance of different building materials and is expressed as the heat transmitted in watts per square metre of material per degree Celsius ($W\,m^{-2}\,°C^{-1}$). For the purposes of the heat loss calculation, it is termed the 'transmittance coefficient U', or 'U' value for short. It is expressed as a number for various structural and non-structural building materials and is obtained by the reciprocal (division of unity by quantity) of the sum of all the thermal resistances of the individual material components that make up the numerous structural compositions. Buildings are constructed from many materials brought together to form different elements of the structure such as a simple cavity wall as depicted in Figure 7.1. The wall is constructed from an inner block material which has an internal plaster finish, an insulated space between the inner block and an outer wall that may be either a brick finish or a block, with an external rendered finish. Each of these individual components are joined together to form the wall and the 'U' value for the wall is the measurement of its resistance to the flow of heat through that composition that has been obtained by the sum of the individual thermal resistances of the materials that form the wall.

This may be calculated by the following equation.

$$U = \frac{1}{Rsi + R1 + R2 + Ra + R3 + R4 + Rso}$$

Whereby:

U	'U' value thermal transmittance	$W\,m^{-2}\,°C^{-1}$
Rsi	Internal surface resistance	$m^2\,°C\,W^{-1}$
R1, R2, R3, R4 etc	Thermal resistance of individual structural materials	$m^2\,°C\,W^{-1}$
Ra	Airspace/insulation resistance	$m^2\,°C\,W^{-1}$
Rso	External surface resistance	$m^2\,°C\,W^{-1}$

The values for the surface resistances R1, R2 etc will vary with the degree of surface roughness and the thickness of the individual material component.

The surface resistance encountered by the flow of heat through the wall resembles an invisible film on both the internal and external wall surfaces, and will vary with the degree of roughness and the air movement over the surface. This is illustrated in Figure 7.1.

Fortunately, for most common building structural compositions, the 'U' value has previously been calculated and can be obtained from recognised engineering data published by professional bodies, such as the CIBSE Guide 'A', the CIPHE Engineering Design Guide, and the Domestic Heating Guide. For new building developments the structure has to have minimum 'U' values conforming to Building Regulations Part 'L'.

Tables 7.4, 7.5, 7.6 and 7.7 list a number of common building material compositions with their 'U' values. The list is not intended to be exhaustive and the reader should refer to the publications previously mentioned for more detailed lists.

It should be noted that the recognised way of expressing 'U' values is $W/m^2\,K$, which translates as 'watts per square metre of building composition material for every one degree Kelvin temperature difference'. As one degree Kelvin is the same as one degree Celsius, this is the same as stating the 'U' value as $W\,m^{-2}\,°C^{-1}$, which means 'watts per square metre of building composition material for every one degree Celsius temperature difference'. As degrees Celsius are more familiar to students studying heating, the 'U' values used in this text are expressed as °C.

The 'U' values listed for solid ground floor constructions are average as the 'U' value will vary with the physical size of the floor: the greater the floor area, the lower the 'U' value. This difference is due to the insulating properties of the compacted earth beneath the floor construction, as the heat is lost mainly around the edge of the floor slab – hence the larger the floor area, the greater the insulated area.

The 'U' values for floors are also based on the assumption that there are two or three exposed edges to the floor: the value will change if there are more or fewer exposed edges.

The 'U' values given for double and triple glazed windows are for hermetically sealed units and should not be selected for secondary glazed windows, where the 'U' value is much higher.

It should be noted that there are two 'U' values given for intermediate floors depending upon whether the flow of heat is upwards to the room above through the ceiling, or downwards to the room below through the floor.

AIR INFILTRATION

The transmission of heat considered so far has been through different elements of physical building materials compositions, but there is also an additional heat loss which frequently accounts for a third or more of the total heat loss from the building. This is caused by air entering the building from the outside. This infiltration of air is due to the movement of air over and through the building, and the stack effect created

Table 7.4 Thermal transmittance coefficients ('U' values)

CONSTRUCTION MATERIAL OR FABRIC		'U' Value W/m²/°C	
External Walls			
Brickwork, Solid, Unplastered	102 mm	3.35	
Brickwork, Solid, Unplastered	228 mm	2.39	
Brickwork, Solid, Unplastered	342 mm	1.7	
Brickwork, Solid, Internally Plastered	102 mm	2.97	
Brickwork, Solid, Internally Plastered	228 mm	2.1	
Brickwork, Solid, Internally Plastered	343 mm	1.65	
Brickwork, Solid, Internally Lined Plasterboard	102 mm	2.84	
Brickwork, Solid, Internally Lined Plasterboard	228 mm	2.05	
Brickwork, Solid, Internally Lined Plasterboard	343 mm	1.6	
Solid Performance Block, Externally Rendered, Internally Plastered	215 mm	0.44	
Solid Performance Block, Externally Tiled, Internally Plastered	215 mm	0.43	
Solid Stone Wall Unplastered	305 mm	2.78	
Solid Stone Wall Unplastered	457 mm	2.23	
Solid Stone Wall Unplastered	610 mm	1.68	
Solid Concrete Wall, Dense Plaster	102 mm	3.51	
Solid Concrete Wall, Dense Plaster	152 mm	3.12	
Solid Concrete Wall, Dense Plaster	205 mm	2.82	
Solid Concrete Wall, Dense Plaster	255 mm	2.55	
		Uninsulated	**50 Insulation**
Cavity Wall, Both Leaves 102 mm Brick, Internally Plastered	254 mm	1.37	0.56
Cavity Wall, Both Leaves 102 mm Brick, Internally Lined Plastered	254 mm	1.23	0.54
Cavity Wall, Both Leaves 102 mm Brick, Internally Plastered Externally Rendered		1.25	0.53
Cavity Wall, Both Leaves 102 mm Brick, Internally Lined Plastered Externally Rendered		1.11	0.51
	Block Thickness	**100 mm**	**125 mm**
Cavity Wall, 102 mm Outer Brick, Inner Aerated Block, Plastered		0.87	0.77
Cavity Wall, 102 mm Outer Brick, Inner Aerated Block, Plasterboard		0.8	0.72
Cavity Wall, 102 mm Outer Brick, Inner Aerated Block, Plastered, Insulated		0.45	0.42
Cavity Wall, 102 mm Outer Brick, Inner Aerated Block, Plasterboard Insulated		0.43	0.41
Cavity Wall, 102 mm Outer Brick, Inner High Performance Block, Plastered		0.68	0.59
Cavity Wall, 102 mm Outer Brick, Inner High Performance Block, Plasterboard		0.64	0.56
Cavity Wall, 102 mm Outer Brick, Inner High Performance Block, Plastered, Insulated		0.39	0.36
Cavity Wall, 102 mm Outer Brick, Inner High Performance Block, Plasterboard Insulated		0.38	0.35

by the temperature difference between inside the building and the external conditions. The air generally enters the building from outside through the lack of seal in the building structure via natural openings, such as air bricks, ventilators, chimneys and flues. If the building has an open flue the warm lighter air created by the heating system will be forced out of the building by the colder and heavier air entering from outside, sometimes at an excessive rate. This is known as the stack effect and is illustrated in Figure 7.2. If

Table 7.5   Thermal transmittance coefficients ('U' values)

CONSTRUCTION MATERIAL OR FABRIC		'U' Value W/m²/°C	
External Walls continued			
	Block Thickness	**100 mm**	**125 mm**
Cavity Wall, 100 mm Outer Block Rendered, Inner High Performance Block Plastered, Insulated Cavity 50 mm		0.33	0.31
Cavity Wall, 100 mm Outer Block Tiled, Inner High Performance Block Plastered, Insulated Cavity 50 mm		0.36	0.34
Internal Partition Walls			
Brickwork Plastered Both Sides	102 mm	3.6	
Brickwork Plastered Both Sides	228 mm	2.3	
Lightweight Block, Plastered Both Sides	80 mm	2.6	
Lightweight Block, Plastered Both Sides	100 mm	2.3	
Timber Studding, Plasterboard 12.5 mm, 75 mm Air Gap		1.72	
Clinkered Breeze Block, 1930s Properties and Earlier	100 mm	1.58	
Doors			
Solid Timber Doors, 25 mm Thick		2.6	
Solid Timber Doors, 30 mm Thick		2.4	
Glazed Doors: As Windows			
Windows & Glazed Doors			
Single Glazed, Timber or PVC-U Frames		5	
Double Glazed Hermetically Sealed, Timber or PVC-U Frames		2.8	
Triple Glazed Hermetically Sealed, Timber or PVC-U Frames		2.1	
Single Glazed, Metal Frame		6	
Double Glazed Hermetically Sealed, Metal Frame		3.6	
Triple Glazed Hermetically Sealed, Metal Frame		2.6	
Roof Lights To Be Taken As Windows			

the heating system is to be installed in an older existing building, the air infiltration rate will be higher if the windows and doors are in poor repair, and it would prove beneficial to increase the insulation of the building as well as to replace windows and doors.

The air infiltration rate is difficult to assess and its indeterminacy renders any temperature guarantees for buildings difficult to sustain, as any test is more of a test of the building's condition rather than of the heating system.

The air change figures listed in Table 7.2 have been arrived at by empirical means and may be used with confidence for well maintained buildings. For buildings that are less well maintained, the air change rates can be expected to be higher than those listed.

The amount of heat for every m³ of infiltration air may be assessed using the specific heat of air capacity at a constant pressure and at a temperature of 20°C, which equals $1.01 \, \text{kJ kg}^{-1}\,^{\circ}\text{C}^{-1}$, and the mass per m³ which equals 1.2 kg.

Table 7.6   Thermal transmittance coefficients ('U' values)

CONSTRUCTION MATERIAL OR FABRIC	'U' Value W/m²/°C				
Solid Floors in Contact With Earth Average Values	**Insulation Thickness 0.04W/mK**				
	0	25	50	75	100
Solid concrete ground floor with floor screed, 25mm thick edge insulation and tile or block floor finish. Two/three edges of floor slab exposed					
Width & Length of Floor Area					
3m × 4m	1.15	0.65	0.45	0.36	0.27
3m × 6m	1.05	0.6	0.43	0.33	0.26
3m × 8m	1	0.58	0.4	0.33	0.25
3m × 10m	0.96	0.56	0.4	0.3	0.25
4m × 6m	0.95	0.56	0.4	0.3	0.25
4m × 10m	0.85	0.53	0.38	0.3	0.24
5m × 7m	0.81	0.52	0.37	0.29	0.24
5m × 10m	0.74	0.5	0.35	0.29	0.23
6m × 8m	0.71	0.48	0.35	0.28	0.23
6m × 10m	0.65	0.45	0.33	0.28	0.22
Suspended Ventilated Ground Floors Average Values	**0**	**25**	**50**	**75**	**100**
Suspended timber ventilated floor on joists with insulation slabs laid between joists on polypropylene net. Two/three edges of floor construction exposed. Unheated underside					
Width & Length of Floor Area					
3m × 4m	1.2	0.68	0.48	0.38	0.32
3m × 6m	1.1	0.65	0.46	0.35	0.3
3m × 8m	1	0.6	0.45	0.35	0.3
3m × 10m	0.98	0.6	0.43	0.34	0.28
4m × 6m	0.95	0.58	0.4	0.34	0.27
4m × 10m	0.87	0.57	0.38	0.33	0.26
5m × 7m	0.83	0.55	0.38	0.32	0.25
5m × 10m	0.8	0.53	0.37	0.32	0.25
6m × 8m	0.75	0.52	0.36	0.3	0.24
6m × 10m	0.72	0.51	0.36	0.3	0.24
Intermediate Floors	**Upward**		**Downward**		
Timber Floor on Joists, Plaster Ceiling	1.7		1.5		
Timber Floor on Joists, Plaster Ceiling – 100mm Insulation	0.36		0.32		
Timber Floor on Joists, Plaster Ceiling – 150mm Insulation	0.25		0.23		
Concrete Slab with 50mm Floor Screed	2.7		2.2		
Hollow Tile & Concrete 150mm	2		1.7		
Hollow Tile & Concrete 200mm	1.9		1.6		
Hollow Tile & Concrete 250mm	1.8		1.5		

Therefore, the heat required to raise $1\,m^3$ of air by $1°C = 1.01 \times 1.2 = 1.212\,kJ$.

As the SI system of measurement uses seconds as a unit of time, this must be converted to hours so as to obtain more practical and sensible figures.

Therefore:

$$\text{Volume } m^3 \times \frac{1.212 \times 1000}{3600} = 0.33\,J/s\,m^3\,°C, \text{ or } W/m^3\,°C$$

Therefore the 'U' value thermal coefficient for the air change rate in the heat loss calculation may be taken as $0.33\,W/m^3\,°C$, which may be regarded as a constant for heat loss calculations.

Table 7.7   Thermal transmittance coefficients ('U' values)

CONSTRUCTION MATERIAL OR FABRIC	'U' Value W/m²/°C
Roofs	
Pitched Roof, Tiles or Slates on Battens, Sarking Felt, Ventilated Roof Space Plasterboard – No Insulation	2.2
Pitched Roof, Tiles or Slates on Battens, Sarking Felt, Ventilated Roof Space Plasterboard – 100 mm Insulation	0.34
Pitched Roof, Tiles or Slates on Battens, Sarking Felt, Ventilated Roof Space Plasterboard – 150 mm Insulation	0.24
Pitched Roof, Tiles or Slates on Battens, Sarking Felt, Ventilated Roof Space Plasterboard – 200 mm Insulation	0.19
Pitched Roof, Tiles or Slates on Battens, No Felt, Ventilated Roof Space Plasterboard – No Insulation	3.3
Pitched Roof, Tiles or Slates on Battens, No Felt, Ventilated Roof Space Plasterboard – 100 mm Insulation	0.38
Pitched Roof, Tiles or Slates on Battens, No Felt, Ventilated Roof Space Plasterboard – 150 mm Insulation	0.26
Pitched Roof, Tiles or Slates on Battens, Sarking Felt, Air Space, Plasterboard on Underside of Pitched Rafters – No Insulation	2.1
Pitched Roof, Tiles or Slates on Battens, Sarking Felt, Air Space, Plasterboard on Underside of Pitched Rafters – 100 mm Insulation	0.33
Pitched Roof, Tiles or Slates on Battens, Sarking Felt, Air Space, Plasterboard on Underside of Pitched Rafters – 150 mm Insulation	0.23
Flat Concrete Roof 150 mm, Waterproof Covering, Screed, Plaster	2.1
Flat Concrete Roof 150 mm, Waterproof Covering, Screed, Plaster, 35 mm Polyurethane Insulation, Vapour Barrier	0.55
Flat Concrete Roof 150 mm, Waterproof Covering, Screed, Plaster, 100 mm Polyurethane Insulation, Vapour Barrier	0.25
Flat Concrete Roof 150 mm, Waterproof Covering, Screed, Plaster, 150 mm Polyurethane Insulation, Vapour Barrier	0.18
Flat Timber Roof, 3 Layers of Felt, Air Space, Plasterboard – No Insulation	1.64
Flat Timber Roof, 3 Layers of Felt, Air Space, Plasterboard, Vapour Barrier – No Insulation	1.87
Flat Timber Roof, 3 Layers of Felt, Air Space, Plasterboard, Vapour Barrier – 50 mm Insulation	0.52
Flat Timber Roof, 3 Layers of Felt, Air Space, Plasterboard, Vapour Barrier – 100 mm Insulation	0.33
Flat Timber Roof, 3 Layers of Felt, Air Space, Plasterboard, Vapour Barrier – 150 mm Insulation	0.24
Flat Timber Roof, 3 Layers of Felt, Air Space, Plasterboard, Vapour Barrier – 200 mm Insulation	0.16

Alternatively, the figures in Table 7.8 may be used for calculating the heat requirements for the air change rates.

Figure 7.2 illustrates the effect that wind pressure can have on a poorly maintained building in so far as there will be an entry of air into the building on the windward side, but a loss of warm air on the leeward side. This is caused by positive and negative pressure effects on the sides of the building, the degree of which is dependent upon the strength of the wind pressure and the angle of the wind direction in relation to the building. This subject is explored further in Chapter 17, Combustion Efficiency Testing.

Table 7.8   'U' values for air change infiltration rates

Number of Air Changes/Hour	'U' Value W/m³/°C
½	0.16
1	**0.33***
1½	0.49
2	0.66
3	0.99
4	1.32
6	1.98
10	3.3

*'U' value for infiltration air.

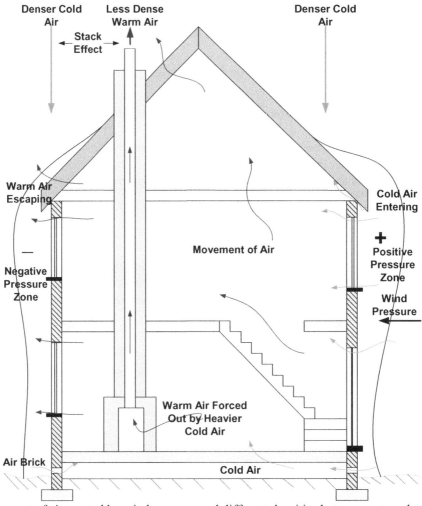

Movement of air created by wind pressure and different densities between external cold heavier air and internal warmer lighter air: stack effect.

Figure 7.2   Movement of air contributing to infiltration

Figure 7.2 also illustrates the stack effect caused by differentials in pressure, the colder and denser external air forcing entry into the building and ejecting the warmer and less dense air through natural openings. The air infiltration rate caused by this stack effect will occur in the same way in a well maintained building as for a poorly maintained building, as all buildings require ventilation in order for humans to function normally.

The stack effect can be best visualised by imagining the internal warmer air being less dense, and therefore lighter, or at a lower pressure than the external colder more dense heavier air, which would be at a higher pressure. Therefore the external higher pressure air will want to try and enter the building to stabilise the pressure difference.

HEAT GAINS

Unless the heat gains to the building or any part of the building are significant and can be relied upon to be constant, they are normally ignored so far as the heat loss assessment calculations are concerned. However, knowledge of any potential heat gain is desirable when specifying the mode of control for the heating system, although it is not taken into account when assessing the heat loss.

Fortuitous heat may be obtained from a number of sources including the following:

- Solar heat gains
- Heat from occupants
- Heat from lighting and illuminations
- Heat from machines
- Heat from electrical appliances

The heat obtained from any of these sources is usually too small and insignificant to have any real contributory effect on the heating system for most applications, with a few exceptions. Therefore, unless these heat gains are considerable they are ignored for the purpose of the heat loss calculation, but should be noted and considered when designing the heating controls system to take full advantage of making possible savings on energy.

Solar gains

Solar heat gains, not to be confused with solar heating, can contribute large quantities of heat during the warm seasons at certain times, especially through glazed elements. However, as sunshine cannot be guaranteed we cannot include this heat gain in our heat loss calculations.

The amount of heat received from the sun through windows and other glazed elements can give an almost instant effect as there is no time-lag involved with glass, even when the external ambient temperature is below zero degrees Celsius. Therefore it is worth considering creating separate heating zones for buildings with large expanses of glass – particularly southerly aspects and conservatories – so as to take advantage of this heat gain when it occurs, by thermostatically turning the heating off in these areas whilst the remainder of the building is still being heated.

Occupants

The human body will emit heat, the amount of which is dependent upon the physical activity involved and the surrounding air temperature.

Table 7.9 states heat emission values for various activities based on an ambient temperature of 20°C.

The colder the ambient air temperature, the greater the amount of heat emission from the human body. Subsequently, the higher the ambient air temperature, the lower the heat emission until the air temperature reaches 37°C or body temperature, when there will be no heat emission.

Table 7.9   Heat emissions from occupants

Activity	Heat Emission/Occupant (W)
Resting	115
Sedentary Worker	140
Walking Slowly	160
Light Manual Work	235
Medium Manual Work	330
Heavy Manual Work	440

This aspect is especiallly important when designing for high occupancy rooms, such as meeting rooms or conference rooms, where individual control of this area will be required.

Lighting

All electrical lighting will produce heat relative to the amount of energy consumed, the rate of heat emission being dependent upon the type of lamp used. Standard tungsten lamps will emit the most heat – approximately four times as much as fluorescent tubes – however, unless there is a high intensity of lighting, the quantity of heat dissipated into the room will be negligible and can be ignored for the purposes of the heat loss calculations.

Machines and electrical appliances

All machines and electrical appliances will convert a certain amount of energy consumed by the work done into heat. For domestic residential dwellings this heat gain may be ignored as the heat produced would be considered insignificant. For industrial applications the heat produced from machinery may be quite large and the heating control system specified should take this into account. Likewise, in commercial buildings the increase in computer equipment is increasing all the time, with the heat being emitted making a reasonable contribution to the heating requirements, although the heating system should be capable of satisfying the heat load without this contribution and the heating control system should be arranged accordingly. It should be noted that future of computer equipment is to be developed with more efficient appliances that will emit less heat.

BUILDING TIME-LAG

The time-lag of a building has no direct consequence on the heat loss calculation, but an explanation of this phenomenon has been included as it is important to have an understanding of this matter because it may influence the selection of the type of heating system.

The structural material of the building has to be warmed up before any effect or increase in the internal room air temperature is noticed. The heavier the construction of the building, the longer the time-lag is between the heating system being turned on and an increase in the room temperature being observed.

Buildings having a very heavy construction with thick walls and negligible window areas, can have a time-lag as long as a week. The opposite form of construction, such as a lightly constructed structure or a hut with a relatively large window area, will have an almost instant response by way of a temperature rise when the heating system has been turned on, with hardly any time-lag at all. But it should also be appreciated that a building that has a long heat-up time-lag will also have an equally long time-lag when cooling down, and in the case of the lightly constructed building or hut, the temperature drop will be noticed almost immediately when the heating system is turned off.

The time-lag of a building should be taken into account when considering the form and method of operation intended for the heating system.

HEAT LOSS CALCULATIONS

Methods of calculation

Having considered the previous information in this chapter, the heat losses from a building may now be calculated. There are a number of methods in use for undertaking this task, all of which have a purpose if used correctly. These methods are:

- Cube x number
- Whole house boiler sizing method
- Slide rule
- Worked engineering calculation

The cube x number method is purely a 'rule of thumb' method as the result is not accurate, but it can serve a useful quick and inexpensive purpose in assessing budget figures for heating loads, particularly when used for estimating prices for domestic heating systems. It is explained further towards the end of this chapter.

The whole house boiler sizing method is a method used in the energy efficiency procedure to assess boiler replacement heating capacity. It should not be used in the design of a heating system. The author recommends that an engineering calculation is conducted to assess boiler capacity when the boiler needs to be replaced.

The slide rule method has been in existence for many years and is one of many in a series of similar slide rule calculators; it is often used to assess heat loss values in the domestic heating sector of the heating industry, as it is simpler and faster to use than the worked engineering calculation method, but it should be used with caution.

The slide rule calculator consists of a plastic circular disc (see Figure 7.3) and enables examples to be worked through using set conditions, but it is the engineer's responsibility to ensure that those set conditions are correct for the particular project in question.

The slide rule calculator is a useful tool for providing fairly accurate and expeditious estimates of heating loads, heat emitter sizes and pipe sizes. It is particularly useful for the small contracting company in the domestic heating market, but if the tender is accepted then a full worked heat loss calculation should be undertaken before any work commences.

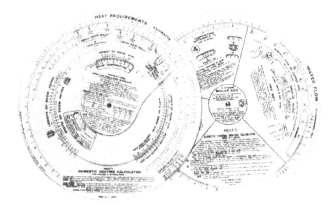

Figure 7.3   Slide rule heating calculator. Reproduced by kind permission of M. H. Mear & Co. Ltd.

WORKED ENGINEERING HEAT LOSS CALCULATION

To determine the heat losses from a building, the only fully accurate method that is recognised and approved by industry is the worked engineering heat loss calculation explained here.

The heat loss calculation is comparatively simple but can be somewhat laborious for a large development, although one of a number of computer programs available can make this task easier. For single domestic dwellings, all that is required is an electronic calculator.

Figures 7.4 and 7.5 illustrate a typical 1930s semi-detached house that is used for the example heat loss calculation, where most dimensions have been conveniently made to equal whole metres to simplify the mathematics.

It will be noticed that the ground floor has been shown as being part solid concrete floor and part ventilated timber floor. This is not a typical construction but has been included to demonstrate the difference in heat loss between the two floor structures.

The adjoining semi-detached property may be known to have, or not to have, its own heating system; either way it cannot be guaranteed to be operating at the same time as our example heating system so we must provide a heating system that is capable of heating our example property when the adjoining dwelling is unoccupied, with no heating on. There have been many opinions expressed regarding the temperature that should be selected for adjoining properties, but it is normally recommended to regard it the same as the external temperature, to ensure that the heating installation is capable of achieving the desired room temperatures selected.

The first step in our calculation procedure is to obtain the information discussed earlier in this chapter, and this is presented in Box 7.1.

The example calculations are shown recorded on a standard heat loss calculation sheet. This enables the calculation to be worked through in a logical manner, recording each step and permitting the calculation to be checked, or fully understood by any third party.

The calculation sheet (Table 7.10) is arranged in such a way that each step of the calculation is recorded. This makes mistakes less likely as a blank space on the calculation sheet make it obvious if a step has been missed out.

The following illustration serves to explain the calculation sheet procedure, showing which column is multiplied by which to equal what column.

EXPLANATION OF EXAMPLE CALCULATION

The dwelling depicted in Figures 7.4 and 7.5 has been arranged to include a number of details designed to make the student think in terms of engineering rather than of the building's occupant.

Separate calculations are made for each wall, window, door, floor, ceiling, roof and air infiltration rate. There are no heat losses through internal partition walls where the temperature in the adjoining room is

Table 7.10 Calculation sheet procedure

A	B	C	D	E	F	G	H	I	J
Material or Room	Dimensions (m)			Cu or Sq m	Air Change	U Value Coeff	W/°C	Temp Diff °C	Heat Loss Watts
	L	W	H						
	5	5	3	75	2	0.6	90	22	1980
				Column B × C × D = E × F × G = H × I = J					

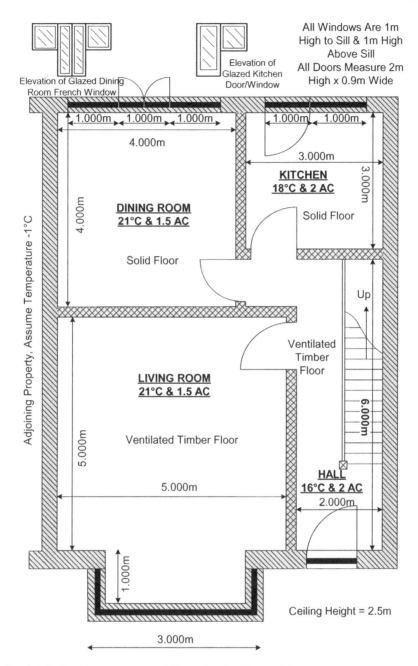

Figure 7.4   Semi-detached house – ground floor plan (not to scale)

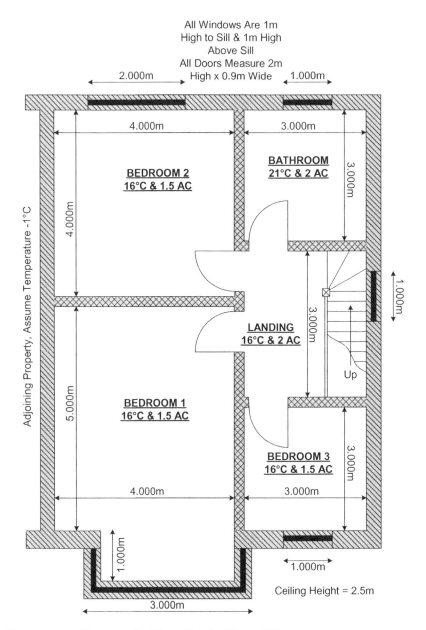

Figure 7.5   Semi-detached house – first floor plan (not to scale)

BOX 7.1   Calculation data

External temperature selected. Room temperatures and air change rates shown on drawings	−1°C.
All external walls and wall between adjoining property, solid brickwork 228 mm internally plastered	U = 2.1 W/m² K
All doors being solid timber 25 mm thick except kitchen external door & dining room French windows	U = 2.6 W/m² K
All intermediate internal partition walls 100 mm thick lightweight blocks plastered both sides	U = 2.3 W/m² K
Ground floors – solid concrete 75 mm insulation	U = 0.36 W/m² K
Ventilated timber floor 25 mm insulation	U = 0.68 W/m² K
Intermediate timber floors no insulation Upward heat transmission	U = 1.7 W/m² K
Downward heat transmission	U = 1.5 W/m² K
Windows & glazed doors, single glazed, timber frames	U = 5.0 W/m² K
Pitched roof with tiles on battens, roof space and 100 mm insulation, no felt	U = 0.38 W/m² K
Thermal coefficient of air	U = 0.33 W/m³ K

the same or higher. The same applies to intermediate floors regarding the room above or below. External walls in corner rooms are added together to make one single external wall, with just one calculation.

The wall surface area is calculated in square metres with the total surface area occupied by doors and windows subtracted before calculating the heat loss through the wall area. The window and door areas must be calculated individually as their 'U' value will be different.

The wall between any adjoining properties is considered to be the same as an external wall, although for large wall areas in non-domestic dwellings the 'U' value may be slightly lower.

Ceiling and floor areas are also squared up, but the 'U' value for upward heat transmission through ceilings is different to that for a downward heat transmission for an intermediate floor.

Heat loss calculations are only conducted through a building element when the temperature on the other side of the structural element is intended to be lower than in the room being calculated.

The total heat loss for the example living room area amounts to 3299.38 watts, rounded up to 3300 watts, or 3.3 kW. This means that 3.3 kW of heat are required to maintain that room at 21°C when the external temperature is at −1°C.

When calculating the heat loss through a pitched roof, the ceiling area or floor plan area is calculated in square metres, as the 'U' value selected incorporates the roof construction including roof void, insulation and angle of pitch.

Heat gains from adjoining rooms that are at higher temperatures are ignored for the purposes of the heat loss calculation, but are considered when selecting the system control scheme.

Generally, each construction element should be included on the heat loss calculation sheet even if there is no heat loss through it so as to demonstrate that it has not been forgotten. The calculation is conducted by multiplying the L × W × H of the element to = cubic or square metres × the air change where applicable × 'U' value = W/°C × temperature difference = heat loss in watts.

On completion, the total heat loss for each room should be added together to obtain the building's total heat loss and enable the boiler rating to be obtained.

Charts for each room are given in Figure 7.6 and the data are summarised in Table 7.11.

The heat losses calculated for the dwelling totalled 13.6 kW, which will be the amount of heat required to be put into the building to maintain the desired room temperatures when the external temperature is not lower than −1°C.

| Heat Loss Calculation Sheet | | | | | | | | | | Sheet _____ 1 Of _____ 9 |

Project   Heat Loss Example Calculation   Contract No. ▓▓▓▓▓▓▓▓▓▓ Date _____

Room   Living Room   Temperature   21°C & 1.5 AC

    Engineer's Name   ▓▓▓▓▓▓▓▓▓▓▓▓▓▓▓▓▓▓▓   ▓▓▓▓▓▓▓▓▓

Material or Room	Dimensions (M)			Cu or Sq M	Air Change	U Value Coeff	W/°C	Temp Diff °C	Heat Loss Watts	Comments
	L	W	H							
Wall Btwn Dining Rm										No Loss
Wall Between Hall	6		2.5	15						
Less Door		0.9	2	1.8						
Wall				13.2		2.3	30.36	5	151.8	
Door				1.8		2.6	4.68	5	23.4	
Wall Between Houses	5		2.5	12.5		2.1	26.25	22	577.5	
External Wall	7		2.5	17.5						
Less Window	5		1	5						
Wall				12.5		2.1	26.25	22	577.5	
Window				5		5	25	22	550	
Ceiling	5	5		25						
	3	1		3						
				28		1.7	47.6	5	238	
Floor				28		0.68	19.04	22	418.88	
Infiltration	5	5	2.5	62.5						
	3	1	2.5	7.5						
				70	1.5	0.33	34.65	22	762.3	
									3299.38	
									Total Say	3300 Watts

Figure 7.6   Example heat loss calculation

Heat Loss Calculation Sheet

Sheet _____ 2 Of _____ 9

Project Heat Loss Example Calculation _____ Contract No. _____ Date _____

Room Dining Room _____ Temperature 21°C & 1.5 AC _____

Engineer's Name

Material or Room	Dimensions (M)			Cu or Sq M	Air Change	U Value Coeff	W/°C	Temp Diff °C	Heat Loss Watts	Comments
	L	W	H							
Wall Btwn Living Rm										No Loss
Wall Between Houses	4		2.5	10		2.1	21	22	462	
External Wall	4		2.5	10						
Less Window/Door				4						
Wall				6		2.1	12.6	22	277.2	
Window/Door				4		5	20	22	440	
Wall Between Kitchen	3		2.5	7.5		2.3	17.25	3	51.75	
Wall Between Hall	1		2.5	2.5						
Less Door		0.9	2	1.8						
Wall				0.7		2.3	1.61	5	8.05	
Door				1.8		2.6	4.68	5	23.4	
Ceiling	4	4		16		1.7	27.2	5	136	
Floor				16		0.36	5.76	22	126.72	
Infiltration	4	4	2.5	40	1.5	0.33	19.8	22	435.6	
									1960.72	
									Total Say	1961 Watts

Figure 7.6 (*Continued*)

Heat Loss Calculation Sheet

Sheet _____ 3 Of _____ 9

Project   Heat Loss Example Calculation _____   Contract No. _____   Date _____

Room   Kitchen _____   Temperature   21°C & 2AC _____

Engineer's Name

Material or Room	Dimensions (M)			Cu or Sq M	Air Change	U Value Coeff	W/°C	Temp Diff °C	Heat Loss Watts	Comments
	L	W	H							
Wall Btwn Dining Rm										No Loss
Wall Between Hall	3		2.5	7.5						
Less Door		0.9	2	1.8						
				5.7		2.3	13.11	2	26.22	
Door				1.8		2.6	4.68	2	9.36	
External Wall	6		2.5	15						
Less Window/Door				3						
Wall				12		2.1	25.2	19	478.8	
Window/Door				3		5	15	19	285	
Ceiling										No Loss
Floor	3	3		9		0.36	3.24	19	61.56	
Infiltration	3	3	2.5	22.5	2	0.33	14.85	19	282.15	
									1143.09	
									Total Say	1144 Watts

Figure 7.6   (*Continued*)

Heat Loss Calculation Sheet

Sheet _____ 4 Of _____ 9

Project   Heat Loss Example Calculation   Contract No. _____ Date _____

Room   Hall _____ Temperature   16°C & 2AC _____

Engineer's Name

Material or Room	Dimensions (M)			Cu or Sq M	Air Change	U Value Coeff	W/°C	Temp Diff °C	Heat Loss Watts	Comments
	L	W	H							
Wall Between Kitchen										No Loss
Wall Btwn Dining Rm										No Loss
Wall Btwn Living Rm										No Loss
External Wall	8		2.5	20						
Less Door		0.9	2	1.8						
Wall				18.2		2.1	38.22	17	649.74	
Door				1.8		2.6	4.68	17	79.56	
Ceiling										No Loss
Floor	6	2		12						
	1	1		1						
				13		0.68	8.84	17	150.28	
Infiltration	6	2	2.5	30						
	1	1	2.5	2.5						
				32.5	2	0.33	21.45	17	364.65	
									1244.23	
									Total Say	1245 Watts

Figure 7.6   (*Continued*)

Heat Loss Calculation Sheet

Sheet _____ 5 Of _____ 9

Project   Heat Loss Example Calculation _____   **Contract No.** [redacted] **Date** _____

Room   Bedroom 1 _____   **Temperature**   16°C & 1.5 AC _____

Engineer's Name

Material or Room	Dimensions (M)			Cu or Sq M	Air Change	U Value Coeff	W/°C	Temp Diff °C	Heat Loss Watts	Comments
	L	W	H							
Wall Between Bed 2										No Loss
Wall Between Bed 3										No Loss
Wall Between Landing										No Loss
Wall Between Houses	5		2.5	12.5		2.1	26.25	17	446.25	
External Wall	6		2.5	15						
Less Window	5		1	5						
				10		2.1	21	17	357	
Window				5		5	25	17	425	
Roof	5	4		20						
	3	1		3						
				23		0.38	8.74	17	148.58	
Floor										No Loss
Infiltration	5	4	2.5	50						
	3	1	2.5	7.5						
				57.5	1.5	0.33	28.46	17	483.82	
									1860.65	
									Total Say	1861 Watts

Figure 7.6   (*Continued*)

Heat Loss Calculation Sheet

Sheet _____ 6 Of _____ 9

Project   Heat Loss Example Calculation _____   Contract No. _____ Date _____

Room   Bedroom 2 _____   Temperature   16°C & 1.5 AC _____

Engineer's Name

Material or Room	Dimensions (M)			Cu or Sq M	Air Change	U Value Coeff	W/°C	Temp Diff °C	Heat Loss Watts	Comments
	L	W	H							
Wall Between Bed 1										No Loss
Wall Between Landing										No Loss
Wall Between Bath										No Loss
Wall Between Houses	4		2.5	10		2.1	21	17	357	
External Wall	4		2.5	10						
Less Window	2		1	2						
				8		2.1	16.8	17	285.6	
Window				2		5	10	17	170	
Floor										No Loss
Roof	4	4		16		0.38	6.08	17	103.36	
Infiltration	4	4	2.5	40	1.5	0.33	19.8	17	336.6	
									1252.56	
									Total Say	1253 Watts

Figure 7.6   (*Continued*)

Heat Loss Calculation Sheet

Project　Heat Loss Example Calculation　　　Contract No. ▨▨▨▨▨▨ Date _____

Room　Bedroom 3　　　　　　　　　　　Temperature　16°C & 1.5 AC

Engineer's Name ▨▨▨▨▨▨▨▨▨▨　　　　　　　　▨▨▨▨▨

Material or Room	Dimensions (M)			Cu or Sq M	Air Change	U Value Coeff	W/°C	Temp Diff °C	Heat Loss Watts	Comments
	L	W	H							
Wall Between Bed 1										No Loss
Wall Between Landing										No Loss
External Wall	6		2.5	15						
Less Window	1		1	1						
				14		2.1	29.4	17	499.8	
Window				1		5	5	17	85	
Floor										No Loss
Roof	3	3		9		0.38	3.42	17	58.14	
Infiltration	3	3	2.5	22.5	1.5	0.33	11.14	17	189.38	
									832.32	
									Total Say	833 Watts

Figure 7.6　(*Continued*)

BOILER SIZING

Having completed the process of calculating the total heat loss from the building, the heat output capacity of the boiler can now be assessed.

The heat output rating of the boiler must be capable of satisfying the following requirements:

$$Q = q_1 + q_2 + q_3 + q_4$$

Where Q equals boiler heat output capacity expressed in kW

　　q1 equals the total heat loss calculated for each room or area

　　q2 equals distribution piping heat losses from unheated areas that the piping passes through. For domestic dwellings the pipework is normally installed within the areas to be heated and can therefore be ignored. For other applications, even though the piping may be insulated, an allowance of 10% may be added. (For large heating projects the heat loss from the insulated pipework should be calculated)

　　q3 equals any primary heat requirements for producing domestic hot water. This requirement will be dependent upon the heating system control scheme specified, on whether the boiler has to satisfy both the heating system's requirements as well as the domestic hot water primary heat load at the same time.

Heat Loss Calculation Sheet									Sheet	8 Of	9

Project Heat Loss Example Calculation   Contract No.   Date

Room Landing   Temperature   16°C & 2 AC

Engineer's Name

Material or Room	Dimensions (M)			Cu or Sq M	Air Change	U Value Coeff	W/°C	Temp Diff °C	Heat Loss Watts	Comments
	L	W	H							
Wall Between Bed 1										No Loss
Wall Between Bed 2										No Loss
Wall Between Bed 3										No Loss
Wall Between Bath										No Loss
External Wall	3		2.5	7.5						
Less Window	1		1	1						
Wall				6.5		2.1	13.6	17	231.2	
Window				1		5	5	17	85	
Floor										No Loss
Roof	3	3		9		0.38	3.42	17	58.14	
Infiltration	3	3	2.5	22.5	2	0.33	14.85	17	252.45	
									626.79	
									Total Say	627 Watts

Figure 7.6   (*Continued*)

q4 equals the overload margin. The overload margin is a factor applied to the heating capacity to take into account deterioration of the boiler and pipework over its expected lifetime. This factor is not a substitute for lack of maintenance to the heating system, or a safety margin for inaccurate heat loss calculations, but is required to compensate for internal fouling of the boiler heat exchanger and the distribution tubing. The exact amount to be added to the heating capacity required will depend upon the engineer's experience; normally the following would be applied:

Gas or oil fired boilers   10%
Solid fuel including biomass fuel   15%

Table 7.12 is used to summarise the above boiler sizing calculation using the results from the example heat loss calculation shown in Figure 7.6.

The overload margin will also compensate for providing an additional heating load during extreme cold external temperatures when the external temperature falls below the base design temperature, providing the heat emitters have a similar extra heat emission capacity.

Heat Loss Calculation Sheet

Sheet _____ 9 Of _____ 9

Project   Heat Loss Example Calculation   Contract No. ▓▓▓▓▓▓▓▓ Date _____

Room   Bathroom   Temperature   21°C & 2 AC

Engineer's Name   ▓▓▓▓▓▓▓▓▓▓▓▓▓▓▓▓▓▓▓   ▓▓▓▓▓▓▓

Material or Room	Dimensions (M) L	W	H	Cu or Sq M	Air Change	U Value Coeff	W/°C	Temp Diff °C	Heat Loss Watts	Comments
External Wall	6		2.5	15						
Less Window	1		1	1						
Wall				14		2.1	29.4	22	646.8	
Window				1		5	5	22	110	
Wall Between Bed 2	3		2.5	7.5		2.3	17.25	5	86.25	
Wall Between Landing	3		2.5	7.5						
Less Door		0.9	2	1.8						
				5.7		2.3	13.11	5	65.55	
Door				1.8		2.6	4.68	5	23.4	
Floor	3	3		9		1.5	13.5	3	40.5	
Roof	3	3		9		0.38	3.42	22	75.24	
Infiltration	3	3	2.5	22.5	2	0.33	14.85	22	326.7	
									1374.44	
									Total Say	1375 Watts

Figure 7.6   (*Continued*)

Table 7.11   Summary of room heat losses

Room	Heat Loss in Watts
Living Room	3300
Dining Room	1961
Kitchen	1144
Hall	1245* }1872
Landing	627* }
Bedroom 1	1861
Bedroom 2	1253
Bedroom 3	833
Bathroom	1375
Total	**13 599 Watts (13.6 kW)**

*Landing and hall heat losses may be added together if a single heat emitter is to be provided located in the hall.

Table 7.12 Boiler heat output calculation summary

Ref	Description	kW
q1	Calculated heat losses from building	13.6
q2	Heat losses from pipework	0
q3	Domestic hot water allowance	4.0*
q4	Overload margin (assume gas fired)	1.36**
Q	**Total heat output required from boiler**	**18.96**

*The quantity of energy in the form of heat required to produce the domestic hot water should be calculated correctly and the method is demonstrated later in Chapter 19. Domestic Hot Water. For the purposes of the above example a figure of 4.0 kW has been allowed.
**A factor of 10% has been applied for the overload margin for a gas fired appliance for the heat loss only, whereby 13.6 kW × 10% which equals 1.36 kW + 4.0 kW = 18.96 kW. It would be normal practice to install a boiler with a heat output capacity of 19 kW, thus having the effect of reducing the overload margin to approximately 8.5% of the total heat output capacity.

Table 7.13 Rule of thumb cube x number values

Room Temperature °C	Cube x Number W/m³
16	32
18	40
21	47
22	49
Average for Whole Building	43

Note – figures given in Table 7.13 should be used with caution for assessing budget heating loads only; full worked engineering calculations should be undertaken before installation commences

CUBE X NUMBER METHOD

As stated earlier, the cube x number method is a rule of thumb method of assessing budget heating loads and requires a certain amount of experience in the engineer applying it. It consists of calculating the heat loss from one room in the correct worked engineering calculation method and dividing the cubic volume of the room into the room's heat loss total, thereby attaching a value in watts per cubic metre for the volume. This value can then be used to multiply the cubic volume of rooms of similar construction, design room temperature and air infiltration to obtain indicative heat loss totals.

Applying this method to our worked examples in Figure 7.6, we have:

Bedroom 1, heat loss total calculated = 1861 watts divided by $57.5\,m^3$ = $32.3\,W/m^3$. Round this figure to nearest whole number, say $32\,W/m^3$.

Applying this to bedroom 2, room volume = $40\,m^3 \times 32\,W/m^3$ = 1280 watts, whereas our worked calculation produced the accurate total of 1253 watts; reasonably close for a budget figure used for estimating purposes, but not for installing. The closeness of the figures is because the rooms and temperature conditions are very similar, but if we apply this method to the living and dining rooms, we find:

Living room, heat loss total calculated = 3300 watts divided by $70\,m^3$ = $47.15\,W/m^3$ rounded off to $47\,W/m^3$.

Dining room volume = $40\,m^3 \times 47\,W/m^3$ = 1880 watts, when our calculation for this room totalled 1961. This reduced accuracy is due to the difference in floor construction, and a figure of $49\,W/m^3$ would have been more accurate.

For buildings of similar construction, the multiplication factors in Table 7.13 can be used to assess quick budget figures that can be used for estimating purposes.

8

Heat Emitter Selection and Sizing

To enable the heat emissions from the heat emitters to be sized, the heat loss calculation from each room or building area has to be completed. The heat loss calculation exercise established the amount of heat that will be lost from each room when maintaining the design room temperature, based on the selected external base design temperature: this will be the quantity of heat that the heat emitters will be required to emit back into the room, as the heat lost will equal the heat input required.

The choice of heat emitters, together with their various attributes and their location requirements is explained in Chapter 5. It is important that these are understood before commencing with the procedure of sizing the heat emitters.

Manufacturers of heat emitters publish data regarding the heat emission values for each size of radiator or convector within their manufactured range, based upon certain conditions that they use during their heat emission testing process. If the conditions used to establish the emission values are the same as the conditions that exist in the heating system that we are designing, then it is a simple matter to select them direct from the manufacturer's data published in their catalogue. If not, then it is essential that our conditions are compared to the heat emitter manufacturer's testing conditions and a factor applied to convert them to the same.

Prior to the introduction of the European Standard for testing the heat emissions from heat emitters, British and Irish manufacturers of heat emitters published heat emission data based on a testing procedure whereby the radiator had pipework connections arranged as 'top, bottom, opposite ends' (TBOE), with a water flow temperature of 90°C and a water return temperature of 70°C, resulting in a mean water temperature of 80°C. This testing procedure was conducted in a test room with an air temperature of 20°C, giving a temperature difference between the mean water temperature of the radiator and the room air temperature of 60°C. The radiator was then used in service, usually connected up with 'bottom, opposite ends' (BOE), and having in most systems a water flow temperature of 82°C and a water return temperature of 71°C. This achieved a mean water temperature of 76.5°C and was used in a room with a design temperature of 21°C, all resulting in a temperature difference between the radiator MWT and room temperature of 55.5°C (see Figure 8.1).

Heating Services in Buildings: Design, Installation, Commissioning & Maintenance, First Edition. David E. Watkins.
© 2011 John Wiley & Sons, Ltd. Published 2011 by John Wiley & Sons, Ltd.
As an aid to lecturers and students, full colour versions of the figures in this chapter may be found at www.wiley.com/go/watkins
These figures are © 2011 by John Wiley & Sons, Ltd.

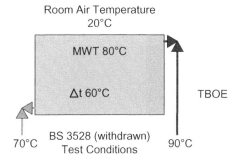

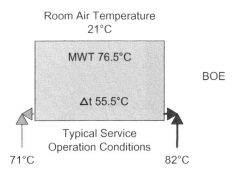

Figure 8.1 Non-European standard testing conditions

Because the testing conditions used a higher mean water temperature and Δt, the test radiator emitted more heat under these conditions than it did when installed in the typical service arrangement shown. For example, a radiator having a published heat emission output of 1000 watts would only emit 874 watts when being used in the above application; therefore a larger radiator would have to be selected, capable of emitting 1145 watts in order to achieve the 1000 watts required.

It was also found that radiators and other heat emitters imported into the United Kingdom from other parts of Europe had published heat emissions that were based on different test conditions to those used by British and Irish manufacturers, which made heat emission comparisons difficult and sometimes misleading to unqualified specifiers, resulting in incorrectly sized heat emitters.

As part of the Europeanisation of the standards programme, the BS EN442 specification for radiators and convectors was introduced in 1996, which served to harmonise the testing procedure and conditions for all heat emitters manufactured in Europe. This had the effect of achieving common emissions criteria published by European heat emitter manufacturers, enabling the data to be easily compared and overcoming the confusion that previously existed regarding heat emission values.

This current standard, illustrated in Figure 8.2, utilises water flow temperatures of 75°C and a water return temperature of 65°C, achieving a mean water temperature of 70°C. The test room air temperature remains at the previous standard of 20°C which produces a Δt of 50°C. It should also be noted that the flow and return piping connections are arranged as 'top, bottom, same end' (TBSE).

If a radiator was selected using these test conditions it would be found to be too large, as the mean water temperature and Δt are lower than that usually employed in our typical service arrangement, which is the opposite to that experienced with the previous standard. For example, a radiator selected on a published heat emission output of 1000 watts would, in this situation, emit 1145 watts, resulting in an over-sized radiator. A radiator selected on a published heat emission of 874 watts would in fact emit 1000 watts under current service conditions.

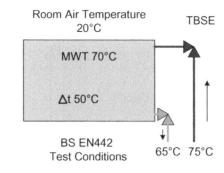

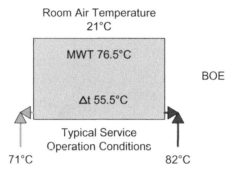

Figure 8.2   European standard testing conditions for heat emissions

Table 8.1 depicts a typical heat emission data schedule for a fictitious radiator manufacturing company. It illustrates the heat emissions for the different sizes of radiators at four heights and varying lengths, including the options of single panel, double panel, single panel single convector, double panel single convector and double panel double convector. The schedule has been limited to a reduced number of sizes, whereas most manufacturers' ranges would be more extensive.

MEAN WATER TEMPERATURE AND DELTA T

From the sample radiator emission schedule illustrated in Table 8.1, it will be noticed that the radiator manufacturer's emission outputs have been qualified on operating temperatures of 75/65/20°C, which is in conformance with BS EN442. These are the design parameters shown in Figure 8.2, where the water flow inlet temperature to the radiator is 75°C, the outlet water flow temperature from the radiator is 65°C and the room air temperature is 20°C. This results in a mean water temperature of 70°C and a Δt of 50°C.

To find the mean water temperature and temperature difference in the standard operating conditions for a typical heating system using conventional flow and return temperatures, we would calculate as follows:

The mean water temperature is the average water temperature inside the heat emitter and is found by adding the flow and return temperatures together and dividing by two:

Standard flow and return temperatures

$$
\begin{array}{ll}
\text{Flow temperature} & 82°C \\
\text{Return temperature} & \underline{71°C} \\
& 153 \div 2 = \text{MWT } 76.5°C
\end{array}
$$

Table 8.1   Acme Radiator Company – Emission outputs based on operating temperatures of 75/65/20°C

Height	Length	Single Panel	Double Panel	Single Convector	Double Panel Single Convector	Double Panel Double Convector
	mm	Watts	Watts	Watts	Watts	Watts
300	400	150	220	220	240	380
	500	190	285	250	386	500
	1000	340	510	510	770	1000
	1500	550	825	770	1160	1500
	2000	760	1140	1032	1554	2020
450	400	190	285	300	440	560
	500	230	345	380	550	700
	600	284	430	460	660	840
	700	330	490	530	770	980
	800	380	570	610	880	1120
	900	420	630	690	990	1260
	1000	470	700	760	1100	1400
	1100	520	780	840	1200	1550
	1200	570	855	920	1300	1690
	1400	660	990	1060	1540	1970
	1600	760	1140	1220	1770	2250
	1800	820	1230	1380	1990	2530
	2000	950	1425	1520	2200	2800
600	400	240	360	400	560	700
	500	300	450	500	700	880
	600	360	540	600	845	1060
	700	420	630	700	980	1245
	800	480	720	800	1125	1420
	900	550	825	900	1260	1600
	1000	610	915	1000	1400	1770
	1100	670	1000	1100	1550	1950
	1200	730	1090	1200	1690	2130
	1400	850	1275	1400	1960	2480
	1600	970	1455	1600	2250	2845
	1800	1090	1630	1800	2535	3200
	2000	1220	1830	2000	2800	3550
	2200	1360	2040	2200	3100	3900
700	600	400	600	680	960	1200
	700	480	720	800	1120	1400
	800	550	825	900	1280	1600
	900	620	930	1020	1430	1800
	1000	700	1050	1140	1600	2000
	1100	770	1155	1250	1750	2200
	1200	840	1260	1370	1900	2400
	1400	980	1470	1600	2235	2800

Therefore a heat emitter with a MWT of 76.5°C installed in a room with a required temperature of 21°C would result in a temperature difference of:

MWT	76.5°C
Room air temperature	21.0°C
	55.5° Δt

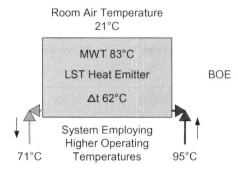

Figure 8.3   LST heat emitter with higher operating F&R temperatures

However, in a situation when higher operating temperatures are employed, such as a sealed heating system incorporating low surface temperature heat emitters, the MWT and Δt would be:

Flow temperature	95°C
Return temperature	71°C
	166 ÷ 2 = MWT 83°C

and the Δt would be:

MWT	83°C
Room air temperature	21°C
	62° Δt.

This example shows that a heating system employing standard flow and return water temperatures of 82°C and 71°C is not consistent with the test conditions used to measure heat emissions from heat emitters. Also, when higher operating temperatures are used, the MWT and Δt are higher than those in either the test conditions or the standard arrangement, which results in a smaller heat emitter being selected.

Figure 8.3 demonstrates the higher operating temperature MWT and Δt.

CONDENSING BOILER SYSTEMS MODE

The subject of condensing boilers is discussed later in the book together with the operating constraints, but it is considered appropriate to include here the considerations that should be applied to sizing and selecting heat emitters that are to be installed in a heating system designed to operate in a condensing mode.

For a condensing boiler to operate in its condensing condition, the return temperature of the water back to the boiler must be below 54°C. If the water was returned at the standard temperature of 71°C, the boiler would operate in the same way as a normal non-condensing boiler and would only be in condensing mode when the boiler started up from cold. The effect that this condition has on the heat emitters is that they should be sized to have a return water temperature of 54°C in order for the condensing boiler to function in a condensing mode most of the time.

Therefore the mean water temperature and temperature difference may be calculated as follows:

Flow temperature	82°C
Return temperature	54°C
	136 ÷ 2 = MWT 68°C

and

MWT	68°C
Room air temperature	21°C
	47°C Δt

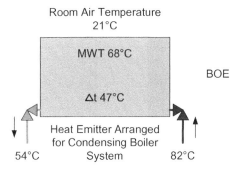

Figure 8.4   Selection parameters for condensing heating system

Table 8.2   Δt for different flow and return temperature conditions

Flow °C	Return °C	MWT °C	Δt at Room Temperatures °C					
			16	18	20	21	22	23
98	71	84.5	68.5	66.5	64.5	63.5	62.5	61.5
95	71	83	67	65	63	62	61	60
95	60	77.5	61.5	59.5	57.5	56.5	55.5	54.5
95	55	75	59	57	55	54	53	52
95	54	74.5	58.5	56.5	54.5	53.5	52.5	51.5
82	71	76.5	60.5	58.5	56.5	55.5	54.5	53.5
82	60	71	55	53	51	50	49	48
82	55	68.5	52.5	50.5	48.5	47.5	46.5	45.5
82	54	68	52	50	48	47	46	45
82	50	66	50	48	46	45	44	43
75	65	70	BS EN442		50	Test Parameters		

50   May be selected direct from manufacturer's published heat emission data
   Will result in smaller heat emitters than contained in published data
   Will result in larger heat emitters than contained in published data

From this calculation, it can be seen that the MWT and Δt are lower than that used for the heat emission test conditions; this would result in selecting larger heat emitters to achieve the desired room temperatures whilst maintaining the condensing boiler in its condensing mode.

Figure 8.4 demonstrates the different operating temperatures utilised on a condensing heating system.

Before selecting heat emitters for any given heating system, the mean water temperature (MWT) and temperature difference (Δt) must be calculated.

The Δt for common flow and return water temperatures and the MWT may be selected from Table 8.2.

CORRECTION FACTORS

Having established the operating conditions of the heating system, if the Δt = 50°C we can select the heat emitters direct from the manufacturer's tabulated heat emission data shown in Table 8.2; if however, the

Table 8.3 Correction factors for values of Δt other than at 50°C

Δt	Correction Factor	Δt	Correction Factor
5	0.05	55.5	1.145
10	0.123	56	1.159
15	0.209	56.5	1.172
20	0.304	57	1.186
25	0.406	57.5	1.200
30	0.515	58	1.213
35	0.629	58.5	1.227
40	0.748	59	1.240
43	0.773	59.5	1.253
44	0.797	60	1.267
45	0.872	60.5	1.281
45.5	0.812	61	1.295
46	0.897	61.5	1.308
46.5	0.910	62	1.322
47	0.923	62.5	1.336
47.5	0.936	63	1.350
48	0.949	63.5	1.364
48.5	0.962	64	1.378
49	0.975	64.5	1.392
49.5	0.987	65	1.406
50	1.000	65.5	1.420
50.5	1.013	66	1.434
51	1.026	66.5	1.448
51.5	1.040	67	1.462
52	1.053	67.5	1.476
52.5	1.066	68	1.491
53	1.080	68.5	1.505
53.5	1.093	69	1.519
54	1.106	69.5	1.534
54.5	1.119	70	1.549
55	1.132	75	1.694

Δt of the heating system is anything other than 50°C, we must apply a correction factor to convert the manufacturer's heat emission data to suit our operating conditions.

Table 8.3 provides correction factors that may be used to convert the heat emissions from manufacturers' published data to suit heating systems with an operating Δt other than 50°C.

EXAMPLE OF CORRECTION FACTOR USE

Using the conditions stated earlier in this chapter:

Heat loss calculated and heat required from heat emitter	1000 watts
Desired room temperature	21°C
Water flow temperature	82°C
Water return temperature	71°C
Mean water temperature	76.5°C
Mean water temperature of	76.5°C
Room air temperature of	21.0°C
Δt from table given in Table 8.2	55.5°C

Therefore, select the correction factor from Table 8.3 equating to a Δt of 55.5°, which is 1.145. Divide the heat emission required from the heat emitter by the correction factor obtained.

$$\frac{1000}{1.145} = 874 \text{ watts}$$

This permits the heat emitter to be selected from the manufacturer's tabulated data sheet that is based on 75/65/20°C conditions to emit 874 watts, but which will emit 1000 watts with our operating conditions.

If we selected the nearest commercially available size heat emitter from the sample manufacturer's heat emission data sheet, we would find that a 450 high × 800 long, double panel, single convector has a stated output of 880 watts. If we multiply this emission by our correction factor we would find its actual heat emission based on our operating conditions:

$$880 \times 1.145 = 1007.6 \text{ watts}$$

Therefore, the actual emission of heat from the heat emitter selected would be just over seven watts above the requirement.

The above example demonstrates that the correction factor can be used by dividing the factor into our required heat emission to enable the heat emitter to be selected on the result, or if the manufacturer's stated heat output is multiplied by the correction factor, then the actual heat emission can be found.

SIZING AND SELECTION PROCEDURE FOR HEAT EMITTERS

The selection and sizing of the heat emitters should be recorded on a worksheet for easy reference. This is essential for projects involving hundreds of heat emitters and should contain all relevant information, including a unique reference number used to identify the heat emitter on the installation drawings. The worksheet should also contain the heart emitter's room location, room temperature, MWT, Δt, heat loss calculation for the room, correction factor from Table 8.3, a fixing factor if applicable for recesses and shelves, etc, a pipe connection factor, if different from the test conditions (see Figure 5.22), the corrected heat emission to be selected from manufacturer's data, size selected and stated output by manufacturer at the test conditions, the actual heat emission achievable when converted back to the system's operating conditions and a descriptive note regarding the configuration of the heat emitters selected.

The heat emitter schedule/worksheet shown in Figure 8.5 shows the working procedure for selecting the heat emitters for the same dwelling used in Figures 7.4 and 7.5 to calculate the heat losses and select the radiators from the fictitious radiator manufacturer's heat emission data sheet shown in Table 8.1.

The selection principle adopted is to try and proportion the radiators below windows and on walls – as explained in Chapter 5 – with a high level fan convector being selected for the kitchen due to lack of usable wall space.

It will also be noticed that the heat requirements for the hall and landing area have been added together and a single radiator, capable of heating both areas, has been located at the foot of the stairs.

Also recorded in Figure 8.5 are both the totals of the heat required to be put back into the building to maintain the desired room temperatures – obtained from loss calculations – plus the actual emission that can be achieved from the heat emitters when they have been corrected to suit the operating conditions of the heating system being designed.

It will be seen that heat loss required matches the total from the heat loss calculation exercise and recorded in Table 7.11, whereas the actual emission is higher:

Schedule/Worksheet of Heat Emitters

Project-Semi-Detached House Fig 8.13/14   MWT-763.5   Date   Sheet 1 of 1

Ref	Location & Room Temp	Δt	Heat Loss W	Emitter Type	Correction Factor	Fixing Factor	Conn Factor	Heat Emission Selected W	Radiator Size HxL & Cat Emis	Actual Emission W	Comments
R01	Living Room 21	55.5	3300	Radiator	1.145	0	5%	3026	600 × 800 – 3200 w	3664	D Panel/D Conv
R02	Dining Room 21	55.5	1961	Radiator	1.145	0	5%	1798	700 × 900 – 1800 w	2061	D Panel/D Conv
C01	Kitchen 18	58.5	1144	F/Convtr	1.227	0	0	933	1000 w	1227	Fan Convector
R03	Hall/Landing 16	60.5	1872	Radiator	1.281	0	5%	1535	600 × 1100 – 1500 w	1985	D Panel/S Conv
R04	Bedroom 1/16	60.5	1861	Radiator	1.281	0	5%	1525	600 × 1600 – 1600 w	2050	Single Convector
R05	Bedroom 2/16	60.5	1253	Radiator	1.281	0	5%	1027	600 × 1800 – 1090 w	1397	Single Panel
R06	Bedroom 3/16	60.5	833	Radiator	1.281	0	5%	683	700 × 1000 – 700 w	897	Single Panel
R07	Bathroom 21	55.5	1375	Radiator	1.145	0	5%	1261	700 × 800 – 1280 w	1466	D Panel/S Conv
	Totals		13599							14747	

Figure 8.5 Heat emitter schedule/worksheet

Actual emission achievable	14 747 watts
Actual emission required	<u>13 599 watts</u>
Emission difference	1 148 watts

This additional heating load can be accommodated by the boiler selected due to the overload margin of 10% that was added (equating to 1760 watts), which will also serve as an additional margin for heating systems operating in an intermittent mode as opposed to continuous operation.

An example of the radiator locations based on the schedule/worksheet has been added to the drawing of the house used to calculate the heat losses, and is shown in Figures 8.6 and 8.7.

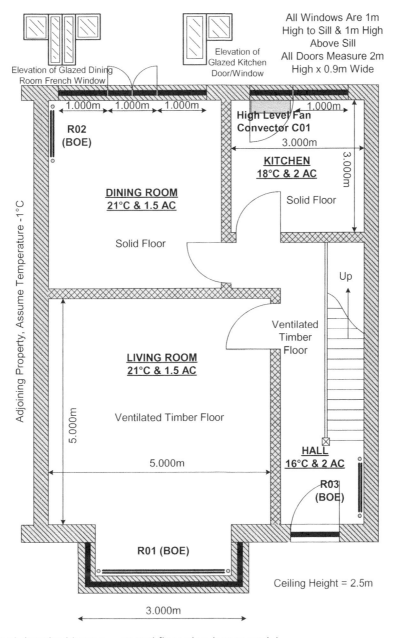

Figure 8.6   Semi-detached house – ground floor plan (not to scale)

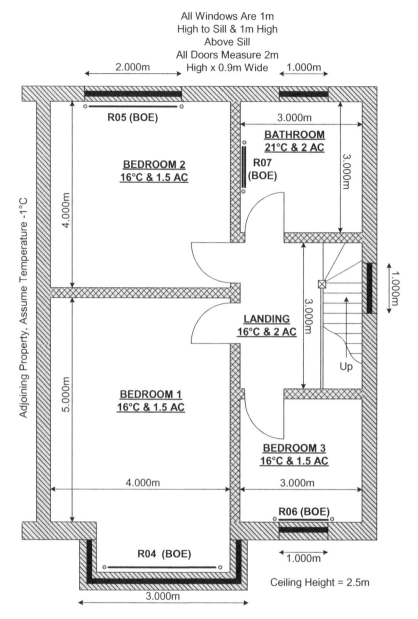

Figure 8.7   Semi-detached house – first floor plan (not to scale)

It should be emphasised that the drawings in Figures 8.6 and 8.7 are included as an example and there is no single solution to the location of the heat emitters as they will depend on the criteria discussed in Chapter 5.

HEATING SURFACE

It was once common practice for radiator manufacturers to publish their heat emissions for each radiator type and height expressed in square metres of heating surface; this was not the overall dimensions of the radiator, but the useful heating surface measured over the corrugations and particular shape and style of the radiator.

Using this emission value, it is possible to size radiators on heating surface alone.

Thus, assume heat emission from manufacturer's data of 624 W m^{-2} and required heat emission is 1560 watts.

Therefore

$$\frac{1560}{624} = 2.5\,\text{m}^2 \text{ heating surface}$$

From this, a radiator having a heating surface equal to or greater than 2.5 m^2 may be selected from the manufacturers' data.

The heat emission per square metre varies from one manufacturer to another, depending upon the particular style of their product, and only the emission value for the radiators proposed to be used should be adopted.

9

Pipe Sizing

Having established the heat requirements for the building by completing the heat loss calculations, arriving at the correct boiler capacity and selecting the heat emitters, the piping layout and sizes may now be considered.

This chapter considers forced circulation piping systems only, which are piping systems where the water is circulated by the aid of a circulating pump. The main object of this exercise is to arrive at economical and suitable pipe sizes to convey the water carrying the correct amount of heat from the boiler to the heat emitters and returning to the boiler.

It is at this stage that the piping arrangement described in Chapter 2 has to be decided upon. It should be laid out on a drawing, and take into account the merits and requirements of the piping system chosen. By laying the piping system out on a drawing, pipe lengths and changes in direction can be established and tabulated with other relevant design information, such as the heat emission requirements from each heat emitter, and recorded on the drawing. Any control valves and commissioning valves can also be located on the drawing.

Before commencing the procedure for calculating the sizes of the pipework, it is important to grasp a basic understanding of the fundamental principles relating to the flow of water in pipes and their heat carrying capacity, in order to determine correct and cost effective pipe sizes. Although the main objective in this exercise is to determine suitable pipe sizes, the procedure that is adopted for this task also results in establishing the operating performance duty required for the circulating pump as these two tasks are closely related.

It is also important to recognise the effects that the velocity or speed of the water flow has through pipes on the system and its reaction to the internal wall surface of the pipes.

FLOW OF HEAT IN PIPES

The objective is to supply sufficient heat from the boiler via the medium of water, through a network system of piping to the heat emitters, whereby it transfers part of that heat to the room or space to be heated before returning to the boiler for the process to be repeated.

Heating Services in Buildings: Design, Installation, Commissioning & Maintenance, First Edition. David E. Watkins.
© 2011 John Wiley & Sons, Ltd. Published 2011 by John Wiley & Sons, Ltd.
As an aid to lecturers and students, full colour versions of the figures in this chapter may be found at www.wiley.com/go/watkins
These figures are © 2011 by John Wiley & Sons, Ltd.

The procedure for calculating pipe sizes for a given flow of water is straightforward when the velocity and pressure are known, but in a circulating piping arrangement such as a hydronic heating system, the water is only the medium employed to convey the heat. Therefore it is necessary to establish the amount of water that will be required to carry the heat that is required to be transferred by the heat emitters to enable pipe sizes to be calculated.

To convert the heat emission requirement to a unit of water flow, the heat losses from the building must have been completed and the individual heat emission values from the heat emitters correctly attributed and proportioned. Utilising this information, the following formula may be used to calculate the mass flow rate of water required, both individually at each heat emitter and collectively, for assessing part of the circulating pump duty:

$$\text{MFR} = \frac{Q}{\text{SpHt} \times \Delta t}$$

where:

MFR = Mass flow rate of water in kg s^{-1}
Q   = Quantity of heat required in watts
SpHt = Specific heat of water in joules
Δt = Temperature difference between flow and return

The mass flow rate is the amount of water required to be conveyed expressed as a weight (kg s^{-1}) of water at a specific heat of 4200 joules, as explained in Chapter 4 under the heading 'Specific Heat Capacity', when supplying water at the operating flow and return water temperatures to each heat emitter.

Applying this formula in practice:

- Assume, for example, a heat loss calculated for a given space of 3000 watts with a heat emitter sized to suit.
- Specific heat of water taken as 4200 joules; this may be taken as a constant for low pressure/temperature heating systems.
- Standard low pressure/temperatures of 82°C for the flow and 71°C for the return for each heat emitter, giving a temperature difference of 11°C. Other flow and return temperatures may need to be considered for condensing boiler applications, or underfloor heating systems.

Example:

$$\frac{3000}{4200 \times (82-71)} = \frac{3000}{46200} = 0.065 \text{ kg s}^{-1}$$

Alternatively, for a Δt of 11 the heat emission in kW may be multiplied by a factor of 0.0216. Therefore:

$$3 \times 0.0216 = 0.065 \text{ kg s}^{-1}$$

From the above example it can be seen that a heat emitter sized to emit 3000 watts will require a flow rate of 0.065 kg s^{-1}, or litres per second (one litre of water weighs one kilogram), when supplied at a temperature of 82°C and transferring 11°C of its heat before being returned to the boiler.

The flow rates for the total number of heat emitters to be installed should be calculated to ascertain the total flow rate for the system.

VELOCITY

The velocity is the speed with which water will travel through a pipe. It has an important bearing when calculating pipe sizes.

Using two extreme examples, it can be best visualised when considering a given quantity of water flowing through two pipes of different sizes: one, a pipe with a fairly large diameter, will result in a relatively slow velocity flowing through it; the other, a pipe with a much smaller diameter, may be capable of passing the same quantity of water in the same time but at a much faster speed. Therefore, although both pipes are capable of supplying the same given amount of water over the same time period, the velocity in each will have completely different effects. This is shown in Figure 9.1.

A pipe designed to have a low velocity for the flow required will result in it being oversized and in turn more expensive than it need be; therefore, it appears reasonable to size the pipes to have a higher velocity by selecting smaller diameter and more cost effective pipes. However, the practice of undersizing pipework by creating a faster velocity will result in much higher frictional resistances being encountered by the flow, with the added effect of developing noise within the pipework at extremely high velocities. As the velocity is increased in the pipework, the level of noise experienced in the system will rise and if the velocity is increased further, generating more noise, this can have the further effect of causing erosion at sharp changes in direction of the piping system, such as at elbows and valve seatings. If the velocity is very high the erosion occurring will eventually wear parts of the pipe or fitting wall away, leaving small pinholes with the obvious disastrous effects.

Therefore, a compromise must be made between the high and low velocities to find some intermediate, more acceptable speed that will not give rise to noise and encounter a reasonable degree of frictional resistance, without resulting in oversized and uneconomical sized pipes.

Traditional pipework arrangements for circulating systems such as heating and cooling piping networks, including domestic small bore heating systems constructed using pipes with diameters of 15 mm (½ inch) and greater and employing standard pipe fittings, would normally be designed to have a velocity on the main pipe runs of between 0.75–1.2 m sec^{-1}, which has been found through experience to achieve acceptable flow velocities and obtain economical pipe sizes.

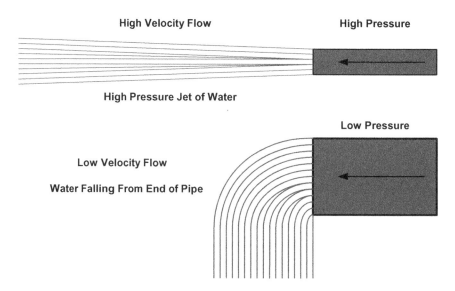

Figure 9.1 Comparison of high and low velocity flow

Therefore, it is normal practice to aim for a design velocity of approximately 1.0m/sec when calculating pipe sizes with lower velocities acceptable for branch connections to heat emitters, and higher velocities for longer but straight pipe runs.

Regarding radial piping systems, including micro bore piping arrangements a higher design velocity may be employed without any adverse effects due to the differing principles with this type of circulating system.

Radial piping systems and micro bore tubing are manufactured in soft copper tube or smooth bore thermoplastic pipe supplied in a continuous coil form. This allows it to be installed between the manifold and each heat emitter without any sharp changes in direction, such as those found in conventional pipe fittings. See Chapter 2 for detailed information on the operating fundamentals.

Due to the absence of any sudden changes in direction with this continuous form of tubing, it is possible to reconsider the design velocity, which may be safely increased from 1.0m/sec without generating noise or causing erosion problems. Thus, the same quantity of water can be conveyed through smaller diameter pipes than would be used for conventional small bore tubing systems.

By removing all tight changes in direction on the piping system, the flow velocities may be safely increased to within a range of $1.4–1.8\,\mathrm{m\,sec^{-1}}$ without any adverse effects, and therefore a design velocity of $1.5\,\mathrm{m\,sec^{-1}}$ may be used for pipe sizing calculations for water being conveyed through continuous smooth bore piping systems.

The design flow velocity through the main conventional flow and return pipework between the boiler and the manifold using conventional pipe fittings should remain at $1.0\,\mathrm{m\,sec^{-1}}$ and is treated the same as conventional circulating piping systems.

FRICTIONAL RESISTANCE

As previously explained in Chapter 4, a circulation of water will naturally occur within a heating circuit due just to the differing densities of water between the flow and return water temperatures. However when this movement of water begins, the flow will encounter a resistance from the walls of the pipe and changes in direction caused by fittings and valves. To overcome these accumulated pipework resistances, a circulating pump is introduced into the circulating piping system to force the circulation to a proportion equal to that of the total resistance. These resistances must be kept to a reasonable level within the economic duty range of circulating pumping equipment, and – in the case of domestic circulating pumps – within the duty range of standard domestic circulators.

When there is a movement of any kind – machinery fluids etc – there will be a resistance encountered due to friction between the two parts. This friction will try to slow the movement down and eventually stop it.

The flow of water within tubes is no exception to this rule, the only variation in the resistance will be from the degree of roughness of the pipework material and the sharpness and number of any changes in direction from the fittings, valves and components.

As previously mentioned, the degree of resistance encountered by the water within the tube will vary in direct proportion to the velocity of the flow: when the velocity is very low, the frictional resistance will also be low; as the velocity is increased, so will be the amount of resistance encountered.

For the purposes of the pipe sizing calculation exercise, this resistance within the piping system is expressed as a unit of pressure, as that amount of pressure will be required to overcome the resistance developed.

The unit of force is the Newton (N) and pressure is expressed in 'Newton's per square metre' $(\mathrm{N\,m^{-2}})$ for circulating piping systems such as heating systems. This unit of pressure is derived from the technical metric unit 'kilogram force per square centimetre' $(\mathrm{kg\,f\,cm^{-2}})$. The SI name given to the unit of pressure is Pascal

(Pa), which is equal to $N\,m^{-2}$. It is not uncommon for technical information to be published in either $N\,m^{-2}$ or Pa:

$$1\,Pa = 1\,N\,m^{-2}$$

and

$$1\,kPa = 1000\,N\,m^{-2} \text{ or } 1\,kN\,m^{-2}$$

or

$$100\,kPa \text{ or } 100\,000\,N\,m^{-2} = 1\,bar.$$

From the above it can be seen that the unit of pressure 'bar' is too large to be of practical use when calculating pipework pressure resistances, and the smaller units Pa and $N\,m^{-2}$ are found to be more practical and are used to express pressure and pressure losses.

The frictional resistance for the flow of water within the pipework is expressed as a unit of pressure, and although it is fairly accurate for straight lengths or runs of piping, where the flow inside the pipework has to negotiate sharp changes of direction – such as at elbows, bends, tees or valves etc – the flow will encounter a greater resistance than through a straight pipe.

It is normal to compare this frictional resistance encountered at changes of direction for the flow to that encountered by an equal length of pipework of the same diameter. This equal length of pipework for each fitting type or valve is termed 'equivalent pipe length' and is added to the actual pipe length to give the total pipe length. For example, from Table 9.1, the resistance offered to the flow of water through a 28 mm diameter angle valve fitted to a radiator is equivalent to 5.5 m of straight 28 mm diameter copper tube.

It has previously been stated that the design velocity used for sizing circulating piping systems is 1m/ second, but as a general guide for domestic heating systems it would also be restricted to a pressure drop,

Table 9.1   Water flow resistance through pipe fittings expressed as equivalent pipe length in metres

Equivalent Pipe Length in Metres

Pipe Size	Fitting Type							
	Tee	Elbow	Labour Bend	Reducer	Return Bend	Angle Valve	Gate Valve	Globe Valve
6	0.2	0.2	0.1	0.1	0.2	0.7	0.1	1
8	0.3	0.3	0.1	0.1	0.3	1	0.1	1.5
10	0.4	0.4	0.15	0.1	0.4	1.5	0.15	2.5
12	0.55	0.5	0.2	0.1	0.6	2	0.2	3.5
15	0.7	0.6	0.25	0.15	0.75	2.5	0.3	4.5
22	1	0.9	0.4	0.25	1.2	4	0.5	7.5
28	1.6	1.2	0.6	0.35	1.7	5.5	0.6	10
35	2	1.5	0.7	0.45	2	7	0.8	13
42	2.6	2	0.9	0.55	2.8	9	1	16
54	3.6	2.5	1.2	0.9	3.5	12	1.5	22
67	4.6	3	1.5	1	4.5	15	2	28
76	5.6	3.5	1.8	1.2	6	19	2.5	34
108	8.5	5	2.5	2	10	28	3.5	52
159	14	7	3.8	3	14	45	6	82

or frictional resistance, of between 100 Pa or N m^{-2} per metre run of pipework to 300 Pa or N m^{-2} per metre run of pipework. This rule of thumb generally achieves an overall total resistance that may be overcome by a standard domestic circulating pump duty range, although this rule of thumb only applies to main runs of pipework, not short branch runs to single heat emitters. It should be noted that this general guide is only applicable to conventional single or two pipe circulating arrangements and does not apply to micro bore piping systems.

INDEX PIPE CIRCUIT

The definition of the 'index pipe circuit' is the pipe circuit having the greatest amount of frictional resistance to the flow of water. It is the pressure element of the circulating pump duty required for the circulating pump to overcome. If the pump is capable of circulating the water through the pipe circuit with the highest amount of frictional resistance, then it will be able to circulate water through the pipe circuits with less frictional resistance.

The index pipe circuit is not necessarily the pipe circuit having the furthest heat emitter from the boiler, although this is quite often the case; it purely depends on the accumulated resistance along the pipeline circuit length including all fittings, valves and other components.

PIPE SIZING DATA

Having previously established the method of converting the flow of heat into a practical flow rate of kg per second and discussed and understood the importance and implications of velocity and frictional resistance, the design engineer will require certain engineering data on the flow characteristics for water through different commercially available pipe materials in order to commence the pipe sizing calculation exercise.

The frictional resistance encountered by the flow of water through pipework has already been established, together with the need to convert all changes in direction of the piping into an equivalent pipe length of the same diameter.

Table 9.1 gives these frictional resistances for various pipe fittings and valves, expressing them as equivalent pipe length in metres of straight pipe of equal diameter that is added to the actual pipe length to obtain total pipe length.

Figure 9.2 shows how a pipe sizing chart is used.

The pipe sizing chart shown in Figure 9.3 has been developed for use in assessing the pressure drop due to frictional resistance for various pipe diameters at different velocities when the mass flow rate is known. This chart can be used for smooth bore tubes such as copper tube and light gauge stainless steel tube.

Using the known flow rate figure, follow the line up vertically until it dissects the selected pipe size diagonal line, then follow the line horizontally to the left, to read the pressure drop that would be encountered. If the pressure drop is considered too high, select a larger tube size and repeat the exercise. The velocity can also be read from this chart and on the example shown above, it falls between two velocity diagonal lines, which means the velocity would be 0.9 m s^{-1}.

An alternative method of presenting the information displayed on the pipe sizing chart in Figure 9.3 is the tabulated version depicted in Figure 9.4. Here, the pressure drop, velocity and mass flow rate can be obtained for a selected pipe size by scrolling down until the flow rate required is reached, then reading off the corresponding pressure drop to the left and – on the right the velocity zone that the flow rate falls between.

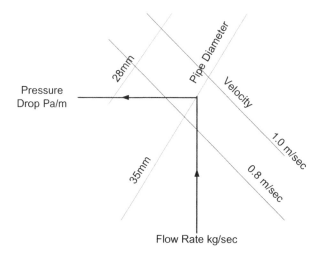

Figure 9.2 Example for using a pipe sizing chart

Figure 9.5 has been included to illustrate in chart form pipe sizing information for thermoplastic pipes such as cross-linked polyethylene (PEX) and polybutylene (PB). Both are also smooth bore tubes and becoming more popular for underfloor heating or radial piping systems.

Alternatively, the pipe sizing chart illustrated in Figure 9.6 may be used for sizing rough bore low carbon mild steel tube, commonly used for commercial building heating applications when larger pipe sizes are required, and having a greater frictional resistance to flow.

PIPE SIZING PROCEDURE

Before the pipe sizing exercise can commence, the following design calculations and decisions must have been concluded:

- Building heat loss calculations completed.
- Individual room heat emitters sized and locations selected.
- Piping arrangement and heating system layout chosen. (The building structural constraints must be taken into account when making this decision.)
- Piping materials selected for the heating arrangement to be installed.

When the above activities have been completed the exercise of establishing suitable pipe sizes may commence by the following procedure.

Explanation of pipe sizing procedure

For continuity with the design calculations previously employed, the same two-storey house has been selected that was used to assess the heat loss calculations followed by the procedure for sizing the heat emitters. The floor plans have been reproduced in Figures 9.7 and 9.8 with the chosen pipe routes added: a two-pipe direct return piping system has been chosen, using copper tube as the preferred piping material.

The pipework has been laid out taking into account the structural constraints of the building, including the fact that the rear half of the building has been constructed with a solid concrete ground floor, with the front portion of the property having a ventilated raised timber floor construction.

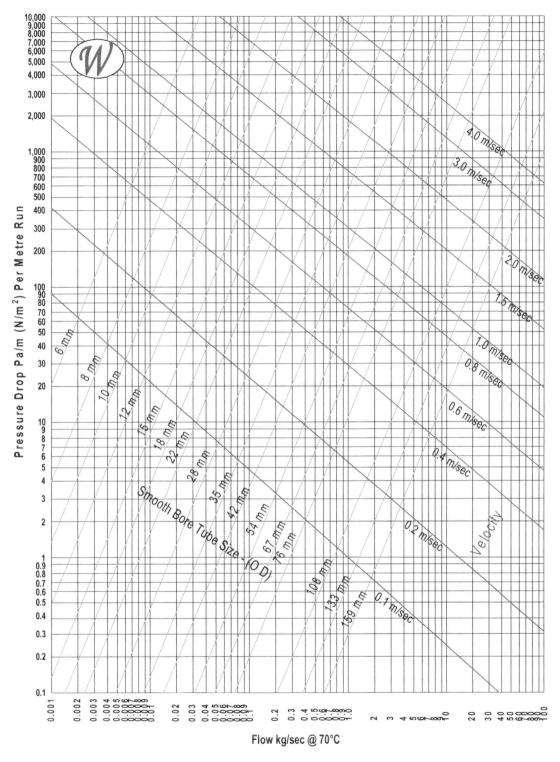

Figure 9.3   Pipe sizing chart for copper tube BS EN1057 R250 and stainless steel tube BS 4127

Pressure Drop Pa/m	Pipe Diameter mm / Mass Flow Rate kg/second													Velocity m/sec
	6	8	10	12	15	22	28	35	42	54	67	76	108	
10				0.01	0.01	0.03	*0.15*	0.12	0.21	0.43	0.77	*0.3*	2.9	
15		*0.1*		0.01	0.01	0.04	0.09	0.15	0.26	0.54	0.97	1.4	3.65	*0.5*
20			0.01	0.01	0.02	0.05	0.1	0.18	0.31	0.63	1.14	1.6	4.3	
25			0.01	0.01	0.02	0.06	0.11	0.2	0.35	0.72	1.3	1.84	4.82	
30			0.01	0.01	0.02	0.06	0.12	0.23	0.39	0.8	1.44	2	5.35	
35			0.01	0.01	0.02	0.07	0.14	0.25	0.42	0.87	1.56	2.22	5.8	
40			0.01	0.01	0.03	0.07	0.15	0.27	0.46	0.93	1.18	2.4	6.3	
45			0.01	0.01	0.03	0.08	0.16	0.29	0.49	1	1.9	2.55	6.7	
50			0.01	0.02	0.03	0.08	0.17	0.3	0.52	1.06	2	2.7	7	
55			0.01	0.02	0.03	0.09	0.18	0.32	0.55	1.12	2.02	2.85	7.5	
60		0	0.01	0.02	0.03	0.09	0.19	0.34	0.57	1.18	2.1	3	7.83	
65		0	0.01	0.02	0.03	0.1	0.2	0.36	0.6	1.23	2.2	3.13	8.2	*1*
70		0	0.01	0.02	0.04	0.1	0.2	0.37	0.63	1.28	2.3	3.25	8.53	
80		0	0.01	0.02	0.04	0.11	0.22	0.4	0.67	1.38	2.5	3.5	9.2	
90		0	0.01	0.02	0.04	0.12	0.24	0.43	0.72	1.48	2.64	3.75	9.8	
100		0.01	0.01	0.02	0.04	0.13	0.25	0.45	0.77	1.55	2.8	4	10.4	
120		0.01	0.01	0.03	0.05	0.14	0.28	0.5	0.85	1.73	3.3	4.4	11.5	
140		0.01	0.02	0.03	0.05	0.15	0.3	0.55	0.93	1.00	3.4	4.8	12.5	*1.5*
160		0.01	0.02	0.03	0.06	0.16	0.33	0.6	1	2	3.63	5.12	13.4	
180		0.01	0.02	0.03	0.06	0.18	0.35	0.64	1.1	2.16	3.9	5.5	14.3	
200		0.01	0.02	0.03	0.06	0.19	0.38	0.67	1.13	2.3	4.1	5.8	15.2	
220		0.01	0.02	0.04	0.07	0.2	0.4	0.71	1.2	2.4	4.35	6.12	16	
240		0.01	0.02	0.04	0.07	0.2	0.41	0.75	1.25	2.54	4.55	6.42	16.7	*2*
260		0.01	0.02	0.04	0.08	0.22	0.43	0.8	1.3	2.65	4.75	6.7	17.5	
280		0.01	0.02	0.04	0.08	0.22	0.45	0.81	1.37	2.76	5	7	18.2	
300		0.01	0.02	0.04	0.08	0.23	0.46	0.85	1.42	2.87	5.15	7.26	18.9	
320		0.01	0.03	0.04	0.08	0.24	0.49	0.87	1.47	3	5.32	7.52	19.6	
340		0.01	0.03	0.05	0.09	0.25	0.5	0.9	1.52	3.1	5.5	7.8	20	
360	0	0.01	0.03	0.05	0.09	0.26	0.52	0.94	1.57	3.18	5.7	8	21	
380	0	0.01	0.03	0.05	0.09	0.27	0.54	0.96	1.62	3.27	5.85	8.25	21.6	
400	0	0.01	0.03	0.05	0.1	0.27	0.55	1	1.66	3.36	6	8.5	22	
420	0	0.01	0.03	0.05	0.1	0.28	0.57	1.03	1.7	3.45	6.2	8.75	22.7	
440	0	0.01	0.03	0.05	0.1	0.29	0.58	1.04	1.75	3.56	6.35	8.95	23.3	
460	0	0.01	0.03	0.06	0.1	0.3	0.6	1.07	1.8	3.64	6.5	9.17	23.9	
480	0	0.01	0.03	0.06	0.1	0.3	0.61	1.1	1.84	3.72	6.65	9.4	24.4	
500	0	0.01	0.03	0.06	0.11	0.31	0.62	1.12	1.88	3.8	6.8	9.6	25	
520	0	0.02	0.03	0.06	0.11	0.32	0.64	1.15	1.9	3.9	6.95	9.83	25.5	*3*
540	0	0.02	0.03	0.06	0.11	0.32	0.65	1.17	1.96	3.98	7	10	26	
560	0	0.02	0.03	0.06	0.11	0.33	0.66	1.2	2	4	7.2	10.2	26.6	
580	0	0.02	0.03	0.06	0.12	0.34	0.68	1.22	2.05	4.12	7.4	10.4	27	
600	0	0.02	0.03	0.06	0.12	0.34	0.69	1.24	2.08	4.2	7.5	10.6	27.6	
620	0	0.02	0.03	0.07	0.12	0.35	0.7	1.26	2.12	4.28	7.6	10.8	28	

Figure 9.4   Flow of fluid in smooth bore pipes

Pressure Drop Pa/m	Pipe Diameter mm / Mass Flow Rate kg/second													Velocity m/sec
	6	8	10	12	15	22	28	35	42	54	67	76	108	
620	0	0.02	0.03	0.07	0.12	0.35	0.7	1.26	2.12	4.28	7.6	10.8	28	
640	0	0.02	0.03	0.07	0.12	0.36	0.72	1.28	2.16	4.35	7.8	11	28.5	
660	0.01	0.02	0.03	0.07	0.13	0.36	0.73	1.3	2.2	4.43	7.9	11.2	29	
680	0.01	0.02	0.04	0.07	0.13	0.37	0.74	1.33	2.23	4.5	8	11.4	29.5	
700	0.01	0.02	0.04	0.69	0.13	0.38	0.75	1.35	2.26	4.57	8.2	11.6	30	
720	0.01	0.02	0.04	0.07	0.13	0.38	0.76	1.37	2.3	4.64	8.3	11.8	30.4	
740	0.01	0.02	0.04	0.07	0.13	0.39	0.77	1.4	2.33	4.7	8.4	12	30.9	
760	0.01	0.02	0.04	0.07	0.14	0.39	0.79	1.41	2.37	4.8	8.5	12	31.2	
800	0.01	0.02	0.04	0.08	0.14	0.4	0.8	1.45	2.45	4.9	8.8	12.4	32.2	
850	0.01	0.02	0.04	0.08	0.15	0.42	0.85	1.5	2.5	5.1	9	12.8	33.4	4
900	0.01	0.02	0.04	0.08	0.15	0.43	0.86	1.55	2.6	5.24	9.35	13	34.3	
950	0.01	0.02	0.05	0.08	0.15	0.44	0.89	1.6	2.68	5.4	9.6	13.5	35	
1000	0.01	0.02	0.05	0.09	0.16	0.46	0.92	1.64	2.75	5.6	10	14	36	
1100	0.01	0.02	0.05	0.09	0.17	0.48	0.97	1.73	2.9	5.85	10.4	14.7	38	
1200	0.01	0.02	0.05	0.09	0.18	0.5	1	1.8	3	6.1	11	15.5	40	
1300	0.01	0.03	0.05	0.1	0.18	0.53	1.06	1.9	3.2	6.4	11.4	16	42	
1400	0.01	0.03	0.05	0.1	0.19	0.55	1.1	1.98	3.3	6.7	12	16.8	43.6	5
1500	0.01	0.03	0.05	0.11	0.2	0.57	1.15	2	3.44	7	12.4	17.4	45	
1600	0.01	0.03	0.05	0.11	0.21	0.6	1.2	2.13	3.6	7.2	13	18	47	
1700	0.01	0.03	0.06	0.11	0.21	0.61	1.23	2.2	3.7	7.5	13.3	18.7	48	
1800	0.01	0.03	0.06	0.12	0.22	0.63	1.27	2.27	3.8	7.6	13.6	19.2	50	
1900	0.01	0.03	0.06	0.12	0.23	0.65	1.3	2.34	3.9	7.9	14	20		
2000	0.01	0.03	0.06	0.13	0.23	0.67	1.34	2.4	4	8.1	14.4	20.4		
	0.5		1		1.5	2			3		4			

Figure 9.4   (*Continued*)

To help illustrate the pipe sizing calculations and enable them to be followed more easily, the piping arrangement has been developed into an isometric projection which includes all the necessary information, such as pipe length dimensions, heat emissions and mass flow rates. The pipework has also been sectioned into a number of circuit zones to enable the pipe sizing calculations to be worked through in a logical order; these circuits have been denoted by the letters A, B, C, etc, in Figure 9.9.

The pipe sizing calculations are recorded on a worksheet depicted in Figure 9.10, which enables the procedure to be followed from the starting point of recording the circuit reference, followed by the heating load for that section stated in watts and converted into the mass flow rate in $kg\,s^{-1}$.

The calculation exercise then proceeds by selecting a suitable pipe size based on the velocity and pressure drop recommendations previously given for small bore low pressure heating systems.

For circuit A–B, selecting a 15 mm diameter copper pipe for a mass flow rate of $0.071\,kg\,s^{-1}$, a pressure drop of $240\,Pa\,m^{-1}$ at a velocity of $0.5\,m\,s^{-1}$ can be read from either the pipe sizing chart on Figure 9.3, or the tabulated version on Figure 9.4.

Having selected a pipe size and tested that it conforms to the general rules on the flow rate and velocity from the pipe sizing chart, the actual pipe length of the flow and return is measured at 8 metres.

The equivalent length of 15 mm pipe for the fittings, including the radiator angle valves, is then assessed using Table 9.1 (5 × 15 mm elbows at 0.6 m = 3 m, plus 2 × 15 mm angle valves at 2.5 m = 5 m, totalling 8 m). This is then added to the actual pipe length to give a total pipe length of 16 m which is then multiplied by the pressure drop of $240\,Pa\,m^{-1}$, resulting in $240 × 16 = 3840\,Pa$.

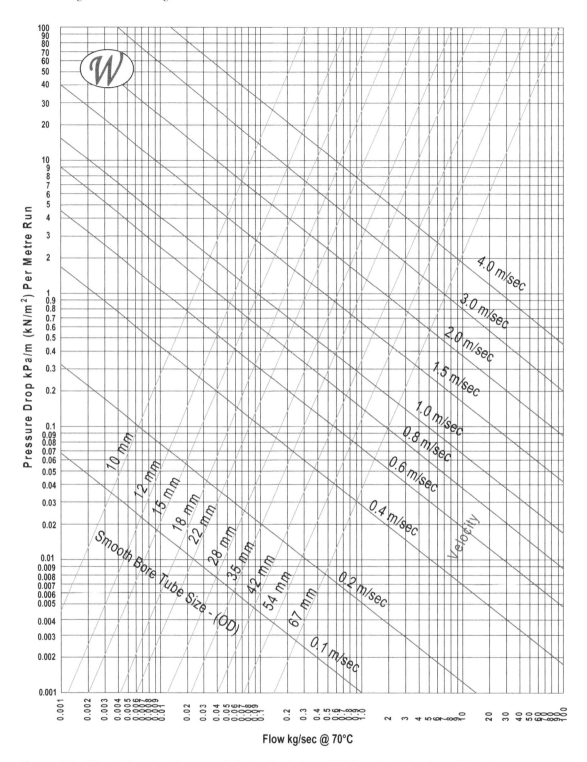

Figure 9.5   Pipe sizing chart for cross-linked polyethylene (PEX) and polybutylene (PB) tube

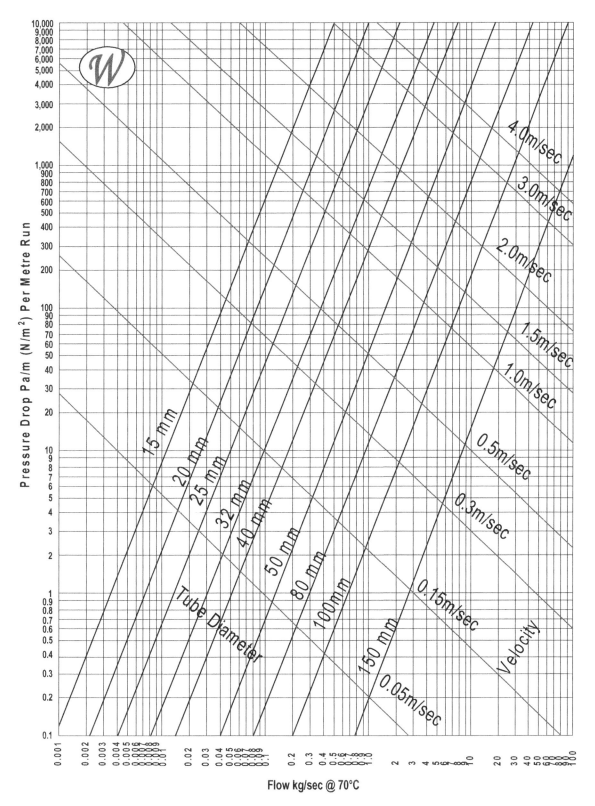

Figure 9.6   Pipe sizing chart for low carbon mild steel tube BS EN10255 (BS 1387) medium weight

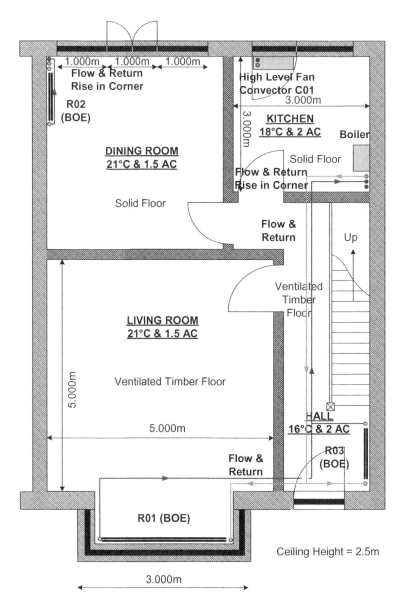

Figure 9.7 Pipe sizing exercise – ground floor plan (not to scale)

The procedure is continued for pipe section circuit C–B with the pressure drop added to the result of the pressure drop for circuit A–B.

As the piping arrangement meets the piping circuit from the first floor, the procedure then should start again, calculating the resistance from circuit D through to L and then from circuit G through to L. From the results it can be seen that the pressure drop G–L is greater than the pressure drop from D–L, so pressure drop from D–L can be disregarded but the flow rate is retained and the procedure continued using the total flow rate from circuit L through to N.

At this stage it is clear that the index circuit will be from circuit A, therefore the procedure continues using the total flow rate for the system from N to P, with the resulting pressure drop added to the pressure drop calculated for circuit A–N.

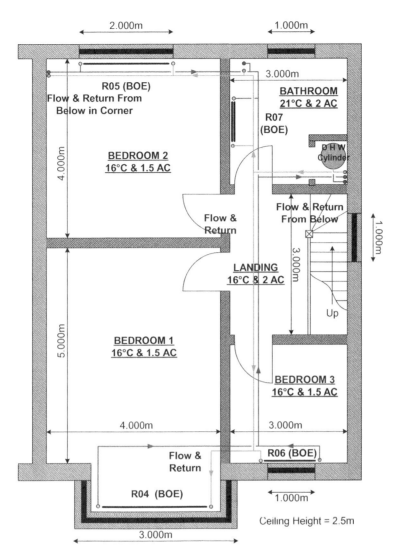

Figure 9.8 Pipe sizing exercise – first floor plan (not to scale)

It can be observed from the pipe sizing calculation worksheet that the resistance for the index circuit amounts to $16\,604\,\mathrm{Pa\,m^{-1}}$; the resistance for the boiler and heat emitter for the index circuit should be added to this figure, which results in a total resistance for the index circuit of $17\,094\,\mathrm{Pa\,m^{-1}}$, which rounds up to 17.1 kPa.

From the results recorded in the calculation worksheet Figure 9.10, the pipe sizes to all pipework circuit sections have been established along with their respective mass flow rates and velocities.

As well as establishing pipe sizes, the circulating pump base duty has also been found, as the circulating pump has to be able to pump the mass flow rate for the entire heating system against the pressure resistance of the index circuit. If the pump can circulate water around the index circuit, it can circulate it around the rest of the heating system.

Therefore the duty required for the circulating pump for the example pipe sizing calculation equals:

$$\text{MFR } 0.38 \text{ kg s}^{-1} \text{ against } 17.1 \text{ kPa.}$$

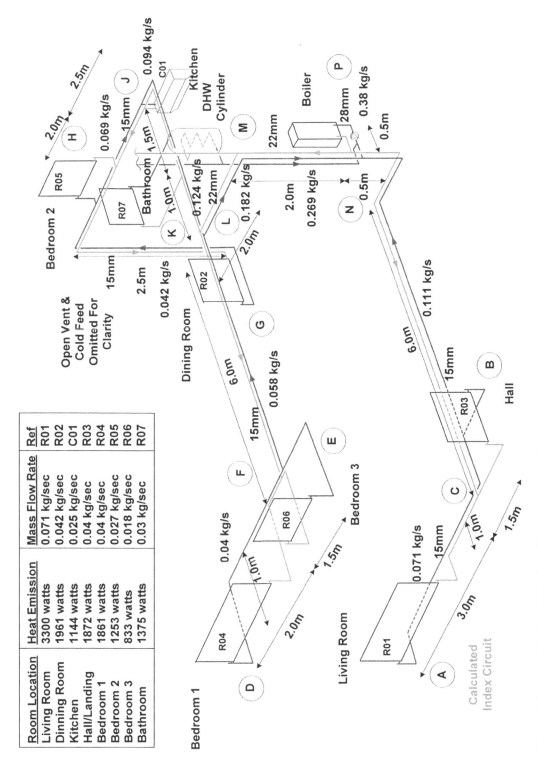

Room Location	Heat Emission	Mass Flow Rate	Ref
Living Room	3300 watts	0.071 kg/sec	R01
Dinning Room	1961 watts	0.042 kg/sec	R02
Kitchen	1144 watts	0.025 kg/sec	C01
Hall/Landing	1872 watts	0.04 kg/sec	R03
Bedroom 1	1861 watts	0.04 kg/sec	R04
Bedroom 2	1253 watts	0.027 kg/sec	R05
Bedroom 3	833 watts	0.018 kg/sec	R06
Bathroom	1375 watts	0.03 kg/sec	R07

Figure 9.9   Isometric layout of pipe sizing arrangement

Pipe Sizing Calculation Sheet

| Project Semi-Detached House Figs 9.7/8 | | | | | | | | Temp Diff Δt 11°C | | | Sheet 1 of 1 |
| Description of Piping System - Two Pipe Small Bore | | | | | | | | Pipe Material – Copper | | | Date |

Circuit Ref	Heating Load Watts	EFR kg/sec	Pipe Size mm	Pressure Drop Pa/m	Velocity m/sec	Pipe Length F & R m	Eq Pipe Length m	Total Pipe Length m	Total Pressure Drop Pa/m	Running Total Pressure Drop Pa	Comments/Notes
A-B	3300	0.071	15	240	0.5	8	8	16	3840		
C-N	5172	0.111	15	560	0.7	16	2.4	18.4	10304	14144	Index Circuit
D-E	1861	0.04	15	90	0.3	6	8	14	1260		
F-L	2694	0.058	15	200	0.4	12	1.2	13.2	2640	3900	
G-H	1961	0.042	15	90	0.3	11	3.6	14.6	1314		
H-J	3214	0.069	15	240	0.5	5	2.8	7.8	1872		
J-K	4358	0.094	15	480	0.7	3	2.6	5.6	2688		
K-L	5733	0.124	22	90	0.4	2	2	4	360	6234	
L-M	8427	0.182	22	280	0.7	4	2	6	1680		
M-N	12427	0.269	22	500	0.9	4	4	8	4000	5680	
N-P	17599	0.38	28	300	0.8	1	7.2	8.2	2460	2460	Add to A-N
Add flow rate for bypass if fitted		0					Index circuit pressure drop			16604	
Total	17599	0.38					Add resistance for boiler			350	
							Add resistance for heat emitter			140	
Total MFR		0.38					Total Pressure Drop			17094	Round total up to 17.1 kPa

Circulating Pump Duty = MFR 0.38 kg/second @ 17.1 kPa

Figure 9.10 Pipe sizing calculation sheet

Added to this base pump duty should be an allowance of 10% to the mass flow rate, to permit a balancing tolerance to the heating system, plus 15% to the pressure generated by the pump for unforeseen pipe installation difficulties, such as additional bends etc, as pipework hardly ever is installed exactly as the working design drawings show.

Therefore the revised duty of the circulating pump will be:

$$\text{MFR } 0.38 + 10\% = 0.418 \text{ kg s}^{-1}, \text{ say } 0.42 \text{ kg s}^{-1}.$$

Pump head, 17.1 kPa + 15% = 19.65 kPa, say 20 kPa.
Therefore the circulating pump should be selected on the duty of:

$$\text{MFR } 0.42 \text{ kg s}^{-1} \text{ against } 20 \text{ kPa.}$$

The selection of the circulating pump is discussed further in greater detail in Chapter 18, Circulating Pumps.

SIMPLER PIPE SIZING EXAMPLE

Figure 9.11 illustrates in a simplistic schematic form a pipe sizing exercise that is easier to follow and understand.

The example portrayed in Figure 9.11 is restricted to just four heat emitters, for the purpose of the exercise and is of unrealistic and exaggerated size. The heating system has been divided into two circuits of equal heat load capacity, but arranged to emit that heat differently. The pipe sizing calculation worksheet records the process, but in this case the equivalent pipe length for fittings and components has been taken as 100% of the actual pipe length for calculation purposes. By following the worksheet through, the index pipe circuit can be found.

If the index circuit proves to have too high a resistance for a standard circulating pump, then revisit the worksheet by selecting a larger pipe diameter at the point of greatest resistance and rework the calculation. This normally has the effect of changing the index circuit to a different circuit.

This process may be repeated a number of times when striving to obtain cost efficient pipe diameters.

BOILER BYPASS FLOW ALLOWANCE

If a bypass is to be fitted to a low water content boiler – which is quite common today for domestic heating systems, to protect the boiler from overheating when the circulation is halted by the control system – then an allowance of 0.1 kg s⁻¹ should be made to the total flow rate during normal operation when sizing the pipes and arriving at the circulating pump duty.

APPROXIMATION OF PIPE SIZES

The pipe sizing procedure can be a time-consuming process, especially on larger heating systems and even on domestic heating systems. This can be costly when in competition to provide an estimate for the heating system.

For budget purposes only, and not as a substitute for conducting an approved engineering pipe sizing calculation, the pipe sizes given in Table 9.2 can be used to assess approximate pipe sizes.

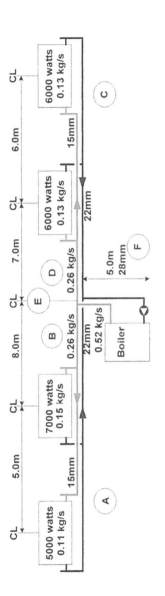

Pipe Sizing Calculation Sheet

Project	Simple Exercise							Temp Diff Δt 11°C			Sheet 1 of 1
Description of Piping System - Two Pipe Small Bore								Pipe Material – Copper			Date
Circuit Ref	Heating Load Watts	MFR kg/sec	Pipe Size mm	Pressure Drop Pa/m	Velocity m/sec	Pige Length F&R m	Eq Pipe Length m	Total Pipe Length m	Total Pressure Drop Pa/m	Running Total Pressure Drop Pa/m	Comments/Notes
A-B	5000	0.11	15	500	0.75	10	10	20	10000		
B-E	12000	0.26	22	400	0.8	16	16	32	12800	22800	
C-D	6000	0.13	15	650	0.8	12	12	24	15600		Index Circuit
D-E	12000	0.26	22	400	0.8	14	14	28	11200	26800	
E-F	24000	0.52	28	420	1	10	10	20	8400	35200	Add to C-E
Total MFR		0.52					Total Pressure Drop – Index Circuit			35200	Total 35.2kPa

Figure 9.11 Simpler pipe sizing exercise

Table 9.2   Budget pipe sizing guide

Pipe Diameter mm	Maximum Heat Load Carrying Capacity kW
8	1.5
10	2.5
15	6.0
22	12.5
28	22.0
35	34.0
42	55.0
54	105.0

Note – The above tabulated heating loads may be exceeded from the given pipe sizes for short pipe runs.

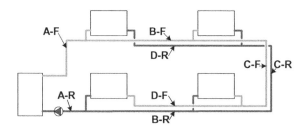

Figure 9.12   Two pipe reverse return pipe sizing example

OTHER PIPING ARRANGEMENTS

A two pipe direct return piping arrangement has been used for the example pipe sizing calculation, but the basic process can also be used for other piping arrangements with the following minor differences.

Two pipe reverse return

The sizing procedure adopted is the same as that for the two pipe direct return system with the exception that the first flow pipe reference A–F is equal to the last return pipe reference A–R, whereby the lengths should be combined for resistance calculation. The procedure continues whereby pipe reference B–F equals B–R, C–F equals C–R, and D–F equals D–R. Apart from this deviation, the procedure is the same.

The branch pipes to each heat emitter are sized in the normal way with the index circuit being the one with the highest resistance added to the main reverse return circuit.

An example is shown in Figure 9.12.

Single pipe system

The single pipe heating system is somewhat simpler to undertake than the two pipe system as it comprises a single calculation for each ring circuit, using the same velocity and pressure drop criteria as previously explained. It is sized to convey the total mass flow rate required by the heating system including the equivalent pipe length for fittings and components.

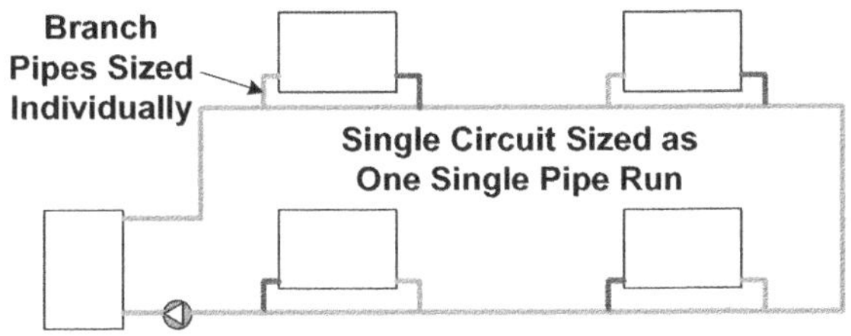

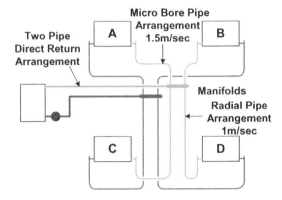

Figure 9.13 Single pipe system pipe sizing

Figure 9.14 Radial or micro bore piping arrangement

The branch pipes are also sized individually with the index circuit being the branch circuit with the greatest frictional resistance, which is added to the resistance from the main single pipe circuit.

An example can be seen in Figure 9.13.

Radial piping systems including micro bore systems

The basic principles of pipe sizing for radial or micro bore heating systems are no different from those of any of the other piping arrangements, with a few minor exceptions.

The piping arrangement from the boiler to the manifold and back again is sized in exactly the same way as for a two pipe direct return system with the velocity not exceeding 1m/sec through this pipe section.

For the micro bore piping heating section, each flow and return piping circuit from the manifold to the heat emitter and back again is sized separately, with an increased velocity of $1.5\,\mathrm{m\,s^{-1}}$ tolerated.

The radial piping heating section is considered in the same way as the micro bore tubing, but the velocity should remain at $1\,\mathrm{m\,s^{-1}}$ unless the piping circuit from the manifold to the heat emitter and back again is free of sharp changes in direction. If this is the case, an increase in velocity to $1.5\,\mathrm{m\,s^{-1}}$ may be considered.

The index circuit will be the circuit having the highest frictional resistance from the manifold to the heat emitter, either circuit A, B, C or D, for both the flow and return, which is then added to the frictional resistance calculated for the two pipe direct return piping section from the boiler to the manifold.

An example is given in Figure 9.14.

For all piping arrangements, the pump duty must be capable of overcoming the frictional resistance encountered by the index circuit, whilst supplying the mass flow rate required for all heat emissions from the heating system.

10

Electricity

Plumbing systems – and in particular heating systems – have become totally dependent upon electricity to provide power for boilers, pumps, water heaters, motorised valves and control equipment. Electrical appliances such as washing machines, dishwashers, tea making boilers and drink dispensing vending machines are also commonplace and are plumbed-in to a water supply and waste pipework system. This interface and dependency on electricity is just as common in domestic dwellings as it is in non-domestic buildings.

During the electrification of the nation in the late 1800s, the first electrical systems installed inside buildings were lighting systems replacing the gas lighting that existed at that time. As electricians had yet to be trained, it was only natural to turn to qualified plumbers to install the conduit and cables required, as plumbers possessed the necessary skills to undertake this new form of work.

The two trades have long parted company, with electricians born from the plumbing industry developing and going their own separate ways, but because of this dependency and interfacing with electricity it is now essential for all plumbers to gain a working appreciation of the subject and become competent in electrical safety.

This necessity to become competent is required for the following reasons:

- For all operatives to remain safe when working on piping systems that may have become live due to an electrical fault occurring.
- For all occupants of the building to remain safe both during their normal living environment and during periods of maintenance to all plumbing and pipework services, particularly if an electrical fault should occur.
- To understand the electrical power requirements for all equipment incorporated into the plumbing and heating system so as to ensure that it is functioning correctly.
- To have a knowledge of electrical wiring controls to understand the control systems.

Heating Services in Buildings: Design, Installation, Commissioning & Maintenance, First Edition. David E. Watkins.
© 2011 John Wiley & Sons, Ltd. Published 2011 by John Wiley & Sons, Ltd.
As an aid to lecturers and students, full colour versions of the figures in this chapter may be found at www.wiley.com/go/watkins
These figures are © 2011 by John Wiley & Sons, Ltd.

The legislation governing electrical safety includes the following:

- The Health and Safety at Work Act
- The Electricity Act
- The Electricity at Work Regulations
- The Building Regulations Approved Document Part P (Part N in Scotland)
- The Electricity Safety, Quality and Continuity Regulations
- The Electricity (Standards of Performance) Regulations

As well as the above, BS 7671: 'Requirements for Electrical Installations' – previously referred to as the IEE Wiring Regulations – is not a statutory requirement itself, but as it is referenced in the Building Regulations, it should be complied with. This is a design standard, and knowledge of its contents, together with its interpretation and application, should be obtained by anyone working on low voltage electrical systems.

The IEE (Institution of Electrical Engineers) is now known as the IET (Institution of Engineering Technology).

It should be emphasised that this chapter is designed to give the reader an understanding and a basic respect for, knowledge and appreciation of electrical systems so that they remain safe. It is far from a comprehensive coverage of the subject and is not a substitute for full electrical training leading to becoming a qualified electrician.

HISTORY AND NATURE OF ELECTRICITY

Electricity is dirived from the Ancient Greek *electron*, meaning amber, leading to the modern Latin name *electricus* and eventually to the English terms *electric* and *electricity*. It may be described as a force created by the movement of subatomic particles (electrons and protons), which interact with electromagnetic fields and cause attractive and repulsive forces between them, whereby the electrons are released and allowed to flow in the form of electricity. Electricity is invisible, odourless and mostly silent, but if touched, it can kill or inflict serious injury and must therefore be treated with respect. It is an energy source that today's world relies on.

Although experiments had been conducted with electricity over the centuries, modern day understanding, generation and distribution of electricity owes a debt to many scientist, physicists and engineers – the most notable include Michael Faraday (1791–1867), André-Marie Ampère (1775–1836), Alessandro Volta (1745–1827), Georg Ohm (1789–1854), Heinrich Hertz (1857–1894), Nikola Tesla (1856–1943), Sebastian Ferranti (1864–1930), Thomas Edison (1847–1931) and George Westinghouse (1846–1914).

DEFINITIONS OF ELECTRICAL TERMS

The concepts in electricity are understood as follows.

Electrical charge

The SI unit for quantity of electricity or electrical charge is the 'coulomb', named after Charles Augustin de Coulomb (1736–1806) and denoted by the symbol 'C', with the symbol Q used to denote the quantity of electrical charge. The coulomb represents approximately 6.24×10^{18} (6 240 000 000 000 000 000) elementary charges (the charge on a single electron or proton) or number of electrons. The coulomb is defined as the quantity of charge that has passed through the cross-section of an electrical conductor carrying one ampere within one second.

Voltage

This is the potential energy per unit or electric potential difference, sometimes described as electromotive force. It is measured as the number of joules of work required to force one coulomb of electrons along a conductor. As the electrical energy moves along the conductor the energy becomes weaker. Voltage is defined as the potential difference across the conductor when the current of a single ampere dissipates one watt of power, expressed as one joule of energy per coulomb of charge.

Voltage may be likened to being the unit of electrical pressure that forces the electrical energy along the conductor whereby one joule/coulomb = 1 volt.

Current

An electric current is a flow of electric charge through a conductor such as a metal wire, and its intensity is measured in amperes, normally abbreviated as amps. An ampere may be defined as a flow of one coulomb of electrical energy in one second.

Electrical current may be either a direct current (DC) which is a unidirectional flow, or an alternating current (AC) which has repeated reversals of the flow direction.

Electrical power

The electrical power is the rate at which electrical energy is produced – or the rate of consumption by electrical equipment – and is measured in watts. All electric motors, appliances and luminaries (light bulbs) have a rating expressed in watts or kilowatts.

Electrical resistance

The flow of electricity in a conductor will encounter a resistance just as the flow of water in a pipe will meet a resistance. The resistance to electrical flow or current is measured in ohms and may be calculated by using Ohm's Law, which states the relationship between the electrical quantities is: voltage will equal the current multiplied by the resistance. This is depicted in 'Ohm's Law triangle' in Figure 10.1.

ELECTRICAL GENERATION, SUPPLY AND DISTRIBUTION

Electricity is generated at power stations by an alternator being rotated by a prime mover. An alternator is an electromechanical device that converts mechanical energy into an alternating current (AC) electrical energy by use of a rotating magnetic field. Alternators comprise a rotating electromagnet termed a rotor, which is turned within a stationary set of three conductors wound in coils on an iron core called the stator.

Where V = volts, I = amps and R = ohms

Figure 10.1   Ohm's Law Triangle

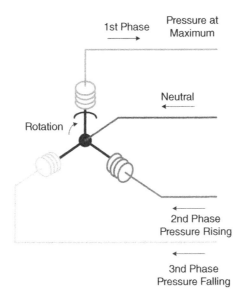

1st Phase   Pressure at
                  Maximum

Neutral

Rotation

2nd Phase
Pressure Rising

3nd Phase
Pressure Falling

Figure 10.2   Three-phase alternator

The mechanical input causes the rotor to turn and pass across the conductors in sequence, thereby inducing an electrical current.

The currents are sinusoidal functions of time all at the same frequency, and the coils are equally spaced around the machine at 120° and the three supplies alternate together. As the electromagnet passes one of the coils the voltage reaches its peak pressure in that conductor, whilst falling away in the other two.

The speed of the alternator is arranged so that each peak in electrical pressure occurs in each coil every $\frac{1}{50}$ of a second, or 50 times every second. The current builds up to a maximum in one direction and drops to zero in the first $\frac{1}{100}$ of a second, it then builds up to a peak in the opposite direction and drops to zero again in the next $\frac{1}{100}$ of a second, making the entire rotation cycle $\frac{1}{50}$ of a second. Figure 10.2 demonstrates the principles of the alternating current.

Traditionally, the prime mover in power stations has been fossil fuels such as coal, gas and oil, which are used to either raise steam to drive a turbine and in turn rotate the alternator, or fuel an engine to turn the alternator. Nuclear power and hydro-electric turbines have also been used to generate electrical energy in large quantities. Alternative energy forms, such as wind power and wave power, have still to prove that they can generate large quantities of electricity, and solar power, geothermal power and photovoltaic power have only been successful for individual buildings. The use of biomass fuels has yet to be proved.

Compliance with the European Harmonisation Directives and Regulations means that the electricity supply authorities declared supply voltages are now 400V/230V plus 10% or minus 6%, which on a single phase electrical supply of 230 volts means that the voltage range can be between 253 volts and 216.2 volts with the declared frequency being within 1%.

The National Grid's transmission system in the United Kingdom is designed to operate at voltages of up to 400kV. Figure 10.3 illustrates the electrical supply transmission network.

The electrical transmission system forms an interlinking National Grid with other power stations, passing through a number of step down transformers that reduce the mains supply voltage from 400kV to 11kV and then to 400V. The 11kV supply would be supplied to buildings requiring a high voltage power requirement, whilst the 400V electrical distribution would supply commercial properties with a 230V supply to each domestic dwelling.

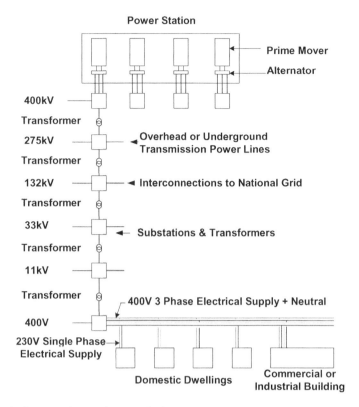

Figure 10.3   Electrical generation and transmission arrangement

The electricity supply authorities' main distribution at the end of the National Grid to commercial and domestic dwellings comprises a cable that incorporates four conductors, three of which are live phases connected to the area substation, whilst the fourth conductor is the neutral and is solidly earthed back to the substation.

The electrical supply to commercial and industrial buildings not requiring a high voltage supply would be a 400 volt, 3-phase, 50 Hertz (400/3/50) supply, and a domestic dwelling would be provided with a single-phase supply from the 3 phase distribution system that achieves 230 volt, single phase, 50 Hertz (230/1/50).

The three phase electrical supply shown in Figure 10.4 is a means of supplying three times as much electricity along the conductors without having to increase the diameter of the conductors, and is used in industry to drive motors and other devices.

The electrical supply to each domestic dwelling is a single-phase 230 volt distribution expressed as 230/1/50. It is achieved by connecting each dwelling to an alternate phase of the main electrical supply, meaning that every third property is connected to the same phased supply, as illustrated in Figure 10.3. The commercial building depicted in Figure 10.3 would require a larger electrical supply and is therefore supplied with all three phases having a voltage between the phases of 400 volts.

Domestic dwellings in Europe, including the United Kingdom, have a harmonised electricity supply of 230 volts at 50 Hertz (Hz), whereas the United States and Canada have a 110 volt, 60 Hz electricity supply.

The single-phase electrical supply is depicted in Figure 10.5, which is similar to the three-phase supply shown in Figure 10.4, except that only one of the phases is used. For a 50 Hz frequency, the current builds up to a maximum oscillation peak of 339 volts in one direction and then drops to zero in the first $\frac{1}{100}$ of a second. It then builds up to a peak of 339 volts in the opposite direction and drops to zero again in the

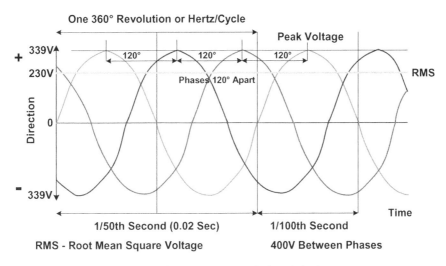

Figure 10.4 Three-phase alternating (AC) electrical supply (400/3/50)

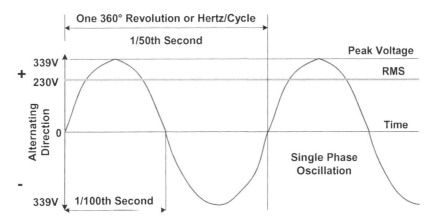

Figure 10.5 Single-phase alternating (AC) electrical supply (230/1/50)

next $\frac{1}{100}$ of a second, making the entire cycle in $\frac{1}{50}$ of a second by flowing first in one direction and then reversing the flow in the opposite direction.

DIRECT CURRENT (DC)

The National Grid electricity supply network delivers electricity utilising an alternating current (AC) supply – which means the flow alternates in direction along the path of the conductor, changing direction at a rate of fifty times a second.

Direct current or continuous current may be described as a constant flow of electrons in a single direction from low potential to high potential, and was formally referred to as 'galvanic current'.

The first commercial electric power transmission, which was developed by Thomas Edison in the late nineteenth century, used direct current to provide a system of electric lighting for approximately 50 residential properties in New York. Because of the advantages of the alternating current over direct current being able to transform the transmission more easily, electric power distribution is almost all alternating current.

All electric batteries discharge electricity by direct current by a continuous flow in the same direction without any oscillations, but electricity may also be generated to provide direct current, or transformed from an alternating electricity supply.

Two of the main users of electricity employing a direct current supply are the London Underground railway and the mainline railway formerly in the ownership of the Southern Railway; both use a direct current supplying electricity at between 650 and 800 volts. The overhead power lines used for the mainline railway use an alternating current electrical supply at 25 kV. Both types of electricity sypplies should be treated with great caution and respect.

CABLE IDENTIFICATION COLOURS

From April 2004 a new system of harmonised cable identification colours has been agreed and implemented throughout the European Union, as shown in Table 10.1.

VOLTAGE DEFINITIONS

Voltages used in industry may be defined by the classifications given in Table 10.2.

Table 10.1   Cable identification colours

Country	Cable Identification				
	Phase L1	**Phase L2**	**Phase L3**	**Neutral**	**Earth**
UK Installed before April 2006 3-Phase Supply	Red	Yellow	Blue	Black	Green/Yellow Stripes, Green before 1970
UK & Europe from April 2004 3-Phase Supply	**Brown**	**Black**	**Grey**	**Blue**	**Green/Yellow Stripes**
UK & Europe Single-Phase Supply	**Brown**	**N/A**	**N/A**	**Blue**	**Green/Yellow Stripes**
United States & Canada 3-Phase Supply	Brown	Orange	Yellow	Grey or White	Green
South Africa 3-Phase Supply	Red	White	Blue	Black	Green/Yellow Stripes
Australia & New Zealand 3-Phase Supply	Red	White	Blue	Black or Light Blue	Green/Yellow Stripes

Table 10.2   Voltage definitions

Classification	Definition
Extra-Low Voltage (ELV)	Normally not exceeding 50 volts AC
Low Voltage (LV)	Normally exceeding extra-low voltage but not exceeding 1000 volts AC. *This includes:* *110 volts AC – Site temporary voltage* *230 volts AC – Single-phase domestic electricity supply* *400 volts AC – Three-phase electricity supply*
High Voltage (HV)	Normally exceeding low voltage

BUILDING WIRING CIRCUITS

As previously established, electricity is supplied through the National Grid, providing a declared 230/1/50 electricity supply to each domestic residential dwelling or a 400/3/50 electricity supply to larger consumers of electricity in commercial or industrial buildings. This supply is taken off the three-phase distribution mains as shown in Figure 10.3, with the live phase being coloured brown, formerly red before April 2006, together with a neutral connection coloured blue, formerly black before April 2006.

The actual method of arranging the incoming electrical supply to each dwelling can vary and is explained further in the 'Earthing' section below. The incoming electrical supply is terminated by the supply authority by the installation of the main fuse, referred to as the 'company fuse', together with the electricity meter. These components remain the property of the electricity supply authority and should not be tampered with. The electrical installation from this point onwards becomes the responsibility of the property owner and should be in complete compliance with the latest edition of the Wiring Regulations for equipment in buildings.

CONSUMER UNIT

The incoming electrical supply is taken from the electricity meter into the consumer unit and terminated at the double pole isolation switch incorporated in the unit – double pole meaning that it switches both the live and neutral conductors. The consumer unit is a plastic constructed box, or a metal box on older consumer units. It comprises a common copper live busbar conductor that miniature circuit breakers (MCBs) are connected to, making the incoming electricity supply to them permanently live when the main isolation switch is made contact with. On older installations these would be rewirable fuses that are intended to be the weakest link to protect the circuit during fault conditions, whereby any sudden surge of electricity would cause the fuse wire to melt and cut the flow of electricity off. The fuse wire would require replacing with the correct wire gauge rating when the fault had been identified and remedied. Miniature circuit breakers have the advantage that the switch will trip during surge conditions and can be simply reset when the fault has been rectified.

Table 10.3 lists the maximum MCB and fuse wire ratings for circuit protection together with minimum cable conductor size.

The number and amperage rating of the MCBs or fuses will depend upon the number of circuits required, which will be determined by the building's electrical requirements, but it is always good practice to allow a couple of spare ways for any future requirements.

Figure 10.6 illustrates the internal arrangement of a typical consumer unit where each circuit is protected by its own dedicated miniature circuit breaker or fuse and can be made safe by switching off the MCB or removing the fuse to any particular circuit.

Table 10.3   Protective circuit breaker and fuse ratings

Circuit Designation	MCB Rating – Amps	Rewirable Fuse Rating – Amps	Minimum Cable Size mm²
13 Amp Ring Mains (1 or 2 per dwelling)	32	30	2.5
Lighting Circuits (1 or 2 per dwelling)	6	5	1.5
3 kW Immersion Heaters	16	15	2.5
Electric Ovens & Hobs	40	30	6.0 or 10.0
Electric Power Showers	45	N/A	10.0
Heating Control System*	N/A	N/A	1.0

*The fuse for the heating control system and power source is not incorporated within the consumer unit.

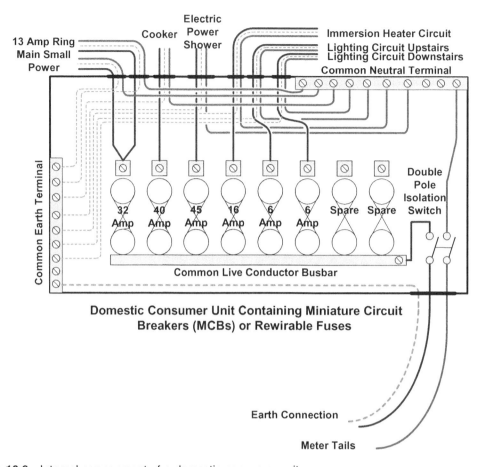

Figure 10.6   Internal arrangement of a domestic consumer unit

It can be seen that all neutrals are returned to a common terminal strip contained inside the consumer unit that in turn is connected to the main isolation switch incorporated in the consumer unit. Similarly, all earth returns are collected together by a common earth terminal strip, which can then be arranged to be connected to the main earth protection system for the building. See the 'Earthing' section below for further details.

The consumer unit can then function as a distribution board to all wiring circuits in the building.

RING CIRCUITS

Figure 10.7 illustrates a typical building wiring system for a domestic dwelling, including the final ring circuit commonly referred to as 'ring mains'. Since its introduction shortly after the Second World War, this wiring arrangement has become the recognised method of providing power to socket outlets to supply electricity for the use of domestic appliances. The socket outlets have a maximum rating of 13 amps and are designed to accept a square three-pin plug which contains a replaceable cartridge fuse.

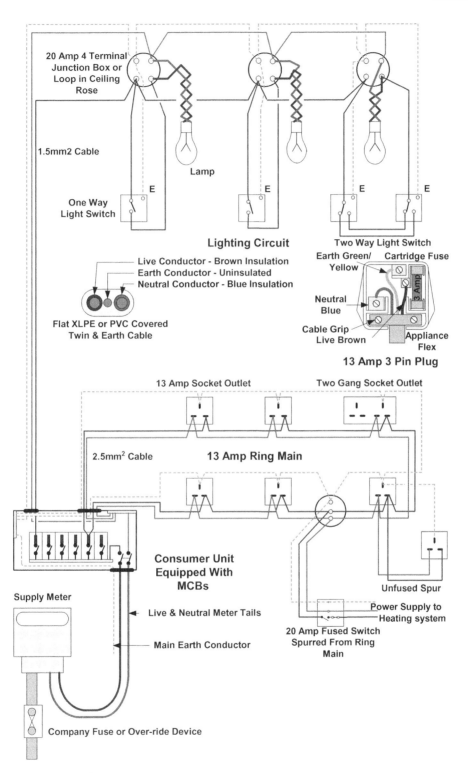

20 Amp 4 Terminal
Junction Box or
Loop in Ceiling
Rose

1.5mm2 Cable

Lamp

One Way
Light Switch

E

E

E

E

Lighting Circuit

Two Way Light Switch

Live Conductor - Brown Insulation
Earth Conductor - Uninsulated
Neutral Conductor - Blue Insulation

Flat XLPE or PVC Covered
Twin & Earth Cable

Earth Green/
Yellow

Cartridge Fuse

Neutral
Blue

Cable Grip
Live Brown

Appliance
Flex

3 Amp

13 Amp 3 Pin Plug

13 Amp Socket Outlet

Two Gang Socket Outlet

2.5mm² Cable

13 Amp Ring Main

Consumer Unit
Equipped With
MCBs

Supply Meter

Live & Neutral Meter Tails

Main Earth Conductor

Company Fuse or Over-ride Device

Unfused Spur

Power Supply to
Heating system

20 Amp Fused Switch
Spurred From Ring
Main

Figure 10.7   Building wiring systems

The weakness of this system is that it is dependent on the correct cartridge fuse rating being placed inside the plug to protect the appliance that it is connected to. The fuse rating may be calculated by the following formula:

$$Watts \div Volts = Amps$$

Therefore, for a 1 kW electric heater the fuse rating required would be:

$$\frac{1000}{230} = 4.35 \, amps$$

Select the nearest standard fuse rating above this figure, which would be 5 amps.

The ring main consists of a twin and earth conductors connected to a 30 amp fuse or 32 amp circuit breaker housed inside the consumer unit and extended around the building. It connects into each socket outlet before returning to the same fuse or circuit breaker, thus forming a ring. Current can therefore reach each socket outlet from two directions, which reduces the risk of any cable section becoming overloaded.

There is no limitation to the number of socket outlets installed on any ring circuit, but the floor area served by a single ring circuit should not exceed 100 m². If the floor area does exceed this figure, then an additional ring circuit should be installed. Normally it is considered good practice to arrange for a ring circuit for each floor, thus having two ring circuits for a two-storey house.

Ring circuits are normally wired in a 2.5 mm² cable, but for exceptionally long circuits a cable size of 4.0 mm² may be required.

The socket outlets are normally fixed to a metal box that has been chased or sunk into the wall to permit a flush finish. The cable to and from the socket outlet should be contained inside a conduit that has also been chased into the wall, which will allow a degree of flexibility in the cable connections when the plate has been removed from the metal box. The socket outlet should also have an earth link from its earth terminal back to the metal box, which should have an earth lug to accommodate this so that the socket outlet plate remains earthed when it has been removed from the metal box.

SPUR POINTS

Socket outlets and appliances may be connected in the form of a spur point, as shown in Figure 10.7 and detailed in Figure 10.8. This is achieved by connecting a single cable from the back of a socket outlet on the ring circuit and connecting to another socket outlet which does not form part of the ring main. This is an acceptable method of adding additional socket outlets to the system, or connecting to single outlets located some distance from the ring main such as outbuildings.

At one time a maximum of two unfused spur points were permitted to be taken from any point on the ring circuit, but now only one single socket outlet, or one two-gang socket outlet is permitted to be taken as a spur from any point on the ring circuit. However, two fused spur points can be taken from any single point on the ring circuit wired up in series. The total number of spur points should not exceed the total number of socket outlets on the ring circuit.

Spur points can also be taken directly off the ring circuit, as shown in Figure 10.8, using of a three pin 30 amp junction box. This is a convenient method of connecting the heating control system and boiler/pump power supply to the electrical system. Here the cable outer sheathing is removed and the conductors bared of insulation to connect into the junction box terminals as shown, along with the spur cable to the fused switched connection unit. This permits individual isolation and control over the heating system's electrical system.

The cable size to the heating control system can be reduced to 1.0 mm² or 1.5 mm² on the outlet of the fused switched connection unit as it is now protected by its own system fuse rated at 3 amp for domestic heating systems.

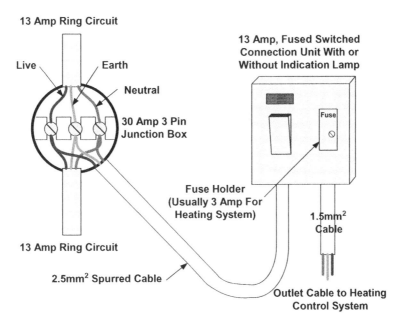

13 Amp Ring Circuit

Live

Earth

Neutral

30 Amp 3 Pin Junction Box

13 Amp, Fused Switched Connection Unit With or Without Indication Lamp

Fuse

Fuse Holder (Usually 3 Amp For Heating System)

1.5mm² Cable

13 Amp Ring Circuit

2.5mm² Spurred Cable

Outlet Cable to Heating Control System

Figure 10.8 Spur connection for heating system

It should be noted that if 4.0 mm² cable has been used for the ring circuit, then trying to fit nine conductors of this size into the junction and wall box, together with the terminals on the socket outlet or junction box is extremely difficult.

LIGHTING RADIAL CIRCUITS

The lighting circuit differs from the small power ring circuit in so far as it takes the form of a radial system rather than a ring circuit. The lighting radial circuit is extended from the consumer unit – where it is protected by either a 5 amp fuse or 6 amp circuit breaker – to each lighting point, which is connected up by using a four terminal 20 amp junction box as shown in Figure 10.7, or by the use of a loop-in terminal ceiling rose as detailed in Figure 10.9.

It can be seen from Figure 10.9 that when two-core twin and earth cable is used, the cable drop to the light switch means that the neutral-coloured conductor is often used as the return back to the ceiling rose. This could lead to confusion to any inexperienced electrician.

IMMERSION HEATER WIRING

An electric immersion heater should be wired up separately from its own dedicated 16 amp circuit breaker or 15 amp rewirable fuse in the consumer unit, and should not form part of any other electrical circuit (see Figure 10.10).

For a 3 kW rated immersion heater, the fuse protection rating can be calculated as follows:

$$\frac{3000 \text{ watts}}{230 \text{ volts}} = 13 \text{ amps}$$

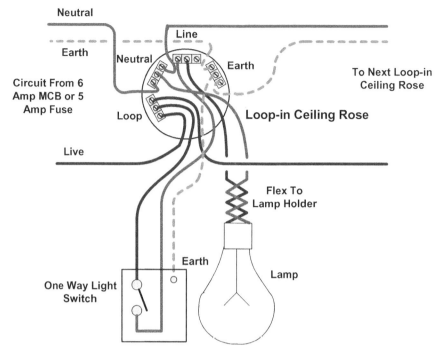

Figure 10.9   Loop-in lighting arrangement

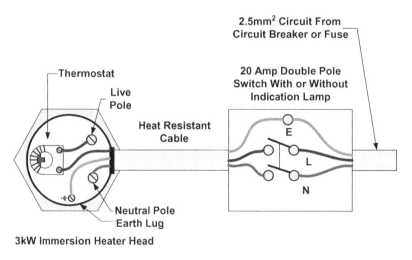

Figure 10.10   Electrical wiring for immersion heaters

The nearest standard fuse rating available would be 15 amp, or a 16 amp circuit breaker.

A single 2.5 mm² twin and earth cable is extended from the consumer unit to a 20 amp double pole switch located adjacent to the immersion heater; the final cable from this switch to the immersion heater should be of the heat resistant type, because it is subjected to heat from the domestic hot water cylinder.

The internal wiring within the immersion heater head is quite simple: the live is connected to the thermostat, which in turn is linked to the terminal on the end of the live pole, with the neutral conductor connected to the other end of the immersion heater to complete the circuit.

The immersion heater will operate under the control of its own thermostat.

EARTHING

All electrical circuits in a fixed installation must have a protective earth conductor, commonly referred to as the circuit protective conductor (CPC). All conductive parts of the electrical installation that may become live and conductive under fault conditions must be connected to the circuit protective conductor.

Earthing may be defined as the process of connecting to the earth by means of electrically conducting materials the conductive parts of an installation which are not normally subject to a voltage or electrical charge.

All exposed conductive components and equipment should be connected to this protective earth and therefore consequently to each other. In the event of an earth fault occurring, the voltage between simultaneously accessible parts will not be of a magnitude and duration sufficient to become dangerous. This type of protection is known as 'earthed equipotential bonding' and causes automatic disconnection of the electrical supply.

In the event of a fault occurring, the current will flow through the live conductor and back to the source of the electricity supply through the protective earth conductor. The extent of this current must be sufficiently large to cause the upstream protective devices, such as the miniature circuit breaker or fuse wire, to trip in the shortest time possible thus shutting off the flow of electricity. The Wiring Regulations require that this protective device should operate within five seconds where the electrical circuit supplies fixed equipment, or 0.4 seconds where it supplies socket outlets.

The function of this protective earth conductor is to provide an easy path for the flow of electrical current in the event of an electrical fault developing, rather than through the body of any unfortunate person who may come into contact with an appliance connected to the system via a socket outlet. The current that will flow is determined by the voltage and the overall resistance or impedance of the circuit along the live conductor and back through the circuit protective conductor, including any resistance of the external connections and the source itself. This resistance is generally termed the earth fault loop impedance.

Electricity will always flow through the circuit with the least resistance and therefore the earthing system should be designed to provide a low-resistance path for the electricity to flow to earth. This includes any exposed installation metalwork so that in the event of a fault occurring, a very high current will take that direct path and almost instantaneously operate the circuit protective devices such as circuit breakers or fuses.

On a 230 volt, 50 Hz alternating current electrical supply it is the voltage or potential difference that drives a current or quantity of electricity around the circuit; the current carrying conductors offer a resistance to the flow of electricity.

This may be calculated by using Ohm's Law which states:

$$\text{Ohms} = \frac{\text{Volts}}{\text{Amps}}$$

Therefore in a heating system consuming a combined total electrical loading of 500 watts for the circulating pumping equipment, boiler and all control components, the amperage would be:

$$\frac{500}{230} = 2.18 \, \text{amps}$$

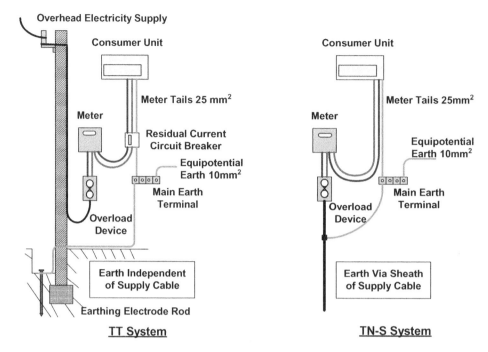

Figure 10.11   TT and TN-S earthing arrangements

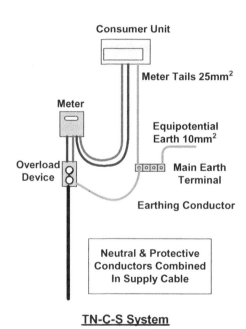

Figure 10.12   TN-C-S earthing arrangement

Therefore the system should be protected by a standard 3 amp cartridge fuse, and the resistance or impedance encountered will be:

$$\frac{230}{2.18} = 105.5 \text{ ohms}$$

Hence, if the earth path provides a lower resistance of, say, 10 ohms, then the current flow will be increased by the following:

$$\frac{230}{10} = 23 \text{ amps}$$

As the heating system is protected by a 3 amp fuse, the increased flow will cause the fuse to trip in the event of a fault and so protect the electrical wiring and equipment from overheating and prevent a dangerous situation occurring.

The common forms of electrical supply entry into the building, together with the method of providing for an earth return, are listed and explained below. They are classified by the use of a letter designation, such as IT, TT, TN-S and TN-C-S. These may be defined as follows.

Definitions

SUPPLY – earthing arrangements are indicated by the first letter

I – Supply system not earthed, one or more points earthed through a fault-limiting impedance.

T – The T stands for the French word TERRA, meaning 'earth'. One or more points of the supply are directly connected to earth.

INSTALLATION – earthing arrangements are indicated by the second letter

T – Exposed conductive parts connected directly to earth.

N – Exposed conductive parts connected directly to the earthed point of the source of the electrical supply.

EARTHED SUPPLY CONDUCTOR – arrangement is indicated by the third letter

S – Separate neutral and protective conductors.

C – Neutral and protective conductors combined in a single conductor.

Earthing classifications

- IT – This system has an isolated or impedance-earthed neutral point with no distributed protective conductor and is not permitted on public supply networks.
- TT – This system has an earthed neutral point but the protective conductor is not distributed. This installation must be earthed by an electrode system at the point of intake. See Figure 10.11.
- TN-S – This system also has the neutral point solidly earthed and has separate neutral and protective conductors. See Figure 10.11.
- TN-C-S – Sometimes referred to as 'protective multiple earthing' (PME). This is the most commonly employed form of protective multiple earthing and is also referred to as the combined neutral and earth (CNE) conductor or protective earthed neutral (PEN) conductor. See Figure 10.12.

11

Controls, Components and Control Systems

The broad term 'controls' covers a wide and varied subject. If we examine the word it may be defined as a means of restraint, regulating and checking – all applicable to the form of control found in a heating system – but the type of controls in a heating system may be classified into two main categories, namely:

- Controls for safety
- Controls for comfort and energy efficiency

Both categories of control have completely different functions, but both are equally important to the operation of the heating system. Therefore this chapter has been divided into two main sections covering these categories.

CONTROLS FOR SAFETY

There are various controls that are applied to a heating system whose only purpose is to perform a critical safety function in the event of a system or component malfunction, in order to keep the heating system safe and prevent a dangerous – or even fatal – situation occurring.

With the one exception of the system safety relief valve, almost all of these safety components or devices are fitted as an integral part of a heating appliance – including the burner safety controls, which are discussed further under the specific headings for the fuels concerned. However, the system designer and installer have a responsibility to ensure that the heating system remains safe at all times and incorporates the means to overcome the risks associated with any component failure.

As the design engineer and installer of the heating system can be held responsible for a possible unsafe condition occurring, then some form of HAZOP risk assessment and analysis – together with any mitigation measures taken or considered – should be undertaken, even for domestic heating applications.

Heating Services in Buildings: Design, Installation, Commissioning & Maintenance, First Edition. David E. Watkins.
© 2011 John Wiley & Sons, Ltd. Published 2011 by John Wiley & Sons, Ltd.
As an aid to lecturers and students, full colour versions of the figures in this chapter may be found at www.wiley.com/go/watkins
These figures are © 2011 by John Wiley & Sons, Ltd.

SYSTEM SAFETY RELIEF VALVES

These are valves for automatically preventing a safe pressure being exceeded, and are categorised as:

- *Safety Valve* – A valve which automatically discharges steam, gases or vapours to prevent a predetermined safe operating pressure being exceeded. These valves usually have a precise action and obtain their rated discharge capacity with a rise in pressure above the operating pressure of 10%.
- *Relief Valve* – A valve which automatically discharges liquid to prevent a predetermined safe operating pressure being exceeded. The term is commonly used for pressure relieving valves in which the lift is proportional to the increase in pressure above the set operating pressure.
- *Safety Relief Valve* – A valve which automatically discharges liquid, gases or vapours to prevent a predetermined safe operating pressure being exceeded. Valves used for heating systems employing liquid as the heating medium should be of this type.

The function of the system safety relief valve is to relieve any increase in system pressure should this rise above normal for any reason. In circumstances such as the cold feed and open vent pipe becoming blocked or frozen, and the boiler thermostat not functioning correctly, the safety relief valve is the last resort to prevent an explosion occurring.

The safety relief valve should be located either directly on the boiler using a spare pipe tapping if it exists, or on the flow pipework as close to the boiler as practical, in order for it to be able to relieve the pressure at the main point of generation.

The safety relief valve should also be accessible so that it may be checked during any annual maintenance or servicing periods to ensure that it is still functioning at the correct pressure.

The safety relief valve set operating pressure should not be higher than the design working pressure of any of the equipment or system components contained on the heating system, but there must be an adequate design margin between this set operating pressure and the system's working pressure.

Figure 11.1 indicates the relationship between the system static head pressure and the safety relief valve's set operating pressure for open vented heating systems. This may be calculated using the following formula:

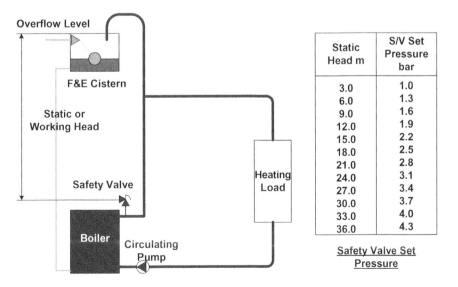

Static Head m	S/V Set Pressure bar
3.0	1.0
6.0	1.3
9.0	1.6
12.0	1.9
15.0	2.2
18.0	2.5
21.0	2.8
24.0	3.1
27.0	3.4
30.0	3.7
33.0	4.0
36.0	4.3

Safety Valve Set Pressure

Figure 11.1 Safety relief valve set pressure/static head

Table 11.1   Minimum safety relief valve sizes

Boiler Rating kW	Minimum Size of Safety Relief Valve
0–39	20
40–99	25
100–249	32
250–399	40
400–499	50
500–599	65
600–799	80
800–999	100

$$\frac{\text{Safety Relief Valve}}{\text{Set Operating Pressure (bar)}} = \frac{\text{System Static Head (m)}}{10.2} + 0.7$$

The static head is obtained by measuring the vertical height above the safety relief valve to the top, or overflowing level, of the feed and expansion cistern.

The static head for safety relief valves for sealed heating system applications is replaced by the compression pressure found by using the following formula as explained in Chapter 2:

$$P_2 = \frac{P_1 V_1}{V_2} + 0.7 \text{ bar}$$

The inlet diameter of the safety relief valve may be obtained from the information given in Table 11.1; the minimum size should not be less than 20 mm (¾ inch), including all domestic heating systems – open vented as well as sealed. This size includes a fouling factor that may reduce the internal diameter over a period of years.

Safety relief valves are normally of the spring type: see Figure 11.2, which illustrates the workings of a typical insurance approved spring type safety relief valve, whereby the set operating pressure can be set and locked in position to prevent any unauthorised tampering. The locking bar also allows the valve to be periodically tested by raising the testing bar, which lifts the spindle against the spring, raising the valve disk off the valve seating to check that it has not become stuck and still functions correctly.

The operation of this type of safety relief valve is quite simple as the tension of the spring holds the valve disk down on the seating at the set operating pressure. When this pressure is exceeded for some reason, it lifts the valve disk off its seating, allowing water to pass into the body and discharge through the side outlet, thus relieving the increase in pressure. The pressure is adjusted by removing the padlock and sliding the locking/testing bar out to allow tightening, or loosening the pressure adjustment screw to tension the spring until the set operating pressure is achieved.

The discharge outlet from the safety relief valve should be piped away to drain, using a pipe of equal diameter to the valve outlet or larger, and it should terminate in a safe manner that cannot cause injury to anyone. This is particularly important regarding the discharge from safety relief valves fitted on pressurised sealed heating systems operating at higher temperatures, as the sudden release of pressure could cause the high temperature water to flash to steam in a violent manner. The discharge outlet from the safety relief valve is normally terminated below the grating of an open top trapped floor gully or drain, but it should be stressed that the drainage material should be suitable for withstanding the possible temperature that may be expected from the valve when relieving the pressure.

Figure 11.3 illustrates a less expensive version of the spring type safety relief valve that is suitable for an open vented domestic heating system and provides the last line of protection in the event of all other provisions failing.

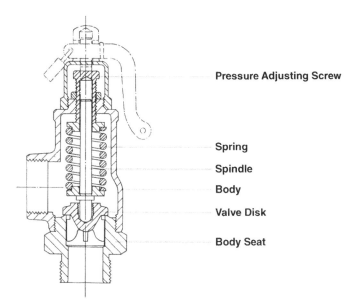

Figure 11.2 Spring type insurance approved safety relief valve

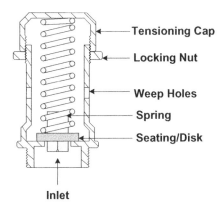

Figure 11.3 Domestic heating version of spring type safety relief valve

The practice of installing these safety relief valves or better quality safety relief valves on domestic open vented heating systems has declined in recent years, leaving the heating system more vulnerable and possibly making the building's insurance policy invalid in the event of any catastrophe occurring.

The reasons given by domestic installers for not fitting safety relief valves are varied but none are valid. Common reasons include:

- *They are not required* – The absence of them not being specifically specified in some organisational guides etc does not mean they are not required. The safety relief valve is a fundamental part of the heating safety system, rarely called upon, but still required by some building insurance policies.
- *They often leak* – They rarely leak, they are either set at the wrong pressure, or they are doing their job of relieving pressure; in either case there is a lack of understanding of these type of valves.
- *They don't work* – This is also a result of not understanding these valves, as they have been incorrectly set at a high pressure.

These domestic versions of the spring type safety relief valve work on the same principles as those approved by insurance companies, shown in Figure 11.2. The spring is tensioned to operate at the system's set relieving pressure by the tensioning cap and then locked in position by tightening the locking nut. Any discharge from this valve is emitted through the series of weep holes around the body of the valve, which is one reason why this form of valve is not suitable for sealed and pressurised heating systems.

The insurance approved spring type safety relief valve can be obtained with the specified operating pressure set at the factory ready for use, but it may need resetting at some time in the future, or set at site when initially installed. This may be achieved simply by attaching the valve to a test pump primed at the required operating pressure, and releasing the valve's tensioning nut until it opens at the primed pressure.

The domestic version of this safety relief valve which does not come factory set must be adjusted on site. This is also a simple and relatively quick operation. The tensioning cap should initially be tightened right down and the heating system hydraulically pressure tested to 1½ times the working pressure. On successful completion of the soundness test, the pressure generated by the test pump should be reset to match the safety relief valve's required operating pressure, and the safety relief valve tensioning cap slowly slackened off until water starts to trickle slowly through the weep holes of the valve body. The tensioning cap can then be locked in that position by tightening the locknut to prevent the tensioning cap from moving. This method can be used to check the pressure setting of all safety relief valves, or this can be done by removing them from the system and attaching them to an adapted connector attached to the pressure test pump.

The setting and checking of the correct operating pressure of the safety relief valve is an essential part of the heating system's commissioning procedure and periodic servicing activities and should be carried out accurately.

Almost all safety relief valves used for protecting heating systems are of the spring type in one form or another, but the tensioning spring will deteriorate and weaken over a period of time and the valve operating pressure will need to be reset or the valve fitted with a new spring. If the valve is in poor condition it should be replaced completely with a new safety relief valve.

Another form of safety relief valve is shown in Figure 11.4. This type was previously commonly used on commercial and industrial heating installations but is rarely used today. It differs from the spring type safety relief valve by the spring being replaced by a lever and adjustable cast iron weight, which should be heavy enough to be capable of holding the valve disk down on its seating until the system set pressure is reached when the pressure acting on the disk will raise the valve lever to relieve the system pressure.

The weights are interchangeable over a range to suit the pressure required. Final pressure adjustment is accomplished by moving the weight along the lever and bolting it down into position when the set pressure is reached.

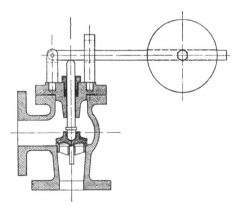

Figure 11.4 Weighted lever type safety relief valve

This type of cast iron safety relief valve has the advantage of not having a spring to maintain and can be easily tested by physically raising the lever to check that the valve disk has not become stuck on the seating.

CONTROLS FOR COMFORT AND ENERGY EFFICIENCY

Of all the many different rudiments that go to form a heating system, the subject of controls is the one that has seen far more developments than any other. Since the 1950s, when controls were fairly basic – either just turning the heating system on or off, if they existed at all, and often offered only as an optional extra in the case of domestic heating systems – they have developed to today's level of sophistication where specialist control companies can now provide a design, installation and commissioning service and where mandatory minimum control requirements contained in the Building Regulations now apply.

These advances can be attributed to a number of developments that have occurred over that time period, including:

- Changes in people's lifestyle, where an automated system of temperature control is desired.
- The desire to have greater flexibility in the degree of control for comfort conditions throughout the building.
- The advancement in control technology including the microprocessor, silicon chip and printed circuit boards.
- The increasing awareness for the need to conserve energy and increase the efficiency of the heating system.

Regardless of the developments and regulations that have occurred to the controls associated with heating systems, the main objectives of the control scheme remain the same today as they did in the 1950s, namely:

- To provide an automatic mode of operation that will permit comfort conditions to be achieved and maintained at all times during the heating season.
- To regulate the heating system and to ensure that it performs exactly as it was designed to.
- To allow the heating system to operate efficiently and economically by conserving fuel and energy.

The control schemes depicted in this chapter include systems and components that have been used during the development of the control of heating systems but which are now considered redundant or have been superseded by advances in technology as well as by current control schemes. They have been included to help the student understand, or remind the practising engineer, how the control schemes used today have been arrived at and how they function.

There are many variations of the control schemes included in this work, but the schemes described are the basic concepts that may be mixed together, or expanded upon.

BASIC SYSTEM OF THERMOSTAT CONTROLLING PUMP

The most basic form of space heating control, from which all other systems have been developed, is a room thermostat arranged to control the circulation of hot water by the circulating pump.

Figure 11.5 illustrates the basic concept of this scheme, which served its purpose as an inexpensive automatic means of controlling room temperature at a time when controls were not commonly used for domestic heating systems. The simple operation involves a room thermostat switching the circulating pump off when the temperature was hot enough, and back on again when heat was required and includes today's unaccepted practice of the boiler cycling on its own thermostat. This practice of allowing the boiler to cycle on its own thermostat was only possible on a traditional large water content system boiler that was also common at the time.

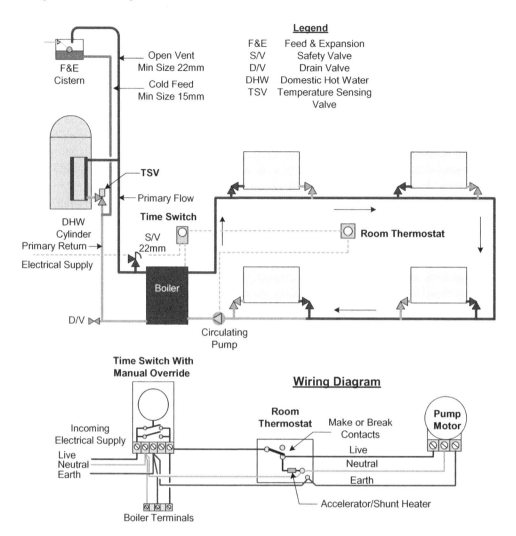

Figure 11.5   Basic control system of thermostat controlling pump

This basic method of control applied to the control of room temperature only; the domestic hot water temperature had to be controlled independently by some other means, if any, such as by the use of a non-electrical mechanical temperature sensing valve.

It should be appreciated that this method of permitting the boiler to cycle on its own thermostat often resulted in a partial thermo-siphon/gravity circulation to occur to the upper floors, and the installation of a check valve on the flow pipe became common practice simply to add frictional resistance to the system to prevent a gravity circulation.

ROOM THERMOSTATS

The basic design concept of the uncomplicated electronic room thermostat has changed very little over the years since its introduction to control space temperature. It comprises a bimetallic spiral strip or corrugated metal expanding bellows, as shown in Figure 11.6.

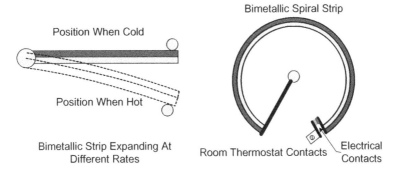

Figure 11.6   Thermostatic bimetallic strip

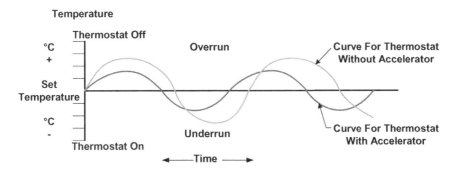

Figure 11.7   Effect of accelerator fitted to a room thermostat

The bimetallic strip comprises a strip of two dissimilar metals joined together. They have contrasting coefficients of thermal expansion and will expand at different rates when exposed to the room temperatures expected for heating systems.

Room thermostats are normally equipped with accelerating shunt heaters, as depicted in the room thermostat shown in Figure 11.5. The action of the shunt heater is to sense – by changes in electrical resistance – when the thermostat is about to operate by making or breaking the electrical contact, and accelerate this function by smoothing out the response time curve.

Figure 11.7 compares a room thermostat fitted with an accelerating shunt heater and one without. The thermostat fitted with an accelerating shunt heater achieves a faster response time and more acceptable comfort conditions. It can be observed from comparing the two performance curves that the temperature time lag is much improved by incorporating an accelerator into the room thermostat, where the fluctuation of the temperature overrun and underrun is restricted to approximately 2–3°C between the thermostat switching on and off, instead of a more prolonged variance spanning about 6–7°C.

Early electrical thermostats incorporated a mercury switch operated by a bimetallic spiral strip that tilted the mercury glass tube to make or break the electrical contacts.

ROOM THERMOSTAT LOCATION

The choice of room location for installing the room thermostat may vary between one building and another, but in all applications it should be fitted within the space being heated and controlled by that thermostat.

When selecting the room location, due regard should be given to the form of control employed and the design of the heating system, as it should be appreciated that when the room thermostat operates by the temperature obtained in the room in which it is located, it will also have an effect on the heating being provided in other rooms that are on the same control zone. This is discussed further in the following section.

In dwellings, it has been normal practice to locate the room thermostat in either the entrance hall, due to its higher air infiltration rate, or the living room, as this is the room that is occupied for more of the time. In buildings where the heating system has been divided into a number of zones, the location of the room thermostats should also be chosen using the same criteria for each thermostat.

Having decided on the location, the thermostat should be fixed on a suitable wall at a height of approximately 1.5 metres. This not only provides an ideal height to allow the occupier to adjust the thermostat, but is also the optimum height to sense the space temperature within the living area of the temperature gradient, as explained in Chapter 5 in the section 'Room Height Effect'.

The decision on what constitutes a suitable wall should not only include the requirement for the thermostat to blend in with the room's décor etc, but also the ability to position it away from any source of false temperature effects, such as from direct sunlight or draughts from doors. They should not be fixed directly above heat emitters or located behind curtains.

MIXING CONVECTORS AND RADIATORS

The difference in heat transference between heat emitters that are predominantly either convective or radiant has been explained in both Chapter 4, Heat and Heat Transfer, and Chapter 5, Heat Emitters. Attention should be paid, when it is desired to mix both forms of heat emitters in the same heating system, to avoid the effects depicted in the transference curves shown in Figure 11.8.

The convection/radiation comparison curve depicted in Figure 11.8 demonstrates that a heating system that is predominantly convective in its heat emission achieves a faster heat-up time than one that is predominantly radiant, but it also has a more rapid cool-down time. For an example, if we explore the situation where the room thermostat has been installed in a room heated by convection type heat emitters, when the thermostat reaches the design set point temperature, other rooms being heated by predominantly radiation type heat emitters would have only reached the lower temperature marked as 'Point A' on the curve. This would result in those rooms being heated by heat emitters that have a high radiation emission value not reaching the design set point temperature and operating at the underheat curve level shown.

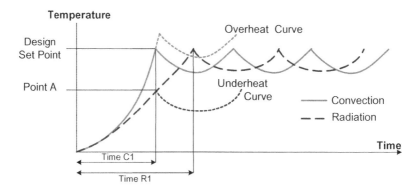

Figure 11.8   Convection/radiation heat-up comparison curve

If the thermostat is relocated to a room that is being heated by predominantly radiation type heat emitters, then providing that the external temperature is the same as the design base temperature used, the system would achieve the design room temperatures selected. However, if the external temperature is higher than the base design temperature, or the convector type heat emitters have an overcapacity, the rooms being heated by the convector type heat emitters will continue to emit heat into the space above the design set point, which would result in a higher temperature as shown by the overheat curve.

This example serves to demonstrate that a single heating zone controlled by a single room thermostat is inadequate for controlling a heating system when mixed heat emitters are to be employed.

TEMPERATURE SENSING VALVE

The control scheme shown in Figure 11.5 depicts a method of controlling the temperature of the domestic hot water independently of the heating temperature control, utilising a non-electric thermostatic temperature sensing valve.

Once commonly used, but less so today, inexpensive thermostatically operated valves are designed to control the temperature of the domestic hot water independently of any other control system. They do not require any electrical power supply, and can still prove useful when updating systems that do not have any temperature control of the domestic hot water.

Figure 11.9 shows the variations in which these type of valves are, or have been, available. They can be classified as being either of the direct temperature sensing type or the remote sensing temperature version.

The earliest developed form of this valve was the direct temperature sensing version, which was fitted to the primary return from the domestic hot water cylinder. There were two varieties: the straight through pattern and the angle pattern. The valve worked by sensing the rise in the temperature of the water passing over the expandable wax filled capsule as the water returns from the heat exchanger in the cylinder. This temperature will eventually equate to the same temperature as the domestic hot water being stored, normally adjusted on site to 60°C.

The temperature of 60°C will gradually be reached, when the valve will be closed, preventing the continual circulation of the primary flow through the heat exchanger, as the return temperature will rise according to the temperature of the domestic hot water being stored in the cylinder. For example, when the water in the cylinder is cold, the primary flow will enter the heat exchanger at 82°C and exit at a temperature of 45–50°C, having transferred part of its heat to the secondary water. When the domestic hot water reaches this return temperature and starts to exceed it, the temperature of the primary return will also rise to equal that temperature until the set temperature of 60°C is reached.

A variation of this type of temperature sensing valve is the remote sensing form which replaces the temperature sensing expandable wax filled capsule, which had a reputation for failing prematurely. The more reliable remote temperature sensing element is connected to the valve by a capillary tube at one end, with the temperature sensing element at the other end, and is strapped to the external surface of the cylinder approximately one third up from the floor (after first having removed any pre-insulation material from the wall of the cylinder so that the sensing element is in contact with the cylinder). Alternatively, the remote temperature sensing element may be inserted into the cylinder if a thermostat pocket or phial has been incorporated in, or fitted to, the cylinder.

A further variation of this type of control valve is the three-way version in conjunction with a remote sensor connected via a capillary tube, where instead of the valve closing off the flow of water, it diverts the flow from the cylinder through a bypass back into the primary return. This three-way variation has to be installed on the primary flow to the cylinder.

Like all the temperature sensing valves described, these valves are thermostatic in their operation, which means they are not an instant action valve and require time to operate from being fully open to fully closed.

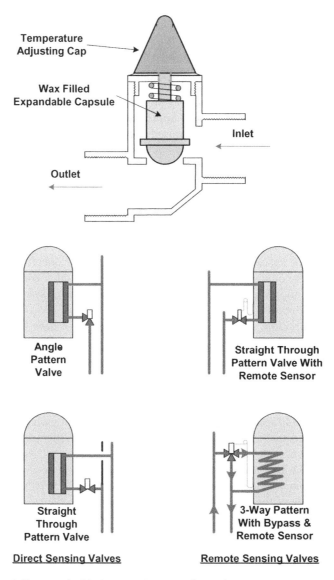

Figure 11.9   Thermostatic non-electric temperature sensing valve

The three-way valve will be fully open to the cylinder heat exchanger when it first operates and the secondary water stored in the cylinder is cold or slightly warm. As the temperature of the secondary water begins to rise, approaching the set temperature of 60°C, the valve will begin to slightly close to the cylinder and open to the bypass, allowing a proportion of primary water to pass through both outlet ports. As the temperature continues to rise, the proportion of primary water to the bypass will increase as the proportion to the cylinder decreases, until all the primary water is diverted through the bypass outlet port and none is being delivered to the cylinder.

Regardless of which version of the thermostatically operated temperature sensing valve is employed, it can provide a simple, less expensive method of providing suitable safe temperature control of the domestic hot water, particularly when upgrading an aged heating system when funds are not available to renew more of the system.

CYLINDER THERMOSTAT ARRANGED FOR DOMESTIC HOT WATER PRIORITY

This method of control is a development of the previous scheme shown in Figure 11.5, and is similar in its operation, except that it incorporates an additional thermostat attached to the domestic hot water storage cylinder and is arranged to prioritise a quick recovery of the water temperature stored in the cylinder.

Figure 11.10 depicts this now redundant control scheme, which was employed to provide a degree of control for the heating system as well as furnishing large quantities of domestic hot water.

This variation of control uses an additional thermostat, which is attached to the side of the domestic hot water cylinder approximately a third of the way up from the floor, and is connected electrically between the room thermostat and the heating system's circulating pump.

The temperature of the dwelling is controlled in exactly the same way as the previous scheme: a single room thermostat regulates the heat output from the heat emitters by switching the circulating pump on and off. This is accomplished by the cylinder thermostat overriding the room thermostat and breaking the electrical supply to the circulating pump and switching it off, allowing the whole of the boiler heat output

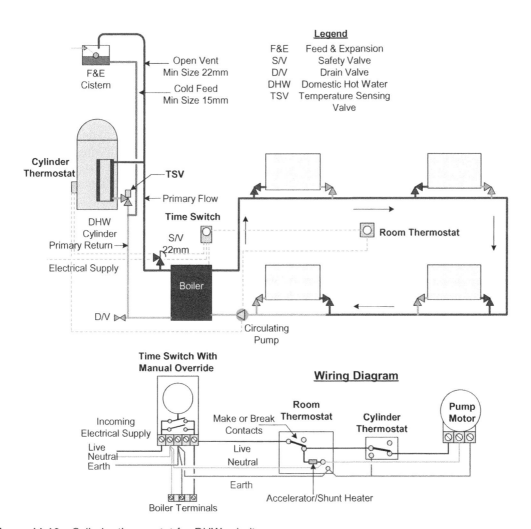

Figure 11.10 Cylinder thermostat for DHW priority

to be directed towards recovering the domestic hot water temperature more quickly. When the cylinder thermostat is satisfied, the heating control is restored to the room thermostat via the electrical supply to the pump.

The cylinder thermostat employed in this control scheme differs from the standard arrangement for cylinder thermostats in so far as the thermostat makes an electrical contact when the temperature is satisfied, instead of breaking electrical contact, as is the case with a standard cylinder arrangement.

There are a number of variations to this scheme that incorporate some method of temperature control of the domestic hot water, including the non-electrical type thermostatic temperature sensing valve, depicted in Figure 11.10. This control scheme, along with the previous version from which it was developed, permitted the boiler to cycle under its own control thermostat when the circulating pump was switched off, which was considered acceptable practice at the time.

CYLINDER THERMOSTATS

Most cylinder thermostats used for domestic dwellings operate on the same make or break bimetallic strip electrical contacts principle as the room thermostat – except that they do not normally incorporate an accelerating shunt heater, as the temperature overrun is not as critical as it is for room thermostats.

Cylinder thermostats incorporated in domestic heating control schemes normally consist of the strap-on contact type, which require any pre-insulated foam applied to the domestic hot water cylinder to be removed to allow the thermostat to come into contact with the cylinder wall, as in Figure 11.11. Only sufficient foam insulation should be removed to permit the thermostat to fit tightly in the area removed, enabling it to sense the temperature of the water stored through the cylinder wall.

Although the strap-on contact type cylinder thermostat can be used on any size of domestic hot water storage cylinder, providing that the metal band that serves as the strap is sufficiently long, most domestic hot water storage cylinders or calorifiers used for providing larger quantities of hot water in commercial or industrial applications use the probe, or phial type, of thermostat. This is inserted into a pocket incorporated in the wall of the cylinder or calorifier.

The fixing location of the thermostat on the cylinder or calorifier is important as it will be at this position that the thermostat will sense the temperature of the domestic hot water being stored. For domestic heating applications, it is normal practice to clamp the contact type thermostat on the wall of the cylinder at a height of approximately one third the way up from the base in relation to the storage cylinder's overall height, and set at a temperature of 60°C. It is at this point of contact that the thermostat will sense the temperature of the domestic hot water, but due to stratification of the water inside the cylinder, the tem-

Cylinder Thermostat
Temperature Adjustment

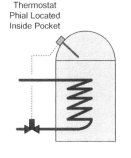

Insertion Type Thermostat

Figure 11.11 Cylinder thermostat details

perature at the top of the storage cylinder will probably be around 65°C. The effects of stratification of domestic hot water are discussed further in Chapter 19 Domestic Hot Water.

However, cylinder thermostats that incorporate a phial inserted into the cylinder or calorifier normally have the pocket arranged at, or near, the top of the cylinder and set at a temperature of 65°C, which means that the temperature of the domestic hot water being stored will be approximately 60°C one third the way up the cylinder – the same as that for domestic applications.

CURRENT CONTROL SCHEMES

The previous control schemes described in this chapter have covered the development of heating control systems, and although there may still be some in existence and coming to the end of their expected life, they are no longer considered acceptable by current standards, both on efficiency grounds and by today's accepted living or lifestyle conditions.

Current control schemes should produce a heating system that is capable of heating the building under the design conditions. However, for most of the time that the heating system is operating, much less heat output is required. The control scheme should be capable of ensuring that the desired room temperatures selected are in fact achieved and maintained in all heated rooms and under all conditions, including mild temperature periods, when little or no heat is required.

Current control schemes are required to provide the following minimum functions:

- Boiler interlock arranged to allow the boiler to fire only when there is a demand for heat
- Thermostatic radiator valves on all heat emitters except in rooms where a controlling thermostat is located
- Thermostatic control of the domestic hot water
- Minimum of one controlling room thermostat controlling heated space temperature
- Controlling programmer with a minimum of two zone channels
- Automatic boiler bypass valve on low water content boilers.

ZONE CONTROL BY TWO-PORT MOTORISED VALVE

There are many variations of zoning by two-port motorised valves forming heating control schemes, too many to include all of them here, but because of its versatility, this method of control is the most commonly used throughout the heating industry – from domestic heating applications to commercial and industrial heating applications.

The versatility of this scheme is depicted in Figure 11.12. The scheme allows the heating installation to be divided into any number of zoned circuits, each controlled independently of the others by a two-port motorised valve activated by its own dedicated thermostat. The scheme shown in Figure 11.12 has three heating zones – one of the zones comprising convector type heat emitters and the other two zones comprising radiator type heat emitters, as this scheme permits heat emitters having differing heat transfer characteristics to be mixed without affecting each other. The domestic hot water temperature control forms the fourth zone.

There are no limitations regarding the number of zones created on a heating system, but most domestic heating applications would normally just have two, one for the heating system itself and the other for controlling the temperature of the domestic hot water.

Figure 11.12 shows a heating system incorporating the provision for domestic hot water divided into four separate zones as a common control system, but each zone is in fact a separate individual control scheme, as demonstrated by the electrical wiring arrangement in Figure 11.13, and each zone can operate

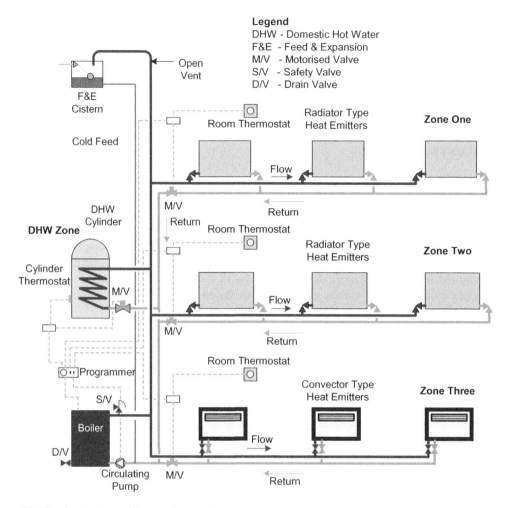

Figure 11.12 Application of two-port motorised zone valves

independently of each other, except if controlled by a common timeswitch. The application of slave timeswitches connected to a master programmer, or programmable room thermostats, would allow each zone to operate totally independently of each other, including the on/off times for the heating to each zone.

Until recently, the provision of the auxiliary switching arrangement on the motorised valve actuator – depicted in Figure 11.14 – was offered as an optional extra by the manufacturers to control the boiler firing and/or the running of the circulating pump only when heat was being called for from any one or more of the thermostats. The Building Regulations Approved Document Part L now requires all new or existing heating systems being upgraded to incorporating the provision of boiler and circulating pump interlocks to automatically prevent the boiler and circulating pump from operating when all thermostats are satisfied, hence saving energy when not required. This method of control – depicted in the wiring diagram in Figure 11.13 – shows that it lends itself very easily to incorporating the auxiliary switching arrangement to provide the interlock required by the Building Regulations. It should be noted that each two-port motorised valve requires this auxiliary switching facility so that if any one thermostat requires heat, the boiler and circulating pump will be energised.

Figure 11.14 shows a typical domestic installation.

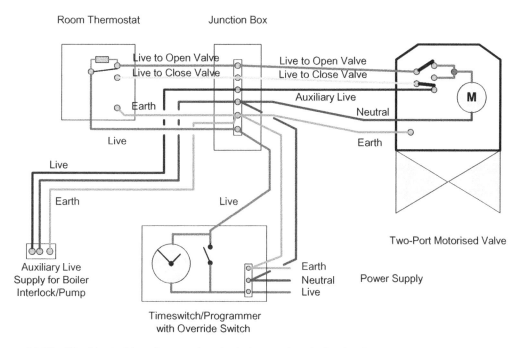

Figure 11.13   Electrical wiring diagram for single two-port motorised valve

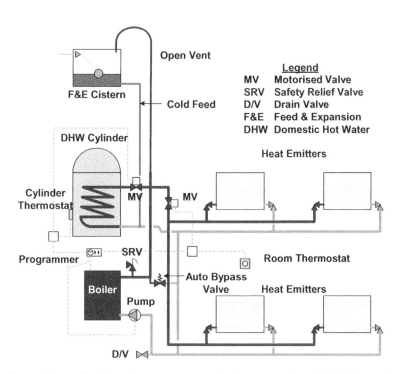

Figure 11.14   Domestic application of two-port motorised zone valve with boiler and pump electrical interlock

The versatility of this control system not only includes permitting heat emitters of different heat emission characteristics to be mixed on the same heating system, but also the zone control system can be used to divide properties into a number of separate dwellings, commercial offices, or dwellings that incorporate a home office – whereby each separate dwelling or office would comprise an individual heating zone.

Buildings may also be zoned into north and south facing zones, whereby the heating system is divided up vertically throughout the building to take advantage of any solar heat gains during the winter months, therefore achieving comfort conditions on both sides of the building and conserving energy.

Rooms containing an open fire can also be treated as a separate heating zone to achieve comfort conditions whilst conserving energy.

The above illustrations are just a few examples of the versatility of the zone control arrangements that have contributed to making it the most popular form of control being employed.

MOTORISED VALVES

Most forms of isolating and regulating valves may be motorised by attaching electric actuators to them, providing the valve head gear has been adapted to suit the actuator type. This permits the valve to be operated remotely either by a manually arranged switching point, or automatically, by means of a thermostat or timeswitch.

Adding a motorised actuator to a valve achieves the means of automating its operation via a switch incorporated within a thermostat, timeswitch or some other temperature, pressure or humidity sensor. Normally, simple quarter turn type valves are chosen for this application, such as butterfly valves or quarter turn spherical plug or ballvalves.

Figure 11.15 illustrates a plan section through a quarter turn spherical plug valve. An electrical motorised actuator is fitted onto the end of the spindle that is arranged to turn the plug through 90°, being either fully open as shown in Figure 11.15, or fully closed.

The motor can be either the double live supply type as depicted in Figures 11.12 and 11.13, or of the less costly single live supply with spring return. The double live supply is arranged through the electrical wiring diagram as portrayed in Figure 11.13, where one live supply energised by the respective thermostat causes the motor to open the valve by turning the valve spindle 90°. When the thermostat is satisfied, the live supply is switched to the second live connection, which energises the motor to turn in the opposite direction to close the valve. The auxiliary interlock live electrical supply to the boiler and pump is only energised when the first live electrical supply arranged to open the valve is activated. When this electrical

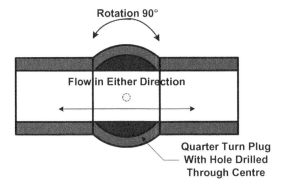

Figure 11.15 Plan section through quarter turn spherical plug valve

supply is de-energised, this electrical supply route to the boiler and pump is no longer available until it is re-energised.

The single electrical live supply with spring return version of this valve works on the principle of the valve being opened when the live supply is energised by the thermostat, and being kept open whilst the electrical supply remains live. When the thermostat is satisfied, the live supply which is keeping the valve open is de-energised, enabling the spring to return the valve plug to the closed position.

Unlike solenoid valves, which are instant action type electrical valves, motorised actuators require a certain amount of time to complete the manoeuvre from when they are first energised to when they are finished operating. Some motorised valves may be arranged to modulate anywhere between being fully open to being fully closed, these are discussed later under other control schemes.

Two-port motorised zone valves are available with either factory assembled actuators, or detachable bolt-on actuators that facilitate maintenance activities in replacing burnt-out motors by allowing the valve body to be left in place and avoiding having to drain down the system.

SELECTIVE DIVERTER CONTROL BY THREE-WAY MOTORISED VALVE

The control schemes previously described may comprise individual control components manufactured by a variety of different manufacturers which can be arranged by the design engineer to form an overall control system that best suits the heating system concerned. They give the designer a degree of flexibility in selecting the best components available to achieve the control required to ensure that the heating system performs as designed.

The three-way diverter valve control method differs from the above in so far as all the main control components are normally obtained as a control package, all manufactured by the same control component manufacturer – including the main item of heating control equipment, the three-way motorised diverter valve.

The first domestic version of this method of control was developed in the 1960s to provide a simple but reasonable degree of control for low cost dwellings. It worked on the principle of providing selective control for the heating and domestic hot water requirements, as depicted in Figure 11.16.

In the early version of this control scheme, it comprises a single two-position three-way motorised diverting valve controlled by both a cylinder thermostat and a single common room thermostat, arranged electrically to provide priority in satisfying the domestic hot water demands at the expense of the heating system; see Figure 11.17 for the electrical wiring diagram. When the domestic hot water temperature has been satisfied, the control scheme then reverts to its energised position by providing heat to the heating system by the diverting valve closing to the cylinder port and opening to the heating port.

The energised position of the diverting valve closes to the port outlet to the domestic hot water cylinder; when the cylinder thermostat calls for heat, the three-way diverting valve is de-energised and the valve closes to the heating outlet port by the assistance of a spring, allowing the full boiler heat output capacity to be directed towards recovering the domestic hot water storage temperature. The heating system will suffer by being off during this period until the cylinder thermostat is satisfied, and re-energises to open to the heating outlet port.

If the primary piping to the cylinder has been sized accordingly and the domestic hot water cylinder incorporates a rapid heat recovery heat exchanger, this recovery period will be fairly quick, with the lack of heating going unnoticed during all but extreme cold weather spells.

The advantage of this selective control system allows the boiler to be sized on the calculated heating system load only, with the domestic hot water requirements being ignored as they are never required to be operated together. This permits a smaller boiler to be installed, making it an ideal heating system for low cost dwellings.

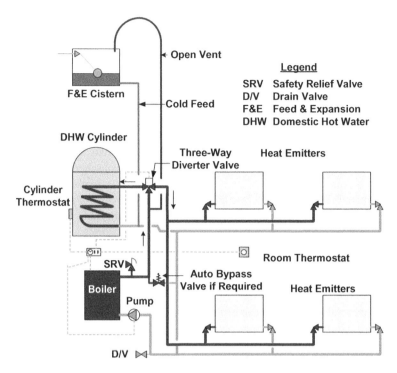

Figure 11.16   Selective diverter control by three-way motorised valve

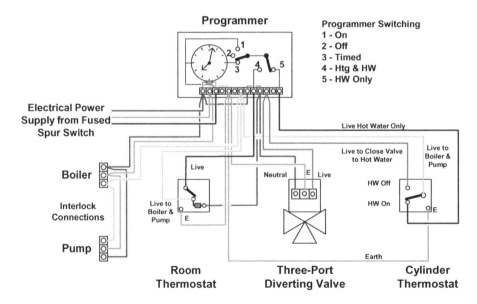

Figure 11.17   Electrical wiring diagram for selective three-way valve control

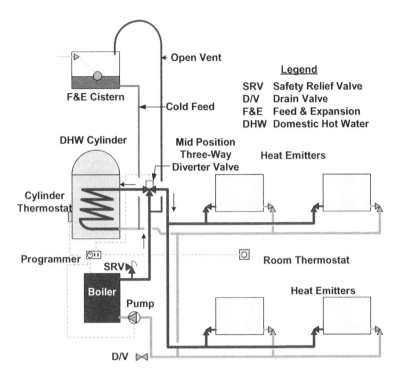

Figure 11.18 Shared flow by mid-position three-way motorised valve

SHARED FLOW BY THREE-WAY MOTORISED VALVE

The selective control system discussed above was often abused; it was developed for low cost dwellings but was frequently specified inappropriately for larger properties, resulting in the heating system being off for longer periods than desired when the heat was being directed towards satisfying the domestic hot water requirements. It was also not capable of providing the degree of control that would be expected in a larger and more expensive dwelling.

A variation developed from this control system is shown in Figure 11.18. From first impressions the mechanical piping arrangement looks just the same as the selective control system shown in Figure 11.16 as it employs a three-way motorised diverting valve. However, this later version of the three-way motorised valve is arranged to adopt any one of three positions. These include the two positions that the previous selective three-way valve is capable of, plus an intermediate position that permits the flow of hot water from the boiler to be directed to both outlet ports to satisfy both the domestic hot water and heating system simultaneously.

This later development of the shared three-way control valve method eliminates the disadvantage experienced by the two-position diverting valve, of the heating circuit being starved of heat when the cylinder thermostat diverts the flow of hot water to recover the contents of the domestic hot water storage vessel. However, the advantage of the smaller boiler size that is possible with the previous system is lost as the boiler now has to be sized to satisfy both the heating circuit and domestic hot water together.

This control system is suitable for slightly larger residential dwellings than the previous two-position three-way valve where it may be used successfully.

Figure 11.19 illustrates a working section through a motorised three-port valve indicating the operational positioning of the valve seating, either shoe or paddle type, applicable to both the two-position selective control valve and the three-, or mid-position motorised three-port valve.

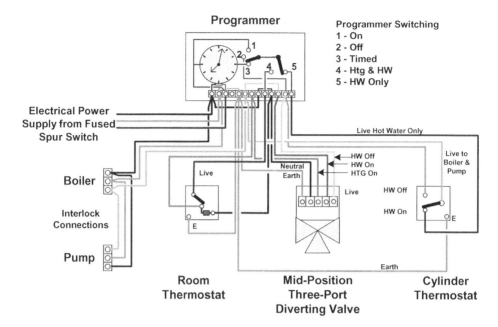

Figure 11.19 Electrical wiring diagram for mid-position three-way valve control

Unlike the two-port motorised valve, the actuator or motor cannot normally be removed from the three-port valve body, as it is usually factory fitted. If the valve motor fails, the whole valve body/motor unit has to be replaced. The motors are normally equipped with a sliding manually operated override lever that allows the valve seating to adopt a neutral mid-position in the event of a motor failure, thus permitting the heating system to remain operational until the valve unit can be replaced.

Figure 11.20 shows a three-way motorised valve and the functions of the three positions.

WEATHER COMPENSATING CONTROL SYSTEM

This system is a form of control for the heating circuit only, but it is often used in combination with other control schemes such as zone control. In the past it has only been used in the larger type of dwelling or commercial building due to its higher financial cost; however, as the emphasis is now on fuel conservation and energy efficiency, this additional control scheme is now being considered for lesser value properties because of its ability to conserve energy which can help justify the higher cost of the system.

The heated space is controlled by a central controller mounted on a three-port mixing valve that is automatically regulated to suit prevailing weather conditions together with internal temperature requirements.

The temperature of the water circulating throughout the heating system is automatically regulated by a three-port motorised modulating mixing valve under the dictates of an external temperature detector, in harmony with an internal temperature detector, as illustrated in Figure 11.21. It should be noted that these are temperature detectors and not thermostats; they detect temperature changes but do not make or break an electrical switch in the same way that thermostats do.

The temperature detectors sense any change in both external and internal temperatures and the results combine to operate the three-port mixing valve, which is modulated to mix the flow water from the boiler with a proportion of returned water from the heating system to obtain an optimum flow temperature to suit prevailing temperatures.

Three-Position Valve	Position	Two-Position Valve
Powered to Close to the Heating Circuit	1	Powered to Close to the Heating Circuit
Neutral Position	2	Manual Override Position
Powered to Close to the Domestic Hot Water Cylinder	3	De-Energised Position

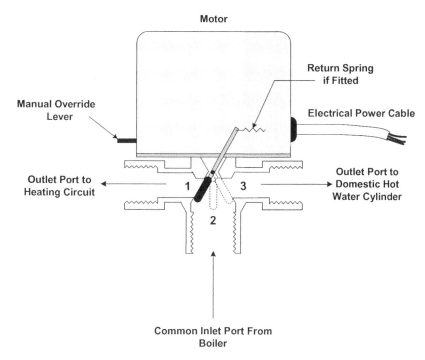

Figure 11.20 Three-way motorised diverting valve

For example, when the external temperature is at or below the base design temperature and the internal temperature is at or below the desired set temperature, 100% of the circulating water will pass through the boiler. As the external temperature and/or internal space temperature begin to rise, the change will be sensed by the detectors which will cause the mixing valve to allow a proportion of the lower temperature return water to mix with the higher temperature water leaving the boiler to achieve an intermediate optimum flow temperature. This condition is constantly monitored and the valve modulated to suit the ever-changing temperature conditions. For this reason the pump must remain running so that the flow temperature remains at the optimum temperature and therefore only the boiler should be interlocked to be switched off when 100% of the return water is being circulated. Some weather compensating systems may incorporate a contact temperature sensor attached to the flow pipe to include this temperature in the monitoring process.

Figure 11.22 shows a suitable wiring diagram.

Figure 11.23 depicts a section through a three-port mixing valve, showing how the rotating shoe seating modulates between the two inlet ports to proportion the flow of water to the heating circuits to match the optimum temperature required. When the valve is totally closed to the boiler, 100% of the return water is re-circulated through the heating circuit; the boiler is then turned off by the boiler interlock to conserve energy when not required. When the valve begins to open to the boiler port, the boiler is then switched back on to sustain the flow temperature and continue to preserve the optimum temperature.

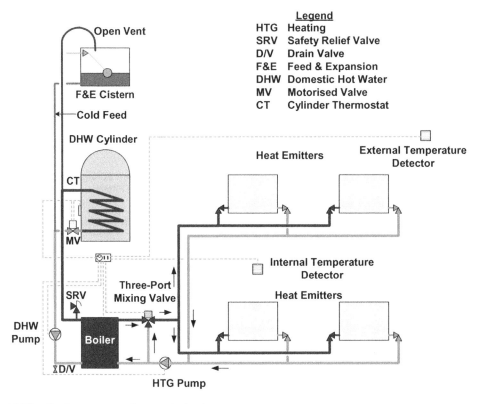

Figure 11.21   Basic compensating control scheme

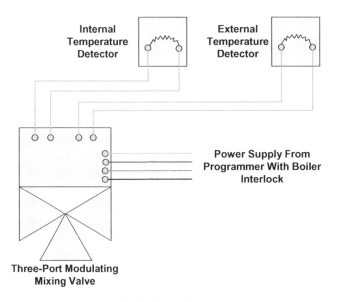

Figure 11.22   Weather compensating electrical wiring diagram

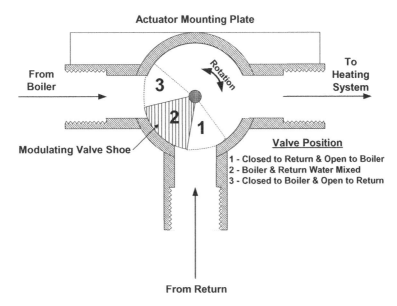

Figure 11.23   Section through three-port modulating mixing valve

The internal room temperature detector should be installed with the same constraints as described for a room thermostat, having due regard for any false sensing of temperature.

The external temperature detector is manufactured with a weatherproof casing making it suitable for external mounting, but the location must also be carefully chosen to avoid places of direct sunlight and other false temperature situations.

As illustrated in Figure 11.21, control of the domestic hot water must be provided by some other control means. The most obvious solution is to treat it as a separate zone using a motorised two-port zone valve, controlled by a cylinder thermostat complete with boiler and pump interlock. The primary circulation to the domestic hot water cylinder would at one time have been arranged to have a gravity circulation, but today the Building Regulations require the circulation to be forced, which results in a second circulating pump being installed and operating independently of the main heating system circulating pump. The problems of duplicate pump arrangements are explained in the next section and in Chapter 18, Circulating Pumps.

The problems discussed earlier in this chapter regarding the mixing of different types of heat emitters must be applied to this control system, as the internal temperature detector will function in the same way as the room thermostat. Figure 11.24 illustrates how this problem is overcome by introducing zone control arrangements.

Care should also be taken to avoid installing fan convectors and air heater batteries on the variable temperature heating circuits controlled by the modulating mixing valves if they require a constant flow temperature to function correctly.

Figure 11.24 illustrates a weather compensating control scheme arranged to incorporate individually controlled zones utilising two-port motorised valves. This control arrangement permits heat emitters that require a constant temperature water flow to be controlled independently of the main heating system that operates using a variable temperature flow.

Because of the greater sophistication of this control scheme, together with the greater number of circulating pumps required, this form of control arrangement is normally only employed in larger dwellings or commercial buildings.

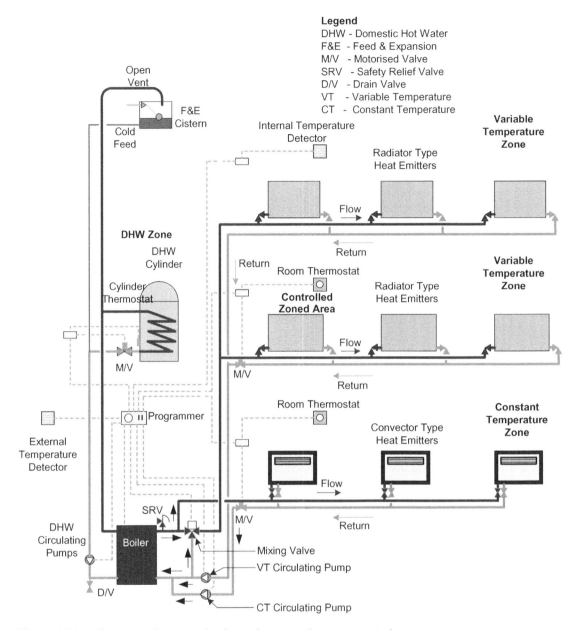

Figure 11.24 Compensating control scheme incorporating zone control

TWIN-PUMP CONTROL SCHEMES

The difficulty associated with incorporating two or more circulating pumps on the same heating system was mentioned in the previous section.

Centrifugal circulating pumps produce both positive pressure at the pump discharge outlet and negative pressure at the pump inlet. If more than one pump has been installed on the heating system, there is a strong possibility that the pressure part of the duty generated by each pump will have an adverse effect on the other pump as they are working against each other.

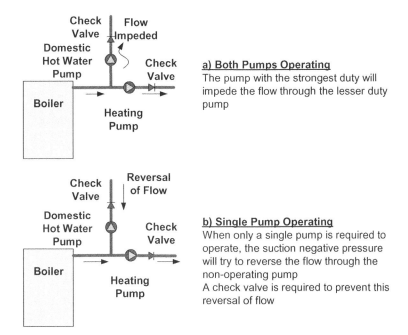

Figure 11.25 Operational effects of circulating pumps on each other

Figure 11.25a illustrates this condition, where both pumps are operating and the negative pressures generated by each pump pull against each other, having the effect of impeding the flow through each circuit. The pump having the greatest pressure reduces the flow through the pumped circuit with the lesser pressure. Depending upon the magnitude of this effect, the life expectancy of each pump will be reduced.

Figure 11.25b illustrates the effect when only one of the circulating pumps is operating. The negative pressure of the operating pump will try to reverse the flow through the circuit containing the non-operating pump, as the negative pressure generated will be uniform on that part of the system. A check valve installed on the pump discharge outlet will prevent this reversal of flow, providing the check valve is functioning correctly.

Figure 11.26 depicts a twin-pump control scheme that is designed to overcome the negative effects created by the pump suction conditions. The scheme comprises two circulating pumps: one for the heating circuit controlled by a room thermostat, and the other installed on the domestic hot water primaries controlled independently from the heating system by a cylinder thermostat. Both pumps are controlled through a common programmer.

The control scheme depicted in Figure 11.26 shows a domestic twin-pump heating system that employs a standard domestic variable head circulating pump to satisfy the space heating requirements. This standard circulating pump is controlled completely independently of the domestic hot water primary heat requirements by the dictates of a single room thermostat, which can also be interlocked back to the boiler; see Figure 11.27.

The circulating pump furnished to satisfy the domestic hot water requirements has to be of a non-standard special construction, in order to make this system work correctly without any interference of the other pump's operating performance. It can be seen from Figure 11.26 that this circulator has three pipe connection ports arranged as one common inlet pump suction port, together with two outlet ports. The inlet connection port is immediately divided into two separate waterway passages at the pump entry. One of these waterway passages is arranged to form a bypass of the pump impeller and continue as a straight

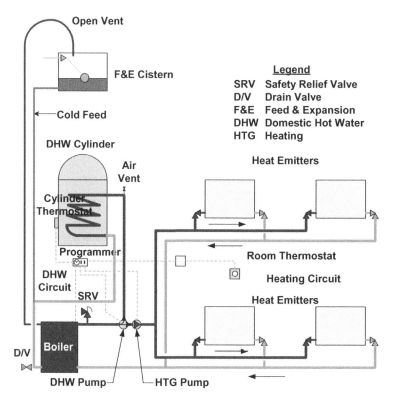

Figure 11.26   Twin-pump heating arrangement

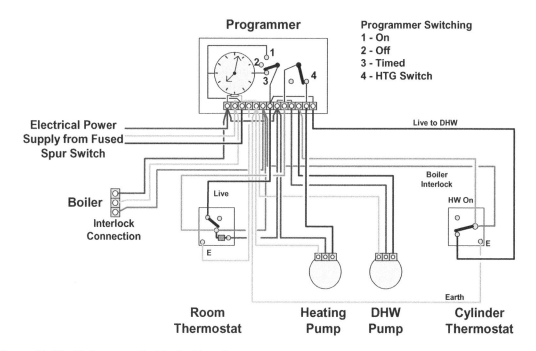

Figure 11.27   Twin-pump electrical wiring diagram

outlet on the opposite side of the pump casing, which can then be piped up to the standard heating circuit pump. The other waterway passage serves as the entry for the domestic hot water primary circulating pump, where the water is boosted in the normal way, except that the discharge outlet port of the pump permits the water to leave at a 90° angle to the pump inlet.

This integral pump bypass has the effect of neutralising each pump's negative pressure effects without interfering on the other's performance.

This special non-standard domestic hot water primary circulating pump is controlled independently by its own cylinder thermostat, and does not have any adverse effect on the water travelling through the casing bypass.

This control scheme provides a reasonably fast recovery time for the domestic hot water regardless of the operating routine of the heating pump.

Figure 11.28 illustrates a more recent development: the twin-pump control scheme, whereby the two individual pumps have been incorporated into a common pump casing. This achieves a more compact arrangement, but if one of the pumps fails, then the whole pump unit has to be changed.

The operating principles of this pump unit are the same: there is a common inlet connection which is split to divert the flow two ways at the entry point; the flow is then directed into each pump, which operates independently of the other.

This more recent pump unit is not restricted to domestic heating systems; it can just as easily be installed in heating systems for small commercial buildings.

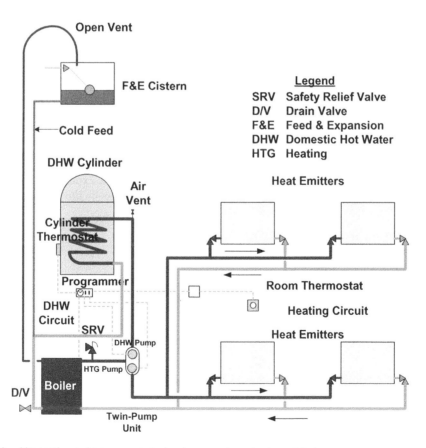

Figure 11.28   Alternative twin-pump control scheme using single unit twin-pump

Multiple pumping arrangements for larger building heating applications are discussed in Chapter 18, Circulating Pumps.

WIRELESS CONTROLS

All the control systems depicted so far in this text have been conventionally hard-wired to and from all control components using electrical cable.

Wireless controls have been developed over a number of years and, as the name implies, they omit sections of cabling from the control scheme. The first generation of these wireless controls concentrated on the room thermostat: by working on the principle of transmitting signals by radio waves, the cable between the room thermostat and programmer becomes redundant.

Figure 11.29 is a copy of the two-port motorised zone valve control system shown in Figure 11.14, but with the standard hard-wired room thermostat replaced by a wireless room thermostat. This transmits a signal of the room temperature status by radio waves to a receiver that is hard-wired to the adjacent programmer. The receiver sends this signal by cable to the programmer, which acts upon the signal and activates the motorised two-port zone valve when required to do so.

The wireless room thermostat is powered by a battery, and dependability has improved since its initial introduction. It is ideal for heating systems requiring upgrading as cable does not have to be chased into the wall – saving installation time, construction mess and money – but it is seldom used for new-build applications.

More recently, wireless technology has been developed to provide a more individual degree of control, but still only sends, and in this system, also receives, signals; it is not capable of providing power. This

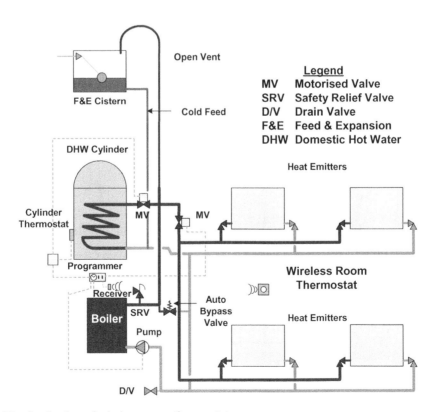

Figure 11.29   Application of wireless room thermostat

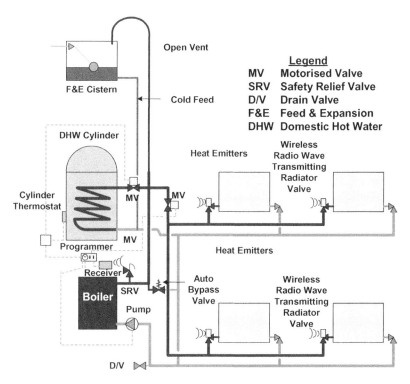

Figure 11.30   Application of wireless thermostatic radiator valves

second generation of wireless control has eliminated the room thermostat and placed the temperature transmitters into the heads of thermostatic radiator valves. These transmitting radiator heads send and receive signals from the central receiver, which is capable of receiving up to four individual signals. This is illustrated in Figure 11.30, which is the same heating scheme as the previous system, altered to depict this second generation of wireless controls. If more than four radiator signals are required to be received, then a further receiver is required.

Each radiator valve can be programmed at the receiver or radiator valve head to operate at different set temperatures. The radiator transmitting head will also display in digital format the actual temperature at any point in time and can be re-programmed easily if required.

Unlike the first generation of wireless controls, this second generation is as suitable for installation in new-build properties as it is for upgrading existing heating systems, as it achieves individual control to each room and can be quickly altered by the occupier of the room if desired.

THERMOSTATIC FROST PROTECTION

The provision of a frost thermostat should be considered if it is likely that the property will be unoccupied, with the heating system off for long periods of time during the heating season.

If the heating system is in the off position and freezing conditions should occur, then the property together with the building's water services are at risk of freezing, particularly during extremely low ambient temperature conditions, with potentially disastrous consequences.

Frost thermostats normally have a temperature range of between −15°C to +10°C and can be easily wired into the heating control system, as shown in Figure 11.31, to override the timeswitch or programmer. It will be set to operate at a temperature of around 1°C to 4°C for normal dwellings. The frost thermostat

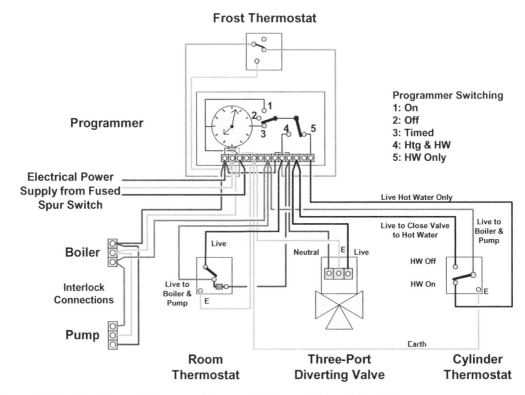

Figure 11.31   Frost thermostat arranged to override timeswitch and off switch on programmer

will automatically switch the heating system on when this temperature is reached, and maintain the system at this low temperature until the ambient temperature rises above the operating temperature.

The frost thermostat should be located in what is considered to be the coldest part of the building that it is protecting; this includes cloakrooms, utility rooms and sometimes loft spaces.

Frost thermostats are normally employed in factory warehouses where the products being stored may have a critical storage temperature above which they have to be kept; in these circumstances the frost thermostat set temperature will be dictated by the products' temperature requirements.

AUTOMATIC SPACE TEMPERATURE REDUCTION

Commonly referred to as a 'set-back facility', the automatic space temperature reduction is an additional control arrangement that can be incorporated into a zone control system to provide a means of reducing space temperature automatically when the area is expected not to be occupied.

This facility is particularly advantagous when the building has more than one use, such as the dual function of a residential dwelling with an office or workspace wing or area. When this part of the building is not occupied the temperature can be arranged to be automatically reduced to a lower, more economical temperature without switching off the main heating system.

This is just one of a number of methods for achieving this function. It requires a room thermostat incorporating a set-back facility, which is an electrical device that, when energised, automatically reduces the room temperature by 6°C lower than that set by the room thermostat's normal setting.

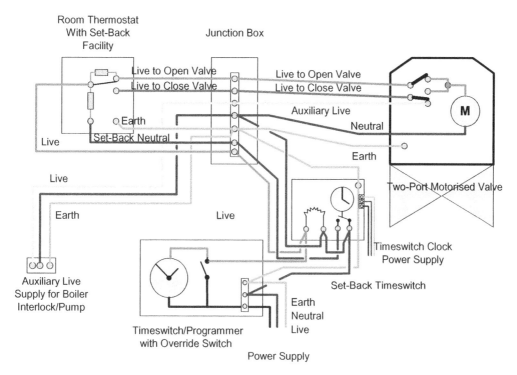

Figure 11.32 Automatic temperature reduction by set-back facility

The automatic space temperature reduction is activated by a separate timeswitch incorporated into the electrical wiring system, as shown in Figure 11.32, so that when the timeswitch activates the set-back switch, the temperature reduction facility comes into operation.

This method for achieving reduced space temperatures is used far less today in favour of alternative methods, but it can still be a useful means of reducing fuel consumption when part of the building is unoccupied.

AUTOMATIC BYPASS VALVE

Since the advent of low water content boilers and the necessity for them to have the circulating pump arranged to overrun – to maintain a minimum flow of water through the heat exchanger to prevent over-heating and boiling occurring – a bypass is required, as indicated on some of the control schemes in this chapter. The boiler manufacturer will advise if a bypass is required and the conditions that may apply that would warrant a bypass being provided.

The automatic bypass valve (ABV) operates on the difference between the inlet pressure and the outlet pressure of the valve; hence it is sometimes referred to as a 'pressure differential bypass valve'. Figure 11.33(a) illustrates, in simplistic form, the normal operation of a standard heating system, in which the water is being heated in the boiler heat exchanger, circulated around the heating system and returned to the boiler for the cycle to continue. The pressure difference (Δp) being created is determined by the pressure drop around the heating system created by the frictional resistance encountered, and the bypass valve should be set at the commissioning stage to be closed during this operation.

When the control system switches the heating system off, such as when all zone valves or all thermostatic radiator valves are closed off – as depicted in Figure 11.33(b) – the pressure difference between the flow

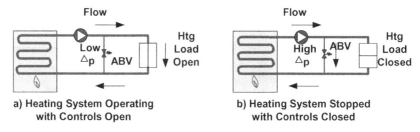

a) **Heating System Operating with Controls Open**

b) **Heating System Stopped with Controls Closed**

Figure 11.33 Heating system effects on bypass valve

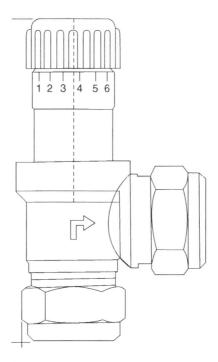

Figure 11.34 Typical automatic bypass valve

and return will increase due to the cessation of water flow. Hence, the circulating pump should be installed on the flow in this application, which will permit the bypass valve to open and maintain a minimum flow of water through the boiler and back again via the bypass. This short-circuiting of the heating system, or circulating pump overrun should be long enough to allow sufficient time for the boiler's burner to shut down before the pump is stopped. It should also be appreciated that it will reduce the life of the circulating pump if it is allowed to continue to run against a closed head, which is what would occur if a bypass valve was not fitted.

The automatic bypass valve is constructed in a similar manner to the safety relief valve, employing a tension spring type seating which is adjustable between approximately 0.2–0.7 bar differential pressure across the seating. The differential pressure setting required can be calculated as part of the pipe sizing exercise. A typical example is shown in Figure 11.34.

Table 11.2 may be used as a guide to selecting the correct size of bypass valve based upon the heat output rating of the boiler, but the boiler manufacturer's instruction manual should be consulted for exact information on the size and positioning of the bypass valve.

Table 11.2 Guide to bypass valve size

Valve Size (mm)	Boiler Heat Output (kW)
15	<19
22	19–35
28	35>

It has been the practice in the past to provide the bypass with a manually operated regulating valve equipped with a lockshield cover, such as the globe pattern valve, where the valve will be regulated to allow a permanent bleed of water through the bypass arrangement, to serve the same purpose as the automatic bypass valve. However, as this manually regulated globe valve permits a small continuous flow of water through it, it is considered to be a less efficient method than the automatic pressure differential bypass valve.

THERMOSTATIC RADIATOR VALVES (TRVs)

It is now a mandatory requirement of the Building Regulations that thermostatic radiator valves are fitted to all radiators as a minimum control requirement, with the exception of radiators installed in rooms that are equipped with a room thermostat that has temperature control over the whole heating system.

The utilisation of thermostatic radiator valves offers the advantage of a degree of individual room temperature control at a relatively reasonable cost. The degree of temperature control achieved will depend upon the quality of the valve specified and its application, in relation to its ability to sense the true temperature existing in the room.

The quality of thermostatic radiator valves has improved since their original conception and the problems associated with valve seatings chattering or becoming stuck is less frequent today. The term 'chattering' was applied to thermostatic radiator valves that made a knocking noise when the seating was in the position of being slightly open to a flow, and the combination of pump pressure together with the velocity of the water passing over it caused the seating to stutter or jump, resulting in the metal-to-metal knocking noise in the valve.

Thermostatic radiator valves are also comparable in cost to using a standard non-thermostatic radiator valve when considering the degree of automatic temperature control attained.

Thermostatic radiator valves work by employing an expanding bellows or capsule filled with gas, liquid or wax that will expand or contract with any change in air temperature surrounding the valve sensing head; see Figure 11.35. The expanding bellows or capsule will cause the valve seating to be lowered or raised, and in turn regulate the flow of hot water into the radiator.

The valve has an element of temperature adjustment by raising or lowering the valve head off or onto its seating against the tension of the return spring; when fully closed, it isolates the heat emitter. The valve may incorporate a 'freezing condition' facility that will automatically open the valve when subzero temperatures are sensed.

Thermostatic radiator valves are available in both the standard angle pattern and the straight through pattern version. They are also available in a vertical sensing head arrangement and a horizontal sensing head arrangement.

The vertically mounted sensing head version is commonly used in the UK, probably because it resembles the appearance of a standard radiator valve and is acceptable in its looks to the building occupier. The sensing head of this vertically mounted thermostatic valve has to be compensated for from the heat received from the warm air convection currents that will flow over the temperature sensing head – and if not temperature compensated, this will affect the valves performance.

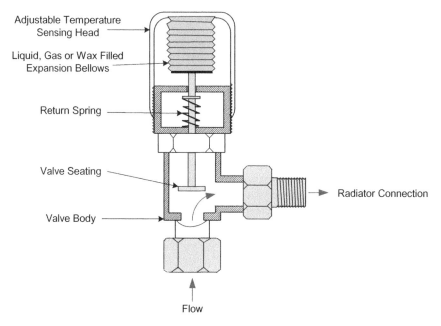

Adjustable Temperature
Sensing Head

Liquid, Gas or Wax Filled
Expansion Bellows

Return Spring

Valve Seating

Valve Body

Radiator Connection

Flow

Figure 11.35   Operating principles of thermostatic radiator valve

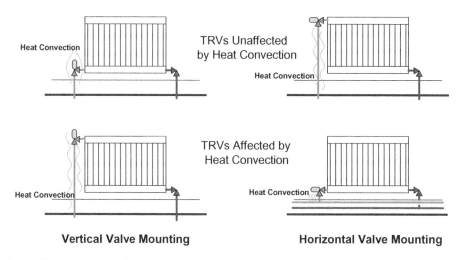

Heat Convection

TRVs Unaffected
by Heat Convection

Heat Convection

Heat Convection

TRVs Affected by
Heat Convection

Heat Convection

Vertical Valve Mounting   **Horizontal Valve Mounting**

Figure 11.36   Thermostatic radiator valve applications

If the flow connection to the heat emitter is required to be top entry, there is a strong chance of this convection air heat being greater than that expected and will give the valve sensor a false temperature reading. In this situation a horizontal pattern thermostatic radiator valve should be considered.

The horizontally mounted temperature sensing head is commonly used in most Northern European countries where it has become the normal arrangement for radiator valves. However, the valve head is more vulnerable to accidental damage in the horizontal mounting and is more sensitive to false temperature sources from heat emitted from warm air convection currents from heating pipework installed below the radiator. If the radiator temperature sensing head is arranged horizontally, mounted projecting forward, the sensing head will not be adversely affected by the rising warm air convection currents.

Figure 11.36 shows typical applications.

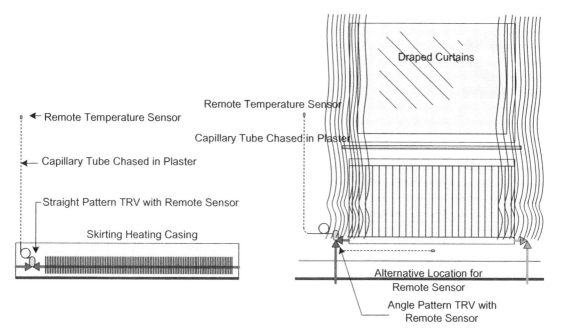

Figure 11.37   Typical TRV remote temperature sensor applications

The performance of the valve's temperature sensing head is not only affected by its position in relation to the heat being convected from the flow and return pipework, but also by its location and ability to sense the true temperature of the heated space. The location of the thermostatic radiator valve is determined by the position of the heat emitter installed in the room. In some instances it will be unavoidable for thermostatic radiator valves to be fitted in locations that are not favourable to sensing the true space temperature, or to be subjected to sources that result in false temperatures being sensed – such as direct sunlight from windows, or hidden in or behind enclosures including long curtains, or installed within the enclosed casings of natural convectors and skirting heating sections.

In such cases the thermostatic radiator valve should be equipped with a remote temperature sensor connected to the valve head containing the expandable capsule or bellows by a factory fitted and sealed small diameter capillary tube, which can be hidden quite easily, or chased within the depth of the plaster. It should be noted that if the capillary tube becomes damaged, it cannot be repaired on site and must be replaced with a new assembly.

Typical remote sensor installations are shown in Figure 11.37.

12

Oil Firing

Oil is a fossils fuel that – in common with all the fossil fuels – is obtained from unsustainable sources and emits carbon into the atmosphere as a product of combustion. It is not only used as a fuel for heating; life has become dependent upon it in every aspect.

This chapter is concerned with oil as a fuel for heating. It is important to understand the characteristics of oil to enable complete combustion to take place whilst everyone, from the installer to the user, remains safe.

Oil is an excellent fuel for heating as it contains a good calorific value, it can be easily stored and distributed, and if treated correctly is safe to use.

Oil as a fuel for domestic heating is second best to natural gas, when the latter is available, but oil is often selected as the main fuel source when a gas supply is not available. However, it is often the first choice for industrial and large commercial applications because the cost is more competitive when compared to natural gas.

Liquid biomass fuels, which are discussed further in Chapter 15, Alternative Fuels and Energy, have similar characteristics to the fossil fuel oils presently used, though it remains to be seen whether they can be produced in the quantities required.

OIL

Crude oil is obtained from sources below ground from various parts of the world, and is a mixture of hydrocarbons built up from the fossilised remains of plants and tiny animals that have been compressed under immense pressure for millions of years.

The products that the heating and associated industries are concerned with are diesel/gas oil and kerosene, which amount to approximately 26% of each barrel.

The term 'barrel' has been internationally adopted by the oil industry as a standard measure for all oil transactions. A barrel is based upon the US standard of 42 US gallons, which converts as follows:

Heating Services in Buildings: Design, Installation, Commissioning & Maintenance, First Edition. David E. Watkins.
© 2011 John Wiley & Sons, Ltd. Published 2011 by John Wiley & Sons, Ltd.
As an aid to lecturers and students, full colour versions of the figures in this chapter may be found at www.wiley.com/go/watkins
These figures are © 2011 by John Wiley & Sons, Ltd.

One barrel of oil equals   42   US gallons

       or   34.98 imperial gallons

       or   158.8   litres

The crude oil is subjected to a refining process to remove any impurities and extract the various petroleum products. The crude oil is passed through a fractioning tower where heat is applied to vaporise the oil and in turn, cool it down into a condensate. The various petroleum products will vaporise at different temperatures, so at each stage of the fractioning the distillate liquid can be drawn off. These oils are referred to as 'distillate oils'. The part of the crude oil that remains is referred to as 'residual fuel oils'. Some petroleum products are a blend of residual and distillate hydrocarbon products.

OIL BURNER FUELS

In the UK, oil burners are manufactured to burn oil complying with BS 2869 Category or Part 2, which covers six classes of oil burner fuels plus the inclusion of a special extra heavy class oil, Category 1, used in diesel driven automotive engines, now BS EN590. The categorisation has now been removed from the British Standard but has been retained here to demonstrate the difference between burner fuels and diesel oils for vehicle engines etc, which attract a different excise duty.

The characteristics of these seven classes of oil are listed in Table 12.1 and it is imperative that the engineer has knowledge of these characteristics to ensure that the oil is stored, distributed and burnt correctly.

Class C1: paraffin

This is a light distillate fuel oil suitable for free standing, flueless domestic oil heaters equipped with a wick burner, but can be vaporised.

Class C2: kerosene

This is also a light distillate fuel oil that is almost identical to paraffin. It is used exclusively for vaporising oil burners and some types of domestic atomising oil burners. Although the sulphur content is higher than that of paraffin, it is still low compared to other fuel oils and therefore requires a flue to convey the products of combustion to terminate externally of the building.

Both Class C1 and C2 oils can be stored and used at the ambient temperatures experienced in the UK without having to apply any special precautions.

Class D: gas oil

This is another distillate fuel oil that is free from residual components. As its name implies, it was at one time mainly associated with the town gas industry, either as a source of gas or for enriching or carburetting water gas.

This class of oil can also be stored at ambient temperatures without any pre-heating. It is used for atomising burners, both domestic and commercial, for heat outputs up to 1000 kW, although some domestic atomising burners are arranged to burn Class C2 oil.

Classes E to H: light, medium, heavy and special fuel oils

These are residual or blended fuel oils suitable for atomising burners only.

Table 12.1 Characteristics of oil burner fuels

Property	Classification of Oil, BS2869						
	C1	C2	D	E	F	G	H
	Paraffin	Kerosene	Gas Oil	Light	Medium	Heavy	Special
Specific Gravity at 16°C	0.79	0.79	0.835	0.93	0.95	0.97	0.98–1.0
Viscosity – Redwood No1	28	28	34	250	1000	3500	5000>
Viscosity – Kinematic Centistokes at 38°C	1.5	1.5	3.4	67	240	800	1700>
Viscosity – Kinematic Centistokes at 82°C	–	–	–	12.5	30	70	115
Flash Point Closed – Abel°C	43	38	–	–	–	–	–
Flash Point Closed – Pensky Martens°C	–	–	55	66	66	66	66
Pour Point °C	–	–	–18	–7	21	21	Varies
Freezing Point	–40	–40	–	–	–	–	–
Cloud Point °C Mar/Sept	0	0	0	–	–	–	–
Cloud Point °C Oct/Feb	–7	–7	–7	–	–	–	–
Water Content % by weight	No measurable amount		0.05	0.5	0.75	1	1
Maximum Ash Content % by weight	–	–	0.01	0.1	0.15	0.2	0.2
Sulphur Content % by weight	0.06	0.2	1	3.5	4	4.5	5
Calorific Value MJ/kg Gross	46.6	46.4	45.5	43.4	42.9	42.5	Varies
Calorific Value MJ/kg Net	43.9	43.6	42.7	41	40.5	40	Varies
kWh/Litre Approx Gross	10.12	10.18	10.57	11.28	11.22	11.42	Varies
Storage Temperature °C Min	Not applicable			10	25	35	Varies
Handling Temperature °C Min	Not applicable			10	30	45	Varies
Sediment Content % by weight	Free from visible sediment		0.01	0.15	0.25	0.25	0.25

- Class E – Light fuel oil may require a small amount of pre-heating during the winter months, and is used for general industrial burners having heat outputs in excess of 1000 kW.
- Class F – Medium fuel oil will require some degree of pre-heating all year during its use, and is normally used for burners with heat outputs in excess of 2000 kW.
- Class G – Heavy fuel oil used for burner plants having heat outputs in excess of 5000 kW, but may be required by some burner manufacturers to be pre-heated by as much as 125°C.
- Class H – Special fuel oil available for large industrial burners to special order only, but its use is often restricted due to its rare availability and problems in disposing of the products of combustion.

EXPLANATION OF CHARACTERISTICS

Table 12.1 lists the typical properties of the oil burner fuels complying with BS 2869. It is important that these properties are fully understood and appreciated when designing an oil fired heating system.

Specific gravity

The specific gravity of an oil represents the ratio of the weight of a given volume of the oil compared to an equal volume of water, at a standard temperature of 16°C.

As previously established in Chapter 4, Heat and Heat Transfer, the specific gravity of water is 1 when at a temperature of 4.4°C, this being the temperature when water is at its maximum density. Any substance with a specific gravity lower than 1 will be lighter than water and will float on the surface.

The specific gravity is also temperature related whereby at a temperature of 16°C the water would have expanded, resulting in the specific gravity of the equal volume of water being reduced from 1 to 0.999. Therefore, at a temperature of 16°C, water will weigh 999 $kg\,m^{-3}$ compared to that of a Class D fuel oil which will weigh only 835 $kg\,m^{-3}$, or 0.999 $kg\,litre^{-1}$ of water compared to 0.835 $kg\,litre^{-1}$ of Class D fuel oil.

Viscosity

The viscosity of an oil or liquid is a measure of its resistance to internal flow: the lower the viscosity, the thinner the oil; and the higher the ambient temperature, the lower the viscosity.

There are a variety of methods for determining the viscosity of an oil using several types of instruments which measure the time taken for a specified volume of oil to flow out of a purpose-made container through a calibrated orifice of a standard size.

The Redwood instrument was commonly used in the UK for establishing the viscosity of a liquid, whilst in America the Saybolt Universal or Furol method was used, with a large part of Europe employing the Engler Degrees method. These methods are now largely obsolete, having been replaced by the internationally adopted technique of using the Kinematic viscometer with its higher degree of accuracy: the other methods were found to be unreliable, particularly with liquids having low viscosities.

The Kinematic centipoise for dynamic viscosity method is based on the measurement of the time taken for a fixed volume of oil to flow by gravity down one leg of the glass 'U' tube viscometer, which is immersed in heated water under prescribed conditions. The time taken for the sample to pass between two timing marks is expressed in stokes or centistokes (cSt); 100 cSt equals one stoke.

The Kinematic viscosity of the fuel oil in centistokes at the controlled test temperature of 38°C, or 82°C is obtained by multiplying the time taken by the calibration constant of the viscometer. The centistokes is equivalent to $mm^2\,s^{-1}$.

The obsolete Redwood method is still used by some to express the viscosity of a liquid in terms of efflux time in seconds, at a temperature of 38°C. There are two Redwood scales, Redwood No I or Redwood Standard, and Redwood No II or Redwood Admiralty – the latter being used for more viscous oils. The

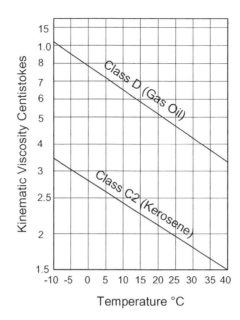

Figure 12.1   Viscosity/temperature relationship

Redwood instrument measures the time in seconds for 50 ml of oil to flow through an orifice of a given standard size at a constant temperature, but becomes inaccurate in the very low viscosity range.

The Saybolt method also has two scales, Saybolt Universal and Saybolt Furol. It is similar to the Redwood instrument except that it measures the time in seconds for 60 cc of oil to flow through an orifice in the bottom of a calibrated tube at a constant controlled temperature.

Figure 12.1 indicates a temperature/viscosity conversion chart that may be used to convert the viscosity of select fuel oils of Class C2 kerosene and Class D gas oil at different temperatures, whereby it can be seen that the viscosity becomes lower as the temperature rises.

Flash point

The flash point of a fuel oil is defined as the lowest temperature at which the application of a source of ignition will cause the oil/air vapour to ignite. It is determined under laboratory controlled conditions where the oil is heated and the resulting vapour mixed with air in an enclosed vessel until it is sufficiently rich to permit ignition on the application of a flame. The flash point should not be confused with the 'ignition temperature', which is much higher.

The flash point gives an indication of the safety of the fuel in storage situations, but has no bearing on its burning quality or performance.

The temperature specified ensures that the fuel oils are safe under proper storage conditions, at temperatures below their flash points.

Pour point

The pour point is determined by a laboratory test and is defined as the temperature at which the oil will cease to flow freely. Therefore, to ensure mobility of the fuel oil, this temperature should be below the normal storage and distribution temperature encountered on site.

Cloud point and freezing point

When distillate fuel oils are cooled to a certain temperature, minute flakes of wax may come out of solution and cause the fuel to become cloudy. This temperature is known as the cloud point and can vary with climatic conditions throughout the year, but providing that a suitable filter of the correct type is used, difficulties with fuel flow arising from low ambient temperatures should not occur.

The term 'cold filter plugging point (CFPP)' is the highest temperature at which Class D gas oil when cooled under controlled conditions will cease to flow through a 45 µm filter due to the formation of wax crystals. Gas oils are manufactured to a lower CFPP during the winter months than during the summer.

Gas oil can contain a blend of a range of hydrocarbons – including paraffin waxes, which when cooled reach a point when the paraffin wax starts to come out of solution by crystallising; this is known as the cloud point.

The freezing point of these fuel oils is below minus 40°C, which means that apart from a few places in the world, this should not present any problem.

Calorific value

The calorific value of a fuel is the amount of heat released on complete combustion when a unit quantity of that fuel is burnt.

There are two values normally given, gross or higher, and net or lower, both normally expressed in MJ kg^{-1}. The gross value includes the heat released by the vaporisation of the water forming part of the products of combustion. The net value is the gross value minus the latent heat of vaporisation. The greater the amount of hydrogen content in the fuel, the greater the difference between the two values.

In the UK, it has in the past, been common practice to use the gross calorific value of the fuel when carrying out tests for boiler efficiency, but more recently the European custom of calculating the boiler efficiency using the net calorific value has been used. This results in a higher efficiency figure and this can be misleading when quoted without qualification.

It is essential to know the gross calorific value of a fuel when making comparisons with other fuels and when calculating running costs.

Calorific value is sometimes referred to as 'specific energy'.

Sulphur content

All fossil fuels contain a degree of sulphur which affects the calorific value: the greater the amount of sulphur, the lower the calorific value.

The main significance of the presence of sulphur in the fuel is its conversion to sulphur dioxide on completion of the combustion process. If this combustion gas is allowed to condense, the sulphur dioxide will convert again to become a highly acidic liquid of sulphur trioxide, which is particularly aggressive to metallic materials.

Sulphur is also aggressive when in solution in the fuel oil and can cause corrosion to metals, with zinc being particularly susceptible to attack from sulphur compounds.

Ash, water and sediment content

These values are of little importance with regard to the quality of a fuel, except that both water and sediment will eventually collect in the bottom of storage tanks over a prolonged period and will require removing periodically.

The ash, water and sediment content of all the classes of oil is quite low, ranging from negligible to less than 1% of the fuel.

Storage and handling temperatures

Certain fuel oils, classes E to H, are required to be stored and distributed above the minimum temperatures stated, explained earlier under the heading of Pour point.

The recommended minimum storage and handling temperatures listed in Table 12.1 include a factor above the pour point temperature to ensure satisfactory flow mobility during operation. Where the ambient temperature is likely to drop below these recommended temperatures, then some form of pre-heating will have to be considered to both the storage tank and distribution pipelines. This pre-heating may take the form of an electrical immersion heater within the storage tank together with trace heating cables attached to the flow distribution pipeline – or where the heating system is arranged to operate continuously and it is practical to do so, a heating pipe coil type heat exchanger can be installed within the storage tank, combined with small diameter heat tracer lines attached to the underside of the oil distribution pipelines, but within the oval shaped thermal insulation.

Classes C1, C2 and D fuel oils used for domestic heating systems do not require any form of pre-heating in the UK, but Class D gas oil has a pour point of −18°C and where this class of oil is stored and used in parts of the world where this ambient temperature is likely to be experienced, then pre-heating should be considered.

OIL STORAGE

The storage capacity of any oil storage tank should be adequate for the boiler and any other oil fired appliances to operate under normal working conditions for a period of approximately eight weeks. It should also have additional spare capacity to allow the oil storage tank to be refilled by delivery road tankers with the eight-week capacity before the storage tank becomes completely empty, normally the tank should still contain one third of it's capacity whilst being refilled.

An example is given in Box 12.1.

Table 12.2 may be used as a guide for assessing the capacity of oil storage tanks based on varying boiler heat output ratings.

Although Table 12.2 may be used as a guide for assessing the minimum capacities of oil storage tanks, it should be appreciated that many fuel delivery companies will offer discounts per unit rate on fuel deliveries in lots of 2270 litres (500 gallons), which serves as an incentive to install larger capacity oil storage tanks – if they can be accommodated – to take advantage of these preferential discounts. An oil storage tank with a capacity of 2700 litres (600 gallons) will allow for bulk purchases of 2270 litres, whilst still maintaining a working reserve or freeboard of 450 litres (100 gallons) to allow the heating system to remain operating whilst waiting for fuel deliveries.

BOX 12.1   Example of fuel oil storage calculation

Eight weeks' normal operation	1000 litres
Oil storage tank's additional freeboard capacity	500 litres
Minimum oil storage tank capacity	1500 litres

It should be noted that the oil industry still retains large elements of the imperial system of measurement in its dealings. The metric values given have in some cases been directly converted from their imperial equivalents and the imperial values are stated alongside the metric figures.

Table 12.2 Recommended minimum oil storage capacities

Boiler Output Rating (kW)	Minimum Tank Capacity	
	Litres	Imp. Gallons
Up to 13	1130	250
13 to 18	1250	275
18 to 25	1400	300
25 to 35	1800	400
35 to 45	2700	600

OIL STORAGE TANKS

The important subject of legislation governing the storage and distribution of fuel oil is discussed in Chapter 24, Regulations, Standards, Codes and Guides. The following regulations apply:

- The Building Regulations Part J in England and Wales and their equivalent documents in Scotland, Northern Ireland and the Republic of Ireland.
- The Control of Pollution (Oil Storage) (England) Regulations.

Oil storage tanks have in the past been traditionally constructed from plain mild steel plates in rectangular or cylindrical form for both domestic and larger heating applications, but more recently non-metallic materials have been used to manufacture single piece oil storage tanks for domestic and small commercial heating installations.

Steel rectangular oil storage tanks

These should be manufactured to BS 799 from plain black mild steel plate of 14BG, 12BG or 10BG with welded seams and internal support bracing. The top should be tented to shed rainwater and prevent the build-up of snow, with the completed construction pressure tested to 0.3 bar.

Figure 12.2 shows a typical rectangular domestic oil storage tank arrangement complete with tank connections and fittings and including a fill point connection, which is a minimum size of 50 mm (2 inch), extended down inside the tank to prevent turbulence when being filled, a vent pipe connection also a minimum size of 50 mm (2 inch), complete with flame arresting gauze mesh, and an oil distribution connection of 15 mm (½ inch) diameter.

The fill point connection is extended from the tank to terminate with a 50 mm (2 inch) BSP threaded hose coupling, blank cap and chain. Where the height of the oil tank prohibits practical access for coupling up to the delivery tanker fill hose, then the fill line should be extended down to the low level and suitably supported, terminating with a check valve and hose coupling gate valve, complete with blank cap and retention chain.

Steel oil storage tanks should be supported on suitable support piers spaced at the maximum dimensions given in the table included in Figure 12.2. The supports may be constructed from either engineering bricks or concrete, and capped with either code 5 sheet lead or a double layer of DPC felt to protect the storage tank from corrosion. Alternatively, some steel oil storage tanks can be obtained complete with steel support legs permanently fixed to the underside of the storage tank.

The oil storage tank should be placed on the supports with a fall of 20 mm per metre of tank length back down to the drain valve – to facilitate any water that may have settled or sludge that may have collected to be drained off and removed.

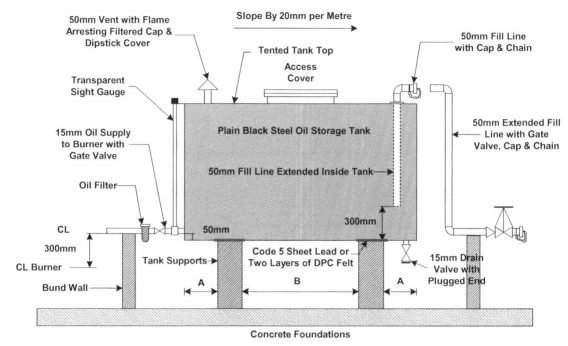

Tank Support Spacing

Tank Thickness	A	B
10BG	300mm	600mm
12BG	300mm	530mm
14BG	300mm	450mm

Figure 12.2   Standard arrangement for rectangular oil storage tanks

The high end of the storage tank should be equipped with a vent pipe of diameter equal to or larger than the fill pipe, to prevent the tank from becoming pressurised when being filled by the delivery tanker's pump, to allow the entry of air when oil is being drawn off, and to generally ventilate the tank to prevent the build-up of a high concentration of flammable oil vapour. The vent pipe should be terminated at a height of less than 600 mm above the roof of the storage tank and be capped with a filtered vent cap, or turned down and fitted with a flame arresting type wire gauze.

All storage tanks should be equipped with the means of determining the amount of fuel oil remaining in storage, so as to allow sufficient time to order and deliver fresh supplies of oil before the storage tank becomes empty.

The simplest and cheapest method of providing a means of indicating the contents of oil remaining in the storage tank for domestic heating applications is shown in Figure 12.2. This method comprises a vertically mounted transparent plastic sight tube, which serves to give a visual indication of the oil contents as the liquid oil will rise up the tube by gravity to a level equal to that in the storage tank. The top of this transparent tube is open to atmosphere to allow the level to rise and fall, but is capped to prevent foreign matter from entering the tube. These visual sight gauges are also available for domestic heating applications incorporating a main oil distribution isolating valve and disposable element oil filter.

The disadvantage of these types of contents gauges is that the clear see-through plastic sight tube can become discoloured after a period of time, making it difficult to ascertain the oil level in the tube and necessitating the replacement of this tube with a new, clean transparent tube.

Table 12.3 Standard rectangular oil storage tank sizes

Capacity		Dimensions			Capacity		Dimensions		
Litres	Gallons	L	W	H	Litres	Gallons	L	W	H
795	175	1050	600	1200	2700	600	2400	900	1200
1130	250	1050	1050	1050	2700	600	2400	600	1800
1130	250	1950	300	1800	3600	800	2400	1200	1200
1130	250	2400	600	1050	4500	1000	2400	1500	1200
1360	300	1800	600	1200	4500	1000	3000	1200	1200
1360	300	1800	900	900	6800	1500	3000	1500	1500
1800	400	1200	1200	1200	9000	2000	2700	1800	1800
2700	600	1800	1200	1200	13600	3000	3000	2400	1800

Other, more sophisticated alternative approaches for determining the contents of oil contained in the storage tank are described later in this chapter.

A list of capacities based upon tank dimensions is given in Table 12.3.

The piping connections made to steel oil storage tanks, as shown in Figure 12.2, should be made using plain black low carbon mild steel tube conforming to BS EN10255 medium weight, formerly 1387, with screwed joints using a jointing compound suitable for oil; galvanised tube should not be used. All copper alloy type valves and fittings should be manufactured from either bronze or gunmetal; zinc containing brass should be avoided as the zinc could be dissolved into the oil to react with the sulphur content of the oil to produce zinc sulphate.

Steel oil storage tanks should not be galvanised, but protected externally with at least two coats of oil resistant paint which should be periodically checked and repainted if necessary.

Steel cylindrical oil storage tanks

Figure 12.3 illustrates a standard storage arrangement for larger oil storage requirements, employing a cylindrical steel oil storage tank that may be mounted either horizontally as shown, or vertically, which although much taller, requires less area of ground.

The cylindrical oil storage tank arrangement is similar to that for rectangular oil storage tanks with regard to pipework connections, except that the fill pipe and vent pipe connections are normally larger and the tank is supplied complete with its own steel support cradles welded to the underside of the tank.

Table 12.4 shows tank capacities based upon dimensions.

The contents gauge indicated in Figure 12.3 is a more sophisticated arrangement comprising a magnetic direct reading gauge calibrated to suit the capacity concerned. It is operated by a rise and fall float fixed on the end of a lever arm. This method of indicating the contents status of the oil storage tank also has the provision to incorporate a second remote located contents gauge that may be mounted inside the fill point cabinet, if the fill point is beyond the sight of the oil storage tank. This method also has the optional provision to incorporate electrical switches that can be set to operate remotely located alarms to give warning of high level conditions occurring at the fill point, or low level alarms to warn the occupier that the contents are becoming low.

Figure 12.4 illustrates other forms of contents gauges that are in common use. These include the 'cat and mouse' contents gauge – see Figure 12.4(c) – whereby a float within the storage tank is attached to a steel wire which passes over a series of pulleys, and operates a pointer which is moved up and down outside the tank on a large calibrated board. These contents gauges are more common on larger storage tanks where the gauge may be visible from a distance, such as from a remote control room, although smaller versions have been made for industrial applications.

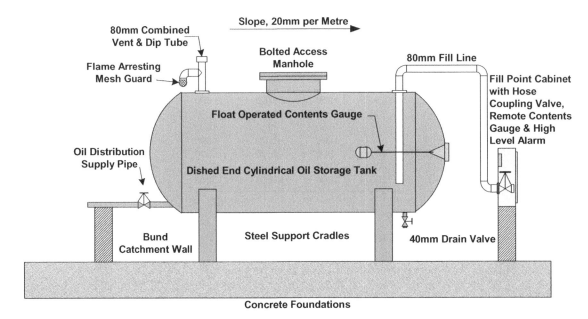

Figure 12.3 Standard arrangement for cylindrical oil storage tanks

Table 12.4 Standard cylindrical oil storage tank sizes

Capacity		Dimensions		Capacity		Dimensions	
Litres	Gallons	Diameter	Length	Litres	Gallons	Diameter	Length
1130	250	1000	1650	13600	3000	2400	3100
2260	500	1350	1650	18000	4000	2400	4050
2700	600	1350	1950	22500	5000	2400	5250
4500	1000	1350	3000	45000	10000	2700	7950
6800	1500	1800	3000	90000	20000	3000	13200
9000	2000	2000	3300	135000	30000	3600	14100

A smaller variation of the 'cat and mouse' contents gauge suitable for domestic heating oil storage tanks is depicted in Figure 12.4(a). This variation employs a float attached to a steel tape inside the storage tank, which in turn is connected through a hole cut in the roof of the tank to a spring-wound reel, similar to that of a retractable steel measuring tape. The status of the oil stored inside the tank can be observed through the window of the contents gauge head by reading the level visible on the tape.

The third version depicted in figure 12.4(b) is a non-electrical hydrostatic contents gauge, which operates on the principle of pressure exerted on a gas filled diaphragm by the weight of the fuel in storage above the pressure sensing diaphragm head. The pressure sensing head is connected to a calibrated contents gauge by a sealed capillary tube. In order for the contents gauge to be calibrated correctly at the factory, the capacity and physical size of the storage tank must be known, together with the specific gravity of the liquid.

Plastic oil storage tanks

Steel has been an admirable material for manufacturing tanks for storing oil as it possesses excellent robust mechanical and fire resisting properties, but it has one main disadvantage of being susceptible to corrosion.

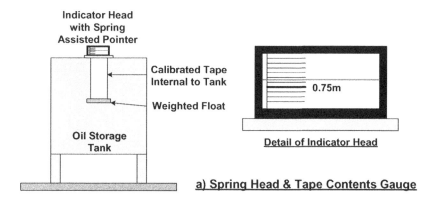

a) Spring Head & Tape Contents Gauge

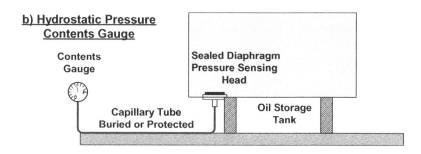

b) Hydrostatic Pressure Contents Gauge

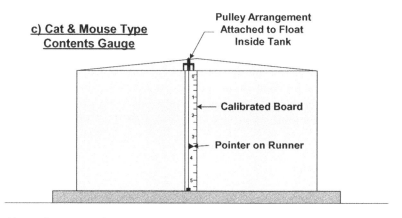

Figure 12.4   Alternative types of contents gauges

Although maintenance in the commercial and industrial heating market has generally controlled this, the situation regarding maintenance to domestic oil storage tanks has often been ignored and resulted in tank failures. More recently, plastic oil storage tanks have been developed for domestic heating and small commercial applications, made from low or medium density polyethylene as a single piece construction.

Polyethylene oil storage tanks have the advantages of not requiring painting and not corroding; they do, however, lack the mechanical strength of steel and can be damaged more easily and are less resistant to fire. They also require firm continuous support beneath the base which must be constructed to have a flat

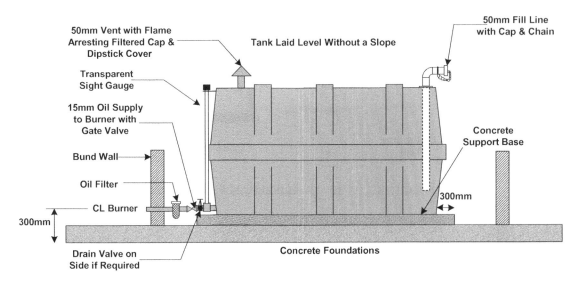

Figure 12.5   Typical domestic heating oil storage arrangement employing a polyethylene storage tank

contact surface that extends 300 mm beyond the tank walls on all sides. The pipe connections must also be fully supported to prevent any stress or strain being exerted on the tank walls, which could cause fracture. Plastic tanks also require restraining straps to prevent them from being moved by wind pressure when they are almost empty.

The moulding of the polyethylene storage tanks is usually ribbed in some form to add strength and rigidity to the walls of the tank, but because they sit directly onto the flat concrete base they do not have the drain-off valve that is common on steel oil storage tanks. Polyethylene oil storage tanks can only be de-sludged by passing a flexible suction tube down the vent or fill connection, or by incorporating a drain valve on the oil distribution piping. Figure 12.5 shows a typical oil storage arrangement employing a polyethylene storage tank with an alternative drain-down facility of a drain valve installed on the oil distribution line. Alternatively by removing the fill point elbow and passing a suction tube down to the bottom of the extended fill pipe and pumping the oil/sludge out, the fuel oil supply company can remove this and return it for refining back into a usable product.

BUNDED OIL CATCHMENT AREAS

A bund is a continuous dyke-like wall that encloses the oil storage tank to form an oil resistant catchment area capable of containing the entire contents of the full oil storage tank, plus a minimum of 10% extra, making a total volume of 110% of the oil storage tank's capacity. If there is more than one oil storage tank contained within the bunded catchment area, then the holding capacity of the bunded area should be at least equal to 110% capacity of the largest oil storage tank enclosed within the catchment area. The 10% additional capacity above the oil storage capacity serves as a safety margin that includes rainfall and any other volume reducing matter.

It has always been considered good practice to construct a bund to all oil storage tank installations to contain the oil in the event of a leak or fracture of the tank, therefore reducing the risk of causing pollution or contaminating water courses, but since 2001, Part J of the Building Regulations and the Control of Pollution (Oil Storage) (England) Regulations has made it mandatory to provide a secondary containment area equal to 110% of the oil storage capacity.

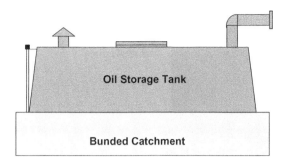

Figure 12.6   Domestic oil storage tank with integral bund

The bund has traditionally been constructed from masonry materials such as brick or concrete, with an oil resistant base being formed to provide a slight gradient to a sump created in one corner. This enables any surface water to be pumped out by the use of a portable sump pump being placed in the sump, or the oil to be reclaimed and returned to the oil supply company for processing back to a usable product in the event of a tank failure.

The bund wall should not have any openings formed in it apart from those for the oil distribution supply pipe and any fire protection pipe, which should be completed with a puddle flange arrangement.

Manufacturers of both steel and polyethylene oil storage tanks now produce a range of tanks that incorporate an integral bund, or twin wall tank construction, manufactured from the same material as the oil storage tank. This achieves a guaranteed oil-/water-tight bund with reduced on-site construction time; this is a particular advantage for domestic applications making for a neater, more acceptable appearance of the oil storage tanks which can be easier to disguise behind shrubs and foliage. An example of such a tank is given in Figure 12.6.

STORAGE TANK LOCATION

There are a number of factors that require consideration when selecting the location for the oil storage tank, including the occupier's desire for it to be unobtrusive and aesthetically acceptable.

Buried oil storage tanks for domestic heating applications should be avoided due to the high cost of the excavation, reinforced concrete raft foundation and restraining steel straps to anchor the tank down to counteract the buoyancy of the tank when empty. Buried oil storage tanks for larger, non-domestic applications should only be considered when no other option exists and planning approval has been granted.

Steel tanks destined for underground applications should be constructed strong enough to withstand external ground pressures acting upon them and have a minimum of 3 mm thick trowelled-on external oil resistant coating to protect them from corrosion. The tank should be pressure tested before the excavation is back-filled with selected material in accordance with the manufacturer's instructions.

The oil storage tank should preferably be sited above ground and in the open air and should never form part of any structure. If the oil storage tank is required to be installed inside a building it should be located in its own dedicated room with the surrounding structure of the walls, ceiling and floor having a 1 hour fire rating. The room that it is located in should also be constructed as a catchment bund with any door opening being placed above the bunded area. The oil storage tank room should also be ventilated with the oil storage tank vent pipe terminating externally to atmosphere.

Wherever the oil storage tank is to be located, it must be accessible for filling by the road delivery tanker without the tanker having to enter the property, unless the access has been properly constructed as

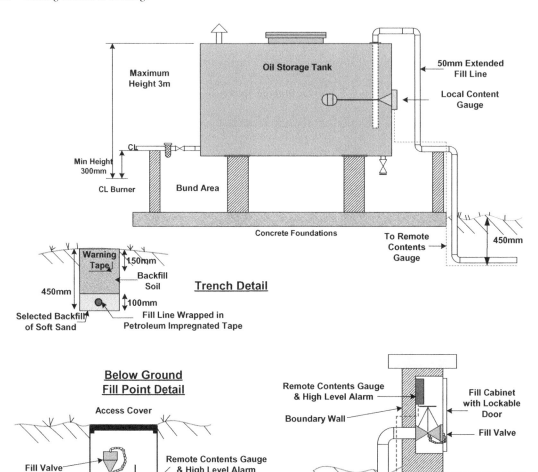

Figure 12.7   Remote oil fill point detail

a road. It should be appreciated that the delivery vehicle can weigh in excess of 14 tonnes fully laden with a large turning circle.

The delivery tankers normally carry a fuel delivery hose of 30 m in length and therefore the oil storage tank fill point should be located within this distance from the road. If the oil storage tank cannot be sited within this distance then the fill line should be extended as indicated on Figure 12.7, and be complete with a remote contents gauge and high level alarm to warn the tanker delivery operator when the oil storage tank is full.

The extended oil fill line should be installed using plain black low carbon mild steel tube conforming to BS EN10255 heavy weight, wrapped in a double layer of petroleum impregnated tape to protect it against corrosion.

As an alternative to the direct buried extended oil fill line indicated in Figure 12.7, the fill point line may be housed in a covered duct constructed in the ground to afford protection.

The oil storage tank should also be located as close as possible to the boiler, 1.8 m being the optimum distance. The height of the oil distribution connection from the oil storage tank should preferably be at least 300 mm above the burner oil inlet connection, but not greater than 3 m. The exact height relationship to the burner will be dependent upon the type of oil burner employed.

OIL HANDLING AND DISTRIBUTION

The oil distribution supply from the bulk oil storage tank to the boiler or boilers should be simple, short and as straightforward as possible and preferably be by gravity, particularly for domestic heating applications. The oil distribution piping system will also be dependent upon the type of oil burner used, as the requirements for atomising burners are different from those required for vaporising burners.

Figure 12.8 illustrates the principles of a single pipe oil distribution gravity supply system, suitable for both vaporising and atomising oil burners alike, although the minimum burner inlet oil pressure for atomising is less than that required for vaporising burners.

Storage tanks serving either vaporising or atomising oil burners employing a single pipe installation by gravity pressure alone, require the oil storage tank to be set up to provide a minimum pressure stated by the boiler/burner manufacturer, usually 300 mm as shown in Figure 12.8. The maximum pressure being exerted on a vaporising burner should not exceed 3 metres, as shown in Figure 12.8.

The distribution supply pipe may be installed using copper tube with Type 'B' manipulative type compression fittings, not Type 'A' compression fittings or soft soldered capillary fittings. If the copper tube is to be installed below ground then it should be complete with a continuous plastic coating. Polyethylene tube can also be used for above ground applications in a single continuous length, thus avoiding the need for intermediate joints. Black low carbon mild steel tube with screwed or welded joints can also be used for both domestic and larger systems, but underground pipes must be wrapped with a double layer of petroleum impregnated tape. Jointing compounds must be suitable for oil applications, or PTFE tape may be used.

All valves should be manufactured from gunmetal or bronze alloy with large diameter valves constructed from cast steel. Cast iron valves should not be used.

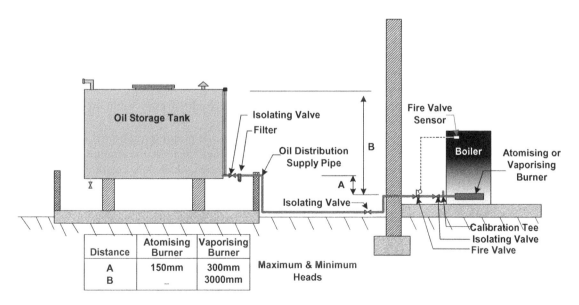

Figure 12.8 Single pipe gravity oil supply system

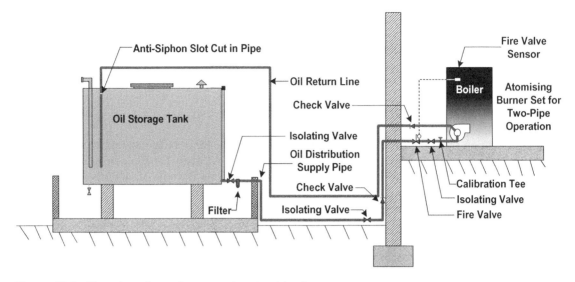

Figure 12.9   Two-pipe oil supply system for atomising burner

The oil supply distribution pipe should be equipped with isolating gatevalves and a fire safety valve, as shown, together with a suitable filter and plugged fuel calibrating tee.

Figure 12.9 illustrates the principles of a two-pipe oil distribution supply arrangement when a gravity flow to the burner cannot be achieved and the minimum pressure head of 150 mm to an atomising burner, shown in Figure 12.8, is not possible.

It is normal practice for the burner manufacturer to supply the burner arranged for a single-pipe flooded suction operation, but where a two-pipe oil lift system is required, as shown in Figure 12.9, it can be arranged on request for the burner to lift the oil from a lower level and any unused surplus oil returned to the oil storage tank, instead of returning it internally back to the burner oil pump suction connection.

The oil supply distribution arrangement is installed in the same way as the single pipe oil supply system, except that a check valve is incorporated on the supply to prevent the loss of oil prime to the atomising burner pump. A return pipe connection is provided on the oil burner pump for returning the excess oil not used by the burner to the oil storage tank, where an additional connection is required and terminated at approximately two-thirds the depth of the storage tank. To prevent this pipe serving as a means of siphoning the oil back to the burner, a slot should be cut into the pipe near the top, inside the tank to allow air into the pipe and thus break the siphon.

The oil supply return pipe should also be equipped with a second check valve to prevent the oil returning back to the burner when the burner pump is switched off.

During the normal working operation of this type of oil burner, the pump incorporated as part of the burner has the ability to raise the oil from the storage tank at a lower level, use the amount required for the burner output and return the unused surplus oil back to the storage tank, thereby creating a circulatory system.

As an alternative to the two-pipe oil supply distribution system shown in Figure 12.9, when the minimum requirements for a gravity flow cannot be obtained – for domestic heating applications only – is a de-aeration device, as depicted in Figure 12.10.

The de-aerator chamber can achieve a higher lift between the oil storage tank and the burner than that achieved by the burner oil circulating pump on the two pipe oil distribution system.

The de-aeration chamber should be fitted close to the oil burner and connected to the oil storage tank by a single supply pipe, as shown. The atomising type oil burner is also connected to the de-aeration chamber by a reduced length looped circuit, where the oil burner pump circulates the oil, in a similar way to the operating principles of the two-pipe system.

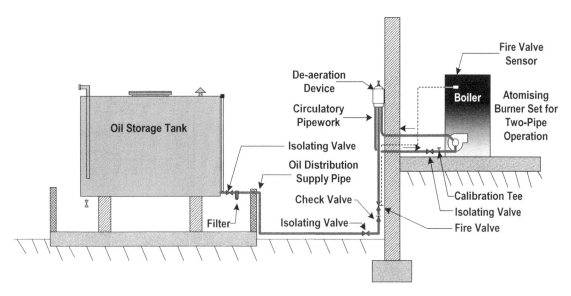

Figure 12.10 Single-pipe oil distribution supply using de-aeration chamber

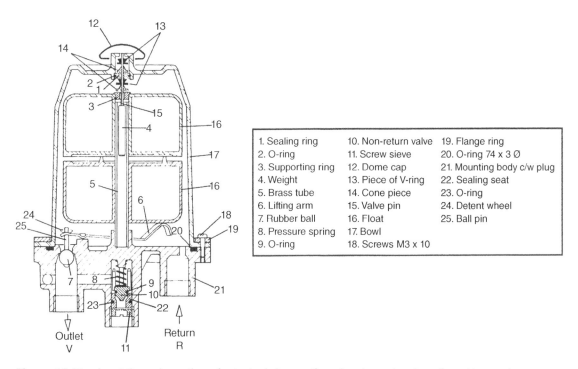

1. Sealing ring	10. Non-return valve	19. Flange ring
2. O-ring	11. Screw sieve	20. O-ring 74 x 3 Ø
3. Supporting ring	12. Dome cap	21. Mounting body c/w plug
4. Weight	13. Piece of V-ring	22. Sealing seat
5. Brass tube	14. Cone piece	23. O-ring
6. Lifting arm	15. Valve pin	24. Detent wheel
7. Rubber ball	16. Float	25. Ball pin
8. Pressure spring	17. Bowl	
9. O-ring	18. Screws M3 x 10	

Figure 12.11 A cut-through section of a typical de-aeration chamber, showing all working parts

The circulatory piping system including the de-aerator should be primed with oil before the oil burner is activated initially, or when the system has been drained down.

The de-aeration chamber serves to release any entrapped air from the oil and acts as a buffer tank for the burner to circulate the oil at a constant rate, irrespective of the burner consumption demand.

A cut-through section is shown in Figure 12.11.

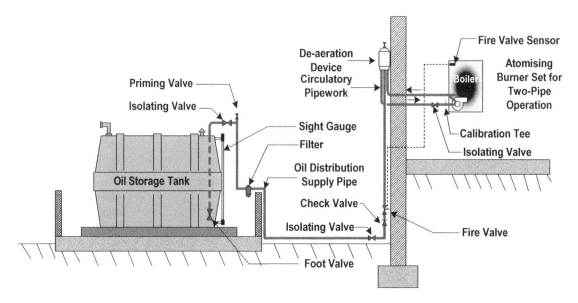

Figure 12.12   Single-pipe oil distribution supply using de-aeration chamber with top outlet oil storage tank

Plastic bodied de-aeration chambers must be installed externally as they are deemed to be at risk if a fire occurs. Metal bodied de-aeration chambers can be installed inside the building, providing that they do not release any oil vapour internally.

All de-aeration chambers are equipped with an automatic air release valve in order for them to function as a de-aeration device and the air that is released will contain a certain amount of oil vapour. Hence, any metal bodied de-aeration chambers located in the building should have the air release valve's outlet arranged to terminate to atmosphere.

Figure 12.12 illustrates another version of the single-pipe oil distribution system. It makes use of a de-aeration device but has a top outlet oil storage tank, and is used for domestic heating applications only.

This method of suppling oil to the boiler/burner unit has found limited use for small domestic heating applications as it suffers from frequent loss of the oil prime when the boiler is on a non-operating period. For this reason the submerged burner suction pipe should be equipped with a reliable foot valve and a convenient means of re-priming the oil supply distribution line.

Figure 12.12 also illustrates the use of a low water content wall mounted domestic heating boiler employing an atomising type burner.

In certain situations it may be required to raise the oil from the oil storage tank to a height beyond the limited capabilities of the burner's pump suction capacity, or that of the de-aeration chamber arrangement, which is normally restricted to between 2 to 3 metres, depending upon the supply pipe diameter used and total length of the distribution pipe. In these situations some form of booster pump equipment will be required.

Figure 12.13 illustrates a method of raising oil that is suitable for low-rise residential flats, or single-dwelling properties where the terrain prevents the previously described methods of lifting being used.

This method employs an electrically operated suction lift pump, commonly referred to as an 'oil lifter', which can be used successfully for raising oil to heights of approximately 7.5 metres above the invert level of the oil storage tank distribution outlet pipe to the top of the oil lifter.

These units consist of an integral miniature oil containing reservoir, housing an electrically driven gear pump and a control float which regulates the level of oil in the reservoir by switching the pump motor on

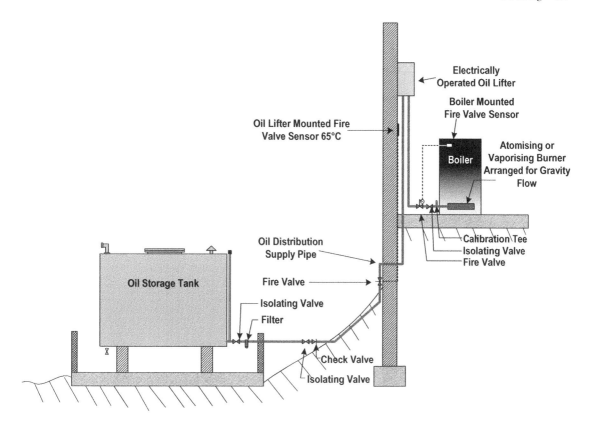

Figure 12.13   Oil supply to boiler raised by oil lifter

and off. The unit incorporates a safety switch operated by the float to switch the pump motor off to prevent overfilling of the reservoir. When the oil level drops below the normal control level, the pump motor is also switched off by a separate mercury switch.

These units are suitable for raising oil to separate dwellings served by a single oil-fired appliance, although larger units are available to serve duplicate appliances.

Fuel oil is fed by gravity from the reservoir of the oil lifter to the appliance's burner as it will function as a constant pressure head device.

As these units contain a store of oil inside them, they are required to have an additional fire valve on the gravity oil supply outlet pipe to the boiler/burner, with the valve sensor located in close proximity to the oil lifter.

Figure 12.14 illustrates in detail the operational components of a standard oil lifting suction lift pump.

In the case of taller buildings with a rooftop boiler house, it would be impractical and undesirable to store large quantities of heavy oil at the top of the building. In these situations a two-stage oil storage and booster transfer system of oil supply, as shown in Figure 12.15, should be considered.

In this system the bulk of fuel oil is stored at ground level – either in an externally located storage tank, or as illustrated in a storage tank housed within its own dedicated room or compartment.

The fuel oil is boosted by a set of oil transfer pumps up to a daily service oil storage tank located in the rooftop boiler house. This 'day tank', as it is termed, would normally be sized to contain approximately 8 hours' normal fuel consumption and equipped with an overflow terminating back into the main bulk oil storage tank. The overflow pipe should be capable of conveying the maximum total oil flow of the transfer

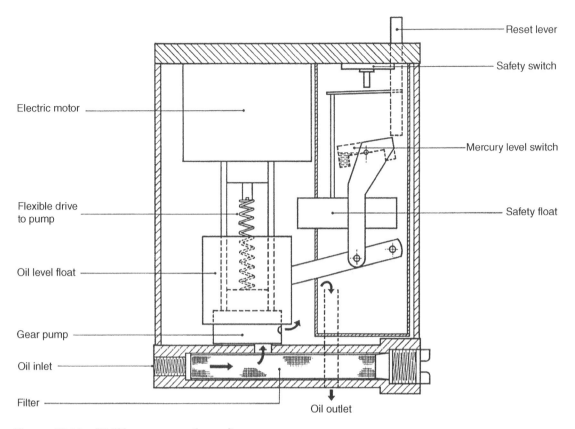

Figure 12.14   Oil lifting pump suction unit

pump duty back to the main bulk oil storage tank by gravity and being sized to be at least one commercial pipe size larger than the boosted transfer main fill line to the daily service tank.

The oil transfer pumps are controlled electrically by a float switch mounted inside the daily service tank, which would be arranged to automatically activate the transfer pumps and boost the fuel oil to fill the daily service tank and in turn, switch them off when full.

The oil booster transfer pumps should be duplicated and arranged to operate as one duty and one standby with the duty/standby being alternated between operations to obtain equal operating time.

OIL FIRE VALVE

An oil fire shut-off valve is an essential safety component installed on the oil supply distribution line from the oil storage tank to the boiler/burner unit. It should be capable of cutting off the flow of oil to the burner in the event of a fire occurring in or near the boiler to prevent further fuel being supplied to the fire.

The fire valve should be located outside the building if possible, but in certain circumstances where this is not feasible, it can be located internally where it should be fitted as close as practical to the external wall where the oil supply pipe enters the building.

Fire valves have in the past traditionally been of the freefall dead-weight drop shut type valve, which is only suitable for internal locations, due to it consisting of a delicately strung steel cable suspended and tensioned on a series of pulley wheels between the boiler casing and the dead-weight fire valve.

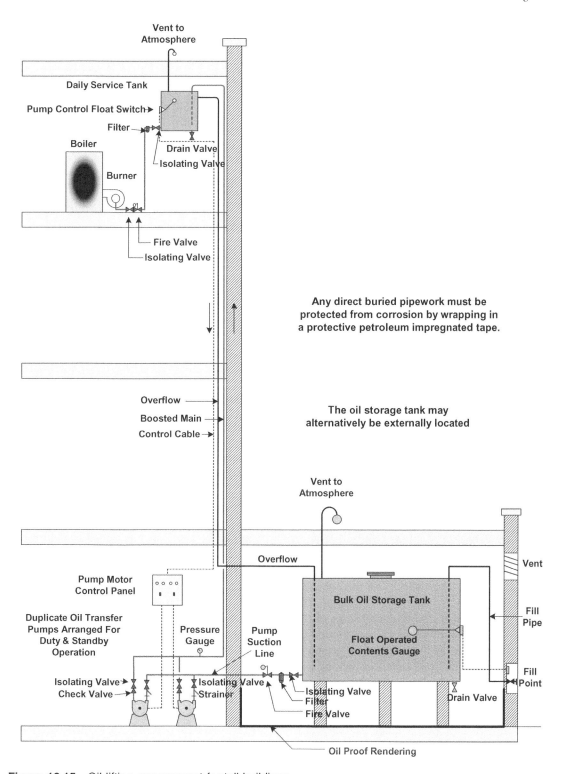

Vent to Atmosphere

Daily Service Tank

Pump Control Float Switch

Filter

Boiler

Drain Valve

Isolating Valve

Burner

Fire Valve

Isolating Valve

Any direct buried pipework must be protected from corrosion by wrapping in a protective petroleum impregnated tape.

Overflow

Boosted Main

Control Cable

The oil storage tank may alternatively be externally located

Vent to Atmosphere

Pump Motor Control Panel

Overflow

Vent

Bulk Oil Storage Tank

Duplicate Oil Transfer Pumps Arranged For Duty & Standby Operation

Pressure Gauge

Pump Suction Line

Float Operated Contents Gauge

Fill Pipe

Isolating Valve

Check Valve

Isolating Valve

Strainer

Isolating Valve

Filter

Fire Valve

Drain Valve

Fill Point

Oil Proof Rendering

Figure 12.15   Oil lifting arrangement for tall buildings

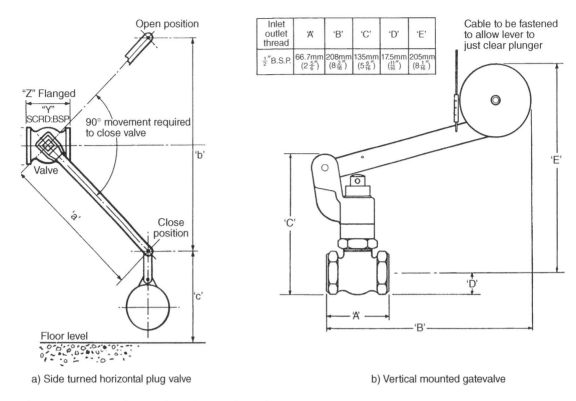

Inlet outlet thread	'A'	'B'	'C'	'D'	'E'
$\frac{1}{2}$"B.S.P.	66.7mm ($2\frac{5}{8}$")	208mm ($8\frac{3}{16}$")	135mm ($5\frac{6}{16}$")	17.5mm ($\frac{11}{16}$")	205mm ($8\frac{1}{16}$")

a) Side turned horizontal plug valve   b) Vertical mounted gatevalve

Figure 12.16   Freefall dead-weight type fire valve

These type of fire valves are available in sizes up to 150 mm diameter and resemble either a quarter turn plug valve with the spindle mounted horizontally with a weighted lever arm attached to it – see Figure 12.16(a), or alternatively, for smaller installations a gate or stop valve bodied version that has a spring-assisted, vertically arranged spindle can be used – see Figure 12.16(b).

The valve is held open by the weighted lever arm attached to the steel cable tensioned to hold it up. The steel cable incorporates a number of release devices that can be activated automatically or manually, which, when released slacken the tension of the cable, allowing the weight of the fire valve lever arm to drop and close the valve tight. It can then only be opened manually after the system has been made safe – see Figure 12.17.

A fusible-link which comprises two strips of copper joined together by a low melting point soft solder, is fitted in the tensioned steel cable and located over the top of the burner. In the unlikely event of a fire occurring at or near the burner, the heat generated will cause the low melting point soft solder to soften and the tension of the steel cable will pull the two strips of copper apart, releasing the weight of the fire valve, thus allowing it to drop, closing the valve and shutting off the oil supply to the fire.

On heating systems installed in larger buildings it may be considered desirable to incorporate a manual quick release mechanism within the boiler house to enable the fire shut-off valve to be manually activated in an emergency. This manual quick release mechanism consists of a wall mounted device incorporating a steel pin with a tension cable attached to it. The pin can be released by pushing in the end button whereby the tension of the cable will pull the pin out; the pin may be reset after it has been activated.

The cable tension may also be released electrically by incorporating an electric quick release device into the tension cable system with the device being activated by part of the main fire alarm system.

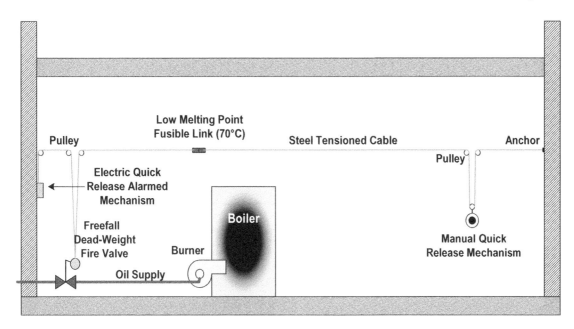

Figure 12.17 Application of the freefall dead-weight fire valve

The working condition and operability of the fire valve system can be easily checked by simply unhooking the lever weight from the cable and physically moving the lever arm up and down to see if it is free to move.

The freefall dead-weight fire valve arrangement has now been almost totally superseded by the use of electrically operated solenoid fire valves, activated by heat or air temperature detectors.

Figure 12.18 illustrates in detail the working principles of the manual quick release mechanism, showing how the release pin is secured inside the quick release device.

Figure 12.19 illustrates in detail a low melting point fusible link copper strip.

DOMESTIC OIL FIRED HEATING FIRE VALVES

The freefall dead-weight fire valve set-up is suitable for a self-contained boiler house where the steel cable strung up around the boiler plant room would not be out of place, but for domestic heating applications it is not practical, especially if the boiler is located in the kitchen, which is the case for the vast majority of cases. For domestic heating applications, a more practical arrangement is required that does not involve a complex tensioned cable being strung across the room where the boiler is located.

A simple fire valve device available for small oil-fired domestic heating arrangements is the 'fusible head' isolating valve. These valves can have a dual purpose as they resemble a standard wheel head isolating valve that can be manually operated to isolate the oil supply to the burner and double up as a fire valve. The valve incorporates a brass ferrule soldered into the head of the valve. In the event of a fire occurring in the vicinity, the solder will melt. This causes the valve to close the supply of oil to the burner, assisted by a spring, and preventing fuel from being fed to the fire. This type of valve cannot be re-set as other types of fire valves can, and must be replaced after the fire has been extinguished.

A more professional engineered arrangement suitable for domestic and small commercial heating applications is the use of an expanding gas capillary tube fire valve, which works on similar principles to that of a thermostatic radiator valve equipped with a remote temperature sensor. The valve is available with

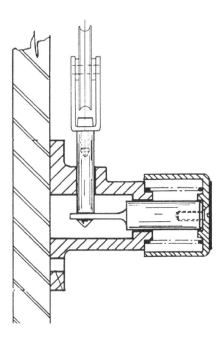

Figure 12.18　Manual quick release device

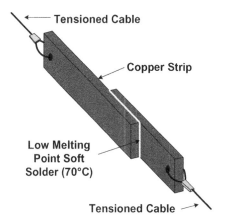

Figure 12.19　Fusible link

varying lengths of capillary tubing factory-attached, which permits the valve body to be located outside the building as shown in Figure 12.20, with the remote sensor fitted inside or adjacent to the boiler casing. When the remote temperature sensor senses a rise in temperature caused by a fire, at a temperature of between 66–70°C, the gas inside the capillary tubing and sensor probe will expand causing pressure to act on the valve head and in turn shutting the valve to the flow of fuel oil to the burner, preventing fresh fuel from being supplied to the area of the fire.

OIL FILTERS AND STRAINERS

Usually the oil burner fitted to the appliance incorporates its own filter or strainer on the inlet, but as they are usually of limited capacity they should be regarded as a secondary filter, and a main oil line filter or

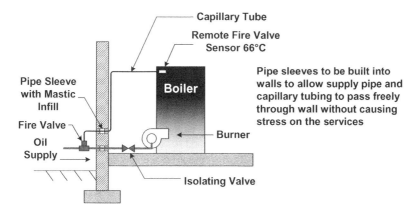

Figure 12.20 Fire valve with remote sensor and capillary tube

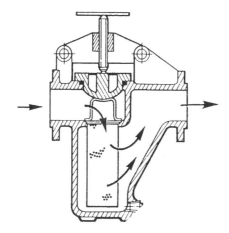

Figure 12.21 Detail of oil strainer for large heating installation

strainer should be fitted on the oil supply immediately after the distribution pipe leaves the oil storage tank, to protect the burner assembly.

Although the oil is delivered in a clean condition, it is still necessary to remove any extraneous matter that may have gained access to the oil storage tank via the vent or deterioration of the system, thus preventing solid particles from reaching the burner or pumping equipment where they could cause mechanical damage.

Filters for domestic oil-fired heating installations are usually a disposable cartridge contained in a removable bowl, whereby the oil flows through the centre of the cartridge element which is made of a material that removes any solid particles from the oil. The filter should not be made from plastic material or contain a glass bowl, but should be made from a suitable metallic material.

Figure 12.21 illustrates a strainer that would be used to filter the oil flow to the burner arrangement; the internal basket consists of a perforated metal cylinder that is designed to filter out any solid foreign particles from the oil flow, so allowing the screened oil to pass into the centre of the basket and out of the strainer. The strainer basket is designed to be cleaned periodically by washing in an oil solution before replacing back into the strainer body. The degree of filtration should be in accordance with the boiler/burner manufacturer's recommendations, but usually the strainer basket's perforations should not be less than 5 mesh per millimetre having a micron rating of 120.

OIL LINE PIPE SIZING

The same criteria and process associated with pipe sizing described in Chapter 9 for water systems applies when selecting suitable pipe sizes for conveying oil to the burner. This includes the length of pipe run, frictional resistance, pressure head available and the flow rate required – obtained from the boiler/burner manufacturer.

In addition to the standard criteria, the viscosity and temperature of the oil will have an effect when sizing piping systems for oil.

The pipe sizing chart shown in Figure 12.22 can be used for assessing pipe sizes for the flow of Class C1, C2 and D fuel oils when either the kilowatt output rating of the boiler or the fuel consumption of the burner is known, by charting them against the total length of the fuel supply pipe.

When more than one boiler is to be supplied with fuel oil, the pipe must be sized to convey the combined maximum flow rate required for all the boilers to be supplied.

OIL PRE-HEATING

As previously established, the viscosity of the oil is dependent upon the temperature of the oil, with the heavier oils usually requiring some degree of pre-heating in both the oil storage tank and the distribution piping.

The oil storage tank heating consists either of hot water or steam if available, circulating through pipe coils located inside the tank at the base, or supplied by an electric immersion outflow heater. Electric heaters are confined to the proximity of the oil outflow so that the oil may be heated at the point of exit from the storage tank rather than in the whole storage tank. The electric outflow heaters are designed to enable the oil to be drawn off through the middle of the heater tube.

In addition, the oil distribution line passes through a heat exchanger heated by hot water or steam, with an electrically heated heat exchanger for initial start-up when the system is cold.

The oil distribution circulation main is arranged to be circulated as close as possible to the boiler/burner assembly so as to avoid dead-legs, with the circulated oil returning excess oil back to the transfer pump's suction line as indicated in Figure 12.23.

The oil distribution lines are kept warm by installing hot water or electric trace heating tapes clipped to the distribution oil pipe under the thermal insulation.

Class C and D oils used for domestic heating applications do not require any pre-heating.

CENTRAL OIL STORAGE

The main disadvantage of and criticism levelled at oil-fired heating for domestic heating applications, is the requirement for on-site storage of the fuel, which can be a particular problem when finding a suitable place for locating the storage tank, especially with the modern trend of building houses with almost no garden area. Another disadvantage is that the consumer is responsible for periodically checking the oil level remaining in the storage tank and arranging for replenishment of the fuel.

Central oil storage answers these criticisms and makes the use of oil as a heating fuel more attractive to domestic consumers, as it eliminates the need for individual oil storage tanks and distribution piping for each property, as well as saving the consumers the hassle of arranging for fuel deliveries.

This method of fuel supply is of particular interest to both private and local government developers of housing estates and blocks of residential apartments, as the fuel delivery supplier takes the responsibility of installing a bulk central oil storage tank together with the piping distribution network to each property.

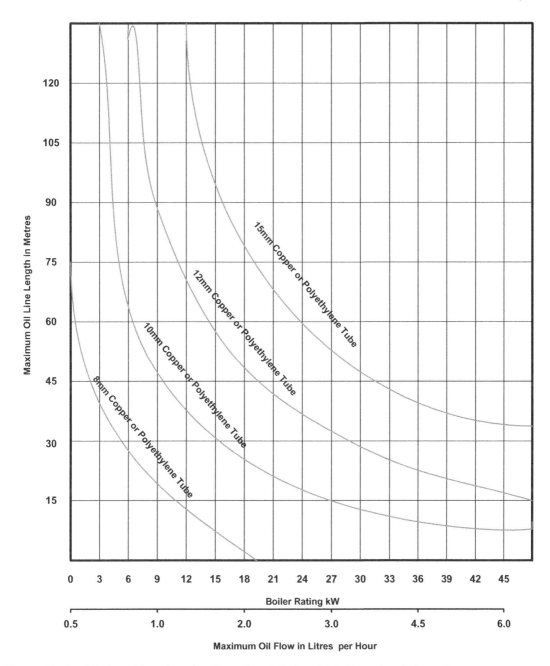

Figure 12.22 Oil pipe sizing chart for Class C and D oils with 300 mm head above burner

The fuel supplier also retains ownership of the storage and distribution facility and responsibility for the maintenance of this system, and undertakes the regular refilling of the storage tank.

To the occupier of the property there is no difference between this method of fuel supply and a natural gas supply, as there is no on-site storage and the occupant receives a bill just for the amount of oil they have consumed, as measured by their individual oil flow meter.

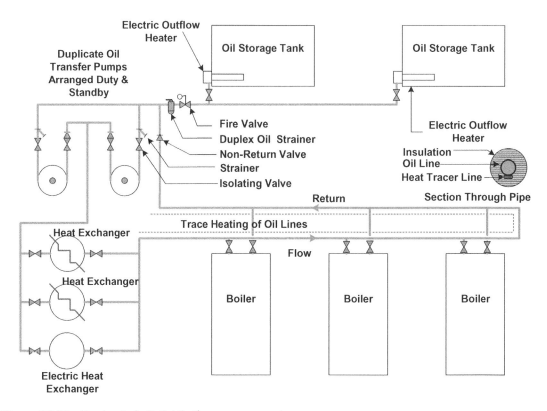

Figure 12.23 Pre-heated oil distribution arrangement

A suitable plot is required to locate the central bulk oil storage tank or tanks; it must be accessible for fuel deliveries, concealed by landscaping and preferably be located on a high point of the area it is supplying so as to achieve a gravity supply and thereby avoiding the need for pumping equipment.

For estate developments, the oil is supplied through an underground network of piping similar to a gas or water supply with individual branch pipes into each dwelling. At the site boundary the oil supply pipe is connected to a meter box containing the oil flow meter, an oil filter, isolating valve, fire shut-off valve and pressure reducing valve to control the oil inlet pressure with the outlet distribution pipe extended on to the boiler. The oil meter may be read directly at the meter, or may be read remotely at a central control office of the fuel supply company. The status of the oil storage tank may also be read remotely by the same telemetry means at the central control point.

The method of supplying oil from a central oil storage tank to low-rise blocks of apartments differs from that of the estate-type layouts, whereby the oil is supplied to each apartment by its own individual oil line from the bulk storage tank. The oil is raised by an electrically operated oil lifter as previously described, with the oil meters located back at the central oil storage tank area; see Figure 12.24.

For situations where the oil lift required is greater than 7.5 m high, then an oil booster employing transfer pumps and day tanks, as shown in Figure 12.15, should be considered.

OIL BURNERS

There are a number of different types of oil burners that have been used for oil-fired heating applications, both domestic and commercial, but they can all be classified as belonging to one of two categories depend-

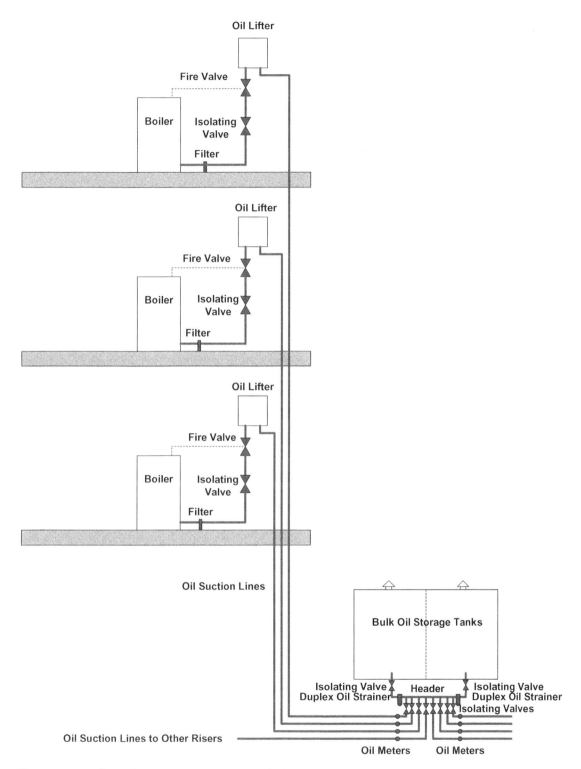

Figure 12.24 Central oil storage arrangement for low-rise apartments

ing upon their method of burning the oil – namely atomising or vaporising, the terms describing the method of achieving combustion.

Atomising oil burners

These type of burners are incorporated almost extensively in larger types of heating appliances, but one type of atomising burner is also used successfully with domestic boilers.

The fundamental principles of operation for atomising burners are that the fuel oil is supplied under controlled pressure which is broken up into a fine spray of oil droplets by a burner nozzle. This spray of oil is then mixed with a measured amount of air to form a combustible mixture, which is then ignited at the point of discharge where it burns as a roaring flame.

Most atomising oil burners are designed to burn Class D gas oil, with larger burners capable of burning Class E oil. Traditionally, domestic oil-fired boilers incorporating an atomising burner have been designed to operate by burning Class D gas oil, but more recently domestic oil-fired boilers employing atomising burners have been arranged to burn Class C2 kerosene oil as well.

Atomising oil burners for commercial heating systems have changed little over the years, but improvements in making them less noisy have made them more popular for domestic heating applications. They are also more responsive to control arrangements, making them an ideal burner for today's heating systems.

Vaporising oil burners

These type of oil burners are exclusively used in domestic heating appliances. They are silent in operation and have been available in many versions with heat outputs up to 50 kW.

The principle of operation for vaporising burners is that the fuel oil is initially heated to form an oil vapour, similar to that of a paraffin blowlamp. The vaporised oil is then allowed to mix with the necessary quantity of air to form a combustible mixture which is then ignited. The heat obtained from the burning flame formed then continues the process of vaporisation, burning as a relatively soft and silent flame.

Vaporising burners burn Class C2 kerosene oil only.

There have been many variations of the vaporising oil burner used for domestic heating systems, but they are limited in their heat output and more difficult to incorporate into a modern control scheme; therefore their use today is not as common as that of the atomising oil burners.

PRESSURE JET OIL BURNER

The pressure jet oil burner, or gun type burner as it is sometimes known, is not only the most commonly used atomising oil burner for both domestic and non-domestic applications alike, but also the most commonly used oil burner for all oil-fired heating applications. It is available with heat outputs up to 2.5 MW, is compact in design and relatively silent in operation, and when used with kerosene oil, can be available in room sealed and conventional flued appliances, which can be either wall- or floor-mounted. Condensing boiler versions are also available.

An example is shown in Figure 12.25.

The pressure jet oil burner comprises an electric motor – which drives a common drive shaft to turn both a low pressure centrifugal fan and oil pump – an atomising nozzle, air directors, ignition electrode and the operating and safety control devices required for automatic operation.

The oil pump is usually of the simple gear type that incorporates a secondary filter on the oil inlet together with a pressure regulating valve for control purposes. The combustion air inlet is controlled by an adjustable air damper on the air inlet to enable the air/oil ratio to be correctly adjusted; see Figure 12.26.

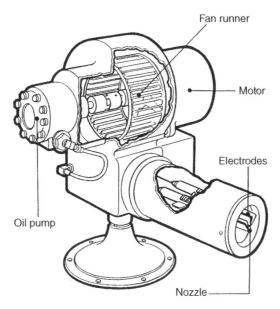

Figure 12.25 Typical pressure jet atomising oil burner

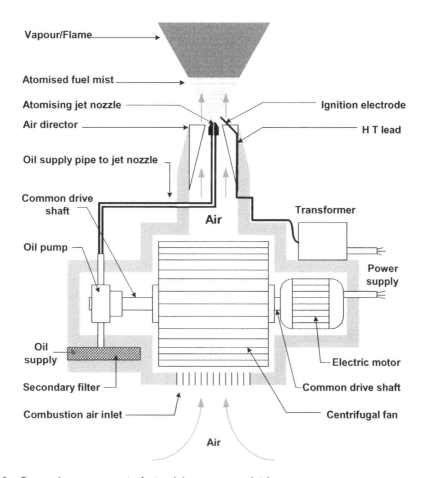

Figure 12.26 General arrangement of atomising pressure jet burner

The fuel oil is delivered under a pressure of 7 bar or more by the gear pump to the atomising jet nozzle, which contains a finely calibrated orifice, or series of orifices – the size and angle of which determine the heat output and spray pattern of the oil burner. This jet nozzle can be interchangeable with jet nozzles of different calibrations to alter the output duty of the oil burner.

The atomising jet nozzle breaks the oil up into a very fine emulsified spray mist, whilst simultaneously oxygen is delivered in the form of air at a controlled rate by the centrifugal fan through a series of swirling vanes of the air director, where it mixes with the oil spray mist. The air director maintains a consistent air pattern, stabilises the flame and creates the required flame shape. Ignition is initially provided by an electric spark from two electrodes located near the nozzle tip, the high-tension supply at about 8.5 kV from the transformer is energised when the boiler is started up. The spark is maintained for a period of time to establish ignition and is cut off by a time delay in the control system. Thereafter the burner operates on an on/off sequence in accordance with the control thermostat's demands for heat.

Pressure jet atomising oil burners burn fuel oil at a fixed rate determined by the combination of the oil pump, forced draught centrifugal fan and atomising jet nozzle, so that the on/off control is dictated by the control thermostat. Sometimes a second high-limit thermostat is incorporated as a safety cut-off for the burner in the event of flame failure or high temperature; this thermostat has to be manually reset after the fault has been rectified.

A solenoid valve, not shown in Figure 12.26, is normally incorporated in the oil supply line and is arranged to hold the oil flow back for a few moments during the start-up procedure, to allow the fan to purge the combustion chamber of any unburnt combustible gases before oil vapour is applied.

An essential component of the pressure jet oil burner is the jet nozzle, which determines the oil vapour spray pattern and angle, together with its function of atomising the oil for combustion to occur. It is believed that one litre of oil being delivered at a pressure of 7 bar will be broken up into approximately 12 billion minute droplets of oil by the action of the atomising jet nozzle. Figure 12.27 illustrates a cross-section through a typical atomising pressure jet nozzle.

Figure 12.27 also shows the variance in the atomised oil spray pattern which is determined by the jet nozzle. This varies between a 90° spray pattern – which achieves a short distance but large area coverage for the burner flame established – to, through a series of discharge angles, a 30° spray pattern, which achieves a longer discharge distance but a smaller area of coverage for the flame.

No attempt should be made to clean the jet nozzle by inserting wire or similar pointed objects into the engineered orifice; this would score the passageway and damage the nozzle beyond repair.

The overall control of pressure jet burners is electrical and is provided by a control box where the ignition, thermostatic and safety control devices are connected. The flame failure safety device is via either

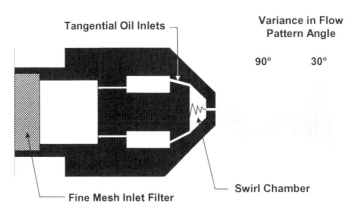

Figure 12.27   Section through pressure jet nozzle

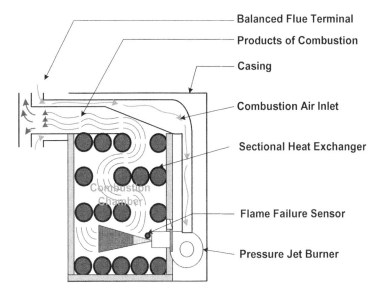

Balanced Flue Terminal

Products of Combustion

Casing

Combustion Air Inlet

Sectional Heat Exchanger

Flame Failure Sensor

Pressure Jet Burner

Combustion Chamber

Figure 12.28 Application of a pressure jet atomising oil burner within a sectional room sealed floor-standing boiler

flue or burner thermostat – or more commonly on larger boilers via a heat sensor or sight type flame sensor employing a photoelectric cell.

Boilers having heat output ratings in excess of 300 kW normally incorporate a modulating control on start-up so that the burner initially operates at a lower reduced heat output rate before the main pressure is applied. This prevents a sudden thermal shock being experienced by the boiler heat exchanger when initial ignition is activated.

Figure 12.28 illustrates the application of a pressure jet atomising oil burner within a room sealed sectional floor-standing system boiler.

ROTARY CUP BURNER

This is another form of atomising oil burner that creates the atomisation by feeding the fuel oil onto a spinning atomising conical shaped cup rotating at high speed. The oil is pumped at low pressure and spreads a film over the rotating surface which is forced by centrifugal action to the periphery of the cup, where it is then flung off the lip as a thin sheet of oil droplets. The atomised oil is then mixed with low pressure air forced by a centrifugal fan over the lip of the rotating cup to form a combustible mixture, which is then ignited.

Rotary cup oil burners are not suitable for domestic applications; they are available for boilers having heat outputs in excess of 150 kW.

Other types of oil burners employing the atomising principles have been developed and used, but the pressure jet burner is by far the most commonly used for all sizes of heating systems and has been adapted to suit all boilers.

In contrast, far more variations of the vaporising oil burners have been developed, but they have only been suitable for domestic heating applications.

The use of vaporising oil burners has declined in recent years in favour of the pressure jet atomising oil burner.

ROTARY VAPORISING OIL BURNERS

The rotary vaporising oil burner – of which there were two forms, commonly referred to under their proprietary names as the 'Wallflame' burner and the Dynaflame burner – was at one time the most popular type of burner for oil-fired domestic heating boilers. It is silent in operation, fully automatic in its control, dependent upon a draught of $5\,N\,m^{-2}$ but limited to heat outputs of up to 40 kW.

Rotary vaporising Wallflame burner

Figure 12.29 illustrates the basic concept of the 'Wallflame' type rotary vaporising oil burner. It consists of a steel base plate supporting a centrally located electric motor, with the armature of the motor wound round a hollow metal shroud, termed the rotor. The bottom of this shroud is immersed into an oil well which supports the journal and thrust bearings of the rotor spindle. Oil is fed into the oil well and its level maintained just above the lower edge of the shroud by a constant level control unit. The shroud is circular in shape with its internal diameter increasing towards the top, from which a pair of diametrically opposed oil distribution tubes are connected and inclined outwards to project through the base plate. Above these distribution tubes and mounted on the rotor spindle, is an air fan. Above the steel base plate there is a circular refractory hearth and within the circumference of this hearth there is a sheet metal band arranged to provide an air gap to insulate and separate the lower part of the boiler from the burner flame. A trough-shaped flame rim is located on the top of the hearth, concentric with the backing band on which is mounted a number of flame grilles. A high tension electrode is fitted into the base of the hearth and projects up to the flame rim grilles; a glow plug rim heater is located in the trough to assist in flame ignition.

   The initial start-up sequence energises the glow plug rim heater and simultaneously drives the motor and rotor clockwise to lift the oil from the well, through the shroud and distribution tubes by centrifugal force, to eject it as a horizontal spray onto the flame rim trough. The oil droplets that fall onto the hot spot of the glow plug immediately vaporise and are ignited by the spark from the high tension electrode against

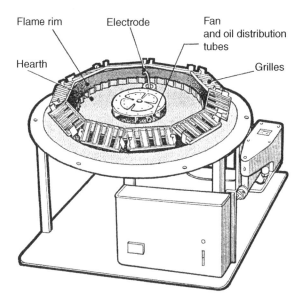

Figure 12.29   General arrangement of 'Wallflame' rotary vaporising oil burners

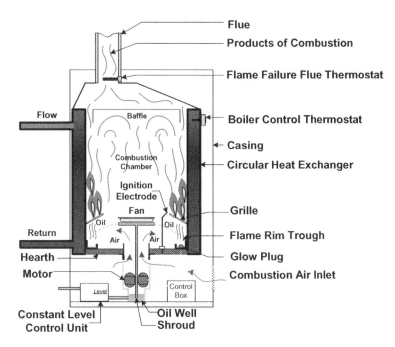

Figure 12.30 Application of rotary vaporising 'Wallflame' burner

the grilles, where the flame is quickly established around the complete circumference of the burner. The grilles on top of the flame rim become red hot and radiate heat back down onto the oil to vaporise it and continue the process. When the burner is operating normally the flame burns through the grilles providing a wall of flame around the circumference, hence the name 'Wallflame' for this type of burner.

An application can be seen in Figure 12.30.

The burner operates on an on/off sequence in accordance with the demands for heat from the control thermostat; when satisfied, the motor is stopped, halting the flow of oil to the burner; when the boiler thermostat calls for heat, the whole process is repeated.

A flame failure device is incorporated to sense if the flame has become established or failed and will shut down the burner, which cannot be restarted until a period of time has elapsed to allow any oil vapour fumes to clear.

The flow of oil to the burner is regulated by a constant level control unit with the combustion air controlled by an air inlet damper on the air inlet grille.

Rotary vaporising oil burners burn Class C2 oil, kerosene.

ROTARY VAPORISING DYNAFLAME BURNER

This is another form of rotary vaporising oil burner. It is a variation of the 'Wallflame' type, but incorporates some unique features.

The Dynaflame burner includes a constant fuel output metering system that ensures that the correct amount of fuel oil is constantly available by way of a vertical spiral rotating tube, which raises the oil up from the well and discharges it by a cone onto the ignition ring around the edge of the burner.

This burner also incorporates a secondary fan, apart from the combustion air supply fan. It recirculates part of the combustion gases and results in a higher efficiency for fuel consumption.

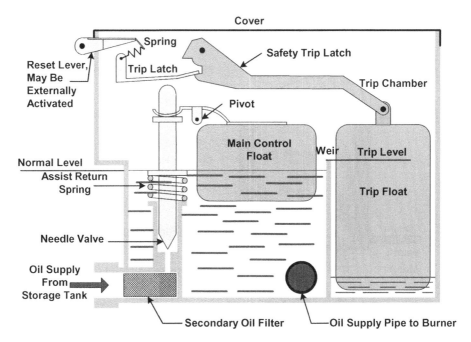

Note - A bellows may be incorporated to operate the constant level float
activated by a non-electric capillary type boiler thermostat

Figure 12.31   Constant level oil control unit

CONSTANT LEVEL CONTROL UNIT

The main fuctions of the constant level control device incorporated on all vaporising oil burners are to maintain a reservoir of oil at a constant level, irrespective of the level of the oil in the main storage tank, and to meter the oil supply to the burner at the correct rate by either manual or thermostatic control. This is depicted in Figure 12.31.

The thermostatic control is by a vapour pressure actuated method, which comprises a heat-sensing phial located in a pocket in the waterways of the boiler heat exchanger and is connected by a capillary tube to a bellows situated on top of the constant level oil control unit. Liquid in the heat-sensing phial reacts to changes in the boiler water temperature and vapour pressure thus produced, and actuates the bellows whose movement is utilised to lower or raise the latch stem. The desired operating temperature is obtained by setting the control adjustment knob at the required temperature.

The constant level oil control unit also functions as a safety device in providing a secondary cut-off, or trip, in case the inlet needle valve becomes stuck and continues to allow oil to flow into the unit when the correct operating level has been exceeded. If not checked, this would eventually lead to the burner becoming flooded with oil.

The detailed design of these units can vary between manufacturers, but in operation they are basically all the same.

Oil under head pressure from the main oil storage tank flows by gravity through a secondary oil mesh strainer and into the inlet needle valve to the level control unit's internal reservoir. The head of oil in the control unit is maintained at a constant level by the action of the main control float, which by the dictates of a lever mechanism, raises and lowers the inlet needle valve as required.

When the needle valve makes contact with its seating, it shuts off the oil supply from the oil storage tank to the burner unit, but in normal operation the inlet needle valve adopts a position such that the inflow of oil is matched to the metered outflow to the burner.

Should the inlet needle valve fail to close off completely due to a faulty valve seat, or foreign matter accumulating between the valve and seating, the oil level will continue to rise until it spills over the weir into the trip chamber. As this space is occupied almost entirely by the trip float, the overflowing oil quickly fills the diminutive space remaining in the chamber and causes the trip float to rapidly rise, which in turn releases the sprung trip mechanism and allows the steel strip to slam down hard on top of the needle valve pin, driving it into the valve seating to dislodge any obstruction and shutting off the oil supply to the burner to prevent the burner from becoming flooded with oil.

The control unit can only be manually reset by pulling the reset lever up, to re-engage the latch mechanism and so drive the trip float back into the trip chamber, displacing the oil within it.

These units may also be coupled to the fire shut-off safety valve arrangement, where a steel trip cable, complete with a soldered fusible link, is attached to the lever to hold it up. This in turn has a weight suspended from it. In the event of a fire occurring, the fusible link solder melts, releasing the tension of the cable, so allowing the weight to pull the lever down, thus activating the trip mechanism and shutting off the oil supply to the fire.

13

Natural Gas Firing

Natural gas, when available, is by far the most popular choice of heating fuel for domestic heating applications and is also competitive for commercial installations. This is probably due to the stability that natural gas has enjoyed since the discovery and online use of North Sea natural gas, as well as the fact that in areas where a gas supply is available, the simplicity of the supply to the consumer does not involve any costly on-site storage and distribution equipment can be very attractive. However, users of large amounts of gas may find that during peak supply periods throughout the day, cheaper tariffs for oil may be available at certain times, compared to gas, in which case dual burners of gas and oil are available to take advantage of these varying fuel costs.

However, natural gas, as a form of energy is also a fossil fuel with limited resources remaining in the world, and its continued popularity will be dependent upon how long these non-renewable, depleting reserves last.

It is not the intention of this volume to provide a textbook on the subject of gas or gases, as there are a number of different gaseous fuels available that may be used as an energy form for heating systems, and this subject would need a volume of equal size on its own to do it justice.

However, as with the case of the previous chapter on the subject of oil, anyone purporting to be a professional in the heating industry and working with gas must have a reasonable knowledge of this energy form so as to obtain the maximum efficiency from it, as well as providing a safe and compliant system.

For this reason, this chapter concentrates on the subject of natural gas, the gaseous fuel that is supplied through a network of distribution piping to its points of use. Therefore, the basic science of natural gas, together with information on gases in general, has been included to give the reader an appreciation of its importance, as well as introducing other gases that may not be so well known.

GAS

The term 'gas' has been used to describe many substances, but the definition of a gas may be defined as being an elastic fluid that is not in either a liquid or solid state, existing at natural ambient temperatures. Air is an example of this, although air is in fact a cocktail of different gases.

Heating Services in Buildings: Design, Installation, Commissioning & Maintenance, First Edition. David E. Watkins.
© 2011 John Wiley & Sons, Ltd. Published 2011 by John Wiley & Sons, Ltd.
As an aid to lecturers and students, full colour versions of the figures in this chapter may be found at www.wiley.com/go/watkins
These figures are © 2011 by John Wiley & Sons, Ltd.

Table 13.1 Typical composition of natural gas

Category	Component	Percentage
Paraffinic	Methane (CH_4) (North Sea value = 92%)	70–98
	Ethane (C_2H_6)	1–10
	Propane (C_3H_8)	Trace–5
	Butane (C_4H_{10})	Trace–2
	Pentane (C_5H_{12})	Trace–1
	Hexane (C_6H_{14})	Trace–0.5
	Benzene (C_6H_6)	Trace
	Other Hydrocarbons	Zero–Trace
Non-hydrocarbons	Nitrogen (N_2)	Trace–15
	Carbon Dioxide (CO_2)	Trace–1
	Helium (He)	Trace–5
	Hydrogen Sulphide (H_2S)	Trace
	Water (H_2O)	Trace–5
	Sulphur Compounds	Trace

NATURAL GAS

The generic term 'natural gas' is applied to gases commonly associated with petroliferous geological formations extracted from underground deposits such as the North Sea and land-based gas fields. In their natural state these gases are combustible, but contain elements of non-flammable constituents such as nitrogen, carbon dioxide and other impurities, some of which may be removed during a refinery process.

Natural gas, like air is made up of a number of different gases (see Table 13.1) but its main energy releasing constituent is predominantly methane, with small quantities of other paraffinic gases and traces of non-hydrocarbon products. The exact composition of natural gas will vary depending on its source and location of extraction, and from one gas field to another, and only typical values are given in Table 13.1.

Natural gas is an ideal fuel for heating applications due to its cleanliness, high calorific value, high flame temperature and ease of distribution, and because it occurs naturally with other hydrocarbon products and decaying organic material.

The gas occurs in the porous rock of the Earth's crust, having been formed millions of years ago either alone or with other accumulations of hydrocarbon products. In the latter case, the gas forms on top of the liquid petroleum to form a gas cap trapped below the impervious rock, forming the cap of the petroleum reservoir. If the underground pressure is sufficiently high, the natural gas will be dissolved in the petroleum liquid and released upon penetration of the reservoir as a result of oil drilling operations.

The natural gas is stored in natural underground deposits of greensand by freezing the ground to keep the potential gas as a liquid, or in constructed storage reservoirs and tanks, prior to distribution in the national piping network.

Table 13.2 gives the operating properties of natural gas, but as gas can vary from source to source, so can the operating properties, therefore the values given are averages.

ODOUR

Natural gas, unlike most other gases, does not contain a characteristic and distinctive odour, and as it is colourless, an aroma has to be added to the gas on the grounds of safety, in order to make it detectable in the event of a leakage.

Following the conversion from manufactured gas to natural gas in the 1960s and early 1970s, the chemical tetrahydrothiophene was added to the gas at an approximate concentration of 1 kg to 60 000 cubic metres

Table 13.2   Operating properties of natural gas

Characteristic	Average Values
Calorific Value – Gross Value	38.6 MJ/m³
Calorific Value – Net Value)	34.9 MJ/m³
Relative Density (Specific Gravity)	0.56
Air Required for Combustion	9.73 m³/m³ Gas
Wobbe Number	51.47
Burning Velocity	0.34 m/s
Maximum Flame Temperature	1930°C
Sulphur Compounds	0–20 mg/m³
Supply Distribution Pressure	17.5–27.5 mbar (1750–2750 Pa)
	Legal requirement, pressure not to fall below 1250 Pa (12.5 mbar)

Relevant gas volume data is based on conditions of 15°C at a pressure of 101.3 kPa (atmospheric pressure)

of gas, as this resulted in a pungent odour that closely replicated that of the manufactured gas that it replaced. This odorant has since been replaced by diethyl sulphide, ethyl and butyl mercaptan.

TOXICITY

Natural gas in itself is not toxic – unlike the manufactured gas that it replaced, which was used by some unfortunate and desperate people to commit suicide. However, it does have an asphyxiating effect if the concentration is high enough, which is another reason for the inclusion of a scent.

Although natural gas is non-poisonous, the products resulting from combustion can be toxic if not completely burnt, due to the production of carbon monoxide (CO), which is discussed further in Chapter 16, Combustion, Flues and Chimneys.

CALORIFIC VALUE

As established in the previous chapter on oil, the calorific value of a fuel is the amount of heat released on complete combustion when a measured volume of that fuel is burnt.

There are two values normally given – gross or higher, and net or lower – both normally expressed in $MJ\,m^{-3}$ for gas. The gross value includes the proportion of latent heat released by the vaporisation of the water forming part of the products of combustion passed up the flue. The net value is the gross value minus the latent heat of vaporisation and is the actual amount of heat energy transferred into the heating water.

In the UK the European standard of using the net value of expressing the calorific value has now been adopted, although the gas supply companies still base their charges on the gross calorific value, which is stated on their gas bills.

RELATIVE DENSITY

The relative density or specific gravity of a gas represents the weight of a given volume of gas compared to an equal volume of air.

Whereas the specific gravity of a liquid or solid is compared to the weight of water, the relative density of a gas is compared to the weight of air at the same temperature. The relative density of air is taken as 1; from

Table 13.2, it can be seen that the relative density of natural gas is stated as 0.56, which means that natural gas is almost half the weight of air. If natural gas escapes, it will rise in the air and readily mix with it.

Gases having a relative density greater than 1 will be heavier than air and will lie close to the ground. Gases having a relative density of less than 1 will rise as in the case of natural gas.

COMBUSTION AIR

For combustion of any substance to take place, oxygen is required. Natural gas, like all fuels, requires an optimum amount of air to provide the oxygen required.

For complete and efficient combustion of natural gas to take place, every one cubic metre of natural gas requires 9.73 cubic metres of air to provide the two cubic metres of oxygen that it needs.

This figure of 9.73 m^3 of air per cubic metre of gas is needed for complete combustion to occur; natural gas can however be ignited at other concentrations and can form an explosive mixture when it is between the flammability range of 5–15% concentration in air. Efficient combustion occurs at a fraction below 10% concentration. A concentration higher or lower than range will result in incomplete combustion or flame loss.

WOBBE NUMBER

The function of the Wobbe number, or Wobbe index, is to ensure that the correct burner assembly is provided for the particular gas being supplied by taking into account the thermal input, gas supply pressure and burner orifice.

The Wobbe number can be determined by the following equation:

$$\text{Wobbe Number} = \frac{\text{Calorific Value (Gross)}}{\sqrt{\text{Relative Density}}}$$

For example

$$\frac{CV = 38.6}{RD = \sqrt{0.56}} \text{ Wobbe No} = 51.47$$

To ensure that the burner assembly operates correctly, the quality of the gas must be maintained within close limits. This is achieved by keeping the gas quality range indicated by the Wobbe number within the family ranges agreed by the International Gas Union.

The family of gases are classified in three ranges, namely:

Family 1	Manufactured Gas	Wobbe No range: 22.5–30
Family 2	Natural Gas	Wobbe No: L=39.1–45; H=45.5–55
Family 3	LP Gas	Wobbe No range: 73.5–87.5

BURNING VELOCITY

Natural gas is said to have a slow burning flame – this was the main criticism that consumers had of natural gas following the conversion from manufactured gas, which was three times as fast.

The velocity of the gas issuing from the burner jets is designed to keep the flame at the tip of the burner: if the speed is too fast the flame could be blown off; if it is too slow the flame could burn its way back inside the burner jets or tube.

GAS MODULUS

The gas modulus is expressed as a number which relates to the heat output from a burner assembly coupled with the gas pressure required to provide sufficient aeration for complete combustion to occur.

The importance of this numerical expression is its application to burners during a change of gas supply or gas type. It is found by dividing the square root of the gas pressure by the Wobbe number:

$$\text{Gas Modulus} = \frac{\sqrt{\text{Gas Pressure}}}{\text{Wobbe Number}}$$

Example: Burner supplied by a manufactured gas supply having a Wobbe number of 27.5 and a supply pressure of 6.25 mbar:

$$\frac{\sqrt{6.25}}{27.5} = \text{Gas Modulus of } 0.09$$

The manufactured gas to be changed to natural gas having a Wobbe number of 51.5:

$$0.09 \times 51.5 = 4.635^2 = \text{Gas pressure of 21.5 mbar.}$$

Therefore, to maintain the same operating conditions the gas supply pressure needs to be increased to 21.5 mbar.

SUPPLY PRESSURE

During transportation of gas from the gas terminals through a national network of pipelines to local storage and distribution stations, pressure can vary between 10 bar, and 30 bar, depending upon length and the geography of the route.

The gas supply pressure through local supply network piping to the consumer's point of connection can also vary between 75 mbar and 2 bar pressure, with the incoming pressure into each domestic dwelling being between 17.5 mbar and 27.5 mbar.

OTHER GASES

The main theme of this chapter is natural gas but there are a number of other flammable gases that can also be used as a fuel for providing heating.

It is not the intention to give full details of each of these gases here other than to include a brief description of them to make the reader aware of their existence.

These gases comprise the following:

Liquefied petroleum gas (LPG)

This gas is explained in detail in the Chapter 14.

Manufactured gas

There are a number of manufactured gases, the most common one being coal gas, or town gas as it was generally known. Town gas is the gas that was in common use to provide street and home lighting, heating

Table 13.3 Typical composition of manufactured gas (town gas)

Component	Percentage
Hydrogen	48
Methane	34
Carbon Dioxide	13
Carbon Monoxide	5

Table 13.4 Operating properties of manufactured coal gas (town gas). Owing to the presence of carbon monoxide in town gas, it is classified as toxic

Characteristic	Average Values
Calorific Value – Gross Value	18.6 MJ/m³
Calorific Value – Net Value	13.6 MJ/m³
Relative Density (Specific Gravity)	0.4–0.6
Air Required for Combustion	4.46 m³/m³ Gas
Wobbe Number	29
Burning Velocity	1.0 m/s
Maximum Flame Temperature	1960°C
Sulphur Compounds	120–390 mg/m³
Supply Distribution Pressure	7.5–20 mbar

Table 13.5 Typical composition of substitute natural gas (SNG)

Component	Percentage
Methane	98.5
Hydrogen	0.9
Carbon Monoxide	0.1
Other Constituents	0.5

and cooking, until it was replaced by the higher calorific value natural gas in the late 1960s. Up until that point almost every town and city had its own gas making plant, hence the term 'town gas'.

Town gas was obtained by the carbonisation of coal at temperatures of between 950°C and 1350°C. The solid carbonaceous residue that remained was sold off as coke for burning as smokeless fuel.

The composition and operating properties of manufactured coal gas are given in Tables 13.3 and 13.4.

Substitute natural gas (SNG)

Sometimes referred to by the contradictory name of synthetic natural gas, substitute natural gas is a manufactured gas made to replicate natural gas to supplement dwindling stocks of natural gas. For this reason it has to be high in methane so that it can be mixed with natural gas entering into the distribution network.

SNG is made using a variety of feed stocks, including liquefied petroleum gas and naphthas, by a process known as the double methanation process. Tables 13.5 and 13.6 illustrate the characteristics and operating properties of substitute natural gas.

Table 13.6   Operating properties of substitute natural gas (SNG)

Characteristic	Average Values
Calorific Value – Gross Value	38.0 MJ/m³
Relative Density (Specific Gravity)	0.55
Wobbe Number	51

Other gases exist which are used in industry rather than as commercial gas available for heating buildings. These include:

- Producer gas: made from the combustion of coal or coke under controlled oxygen supply and used in the steel making industry.
- Coke oven gas: also used in the steel making industry, and made by destructive distillation of coal under conditions that exclude air.
- Blast furnace gas: a form of low-grade producer gas.
- Water gas: produced by the thermal decomposition of coal or coke by the application of steam.
- Blue water gas: obtained by passing steam over red-hot coke. It burns with a characteristic blue flame consistent with the combustion of carbon monoxide, CO.

Other types of gases not mentioned in this chapter are biomass gases, details of which are included in Chapter 15, Alternative Fuels and Energy.

GAS SUPPLY AND DISTRIBUTION

As established earlier in this chapter, natural gas is supplied through a national network of distribution pipelines to most cities and towns, where it is tapped-off to enter each building requiring a gas supply.

Figure 13.1 illustrates a typical gas supply system from the gas main into the building, serving a gas cooker and boiler. Unlike the water supply whereby the water supplier's ownership ends at the site boundary, the incoming gas main supplier's ownership is retained up to and including the main gas supply meter.

The incoming gas supply is a simple system which just brings the low pressure gas supply into the building to the meter – at a pressure ranging between 2 bar to 75 mbar for domestic use, or up to 5 bar for industrial use. From the meter and forming part of the internal piping installation, the low pressure gas

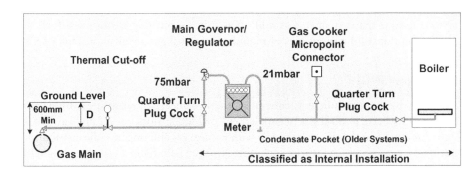

Figure 13.1   Internal gas piping distribution system

distribution piping of a suitable size is extended from the meter to each point of usage at a pressure of 21 mbar. Although this distribution system is simple, it is imperative that it is designed and installed in a safe and professional manner, conforming to all current regulations appertaining to gas.

The piping material has in the past been plain black low carbon mild steel tube – medium weight below ground, wrapped in overlapping petroleum impregnated tape to protect it from external corrosion; and light weight tube above ground, both joined together by BS 21 screwed joints. More recently, flexible polyethylene tube, manufactured in an easily identifiable yellow colour, has been employed for underground tubing. This is joined together by electro-fusion joints resulting in a flexible non-corrosive material ideal for underground applications. The internal piping is normally metallic piping materials, such as copper tube with a tensile strength of R250, or plain black low carbon mild steel tube of medium weight grade with BS 21 screwed joints. Metallic piping materials provide a stronger material and are more resistant to accidental damage or vandalism.

Table 13.7 shows the minimum depth for incoming gas supplies.

GAS METERS

The purpose of the meter is to accurately measure the flow of gas that passes through it, and in most cases to record the quantity of gas to allow charges to be made by the gas supply company.

Gas meters are either of the positive displacement type or the rate of flow type, although there are many variations of meters.

The most common gas meter in use is the positive displacement diaphragm meter, which is available in a number of sizes, depending upon the quantity or volume of gas that it is capable of passing, as listed in the table shown in Table 13.8.

Figure 13.2 illustrates the general operating principles of the diaphragm positive displacement meter, which consists of four chambers – designated 1 to 4 – that fill and empty in sequence as shown. The meter measures the exact quantity of gas passing through it by counting the number of displacements taken; sliding valves control the flow of gas to and from the compartments.

Accompanying the meter and forming part of the assembly, is a gas pressure regulator, which is installed to regulate the downstream gas pressure regardless of any fluctuation in the incoming gas supply.

Table 13.7   Dimension 'D', minimum depth of incoming gas supply main

Pipe Size	Under Roadways	Below Unpaved Ground	Under Footpaths
<50 mm	375 mm Metal or 450 PE	375 mm Metal or 450 PE	375 mm
>50 mm	750 mm	750 mm	600 mm

Table 13.8   'U' series positive displacement diaphragm-type gas meters

Designation	Rating m³/hr	Connection mm
U4	4	25
U6	**6**	**25**
U10	10	32
U16	16	32
U25	25	50
U40	40	50
U65	65	65
U100	100	80
U160	160	100

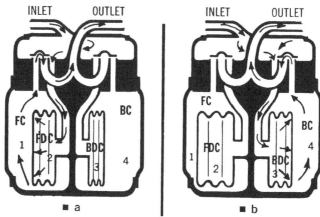

a) Chamber 1 is emptying, chamber 2 is filling, chamber 3 is empty and chamber 4 has just filled.
b) Chamber 1 is now empty, chamber 2 is full, chamber 3 is filling and chamber 4 is emptying.

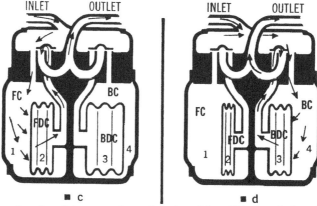

c) Chamber 1 is filling, chamber 2 is emptying, chamber 3 has filled and chamber 4 has emptied.
d) Chamber 1 is now full, chamber 2 is empty, chamber 3 is emptying and chamber 4 is filling.

Figure 13.2   Operating principles of positive displacement diaphragm gas meter

Table 13.9   Diversity factors for gas appliances

Appliance Type	Diversity Factor
Boilers, Warm Air Heaters, Wall Heaters & Circulators	1.0
Combination Boilers & Instantaneous Multi-Point Water Heaters	0.8
Instantaneous Single Point Water Heaters, Gas Fires, Hotplates & Ovens	0.6

It is important that the meter is capable of passing a sufficient volume of gas to satisfy the demand required to operate all gas appliances installed on the supply system. Most domestic gas supplies require a gas flow rate of up to $6\,m^3$/hour, which would be satisfied by the installation of a 'U6' gas meter.

The gas meter size may be ascertained by using the following simple method, using data from Tables 13.9 and 13.10.

The volumetric gas consumption is calculated by multiplying the total heat input rating in kW by 3.6 to convert the kW loading into MJ, then dividing this result by the calorific value to equal the gas volumetric

Table 13.10   Example of volumetric gas consumption

Appliance	Maximum Heat Input in kW	Diversity Factor	Calculated kW Load
Boiler	22	1	22
Cooker Point	20	0.6	12
		Total	34

Total kW	Convert to MJ	Gross Calorific Value	m³/hr
34	3.6	38.6	3.17

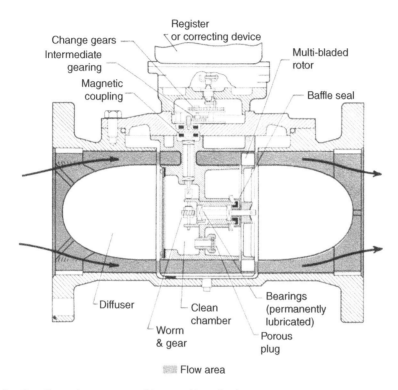

Figure 13.3   Section through a rotary turbine positive displacement meter

consumption rate required to be passed by the meter. Alternatively, multiply kW by $0.093 = m^3/hr$ ($34 \times 0.093 = 3.17$).

Another form of positive displacement gas meter is the rotary displacement meter, which works on the principle of two rotating impellers similar to that of the positive displacement rotary metering pump, explained in the chapter on oil firing.

The positive displacement types of meters are very accurate, particularly at low flows, making them ideal for domestic gas supplies. Other methods of gas metering function on the rate of flow principle – such as the ultrasonic measuring type, which relies on sound being sensed as the flow of gas passes through; and the rotary turbine meter, which is very accurate for large flow rates but not so accurate at low flows. The rotary turbine gas meter is depicted in Figure 13.3. It requires a flow straightening device together with a specified length of straight pipe upstream of the meter to operate correctly. Without this installation constraint, this form of flow meter would not record an accurate gas flow. This form of meter is ideally suited for installations requiring large rates of gas flow that can be relied upon to pass through the meter.

METER COMPARTMENT

It has been common practice in the past to locate the gas meter for domestic installations in the cupboard below the stairs. This practice is now discouraged for reasons of safety, such as lack of suitable ventilation, close proximity to other services such as electricity, and the need to gain access to the property in order to read the meter.

New-build developments require the main primary gas meter to be installed in locations that can be accessed without having to enter the building and conform to current safety regulations.

Acceptable external locations include the following:

- Surface mounted in lockable meter box on external wall
- As above, but built into the wall to form a flush mounted finish in the wall, also contained in a lockable meter box
- Semi-concealed in a ground located lockable meter box

Regardless of which type of compartment is used, the meter box that contains the meter, together with the main isolating valve and pressure regulating governor, should be of an approved type, which can be accessed by the meter reader and for emergency reasons.

Figure 13.4 illustrates the arrangement for a typical meter compartment installation, including the piping entry and exit requirements.

All pipe penetrations through walls should be made via pipe sleeves with the ends sealed with flame-resistant flexible mastic; sleeves passing through cavity walls should be continuous through the cavity gap. Any gas pipe installed within ducts or service cavities should have the duct fully ventilated at both the top and foot of the service duct.

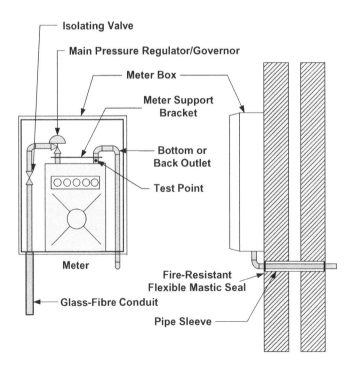

Figure 13.4   Typical surface-mounted meter box compartment

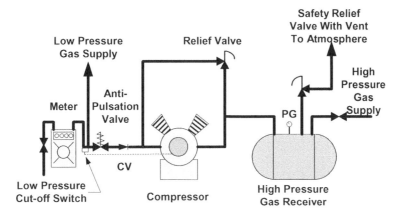

Figure 13.5   High-pressure gas boosting arrangement above 75 mbar

GAS BOOSTING

As natural gas is lighter than air, it will rise naturally and does not require any lifting equipment to raise it to a higher level. However, gas boosters are required when the gas supply pressure is inadequate for the appliance that it serves. Gas boosters are centrifugal fan machines used to increase the gas pressure to approximately 75 mbar. For pressures higher than 75 mbar, positive displacement type compressors are employed, as depicted in Figure 13.5.

Gas boosting is required:

- To provide sufficient pressure at the gas burner nozzle and so ensure the correct flame shape.
- To be at an equal comparable pressure with the pressurised combustion air supply at the combustion mixing point.
- To enable smaller diameter gas supply pipes, valves and controls etc to be used.

The gas booster system shown in Figure 13.5 consists of a low pressure gas supply with the low pressure gas service taken off before entering the gas booster. To prevent the compressor from lowering the main supply pressure below the required minimum, a low pressure cut-off switch is incorporated in the system to stop the compressor; this functions in conjunction with an anti-pulsation valve which prevents any initial start-up surges from falsely activating the low pressure cut-off switch.

The high pressure outlet is normally controlled by a pressure switch integrated within the gas receiver, which either starts or stops the compressor according to demand.

The outlet pipe from the safety relief valve fitted to the high pressure gas receiver must terminate to atmosphere in a safe manner and not in a position close to any source of ignition.

GAS SUPPLY PIPEWORK

As stated earlier, the piping material has in the past been plain black low carbon mild steel tube – medium weight below ground, wrapped in overlapping petroleum impregnated tape to protect it from external corrosion; and light weight tube above ground, and both joined together by BS 21 screwed joints. The screwed joints are made using an approved jointing compound suitable for dry gas; town gas was classified as a wet gas and employed a different formulated jointing compound. Alternatively, a gas-formulated PTFE (polytetrafluoroethylene) tape – which is a heavier duty PTFE tape – can be used on screwed threads up to 50 mm diameter. More recently, flexible polyethylene tube manufactured in an identifiable yellow colour has been employed for underground tubing, joined together by electro-fusion joints. This flexible non-corrosive material is ideal for underground applications. The internal piping is normally metallic piping materials such as copper tube with a tensile strength of R250, or plain black low carbon mild steel tube of medium weight grade with BS 21 screwed joints. Metallic piping materials provide a stronger material and are more resistant to accidental damage or vandalism. Copper and polyethylene tubing should be protected from accidental penetration by nails etc when installed below floorboards or in shallow service ducts.

DOMESTIC PIPE SIZING

Traditionally, the carcass pipework for the domestic gas supply was installed using 20 mm diameter low carbon mild steel tube from the main supply meter up to each appliance, reducing just before the appliance to match the connection line size.

Since the mid-1980s, a simplified pipe sizing procedure, based on limited pressure drop, has been developed for use by gas installers to enable suitable pipe sizes to be obtained for domestic applications.

The simplified pipe sizing procedure is based upon restricting the pressure drop from the main gas supply meter to the appliance pressure test point by an amount not exceeding the maximum of 1 mbar. This is achieved by selecting pipe sizes that restrict the difference between the no-flow static pressure and the dynamic flow pressure to 1 mbar or less.

The static pressure is the pressure that exists in the distribution carcass pipework along its entire length when no flow occurs. When any appliance is operating, a flow of gas will commence and encounter a frictional resistance along the length of the distribution piping resulting in a lower pressure being delivered at the appliance: this is the dynamic pressure and should not be less than 1 mbar lower than the static pressure. The static pressure can be ascertained by connecting a 'U' gauge manometer to the pressure test point nipple on the main gas meter outlet pipe connection.

The pipe layout illustrated in Figure 13.6 has been included to demonstrate the simplified gas pipe sizing procedure and shows pipe lengths and kW loadings of the appliances.

Tables 13.11 and 13.12 tabulate the gas flow rates through copper tube, stainless steel tube and low carbon mild steel tube for domestic size tubes, expressing the flow rates in cubic metres per hour against the maximum permissible length for these tubes.

The gas flow rates given in Tables 13.11 and 13.12 are expressed in straight lengths of tube for each size, but the pipe fittings are totalled on each pipe section and included as an equivalent length of tube which is then added to the measured length of tube to give the total length of tube.

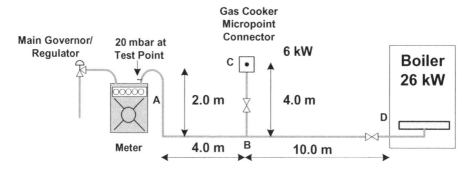

Figure 13.6   Simplified pipe sizing layout

Table 13.11   Gas flow rates through copper/stainless steel tube

Flow of Gas in Copper & Stainless Steel Tube

Tube Size mm	Length of Pipe (m)							
	3	6	9	12	15	20	25	30
	Gas Discharge m³/hr							
10	0.85	0.57	0.5	0.37	0.3	0.22	0.18	0.15
12	1.5	1	0.85	0.82	0.69	0.52	0.41	0.34
15	2.9	1.9	1.5	1.3	1.1	0.95	0.92	0.88
22	8.7	5.8	4.6	3.9	3.4	2.9	2.5	2.3
28	18	12	9.4	8	7	5.9	5.2	4.7

Table 13.12   Gas flow rates through low carbon mild steel tube

Flow of Gas in Low Carbon Mild Steel Tube

Tube Size mm	Length of Pipe (m)							
	3	6	9	12	15	20	25	30
	Gas Discharge m³/hr							
8	0.08	0.53	0.5	0.37	0.3	0.22	0.18	0.15
10	2.1	1.4	1.1	0.93	0.81	0.7	0.69	0.57
15	4.3	2.9	2.3	2	1.7	1.5	1.4	1.3
20	9.7	6.5	5.3	4.5	3.9	3.3	2.9	2.6
25	18	12	10	8.5	7.5	6.3	5.6	5

Table 13.13   Tabulation of simplified gas pipe sizing

Material	*Copper Tube*								
1	2	3	4	5	6	7	8	9	10
Pipe Section	Heat Load (kW)	Gas Flow Rate (kWx0.093) (m³/hr)	Measured Pipe Length (m)	Number of Fittings	Equivalent Length (m)	Total Length (4 + 6) (m)	Pipe Diameter (mm)	Maximum Permitted Length (m)	Pressure Drop (7 ÷ 9) (mbar)
A–B	32	2.98	6	4	2	8	22	15	0.53
A–B	32	2.98	6	4	2	8	28	30	0.26
B–C	6	0.56	4	6	3	7	15	30	0.23
B–D	26	2.42	10	6	3	13	22	25	0.52
A–B + B–C	Pressure drop from meter to cooker = 0.26 + 0.23								0.49
A–yB + B–D	Pressure drop from meter to boiler = 0.26 + 0.52								0.78

Add 0.5 metres of tube for each pipe fitting to express the frictional resistance encountered as an equivalent length of tube.

Table 13.13 tabulates the results of the pipe sizing exercise by calculating each pipe section separately. Section A–B has been calculated twice using both 22 mm and 28 mm copper tube, with the 22 mm size tube being discarded as the pressure drop is considered too high when it is added to section B–D.

The results obtained using this simplified pipe sizing procedure ensure that the maximum pressure drop of 1 mbar between the main supply gas meter and the inlet of each gas appliance is not exceeded for each pipe section.

For all other installations, an engineered pipe sizing calculation must be undertaken.

Table 13.14   Equivalent length of steel tube

Pipe Diameter mm	Fitting Type – Equivalent Length (m)			
	Elbow	**Tee**	**90° Bend**	**Valve**
Up to 25	0.6	0.6	0.3	0.23
35–40	0.9	1	0.45	0.3
50	1.5	1.6	0.6	0.5
80	2.4	2.6	1	0.75
100	4.3	4.5	2	1.7
150	7	7.4	2.6	2

FLOW OF GAS IN PIPES

As for all fluids previously discussed, economic pipe sizes have to be arrived at to convey the full demand of gas quantity required.

Firstly, the quantity of gas required to flow in litres per second has to be established using the following formula:

Gross calorific value of natural gas = 38.6 MJ m^{-3}, which equates to 38.6 kJ/litre.

Therefore, 1 litre per second will equal 38.6 kW.

Example – A boiler having a required input rating of 20 kW will require a gas flow rate of 0.52 litres per second.

$$(20 \text{ kW} \div 38.6 = 0.52 \text{ l/s})$$

Next, the pressure drop has to be determined, after the total pipe run and equivalent length of pipe for fittings have been found.

Table 13.14 may be used to assess the additional length of pipe required to compensate for increased resistance resulting from the pipe fittings.

Having established the total equivalent length of pipe, the next step is to ascertain the available pressure drop. This may be achieved by recording the gas pressure at the meter outlet and noting the pressure required at the boiler; the difference between the two is the available pressure drop.

Therefore, using the previous example of a required flow rate of 0.52 l/s:

Pressure at meter, assume	2500 Pa (25 mbar)
Pressure required at boiler	1250 Pa (12.5 mbar)
Pressure drop available	1250 Pa (12.5 mbar)

Therefore, with a total equivalent pipe length of 50 m

$$\text{Pressure drop per metre run of pipe} = \frac{1250}{50} = 25 \text{ Pa m}^{-1} \text{ run of pipe (0.25 mbar)}$$

From Table 13.15, a pipe having a diameter of 15 mm will be suitable, having a pressure drop of 6.5 Pa m^{-1} = 50 × 6.5 = 325 Pa (3.25 mbar).

GAS BURNERS

There are many types of gas burners, but they are all designed to operate in one of two ways and can therefore be classified as being either:

- Natural draught: subdivided into pre-aerated and post-aerated
- Forced draught

Table 13.15   Flow of natural gas in low carbon mild steel tube

Pressure Drop Pa/m	Flow Rate in Litres/Second for Pipe Sizes mm									
	15	20	25	35	40	50	65	80	100	150
0.5	0.08	0.25	0.49	1.1	1.64	3.16	6.4	10	20.7	60
1	0.15	0.39	0.75	1.62	2.47	4.72	9.6	15	30.6	88.3
2	0.26	0.59	1.12	2.43	3.7	7	14.3	22	45.2	130
3	0.33	0.75	1.43	3	4.66	8.9	18.5	28	57	165
4	0.39	0.89	1.7	3.6	5.5	10.8	22	33	67	198
5	0.45	1	1.92	4.12	6.24	12.4	24.4	37.6	76	218
6	0.5	1.13	2.14	4.57	6.92	13.8	27	41.2	83.6	244
7	0.54	1.23	2.34	5	7.55	14.7	29.3	45	91	266
10	0.67	1.51	2.86	6.1	9.22	18	36	54.8	112	322
12.5	0.76	1.72	3.25	6.9	10.2	20	40.7	62	126	361
15	0.85	1.9	3.6	7.66	11.6	22	44.5	68	139	394
20	1	2.25	4.24	9	13.6	26	52	80	163	460
25	1.14	2.55	4.8	10.2	15.4	29.7	59	90	184	520

There are numerous variations of these types of gas burners but they may all be classified as belonging to one of these categories.

Traditionally, burners for domestic gas boilers have been of the pre-aerated natural draught type, which may or may not be fan assisted, and forced draught burners are used on boilers having larger heat outputs. More recently, developments in some high efficiency boilers have employed a form of forced draught burner that enables the combustion process to be better controlled.

NATURAL DRAUGHT BURNERS

The natural draught or atmospheric gas burner operates at normal gas supply pressure supplied through the gas mains and is quiet in its operation. Its heat output is limited to about 150 kW and is ideal for domestic applications. It requires a flue that is capable of creating a sufficient draught to supply the necessary air for complete combustion and enable the burner to function efficiently. This requirement restricts the physical size of the boiler to allow for the natural draught to occur.

Figure 13.7 indicates the working principles and components of a standard natural draught atmospheric burner depicted in its basic form. It functions by the gas passing through a series of control components on its way to the burner.

The type of burner illustrated is known as an aerated burner, where the action of the gas flowing into the burner induces a calculated amount of primary air into the burner that mixes with the gas before ignition. This mixture of gas and air issues through the jets of the burner, where it is ignited in the presence of the secondary air to complete the combustible mixture, and burns with a medium velocity flame.

The proportion of primary air and secondary air required to provide the necessary quantity of oxygen for combustion is approximately 50% each.

The practice of domestic gas burners employing individual control components, as depicted in Figure 13.7, has long ceased to be the custom, as they have been replaced by a single multi-functional valve, but they have been discussed here to give the reader an understanding of how they work and help illustrate the workings of the single multi-functional valve.

These single-function control devices are still retained for controlling the gas flow to burners fitted to boilers with larger heat outputs.

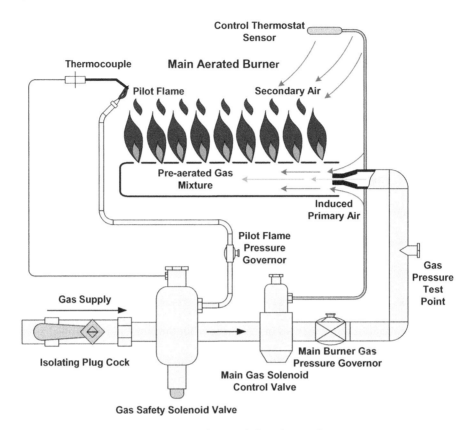

Figure 13.7   Basic component arrangement of natural draught gas burner

Figure 13.7 illustrates how the pre-aeration of the burner is achieved. The gas pressure is increased by the venturi shape of the nozzle entering the burner which induces the primary air into the burner to form the pre-aerated gas mixture. The combustible mixture is completed by the addition of the secondary air at the point of ignition. In this example the ignition is provided by the inclusion of a permanent pilot flame controlled by a thermocouple connected to the gas safety solenoid valve. The main burner is controlled by a boiler-mounted thermostat which operates the main gas solenoid control valve, which has the important task of setting the correct gas supply pressure. This is achieved by connecting a 'U' gauge manometer to the test point and adjusting the burner pressure at the main burner gas pressure governor to suit the heat output required.

In this example, each of the components operates independently from the others, but they combine to provide an effective overall control.

GAS SAFETY SOLENOID SHUT-OFF VALVE

The gas safety solenoid valve is a thermoelectric device operated by a low output thermocouple as shown in Figure 13.8. Its function is to shut off the gas supply to both the main burner and pilot jet in the event of the pilot flame failing, and thus preventing gas from flowing unchecked into the boiler combustion chamber.

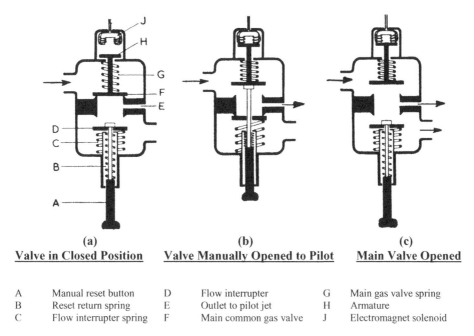

(a)	(b)	(c)
Valve in Closed Position	**Valve Manually Opened to Pilot**	**Main Valve Opened**

A	Manual reset button	D	Flow interrupter	G	Main gas valve spring
B	Reset return spring	E	Outlet to pilot jet	H	Armature
C	Flow interrupter spring	F	Main common gas valve	J	Electromagnet solenoid

Figure 13.8   Thermoelectric gas safety solenoid shut-off valve

Position (a) shown in Figure 13.8 shows the safety valve in the unactivated closed status, preventing the flow of gas past this point. To enable the pilot flame to be established the valve has to be manually opened by pressing in the reset button, which has the effect of raising the main common gas valve to open the passage for gas to flow through to the pilot jet after it has also raised the flow interrupter valve to close off the main flow of gas through the valve, as shown in (b). The reset button has to be held in that position for approximately 60 seconds until the pilot flame has been ignited to enable the thermocouple to sense that the pilot has become established by generating a small electric current sufficient to energise the electromagnet solenoid and keep the interrupter valve open automatically, as depicted in (c).

If the flame fails for any reason, the thermocouple will cool down in about 30 seconds and cease to generate the small electric current. The electromagnet solenoid valve will in turn demagnetise, causing the spring to close the main common gas valve and return the safety valve back to position (a).

The procedure for re-igniting the pilot flame and main burner assembly will have to commence again after the fault has been investigated and rectified.

THERMOCOUPLE ASSEMBLY

The thermoelectric effect was first demonstrated by Thomas Seebeck, a German/Estonian scientist, in 1821, and examined further by Lord Kelvin. Seebeck discovered that heat can produce electrical energy when applied to a metal that is joined together with a different metal. When the junction between the two metals is heated, an electromotive force is produced in the circuit.

The voltage produced will depend upon the temperature of the hot and cold junction between the two dissimilar metals, as well as the dissimilar metals selected for this use. The two metals commonly used for thermocouples employed in domestic boiler applications are copper and constantan, a copper–nickel alloy of approximately 55% copper and 45% nickel.

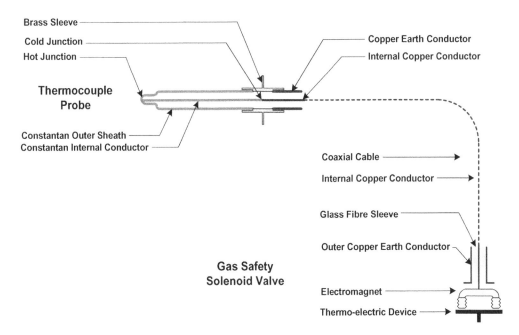

Figure 13.9   Thermocouple arrangement

Other combinations of metals used for higher temperature applications include: copper and iron, chromel (Cr and Ni) and alumel (Al and Ni), platinum and rhodium, and tungsten and rhenium.

The voltage produced by the thermocouple fitted to a domestic boiler ranges between 10 and 30 mV, which is sufficient to hold the valve open, but not sufficient to lift the valve off its seating.

The thermoelectric device is remote from the thermocouple and is connected via a coaxial lead cable, which consists of an internal copper conductor carrying the milli-volt electrical current enclosed in an outer copper sheath that serves as an earth conductor. The two copper conductors are separated by a continuous fibreglass sleeving as shown in Figure 13.9.

BOILER THERMOSTAT AND MAIN GAS SOLENOID CONTROL VALVE

The boiler is the primary heat source for the heating system, and the temperature of the water within the boiler heat exchanger is controlled by an integral thermostat that energises a solenoid valve when calling for heat, and de-energises when satisfied.

The functions of the boiler thermostat and main gas solenoid valve combination are as follows:

• Regulation of the heating water temperature leaving the boiler via the flow pipe to enable the heating system to function as it was designed. Normally, domestic boilers should operate at their highest boiler thermostat setting, which should equate to approximately 82°C for low temperature heating systems, or higher at 95°C if intended to operate as a pressurised heating system. Prolonged operation of non-condensing boilers at temperatures below 82°C can result in condensation occurring in the flue gas, which would result in reducing the boiler's expected life due to corrosion being accelerated in the heat exchanger. Some condensing boilers will be arranged to operate on an upper temperature limit of 75°C.

- By the boiler thermostat regulating the flow temperature to match the heating system's design temperature, this will ensure that the pipes will convey the correct heat capacity and the heat emitters will emit the designed amount of heat.
- To prevent the boiler from overheating and generating steam, some boilers are equipped with a second thermostat known as a 'high limit thermostat', arranged to cut off the fuel supply in the event of the main control thermostat failing. This second high limit thermostat has to be manually reset after the failure of the main control thermostat has been investigated and rectified.

Control thermostats fitted to gas fired boilers can vary in their method of operation. Figure 13.10 depicts a liquid expansion thermostat that has been in common use for controlling the gas supply to the main burner in free-standing traditional boilers with a cast iron heat exchanger, although low water content boilers are normally equipped with clip-on bimetallic thermostats, or rod type thermostats.

The liquid expansion thermostat illustrated in Figure 13.10 comprises a temperature sensing bulb factory assembled to a capillary tube, which in turn is connected to expanding bellows. This unit is completely sealed and filled with a gas or liquid.

The temperature-sensing bulb is located within a pocket cast into the boiler heat exchanger and retained in position by a small screw. The bellows arrangement is connected to an electrical contact switch which makes or breaks the electrical supply under the dictates of the boiler temperature adjustment, and the expansion rate causes the bellows to expand or contract by the expansion of the gas or liquid medium in the capillary tube.

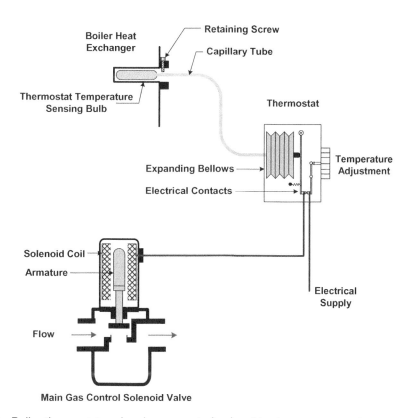

Figure 13.10   Boiler thermostat and main gas control solenoid valve arrangement

The electrical contacts in the thermostat are wired up to the main gas control solenoid valve, which serves to make or break the electrical supply to the solenoid of the main gas control valve as the temperature of the boiler dictates.

A solenoid operated valve is an instant action valve – it is either fully open or fully closed and cannot take up an intermediate position – unlike motorised valves, which are the other type of electrically operated valve used for heating systems and described in Chapter 11, Controls, Components and Control Systems. Motorised valves are slow to react when energised and can adopt an intermediate position if arranged to do so.

Solenoid valves are sometimes referred to as magnetic valves; they operate by creating a magnetic field when energised by the electric current controlled by the boiler thermostat.

The valve is attached to a steel rod which serves as the armature which is free to slide inside a brass tube. The solenoid coil is wound around a bobbin placed over the brass tube and covered by a steel casing which serves to concentrate the magnetic flux. When the electric current is energised, the polarity of the coil creates a magnetic attraction which draws the steel armature up into the brass tube to open the valve. When the electric current is de-energised the valve will fall back down to shut off the gas supply; this may be by a freefall arrangement, or more commonly, spring-assisted.

Solenoid operated valves are suitable for low pressure applications only, as a high pressure medium can prevent the valve opening against the internal pressure that exists.

MAIN BURNER GAS PRESSURE GOVERNOR/REGULATOR

Gas burners are equipped with gas governors that enable a constant gas pressure to be maintained at the burner irrespective of any fluctuations in the gas supply, although the main gas governor or regulator would normally undertake the function of preventing any fluctuations of the gas pressure to the whole gas system. The main function of the governor fitted to the burner assembly is to enable the gas pressure to the burner to be regulated and set to suit the gas pressure corresponding to the heat output required for the boiler.

Figure 13.11 illustrates a low-pressure spring-assisted single diaphragm compensated gas governor; it consists of a double-seated valve attached to a diaphragm that is tensioned by an adjustable spring. The gas pressure is regulated by turning the screw located above the spring, which raises or lowers the main seating; this has the effect of increasing or decreasing the gas pressure to the burner and maintaining it at a steady state. As the pressure acts equally on both the top and lower seating elements, the gas governor is unaffected by pressure fluctuations.

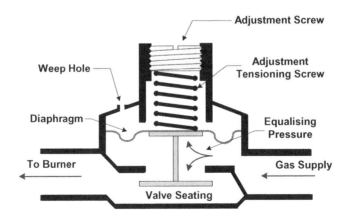

Figure 13.11 Low-pressure gas governor

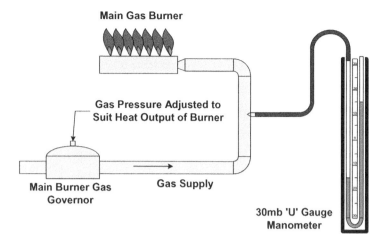

Figure 13.12   Gas burner pressure adjustment at governor

The procedure for adjusting the gas pressure to suit the burner's heat output is depicted in Figure 13.12. The gas pressure can be observed by attaching a 'U' gauge manometer to the test point located immediately prior to the main burner. The main burner should be alight before undertaking this procedure and the adjusting screw on the governor should be turned until the pressure required is registered on the manometer to correspond to the boiler/burner manufacturer's recommended pressure for the heat output required.

Domestic boilers are manufactured with a heat output that varies over a range of rates from its minimum output to its maximum output. The boiler manufacturer's installation literature will give the gas pressure that will correspond to these heat outputs.

This activity forms part of the commissioning and the annual servicing procedures.

MULTIFUNCTIONAL GAS CONTROL VALVE

The individual gas control components explained in the previous sections have been replaced by a single multifunctional control valve on domestic gas fired boilers, but boilers having larger heat outputs still retain the individual equipment items.

The multifunctional valves perform the multiple functions described, and which can be traced through using the sectional view depicted in Figure 13.13.

The multifunctional valve is divided into a number of sub-assemblies that can on some versions be removed from the main body chassis to be replaced if necessary, but on other versions the whole assembly has to be replaced as a complete unit.

The pilot flame can be established and in turn open the main safety valve when the control button is depressed and manually held down. This action causes the power unit armature to come into contact with the pole faces of the solenoid and opens the pilot valve section, allowing gas to flow through to the pilot burner: see the sectional arrangement illustrated in Figure 13.14.

When the pilot flame has been established for approximately 30 seconds, the thermocouple assembly generates a small electrical current of about 30 mV, which is sufficient to energise the power unit solenoid and hold the armature to the magnetic pole faces. When the control button is released, the safety latch pivots under the action of the cantilever latch spring, thereby engaging the safety valve lever and opening the safety valve: see Figure 13.14(II).

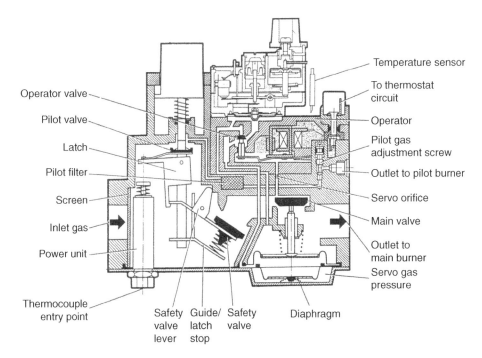

Figure 13.13   Section through multifunctional gas valve

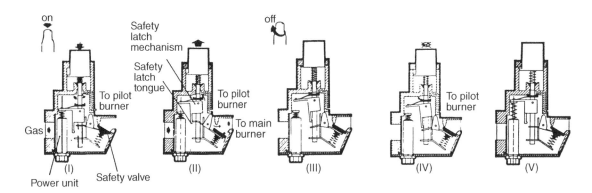

Figure 13.14   Section through pilot valve assembly of multifunctional gas valve

Gas is now allowed to flow via the servo orifice into the operator and regulator system. If there is a call for heat, the main valve will open under the control of the operator assembly and the main burner will be ignited, either electronically or by the pilot flame.

To close the appliance down the control button is normally rotated a quarter of a turn. This causes the safety latch tongue to disengage the safety valve lever which immediately closes the safety valve and prevents the gas flowing to the main burner.

The safety latch now moves upwards and tilts under the action of the cantilever spring, simultaneously closing the pilot valve. The pilot valve is now extinguished but the thermocouple will keep the power unit energised for a further 30 seconds. Should the control button be depressed during this period, the tilted latch mechanism ensures that the safety valve cannot be opened; this permits time for any accumulation of gas in the appliance to disperse before the main gas is re-admitted. When the power unit de-energises, the armature strikes the safety latch mechanism and re-aligns it. The start-up sequence can now be repeated: see Figure 13.14(III, IV and V).

The servo pressure regulator maintains the outlet gas pressure at a constant level regardless of any fluctuations in the inlet pressure. Working gas is taken from the inlet to the main valve for use in the servo pressure regulator and is used to adjust the position of the main valve according to demand.

The minimum gas outlet pressure is set using the adjustment screw; this screw exerts a force on the pressure regulating diaphragm. The maximum rate adjustment is set by using the main adjustment screw located under a plastic dust cap.

The main thermostat operates in the same manner as before, by the gas or liquid fill in the sensor expanding due to the increase in the temperature being sensed. This exerts a force on the diaphragm and shaft against a pivoting lever. This lever pivots until the normally closed micro-switch operates; this switch will then shut down the main burner when opened.

IGNITION METHODS

The conventional method of igniting the main gas burner has been by the presence of a permanent pilot flame, which itself had to be ignited by creating an electric spark generated by manually operating a piezo-electric generator.

The piezoelectric generator employs lead zirconate–titanate crystals which have been exposed to a powerful electric field during their manufacture. This polarises the crystals so that when they are subjected to stress they will produce an electromotive force of about 6000 volts; see Figure 13.15. The term piezo is derived from the Greek word piezein, meaning press, or squeeze.

The igniter consists of two lead zirconate–titanate crystals, each measuring about 10–12 mm in length and 6 mm in diameter, connected in parallel as shown in Figure 13.15. The two crystals are separated by a metal pressure plate connected to the spark electrode by a high tension lead. Operating the applicator by applying pressure in the form of impact or force to the crystals, causes voltage to be built up until it

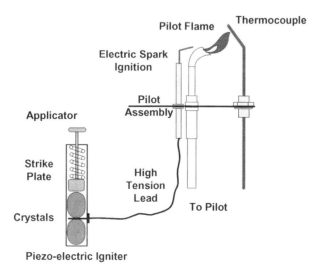

Figure 13.15   Detail of piezoelectric ignition assembly

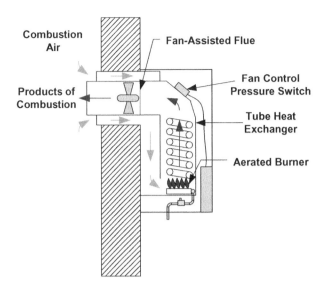

Figure 13.16   Fan-assisted natural draught gas burner

overcomes the resistance of the electrode spark gap and an electric arc is produced. This method of apply-ing pressure to the crystals has the advantage of achieving a consistent force every time it is operated to produce the spark at the electrode.

The use of a permanent pilot flame has now been recognised as an inefficient and wasteful use of gas, as the combined effect of every gas burner in the country that employs a permanent pilot flame wastes a huge amount of a non-renewable fossil fuel that does not contribute to the heating load.

Ignition systems employed on domestic gas boilers now use electronic methods similar to those that have been used for oil-fired boilers for many years. The electronic methods work on the principle of creating a spark caused by an electrode that is energised by the start-up procedure via a high tension lead which ceases to spark when the burner flame has sensed that it has become established, and from thereon works under the dictates of the control thermostat.

FAN-ASSISTED NATURAL DRAUGHT BURNERS

Developments over a number of years in the design of domestic gas fired boilers have moved away from the traditional free-standing cast iron heat exchanger boiler, which is large and heavy, to smaller lightweight low water content boilers employing tube type heat exchangers; see Figure 13.16.

As the physical size of the boiler has shrunk in size, making it easier to install and locate, the reduced size makes it impossible for the correct quantity of combustion air to be supplied by natural draught alone. To compensate for this situation, an electrically operated centrifugal fan is incorporated into the flue, that will supply a forced or assisted draught to the burner for complete combustion to occur.

The boiler has to incorporate a fan failure device to ensure that the burner does not attempt to operate without the fan running, this is achieved by incorporating a pressure switch within the flue passageways. The pressure switch has to be part of a fan-proving procedure during the start-up sequence: the multifunc-tional gas valve will not permit the burner to be ignited until it has sensed that the fan is running by the action of the pressure switch. The multifunctional valve will shut the gas supply down if the pressure switch senses that the fan pressure has been maintained.

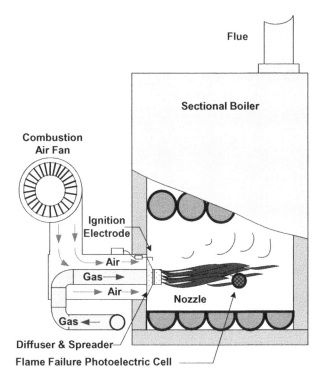

Figure 13.17 Forced draught gas burner

FORCED DRAUGHT GAS BURNER

The forced draught gas burner resembles the pressure jet oil burner in appearance and in some ways is similar in its operation. It has been used on commercial sized gas burners for many years, but more recently has been adapted for use on condensing gas-fired boilers as it offers many advantages in obtaining higher efficiencies.

Figure 13.17 illustrates the working principles of the forced draught gas burner fitted to a medium sized sectional boiler. Its heat capacity is virtually unlimited, but until recently, and because of its noisier operation compared to the natural draught burner, it had seldom been used on domestic gas boilers.

It has the advantages over the natural draught burner of not being dependent upon a flue, or having to be large in order to create the necessary draught to provide the air for combustion. It has always been equipped with electronic ignition.

The forced draught gas burner comprises a gas supply pipe enclosed within an air supply duct, whereby both the gas and forced air supply are mixed together at the diffuser and ignited by an electrode, with the combustion air being supplied under controlled conditions by a centrifugal fan.

These types of gas burners are fully automatic in their operation. A firing sequence occurs after depressing the start button, commencing with a pre-purge period to expel all air out of the burner supply pipe and followed by ignition via an electric spark. Establishment of the flame, together with thermostatic and safety control, is provided by either a photoelectric cell or an ionisation probe set in front of the flame nozzle spreader and which will almost instantly shut the burner down in the event of the burner flame failing.

Because of the similarity of this type of burner to the pressure jet oil burner, it is easy to adapt both of these types of burners to function as a dual fuel burner when it is considered economically beneficial to use both oil and gas and take advantage of lower fuel supply tariffs at certain times of the day and year.

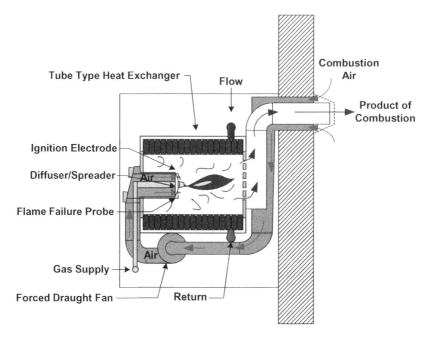

Figure 13.18 Application of domestic version forced draught burner

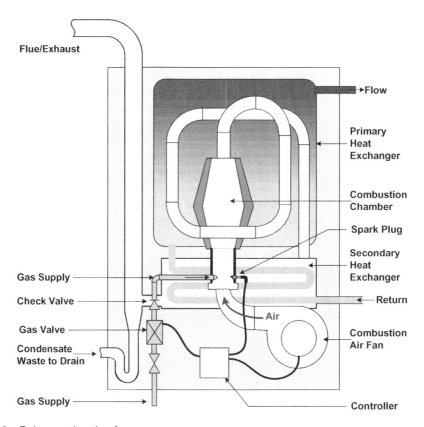

Figure 13.19 Pulse combustion furnace

Figure 13.18 illustrates a domestic version of the forced draught gas burner designed to fit inside a wall-hung domestic boiler.

PULSE COMBUSTION FURNACE

The pulse combustion furnace is an early form of condensing boiler that was developed to produce a more efficient and economical appliance. It was adapted for use in warm air heater applications in the 1970s. The method of producing combustion is fundamentally different from a conventional burner, as illustrated in sectional form in Figure 13.19.

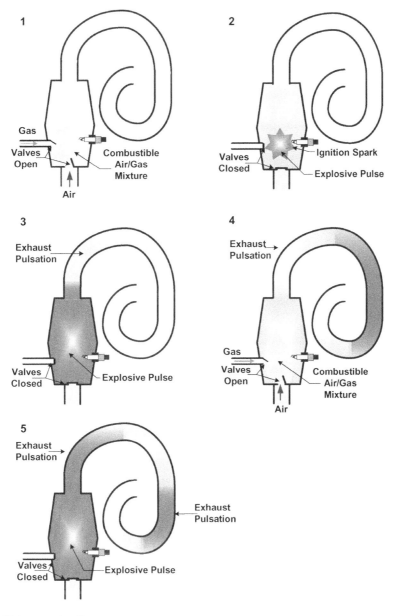

Figure 13.20   Pulse combustion process sequence

As can be seen, the pulse combustion furnace is fundamentally different from other firing methods, whereby the conventional flue arrangement is replaced by an exhaust vent pipe which is terminated above the roof or through a side wall outlet. It also needs to be connected to a drainage system in order to dispose of the acidic condensate waste created by the process.

The combustion process is illustrated in sequential order in Figure 13.20, which commences with (1) a small amount of gas being introduced into the combustion zone through the non-return type flapper valve, by the action of the combustion air being forced in by a blower fan.

This mixture of gas and air (2) is ignited by an automotive type spark plug causing the initial explosive pulse (3), which causes the non-return valve to close, thus forcing the products of combustion out through the tailpipe.

The length of this tailpipe is designed so that as the shockwave from the pulse reaches the end, it is reflected back to the combustion zone. Meanwhile the negative pressure created in the combustion zone (4) has allowed the non-return flapper valve to open again, admitting further quantities of gas and air. When the reflective pulse wave re-enters the combustion zone (5), there is sufficient flame left to ignite the new mixture causing the second explosive pulse to be formed. This initiates self-perpetuating combustion, allowing the spark ignition and combustion air blower to be turned off.

The pulse combustion process burns in separate pulses, occurring 68 times per second, as opposed to a conventional continuously burning flame. Each pulse produces between 0.1 and 0.2 watts.

The burning process generates the pressure to exhaust the products of combustion through a 42 mm diameter vent pipe and at a low enough temperature to allow the use of PVCu tube to be safely used for the vent pipe material.

Because of these lower flue gas temperatures created, condensation will occur within the finned tube heat exchanger, which must be piped away via a trapped waste pipe to drain.

The pulse combustion furnace is classified as a direct vent central furnace and is relatively silent in its operation, achieving a high degree of efficiency comparable to other forms of condensing boilers.

14

Liquefied Petroleum Gas Firing (LPG)

Liquefied petroleum gases are products of the refining of crude oil and are therefore closely related to other hydrocarbon products, which are compounds containing mainly hydrogen and carbon.

Methane is the simplest member of the hydrocarbon family and is the main constituent of natural gas (see Table 13.1 in the previous chapter), but as methane can only be liquefied at low temperatures it is difficult to handle. The main family of liquefied petroleum gases comprises:

Methane	CH_4
Ethane	C_2H_6
Propane	C_3H_8
Butane	C_4H_{10}
Pentane	C_5H_{12}
Benzene	C_6H_6
Hexane	C_6H_{14}

The composition of the liquefied petroleum gases is given in the chemical formula above, which states the make-up of its carbon and hydrogen molecule content; for example, methane is made up of one carbon atom bound to four hydrogen atoms.

The liquefied petroleum gases, or LPGs as they are commonly abbreviated, are used as fuels for providing energy for heating buildings. They are propane and butane, or a mixture of both, with small amounts of ethane, ethylene, butylene, propylene and iso-butane, the exact composition depending upon whether the source is a gas well or oil refinery.

Liquefied petroleum gases normally used for heating fuels are the C_3s commercial propane, or the C_4s commercial butane. Although they are gases at normal temperatures existing at atmospheric pressure, they can be liquefied by the application of pressure and therefore large quantities of gas can be stored as a liquid in relatively small pressure vessels. As the fuel is stored as a liquid under pressure, it will convert back to a gas at ambient temperatures when the pressure is released.

Heating Services in Buildings: Design, Installation, Commissioning & Maintenance, First Edition. David E. Watkins.
© 2011 John Wiley & Sons, Ltd. Published 2011 by John Wiley & Sons, Ltd.
As an aid to lecturers and students, full colour versions of the figures in this chapter may be found at www.wiley.com/go/watkins
These figures are © 2011 by John Wiley & Sons, Ltd.

Table 14.1 Typical Properties of Liquefied Petroleum Gas

Property	Commercial Butane	Commercial Propane
Specific Gravity of Liquid at 15°C	0.582	0.509
Litres per Tonne at 15°C	1718	1965
Specific Gravity of Gas Compared to Air at 15°C	2.1	1.52
Ratio of Gas Volume to Liquid Volume at 15°C at 1013 mb (Atmospheric Pressure)	233	274
Volume of Gas (Litres) per kg of Liquid at 15°C at 1013 mb (Atmospheric Pressure)	429	540
Boiling Point °C at Atmospheric Pressure	−4	−45
Air Required for Combustion m^3/m^3 Gas	30	24
Calorific Value Gross MJ/m^3	121.8	93.1
Calorific Value Net MJ/m^3	112.9	86.1
Calorific Value Gross MJ/kg	49.3	50
Calorific Value Net MJ/kg	45.8	46.3
Ignition Temperature °C	480–537	490–600
Maximum Flame Temperature °C	1987	1980
Weight per m^3 of Liquid at 15°C kg	582	509
Limits of Flammability (Percentage by Volume of Gas in a Gas/Air Mixture to form a Combustible Mixture)	Upper 9.0 Lower 1.8	Upper 10.0 Lower 2.2
Sulphur Content, % by Weight	0.02	0.02
Specific Heat of Liquid at 15°C	2.386	2.512
Vapour Pressure bars at °C		
−40		1.38
0	1.93	5.52
20	3.31	8.95
45	6.89	18.6

Note – Liquefied petroleum gases are colourless and possess anaesthetic properties, but are not in themselves toxic.
As with natural gas, an odorant ethanethiol is intentionally added to liquefied petroleum gases to achieve a perceptible odour that will assist in the detection of any leakage.

PROPERTIES OF LIQUEFIED PETROLEUM GAS

Table 14.1 gives the typical properties of both propane and butane types of liquefied petroleum gases.

EXPLANATION OF PROPERTIES

The definitions of the various fuel property terms have been given in the two previous chapters, but as LPGs have some interesting properties peculiar to themselves, further explanation is included here.

The two commercially available liquefied petroleum gases available have many uses in industry and transport, but as a fuel for providing heating to buildings, propane is the preferred gas due to its more advantageous properties when compared to butane, particularly for domestic heating applications. However, for industrial applications when large quantities of gas are required, butane is chosen due to its price advantage, its cost-effectiveness overcoming the advantages held by propane.

Butane is also used as the fuel for portable blow-torches.

Specific gravity

It can be seen from the properties in Table 14.1 that the specific gravity of both propane and butane is about half that of water when in liquefied form, but when in a gaseous state, butane is over twice as heavy as air and propane is approximately one and a half times as heavy as air. Therefore, because they are both heavier than air, they will tend to displace it at low level in the event of a leak, and if allowed to collect in large quantities they could cause a deficiency of oxygen for human respiration and create a dangerous situation and could be easily ignited if they went undetected. For this reason, distribution piping should be installed in a floor trench constructed with a small sump type area that will permit a gas detector to be housed to raise an alarm in the event of a gas leak.

Ratio of gas to liquid volume

LPG is stored under pressure to keep it below its vaporisation point in liquid form. When that pressure is released, the LPG will vaporise into its gaseous state and expand at the same time – resulting in one cubic metre of liquid butane producing $233\,m^3$ of butane gas, and one cubic metre of liquid propane producing $274\,m^3$ of propane gas, both at 15°C and at atmospheric pressure. Higher temperatures will result in a larger expansion rate. This means that large quantities of gas can be stored in relatively small storage containers.

Air required for combustion

It can be seen from Table 14.1 that butane requires three times more air for complete combustion to occur than that required for natural gas, and propane nearly two and a half times more air. This statistic has a fundamental bearing on the difference between butane, propane and natural gas in the design for the boiler/burner unit.

Calorific value

As previously established, the calorific value of a fuel is the amount of heat released on complete combustion when a unit volume or unit mass of that fuel is burnt.

Table 14.2 compares the calorific values of various common fossil fuels, and it can be seen that the two liquefied petroleum gases compare well and possess a high calorific value.

Limits of flammability

All gaseous fuels will only burn when mixed with a certain quantity of air, which provides the oxygen required for complete combustion to occur. The proportion of air required lies between two defined limits, known as the limits of flammability. As a combustible gas is gradually mixed with air in increasing proportions, a concentration is reached at which the mixture just becomes flammable; this is referred to as the 'lower limit of flammability'. As the concentration of gas in the mixture is increased further, a point is reached at which the mixture ceases to burn; this is known as the 'upper limit of flammability'. Therefore,

Table 14.2 Comparison of net calorific values of common fuels

Fuel	CV in MJ/kg	CV in MJ/m³
Butane	45.8	112.9
Propane	46.3	86.1
Natural Gas	–	34.9
Kerosene	43.6	–
Gas Oil	42.7	–

combustion of the gas can only be achieved if the concentration of the gas in the gas/air mixture lies between these two limits.

Vapour pressure

The vapour pressure of a liquid at a given temperature is defined as the pressure exerted by the liquid in equilibrium with its vapour at a specified temperature.

Regarding liquefied petroleum gases, it determines the pressure exerted by the gas at ambient temperatures and therefore affects the handling requirements and the designed working pressures of the storage vessels.

Since the boiling point of a liquid is the temperature at which its vapour pressure is equal to the applied pressure, then the higher the vapour pressure of a liquid at a given temperature, the lower will be its boiling point. The main difference between propane and butane is the vapour pressure. It can be observed from Figure 14.1 that for propane stored in a vessel at an ambient temperature of 20°C, it will exert a pressure of 6–8.5 bar, while the corresponding vapour pressure for butane is 1.5–2.5 bar.

Boiling point

It can also be observed from both Table 14.1 and Figure 14.1 that the boiling point of both propane and butane are significantly different, with propane vaporising at −45°C at atmospheric pressure, whilst butane will vaporise at −4°C also at atmospheric pressure. This makes propane much easier to store and handle than butane, which could be prone to remaining as a liquid at sub-zero temperatures.

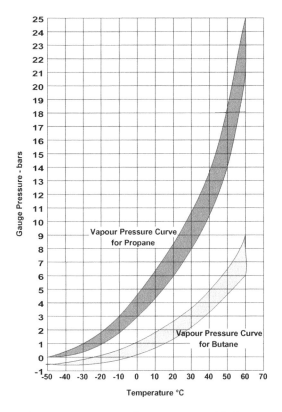

Figure 14.1   Vapour pressure/temperature curves for propane and butane

Table 14.3 Average gas consumption rates for LPG domestic appliances

Type of Appliance	Average Gas Consumption Rates			
	Butane		Propane	
	m³/hr	kg/hr	m³/hr	kg/hr
Boiler/Warm Air Heater	0.42–0.57	1–1.4	0.54–0.71	1–1.35
Oven, Large Domestic	0.57–0.71	1.39–1.74	0.74–0.93	1.37–1.74
Oven, Normal Domestic	0.34–0.42	0.83–1	0.42–0.57	0.79–1.05
Hotplate – 2 Burners	0.1–0.14	0.28–0.35	0.14–0.2	0.25–0.36
Hotplate – 3 Burners	0.14–0.2	0.35–0.5	0.2–0.25	0.35–0.45
Hotplate with Grill	0.2–0.23	0.5–0.56	025–0.28	0.47–0.53
Multipoint Water Heater	0.68–0.9	1.65–2.23	0.88–1.2	1.64–2.22
Storage Water Heater	0.06–0.08	0.14–0.2	0.08–0.1	0.15–0.2
Instantaneous Water Heater	0.3–0.37	0.77–0.9	0.4–0.48	0.7–0.9
Room Heater – Large	0.14–0.23	0.35–0.56	0.2–0.28	0.37–0.53
Room Heater – Small	0.04–0.06	0.1–0.14	0.06–0.08	0.1–0.16

LIQUEFIED PETROLEUM GAS STORAGE

LPG is delivered and stored in liquid form under pressure in either portable cylinders of varying sizes, or where larger volumes of gas are required, in bulk cylindrical or spherical pressure vessels.

As with oil, liquefied petroleum gas should be stored in sufficient quantity to provide normal consumption for a minimum period of one month, together with an additional one month's storage to allow for possible delay in delivery.

In the absence of any specific liquefied petroleum gas consumption rates, which should always be used, the values given in Table 14.3 may be used as a guide.

Example – Assume a domestic dwelling with LPG (propane) heating and cooking (using average values and winter daily use).

Oven (normal) @ 0.93 kg/hr × 1.5 hr daily use = 1.4 kg
Hotplate & Grill @ 0.52 kg/hr × 1.5 hr daily use = 0.78 kg
Boiler @ 1.2 kg/hr × 7 hrs daily use   = 8.4 kg
Total daily consumption in winter   = 10.58 kg/day × 28 days equals 296 kg per month.

Therefore, 296 ÷ 47 = 6.29 cylinders each having a capacity of 47 kg, say 7 cylinders, plus an additional 7 cylinders for reserve standby.

The use of 7 × 2 × 47 kg cylinders would not be considered practical for this application and a bulk pressure vessel should be considered.

CYLINDER INSTALLATIONS

Cylinders are a convenient means of storing and handling liquefied petroleum gas, in either singular or multiple forms, when the expected gas off-take is relatively low. LPG cylinders are available in a variety of sizes which can vary between suppliers, particularly with regard to the small cylinders intended for use in caravans, barbecues, camping etc. Table 14.4 lists the details of the larger cylinders commonly available and suitable for fixed gas installations.

Propane portable gas cylinders are equipped with a ⅝ BSP left hand female thread, whilst butane gas cylinders have a 21.8 mm, 14 tpi left hand male outlet thread.

Table 14.4 Commonly available LPG portable cylinder capacities and sizes

Cylinder Detail		Propane						Butane
Cylinder Colour		Orange with Red Shoulder						Blue with Red Shoulder
Product Capacity	kg	4	13	15	18	19	47	15
Total Weight	kg	13.3	28.5	43.4	46.27	43.5	108	30.4
Liquid Stored	Litres	9	23	30	35	37	92	25
Gas Stored	m³	2.5	6	8	9.8	10	25.5	6
Diameter	mm	280	330	330	330	330	381	330
Height	mm	356	635	712	788	794	1240	610

For normal fixed cylinder installations it is usual to provide at least one month's usage in storage with an equal amount in reserve. This will involve at least one duty cylinder and one standby cylinder connected to each other on a common manifold. Usually more than one duty cylinder is required in order to achieve a month's gas storage. However, unless the gas off-take is very small a single LPG duty cylinder cannot provide sufficient gas for a reasonable length of time and is therefore not usually considered for fixed installations. Single cylinders are normally used for portable rather than fixed appliances.

Multiple cylinder assemblies are used either when the gas consumption rate does not justify the selection of a bulk storage vessel but the expected gas off-take required is greater than any single cylinder can satisfy, or when longer service life is required before cylinder replacement has to be made.

A suitable external location in compliance with the Building Regulations should be found to house the multiple cylinder assembly in a permanently fixed installation; this usually involves making use of an external fire-resistant wall to keep the length of gas distribution piping to a minimum.

The cylinder storage should be shaded from direct sunlight and, if located in an unsecure area, it should be housed in a security cage to prevent any unauthorised tampering with the gas storage.

The storage of LPG cylinders should conform to the Building Regulations (discussed in Chapter 24); however, care should be taken to ensure that there are no openings in the wall of the building in the vicinity of the gas storage assembly, such as air-bricks, vents or windows. Also, open-top drain gullies should be avoided, as the gas is heavier than air and any leakage will enter into the drainage system, creating a potentially dangerous and illegal situation.

Figure 14.2 illustrates a typical two by two multiple LPG cylinder arrangement located against an external fire-resistant wall of the building. There are no limits to the number of LPG cylinders that can be manifolded together, but normally the arrangement is restricted to a maximum of six cylinders each side of the automatic change-over unit for practical reasons.

The automatic change-over valve will automatically change over the standby gas cylinders into service when the duty gas cylinders have become exhausted, permitting the empty cylinders to be replaced with fully recharged gas cylinders. These cylinders then become the designated standby cylinders, ready for when the automatic change-over valve operates again.

The automatic change-over valve also serves as a first stage pressure regulator, that is used to reduce the line pressure exerted by the vapour pressure of the gas contained in the cylinders. This pressure is reduced to either a low service pressure of 0.4–0.85 bar, or when used as a first stage pressure reduction, the line pressure is reduced to a medium pressure of 1.7–2.4 bar.

A low pressure gas distribution system is used for most domestic applications, with the gas storage located reasonably close to its point of consumption. For those distribution systems where the gas storage has to be located further away from the point of consumption, resulting in a longer piping distribution arrangement, a medium pressure system should be adopted that employs a second stage pressure regulator located close to the point of gas consumption. This permits smaller distribution piping to be used that can tolerate a higher pressure drop through the pipework.

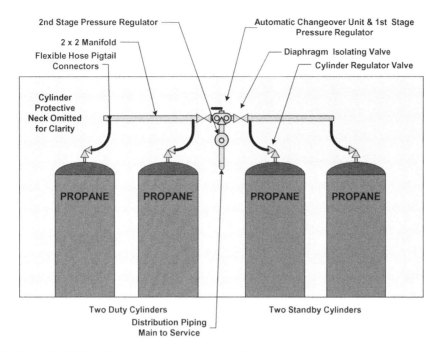

Figure 14.2   Multiple LPG cylinder and manifold assembly

Figure 14.3 illustrates a section through an automatic change-over device that is operating with the liquefied petroleum gas entering from the right-hand side and closed to the standby/reserve gas supply on the left-hand side. When the duty cylinders become exhausted, the service pressure will drop, causing the diaphragm and spring to tilt the opposite way and open to the standby/reserve cylinders and close to the empty cylinders.

Normally the largest commercially available liquefied petroleum gas cylinders are chosen for a manifold arrangement, this is currently 47 kg.

BULK LPG STORAGE INSTALLATION

Bulk liquefied petroleum gas storage pressure vessels may be obtained in both spherical and cylindrical form for above-ground installation in varying capacities.

Figure 14.4 illustrates a typical above-ground cylindrical liquefied petroleum gas storage pressure vessel which should be constructed to conform to the requirements of BS 1515.

The pressure vessel's capacity should be designed to contain the required liquid storage when about 85% full, leaving space for the liquid expansion and a supply of compressed vapour in the freeboard space above the liquid level. The gas vapour can then be drawn off at the top of the vessel without vaporising in the distribution pipe, with further gas being vaporised from the liquid by the release of pressure to take its place.

As previously established, liquefied petroleum gas is stored under pressure to maintain it as a liquid, which, depending upon the ambient temperature and the percentage capacity of the liquid stored in the vessel, will exert a pressure of 2–14 bar for propane and 0–5.5 bar pressure for butane. Therefore, the storage pressure vessels are constructed with a working pressure of 17 bar at 38°C for propane and 7 bar at 38°C for butane. The safety relief valve should be set to operate at 14.25 bar for propane and 6 bar for butane.

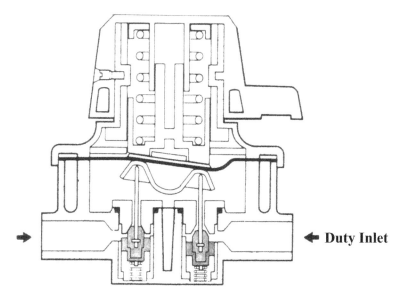

Figure 14.3   Section through automatic change-over valve

Multiple Valve Assembly

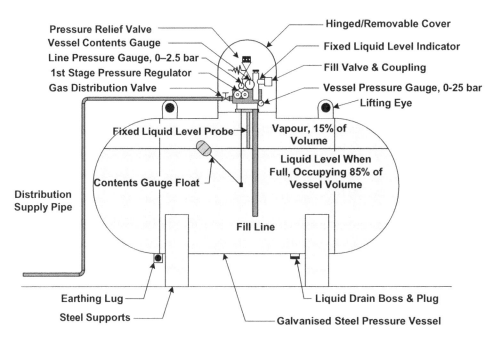

Figure 14.4   Cylindrical LPG storage pressure vessel

Figure 14.4 illustrates a small domestic LPG storage vessel that incorporates all the control components on a single multiple valve arrangement, which includes pressure relief, fill and off-take provision, contents probe and indicator to register when the liquid level has reached its maximum permitted fill point of 85% of the vessel's capacity.

Table 14.5 shows vessel capacities and sizes.

Table 14.5 Standard LPG pressure vessel capacities and sizes

Capacity		Length m	Diameter m
Litres	Tonnes		
380	0.2	1.8	0.9
1200	0.6	2.7	1.2
2000	1	3.1	1
4000	2	3.3	1.4
14000	7	7.6	1.7
24000	12	12.9	1.7
26000	13	7.7	2.3
38000	19	10.8	2.3
50000	25	12.2	2.5
60000	30	7	3.7
100000	50	10.4	3.7
130000	65	13.8	3.7
170000	85	17	3.7
200000	100	20.5	3.7

STATIC ELECTRICITY

As a safety precaution, it is a requirement to provide an effective earthing point on all bulk LPG installations for the discharge of static electricity during the delivery of LPG bulk supplies via road tankers.

This is achieved by providing complete electrical continuity between all pipelines conveying liquids and the bulk storage vessel, including all hoses. All flanged pipe joints should be bridged across the flanges by attaching an earth continuity strip attached to one of the flange bolts. The storage vessel should be earthed by attaching one end of a 12 mm wide copper earthing tape to the earthing tag provided on the storage vessel and the other end to a copper earthing rod driven into the ground to a depth of 1.2 m, alongside the bulk storage vessel's concrete support base.

FIRE PROTECTION

Liquefied petroleum gas installations can be considered a potential fire hazard and should be treated with the highest respect. The risk of an uncontrolled fire occurring with LPG is no higher than with that of any other flammable fuel, but the real hazard comes after the fuel has been ignited as it can only be extinguished by isolating the fuel supply to the fire or starving the fire of oxygen – it cannot be extinguished by applying water, foam or extinguishing gas unless the fire has occurred in a relatively small enclosed space where the fire can be starved of oxygen.

For these reasons some degree of fire protection should be considered for LPG installations, which should warn of a gas leakage at an early stage or cool the storage vessel to protect it from an adjacent fire.

LPG STORAGE LOCATION

The location of any liquefied petroleum gas storage installation, whether it is a multi-cylinder installation or a bulk LPG storage vessel, should comply with the requirements of the Building Regulations Approved Document Part J, which addresses the fire and safety issues and is discussed in detail in Chapter 24.

LPG DISTRIBUTION SYSTEMS

The distribution of LPG can differ between propane and butane due to the different viscosity and the frictional resistance encountered. This should be understood when considering the distribution of the gas to its point of consumption.

Propane

The distribution of propane gas where natural vaporisation exists may be either by low or medium pressure. Most domestic applications are by low pressure distribution, whereby the outlet pressure is controlled by the pressure regulating valve to 27.4–34.9 mbar, suitable for low pressure LPG burners.

Where larger demands for LPG gas are required or where long pipe runs are needed, a medium pressure distribution system should be considered. Medium pressure distribution systems employ a first stage pressure reduction at the storage vessel, controlling the gas outlet pressure to 1.7–2.4 bar, distributing the gas at this pressure, and reducing it at or near the point of consumption by a second stage pressure regulating valve to a low pressure. This system permits smaller pipe sizes to be used on the main distribution pipe run, as a greater pressure loss may be tolerated.

Figure 14.5 may be used to select steel or copper pipe sizes for the flow of propane gas.

Butane

The distribution system for butane gas is similar to that for propane except almost invariably, a vaporiser must be used at the storage vessel. The vaporiser is required to convert the butane liquid into a vapour state due to its higher boiling point compared to propane. This may be achieved either electrically or by the use of steam or high temperature water. It also follows that having converted the butane into a vapour, it must be maintained as a gas whilst in the pipeline, not allowing it to cool to below its dew point, which will involve some form of trace heating.

Because butane needs to be vaporised, it is not normally considered practical for domestic applications.

DISTRIBUTION PIPING MATERIALS

Standard piping materials are used, but due to the particular characteristics and searching properties of LPG, only the following materials are deemed to be satisfactory.

Low carbon mild steel tube

BS 1387 heavyweight tube may be used with either a plain black or galvanised finish. Joints on the pipework up to and including 40 mm may be either screwed or welded, but any screwed joints should be made with tapered threads using an approved jointing compound for LPG gas or PTFE tape.

On liquid LPG lines, only welded joints should be employed.

Copper tube

BS EN1057 R220 or R250 tube should be used on LPG vapour lines only, and never on piping that may become subjected to LPG in liquid form. Normally only low carbon steel tube should be used for this latter situation. Copper tube is mainly used on internal carcass piping, or final pipe connections to gas appliances.

Joints employed on copper tube should be made using capillary fittings or Type 'B' manipulative fittings; alternatively, hard soldering or bronze welding may be used.

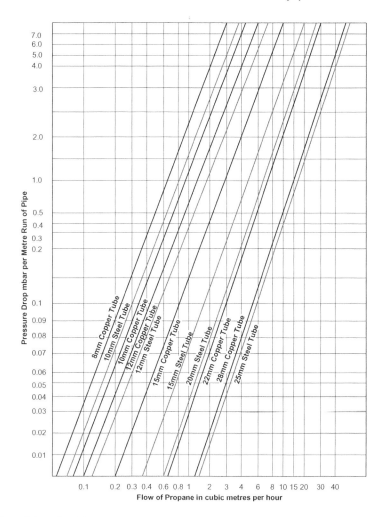

Figure 14.5 Pipe sizing chart for propane gas

Valves

Diaphragm type valves are preferred for the control of LPG vapour lines, due to the absence of any form of valve glands which could be susceptible to gas leakage.

LPG GAS BURNERS

Burners employing liquefied petroleum gas are similar in design to those that burn natural gas, except that they are calculated for the different pressures and calorific value of LPG.

15

Alternative Fuels and Energy

The term 'alternative fuel' has in the past been applied to fuels that were commonly available, but not necessarily sustainable or desirable in normal situations, and were only used when circumstances prevented the use of other commonly available fossil fuels.

Alternative methods include wood burning and electrically powered heat generators, both of which have been available for many decades, and in the case of wood, since prehistoric times; in neither case could they be described as new forms of energy. Neither of these fuels has been the first choice for the vast majority of heating systems – unless the owner/operator of the heating system had a readily available supply of wood, or structural problems with the building that excluded the use of appliances that required a flue, which all fossil fuels do.

This chapter gives brief information on these less popular fuels and energy sources to enable the student to understand them and – on the rare occasions when confronted with the situations described above – be able to correctly specify them.

More recently, the term 'alternative fuel' has become associated with new forms of energy that are either sustainable or more carbon friendly, or both. Sustainability means that the fuel comes from renewable resources, unlike the fossil fuels described in the previous chapters.

The need for viable forms of clean energy has been debated across the world, but it is considered beyond the scope of this technical work to include the political arguments, other than to include information on the use of these fuels or forms of energy together with their environmental impact, to enable systems to be designed and installed correctly and to obtain the maximum benefit from them.

Many of the potential forms of energy proposed, such as wave energy and wind power, are aimed at producing electricity on a large national scale; these are also beyond the scope of this work as we are only concerned with those fuels or energies that are suitable for individual heating systems used for both domestic residential use and larger commercial buildings.

Heating Services in Buildings: Design, Installation, Commissioning & Maintenance, First Edition. David E. Watkins.
© 2011 John Wiley & Sons, Ltd. Published 2011 by John Wiley & Sons, Ltd.
As an aid to lecturers and students, full colour versions of the figures in this chapter may be found at www.wiley.com/go/watkins
These figures are © 2011 by John Wiley & Sons, Ltd.

WOOD BURNING

Although wood or timber may be considered as coal in a raw state, it should not be regarded as a primary fuel, except under certain circumstances, due to its relatively low calorific value (see Table 15.1): fairly large quantities of wood are required to raise the amount of heat needed compared to the primary fuels discussed in previous chapters.

All values given are approximate only, and will vary considerably depending upon the moisture content of the timber concerned, but may be used for guidance purposes. Where a mixture of different timber types is used, an average calorific value of $8500\,KJ\,kg^{-1}$ may be taken as a reasonable representation for guidance purposes in calculating heat outputs and running costs.

In special circumstances, wood may be regarded as an attractive proposition as a heating fuel – such as industrial places like lumber and timber yards, wood mills or any factory/workshop where large quantities of timber waste material are produced, which can then be regarded as a free, or at least a low-cost fuel.

In these situations a hand-fired solid fuel wood-burning boiler should be used, which can be arranged as shown in Figure 15.1, in conjunction with a secondary boiler fired by a conventional primary fuel such as gas or oil, which serves as a back-up supporting fuel and will only operate when there is insufficient wood available.

In this system, the wood-burning appliance is arranged as the primary heat source controlled in the normal way by a central room thermostat with timer control. The secondary back-up boiler should be controlled by a pipe thermostat that activates this back-up facility when the primary flow temperature is less than 82°C, or the designed flow temperature, during a timed-on period. The secondary boiler providing the back-up heat source should be fitted with a high limit non-self-resetting thermostat to shut the boiler down for safety reasons.

Because of the slower response to the control thermostat fitted to the wood-burning appliance, a small diameter weep bypass, complete with a lockshield pattern regulating valve, should be installed to allow a small circulation through the heat exchanger when the motorised valve is in the closed position. All isolating valves should also be of the lockshield pattern to prevent unauthorised operation, which could create a dangerous situation.

The design of the wood-burning boiler is very similar to that for burning coal-type solid fuel, as the combustion process is the same.

Figure 15.2 illustrates a cross-section through a form of wood-burning boiler which incorporates a large fuel storage hopper similar to the solid fuel magazine type boiler. This allows the boiler to be stocked up every 6 to 8 hours for normal continuous usage. The boiler also incorporates a thermostat that controls

Table 15.1   Approximate calorific values for various woods

Type of Wood (In Green Condition)	KJ/kg (Average Values)
Ash	10770
Beech	9075
Birch	8840
Chestnut	6050
Elm	8375
Hickory	9540
Maple	8600
Oak – Red	7910
Oak – White	9305
Pine	9770
Walnut	9540
Willow	5585

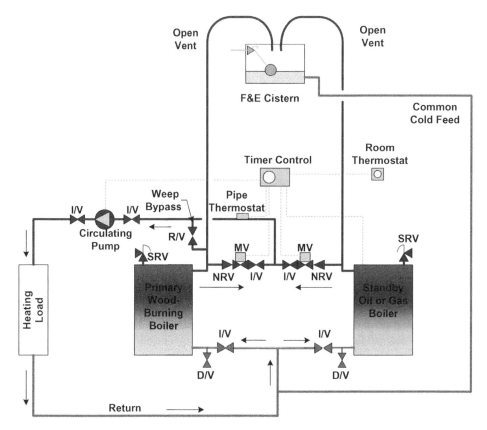

Figure 15.1 Primary wood-burning boiler linked-up to alternative fuel boiler

the combustion air inlet damper by a non-electrical mechanical means, but still requires manual re-stocking with fuel and regular cleaning and emptying of the residual ash.

For domestic residential applications, a multi-fuel room heater, complete with an integral back boiler, can be used as the combustion process, heat transfer and system requirements are the same. This gives the owner/occupier the facility to burn wood logs when they are available.

ELECTRICAL ENERGY

Electric heating systems may be defined as being a system of heating comprised exclusively of electrical components, but this text is only concerned with those items required to provide electrical energy to generate heat at a central source.

Electricity as a source of energy has – like all other fuels and energies – its own particular merits and disadvantages. Each need to be appreciated when selecting the fuel or energy type.

The advantages of electrical energy are:

- The main significant advantage associated with electrical energy is that it does not require any flue or chimney to function correctly and therefore removes this constraint common to all other traditional forms of energy.
- It is a clean fuel supplied directly to all properties and does not require any on-site storage. It should be noted that although electricity does not produce any carbon dioxide or any other products of combus-

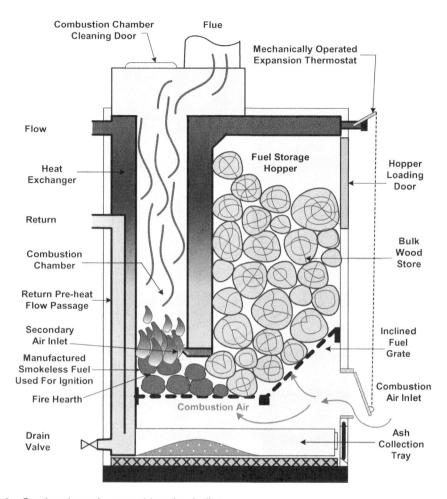

Figure 15.2   Section through a wood-burning boiler

tion on site, it does produce these gases in large quantities at its point of generation at the power station, the exact amount depending upon its method of generation.
• Electric heat generators are quiet and odourless in operation and simple to operate. They can also be installed in less time than that required for the more common forms of fuels.

The disadvantages are:

• The main disadvantage that outweighs all of the merits is the cost of eletrical energy has traditionally been considerably more expensive than the other commonly used fossil fuels. For this reason alone, electricity, as a source of energy for generating heat at a central point, has always been the last choice used only when all other fossil fuels have either not been available, or constraints have precluded them. The future regarding the use of electricity in this aspect will depend upon how quickly the cost of the traditional fossil fuels rises and how successful we are at exploring new ways of mass generation of electricity using sustainable means.
• Unit heaters that operate using a heat store is their weight, the floor may require strengthening to support them.

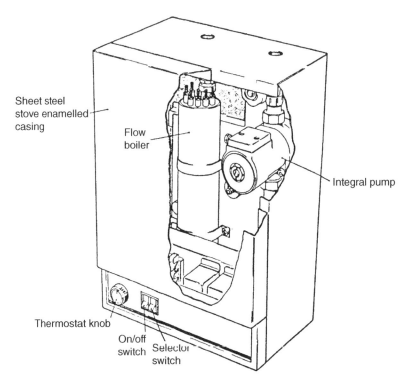

Sheet steel stove enamelled casing

Flow boiler

Integral pump

Thermostat knob

On/off switch Selector switch

Figure 15.3   Electric wall-hung heat generator

ELECTRIC HEAT GENERATORS

Electrically powered boilers, or 'heat generators' as they should be referred to, are installed in exactly the same way as any other boiler on any hydronic heating system regardless of the fuel used.

Electrically powered heat generators have the advantage of not requiring a flue and they can therefore be located anywhere in the building, without any of the constraints that apply to appliances burning fossil fuels.

Two basic forms have been commercially available; the most commonly used type is depicted in Figures 15.3 and 15.4. This form is arranged in a similar way to a conventional wall-hung lightweight boiler. It has heat outputs ranging from 2 kW to 12 kW, it does not have any thermal storage capacity, and it operates on the principle of an almost instantaneous heat-up as the water passes through the electrically powered heat exchanger – termed the 'flow boiler' – which comprises a series of electrode-type resistance elements that rapidly raise the temperature of the water passing through it.

This unit is supplied electrically by a standard tariff 230 volt electrical supply and can therefore be expensive to run at today's cost of electricity.

The other form available is also a low water content model, but is floor standing and much larger as it incorporates an amount of thermal storage. This version operates on similar principles to those of the electric warm air heat storage unit, whereby heat generated overnight by off-peak low tariff may be stored to be released the following day, except that the fan circulates the hot air over a water-to-air heat exchanger, instead of discharging it into a duct. It also incorporates an electrical flow boiler similar to that used in the wall-hung model, to supplement the heat input requirements when the thermal storage bank has become exhausted.

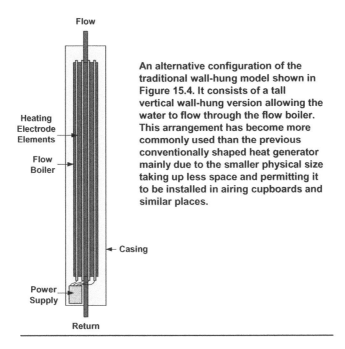

An alternative configuration of the traditional wall-hung model shown in Figure 15.4. It consists of a tall vertical wall-hung version allowing the water to flow through the flow boiler. This arrangement has become more commonly used than the previous conventionally shaped heat generator mainly due to the smaller physical size taking up less space and permitting it to be installed in airing cupboards and similar places.

Figure 15.4   Alternative arrangement for wall-hung heat generator

These types of heat generators are available with heat outputs of up to 24 kW, and are less costly to run kW per kW than the wall-hung low water content version, due to the low-tariff thermal storage element. However, they are considerably heavier in weight and the larger heat output models may require a three-phase 400 volt electrical supply.

Because of the high operating costs of these electrically powered heat generators, they are normally only selected when it is impossible to provide a flue for the more conventional fuels.

NEW ALTERNATIVE FUELS, ENERGIES AND SYSTEMS

The previous text in this chapter has concentrated on those alternative fuels that have been in existence for many years; that is to say they are not new, but are used only when circumstances prevail that made them attractive for a particular application.

This section discusses those fuels, energies and systems that have been identified as possible viable alternatives and are both sustainable and carbon neutral for individual building use.

Research and development into the viability of alternative sustainable fuels and energies is being carried out to find feasible methods of generating electrical power, on both a national supply basis and smaller individual units aimed at single building and dwelling applications. These methods include:

- Wave power
- Wind power
- Photovoltaic

Although these are all important methods of generating electricity, they are outside the scope of this work which concentrates on those fuels and energies that can be adapted directly to raising heat for single-source heating applications. However, if it is ever discovered how to generate large quantities of cheap and clean electricity by sustainable means, then electric heat generators could become the norm.

The alternative fuels, energies and systems that have shown promise in meeting the challenge of providing sufficient energy for heating applications that is sustainable and at least part carbon neutral are:

- Solar heating
- Combined heat and power
- Biofuels
- Heat pumps

The above list is by no means exhaustive as there are other means being investigated, but as yet they have not been produced commercially. The energies and heating methods listed above have been made commercially available, and have proved to be reasonably successful in partly meeting the challenge of being sustainable and carbon neutral.

SOLAR HEATING

The practice of utilising the sun's radiated heat can be traced back to biblical times, and even in its modern form has been around for over a hundred years, although its use in temperate climates is more recent.

Solar heating systems have now become quite common and the general principles that are employed are well known. However, because we now apply a degree of science to the subject rather than the hit or miss approach that has been used in the past, it is covered in detail separately, in Chapter 20.

COMBINED HEAT AND POWER (CHP)

Providing heat and electrical energy from a single source is not a new concept; it has been in commercial use for many decades.

Combined heat and power, or co-generation as it is sometimes known, has in the past been limited to large industrial or commercial applications whereby the waste heat produced by the process of generating electricity is recovered to raise the temperature of water that can be used to provide heating. The main purpose of this process is to generate electricity by consuming any of the fossil fuels – with the recovery of the waste heat produced, very much of secondary importance. The prime movers in these applications have usually been gas turbines, diesel oil engines or coal-fired steam turbine stations, with the waste heat being recovered from a combination of the exhaust gases and cooling systems.

Figure 15.5 illustrates in a simplified schematic layout of one form of co-generation whereby heat is recovered from both the exhaust gases and cooling jacket of the prime mover. There are many versions of this means of recovering heat from the basic object generating electrical power that are normally only applied to applications involving the generation of large quantities of electricity. This has the result of being able to raise quantities of high-grade hot water at the small cost of the parasitic electrical power used in the process.

Figure 15.6 illustrates a section through a exhaust heat recovery combined heat exchanger and exhaust silencer, which has the dual purpose of transferring heat from the exhaust gases to the water and reducing the mechanical noise emitted from the generating process.

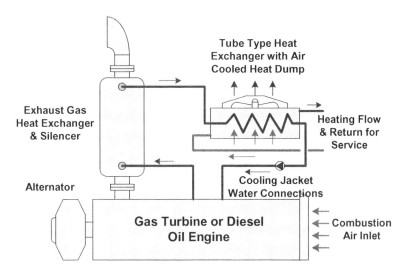

Figure 15.5   Combined exhaust and jacket cooling water heat recovery system

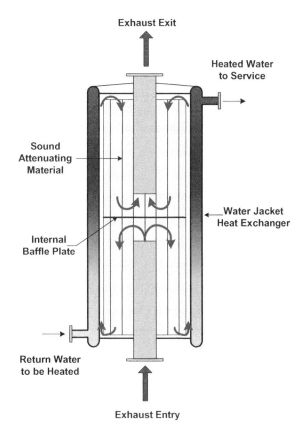

Figure 15.6   Dual-purpose exhaust silencer and heat exchanger

MICRO-COMBINED HEAT AND POWER

The previous text and illustrations of combined heat and power systems have been based on electrical power generating plants, where the opportunity has been taken to recover the waste heat that has resulted from the electrical power generation process.

A more recent and interesting development has been the introduction of micro-co-generation units designed to simultaneously produce both heat and electricity at a decentralised point of use in the home or office. The main difference between micro-combined heat and power units and traditional co-generation systems, is that the micro-co-generation units are basically heat-producing appliances that incorporate the facility to generate a small amount of electrical power.

The first generation of these micro-co-generation units was quite large compared to a traditional floor-standing system boiler fuelled by gas or oil, and required space that is not normally available in most UK homes; it has therefore found less use in the UK compared to other northern European countries, where architectural differences permit more space for installing these units.

These units operate by employing a gas or oil driven internal combustion engine – normally the two-stroke Stirling engine has been selected for this use – which acts as the prime mover to generate the electricity element of the unit. The heat is obtained partly by waste heat recovery, and partly by employing a secondary conventional burner to supplement the heat requirement.

The power to heat generated ratio varies considerably for these micro-combined heat and power units, as depicted in the Sankey energy diagram in Figure 15.7. Current developments of these units are resulting in efficiency improvements by reducing the flue heat loss proportion.

The larger physical size versions of these micro-CHP units can produce approximately 5.5 kW of electricity and approximately 15 kW of thermal energy in the form of heat being transferred into the heating water. This makes them attractive for dwellings that can accommodate their large size.

An even more recent development involving micro-generation is the production of a physically smaller wall-hung version, making it a viable practical proposition for installation in many more dwellings, although the total heat and power being generated has been sacrificed.

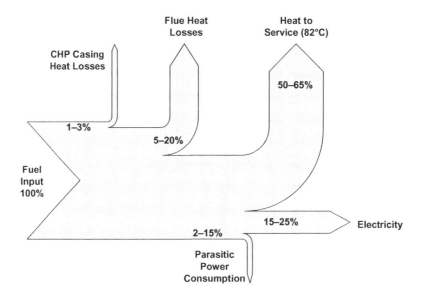

Figure 15.7   Sankey energy diagram for a typical micro-CHP system (not to scale)

The wall-hung versions generate approximately 1 kW of electricity and approximately 6 kW of heat, neither of which would satisfy most dwellings' demand for electricity and heat during the winter months. When the demand for heating surpasses the 6 kW rate, a secondary conventional pre-aerated gas burner is activated that is capable of providing an additional 18 kW of thermal energy for heating that satisfies most domestic dwellings' heating requirements.

The generation of electrical energy remains constant at 1 kW. This can be sold back to the electricity supply company when the full loading is not required and when a special import/export electricity meter has been installed by electricity supply company.

The micro-CHP unit is connected to the heating system in exactly the same manner as any other boiler to provide the primary heat requirements; the only difference is that the electrical supply is arranged to export power as well as import it for its own power requirements.

The introduction of these micro-combined heat and power generating units will require plumbers and maintenance operatives to obtain new skills more commonly associated with automotive mechanics, as well as competencies in installing and maintaining gas- or oil-fired appliances.

The micro-combined heat and power units that have been developed and are in commercial use, are fuelled by conventional fossil fuels such as gas and oil. If these units are to have a long-term future, that will depend upon the availability and ready obtainability of alternative sustainable fuels to make them a viable alternative source of heating and power.

A typical device is illustrated in Figure 15.8.

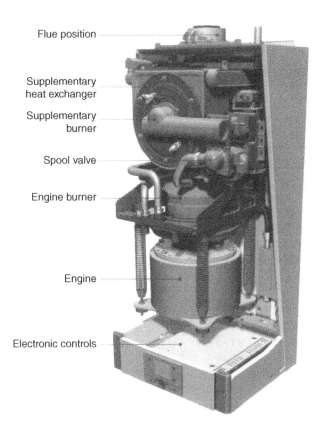

Flue position

Supplementary
heat exchanger

Supplementary
burner

Spool valve

Engine burner

Engine

Electronic controls

Figure 15.8 Illustration of a micro-combined heat and power generating unit. Reproduced by permission of Baxi Heating.

BIOFUELS

'Biofuels' may be defined as being derived from recently living plant or animal life, formed over a relatively short period of time, and are replaceable if managed properly; hence this form of fuel can be considered sustainable. This is in complete contrast to fossil fuels which have been formed over tens of millions of years but which are being consumed in just a few hundred years and cannot be replaced.

Biofuels are thought of by some as being new fuels, but they are the oldest form of fuel known to mankind, having been available since the discovery of fire and having remained in common use until the industrial revolution exploited the newly found fossil fuels.

Biofuels are now being used globally and their use is expanding all the time. They can be produced in any one of three forms, namely:

- biomass   –   solid form
- biodiesel/bio-oil   liquid form
- biogas   –   gaseous form

Biofuels are claimed to be carbon neutral – which should not be confused with being zero carbon rated, as all fuels, including biofuels, when burnt, will release carbon into the atmosphere.

Carbon neutral is the term given to a fuel that extracts the carbon from the atmosphere during its growing time period and releases it back into the atmosphere when burnt, the theory being that there should not be any net increase in carbon dioxide in the atmosphere as a result of converting it into heat. This is depicted in Figure 15.9 as a balance between the carbon dioxide being absorbed by the crops/plants and the carbon being produced as a fuel being released back into the atmosphere.

In reality, most biofuels are seldom as carbon neutral as the carbon cycle suggests because energy is required to grow the crops and process them into a fuel. Cost of carbon-producing energy used in the production of biofuels should include the manufacture of the fertiliser used, the fuel consumed by the machinery to plant, harvest and transport the biofuel as well as the energy required to process the raw material into a biofuel. However, the life cycle analysis used to calculate the carbon emissions – that take into account the total carbon dioxide produced from planting the seeds to the consumption of the biofuel – shows that most biofuels produce between 60% and 80% less carbon dioxide than conventional fossil fuels when using this method.

BIOMASS

Biomass fuels are fuels produced mainly from fast-growing virgin wood or energy crops, food waste, and sometimes industrial waste. Examples are shown in Figure 15.10.

Biomass fuels are normally produced as wood pellets, by chipping waste timber and bark from newly felled trees during forest management routine thinning. Short rotation coppice is fast becoming the main source of wood chips where fast-growing trees such as willow, pine and other evergreen trees are grown as a fuel crop. Wood pellets are made by bonding together sawdust that has been obtained from saw mills and other timber manufacturing concerns, although joinery waste should not be selected as it could contain contaminants such as glue, preservatives or varnish.

Wood chips and pellets are available specified as wet or dry. The moisture content of these fuels adds weight but reduces the energy content when burnt, as the moisture is converted to steam and passes up through the flue as a vapour.

The calorific value of the fuel will vary depending upon the tree from which the timber has been obtained, together with the moisture content: see Table 15.1 for typical calorific values. Wood chips and pellets are normally sold with the heat emission expressed in kW h/kg or kW h/m^3.

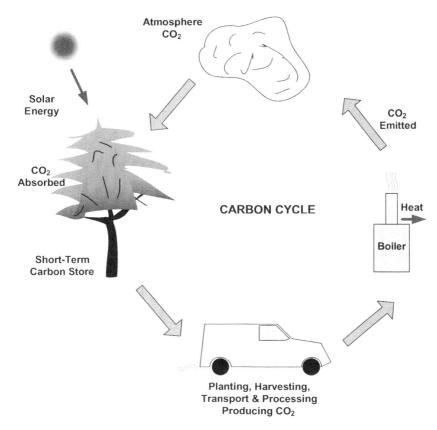

Figure 15.9   Biofuel and the carbon cycle

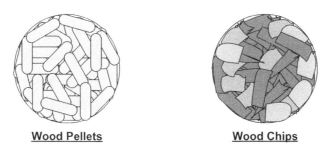

Figure 15.10   Examples of biomass fuels

BIOMASS BOILERS

The storage, handling and combustion of biomass fuels is similar to that employed for coal type solid fuel products, as biomass fuels are also classified as being solid fuel. For this reason the technology and methods developed for coal-burning appliances have been adapted for use with biomass.

Figure 15.11 illustrates the magazine boiler developed from coal type solid fuel and modified for using a biomass pellet form of solid fuel.

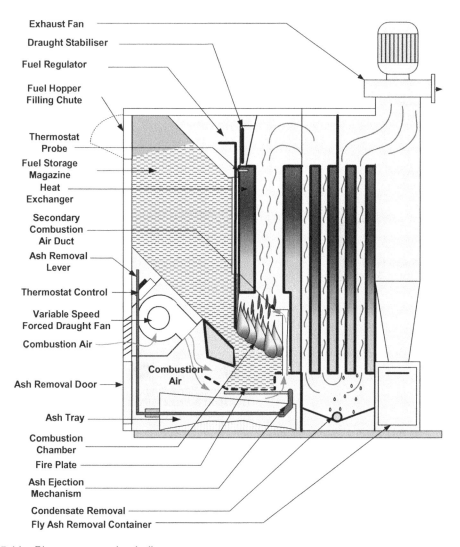

Exhaust Fan
Draught Stabiliser
Fuel Regulator
Fuel Hopper
Filling Chute
Thermostat
Probe
Fuel Storage
Magazine
Heat
Exchanger
Secondary
Combustion
Air Duct
Ash Removal
Lever
Thermostat Control
Variable Speed
Forced Draught Fan
Combustion Air
Ash Removal Door
Ash Tray
Combustion
Chamber
Fire Plate
Ash Ejection
Mechanism
Condensate Removal
Fly Ash Removal Container

Combustion
Air

Figure 15.11 Biomass magazine boiler

The main differences in storing, handling and burning coal type products and biomass solid fuels are that the biomass is less dense and therefore lighter than coal type products, and the combustion of biomass fuels derived from wood produces fly ash. These differences require a different approach in handling the fuel and the prevention of fly ash from being released into the atmosphere, which would cause a nuisance to neighbouring properties.

The method of transporting biomass fuel in pellet form, and to a lesser extent in wood chip form, is by the Archimedean screw method, also developed from coal type solid fuel appliances. However, biomass in pellet form is better delivered by pneumatic means, which eliminates the problem of the Archimedean screw becoming jammed inside its casing.

Figure 15.12 illustrates a packaged biomass boiler, whereby the fuel is fed into the burner at a controlled rate by pneumatic means, which forces slugs of wood pellets along a supply tube to the burner at intermittent intervals. The rate of fuel supply is controlled by the rate of consumption, which is determined by thermostatic sensing. The compressed air is produced by an air compressor that stores it at pressure in an air receiver, which is then released at a rate controlled by the demand.

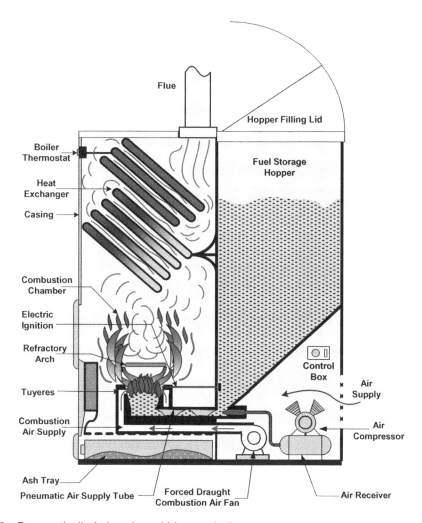

Figure 15.12 Pneumatically fed packaged biomass boiler

Fly ash is present in all flue gases when wood products are burnt, and if not dealt with this will cause a nuisance to neighbouring properties and contribute to localised air pollution. The fly ash in most cases is removed by incorporating a cyclone within the flue outlet that collects the ash at the base, where it can be removed periodically.

There are many variations of biomass boilers available and as can be observed from the drawings, they are developments of the coal burning solid fuel appliances that incorporate further advances in burner control. It will also be noticed that biomass solid fuel boilers still require a degree of manual handling of the fuel and ash removal.

BIO-OIL

The method of producing diesel gas oil from vegetable and mineral oils, including reclaiming it from used cooking oils, has been known for many years. Rudolf Diesel first demonstrated his engine designed to run on peanut oil long before it was adapted to function on hydrocarbon fossil oil.

The sustainable crops used as the source of bio-oil, or biodiesel as it is often referred to, are numerous, but are most commonly soybean, rapeseed, peanut oil, sunflower seed and hemp seed. The oil obtained from these crops is then subjected to trans-esterification, which is the process of exchanging the alkoxy group of an ester compound with another alcohol. These reactions are often catalysed by the addition of an acid base.

The bio-oil, after it has been processed, can be used as a fuel for heating by employing an oil-fired boiler equipped with an atomising type burner that has been adapted to use this oil, either in neat form, or as a mixture with hydrocarbon oil. See Figures 12.25 and 12.26 for details of the pressure jet atomising oil burner using this fuel.

The problem that is associated with producing bio-oil is that it requires enormous areas of arable land to cultivate these crops, which have to compete with the same land that is used to grow food crops. It should also be noted that the oil being produced is mainly aimed at the more profitable automotive industry to provide fuel for road vehicles, leaving little for the heating industry.

Ethanol is another bio-oil that has been known for many years and commercially used in some industrial manufacturing processes. It is extracted by the fermentation of sugars derived from sugar cane, sugar beet, wheat and corn, followed by distillation and drying. Ethanol is seen as a replacement for petrol/gasoline for automobiles, but at present there is no research or development to produce heating oils from these sources.

Ethanol is the red-coloured alcohol that is used in alcohol-based thermometers.

There are other liquid biofuels such as butanol, methanol and other alcohol sources – including the production of biodiesel, ethanol and methane from micro-algae – but none of these are yet a serious contender for producing a fuel for heating.

BIOGAS

Alternative gases to natural gas are discussed in Chapter 13, Natural Gas Firing, but none of those included can be considered as being sustainable.

The production of biogas is normally by the process of anaerobic digestion of organic material from sources such as biodegradable waste material, manure, high-energy crops and waste from lanfill sites. The organic material is fed into anaerobic digesters to accelerate the gas production.

The gas produced by this method is predominantly methane, which – as established in Chapter 13 – is a high calorific value fuel and, as there are numerous supply material sources available, this is a method that has great potential. However, although the production of methane as a bio-gas is a sustainable fuel, it is not carbon neutral and if it escapes into the atmosphere it is a potent greenhouse gas.

HEAT PUMPS

The technology of heat pumps has been known for many years, but more recently the need to develop more energy efficient heating systems has led to the wider use of heat pumps, and the quest to improve on their application.

A heat pump is a thermodynamic device that uses a refrigeration cycle to extract low-grade/-temperature heat from a source and eject a higher temperature heat to a sink. In simple terms, it is said that heat is moved from one place to another by being absorbed by a substance and conveyed to the point of delivery by that substance, where it is released. The intermediate substance used for conveying the heat is known as a refrigerant.

By this definition, all pieces of refrigeration equipment, air conditioners and chillers employing the refrigeration cycle are heat pumps, but in engineering, the term 'heat pump' is generally reserved for equipment that supplies heat as a beneficial end product, rather than that which removes or rejects heat, as in the case of air conditioning cooling systems. However, some heat pumps can be obtained that have a dual-mode service, whereby they can be reversed from supplying heat in winter to provide summer time cooling. This should only be done if the system has been designed to provide cooling, as a standard heating system employing radiators and where there is no provision to collect condensate would not be suitable.

The aim of the heat pump within a heating system is for it to replace the conventional boiler as the source of heat generation. The energy source used to power the heat pump is electricity. The difference between an electrically powered heat generator/boiler and a heat pump, is that the heat pump produces a greater amount of heat energy than the electrical energy consumed by the compressor motor, pump or fan. For this reason, the cost of heat provided by a heat pump is less than that of the heat produced by a boiler fired by any of the fossil fuels, even though the cost of electricity is greater than oil, gas etc. It should also be appreciated that the carbon dioxide produced nationally by the power station using fossil fuels to produce the electrical energy required by the heat pump, is less than that produced by a gas- or oil-fired boiler of an equal heat output. Heat pumps are therefore considered a means of reducing the carbon emission and are recognised by the UK government as a renewable energy technology, although as they are electrically powered they do not use sustainable energy unless the method of power generation does.

Heat pumps use either:

- The vapour compression cycle, or
- The absorption refrigeration cycle

Figure 15.13 illustrates the principles of the vapour compression cycle, which is the most common arrangement employed for heat pumps and used almost exclusively for heat pumps in domestic applications.

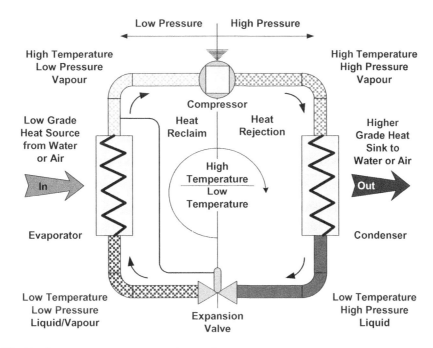

Figure 15.13   Heat pump vapour compression cycle

The thermodynamic cycle consists of four basic components that operate through four basic stages. The components comprise a compressor, condenser, expansion device and evaporator. Superheated refrigerant vapour with low temperature at low pressure enters the compressor where it is compressed to a much higher pressure, and temperature and passed on to the condenser. The condenser is the internal supply heat exchanger which is arranged to give up its high temperature heat to the internal environment, by either water in the case of a hydronic heating system, or air in the case of a warm air heating system. The result of this heat exchange in the condenser is to change the refrigerant vapour into a sub-cooled liquid still at high pressure, which passes onto the expansion device, sometimes referred to as a throttling valve. Then it passes through the valve, which causes a pressure drop to the refrigerant together with an accompanied drop in temperature, so that the refrigerant leaves the expansion device and enters the evaporator as a low pressure, low temperature mixture of liquid and vapour. In this state the refrigerant passes through the evaporator where it absorbs heat from the low temperature source – water or air – before returning as a superheated refrigerant vapour back to the compressor for the cycle to continue.

The absorption refrigeration cycle differs from the vapour compression cycle as it is heat operated and a secondary absorbent fluid is used to absorb the primary gaseous refrigerant that has been vaporised in the evaporator. The equipment is much larger than for the vapour compression cycle, but as it requires a heat input from a secondary source, such as a high temperature exhaust from an engine or burner, it is only considered feasible when that heat is freely available at low cost.

HEAT PUMP REFRIGERANT

The heat pump uses a medium for the refrigeration cycle that is known as a refrigerant. This is a fluid that has the ability to absorb heat from one area and deposit it at another. Many different refrigerants have been used for this purpose over the years. Most refrigerants have been chlorofluorocarbons (CFCs), which have been shown to contribute to the depletion of the Earth's ozone layer and were required by the Montreal Protocol to be phased out by 2010, although the European Union countries phased them out in 1995. Hydrochlorofluorocarbons (HCFCs) also contribute to the depletion of the ozone layer but at a lesser extent than CFCs, and are also scheduled to be phased out by 2030 for developed countries.

Hydrofluorocarbons (HFCs) are said not to be ozone depleting but have been shown to be associated with global warming if allowed to escape into the atmosphere, and they are now being widely adopted as substitute refrigerants. For these reasons, anyone involved in the installation, maintenance, charging or de-gassing of refrigerant systems must be deemed to be a competent person holding a valid approved qualification.

Packaged heat pump units are normally charged with R-134A, R-407C or R-410A, which are HFC refrigerants that are not at present regulated by the Montreal Protocol, but can be assessed by individual governments at any time.

The UK government's position on the use of these refrigerants has been to recognise their potential environmental issues, but to permit their use under controlled conditions providing that there are no suitable alternatives available.

Table 15.2 lists the characteristics of these three refrigerants.

R-134A, R-407C and R-410A are also classified as being low in toxicity levels and non-flammable, making them relatively safe to use under their specified controlled conditions.

HEAT PUMP TYPES AND APPLICATIONS

Heat pumps are classified by their heat source together with their heat sink.

The heat source is the low grade heat that is available for input into the heat pump as depicted in Figure 15.13; this is normally obtained from either air or water through the evaporator. The heat sink is the higher

Table 15.2   Properties and characteristics of common refrigerants

Refrigerant Number	Chemical Name or Composition	Chemical Formula	Freezing Point °C	Boiling Point °C
R-134A	Tetrafluoroethane	CH_2FCF_3	−103	−26
R-407C	R-32/125/134A (23.0/25.0/52.0)	N/A		−43
R-410A	R-32/125 (50.0/50.0)	N/A		−51
R-32	*Difluoromethane*	CH_2F_2		
R-125	*Pentafluoroethane*	CHF_2CF_3		

R-32 and R-125 are not permitted to be used in a 100% concentration or used on their own.

Table 15.3   Heat sources and sinks

Heat Source	Heat Sink
Air	Air
Air	Water
Water	Air
Water	Water

temperature heat that is given up by the condenser as useful heat for input into the heated space – also depicted in Figure 15.13 – and it also employs either air or water, as shown in Table 15.3.

AIR SOURCE

Air as a heat source is a variable temperature medium that is dependent upon seasonal change, altitude, exposure and geographical location – making it difficult to assess the heat quantity that can be obtained. Air is the most widely used heat source for heat pumps for residential dwellings and small commercial premises as it is suited to more buildings than water is, and can have a lower installation cost. When selecting or designing an air source heat pump installation, three main factors should be considered:

1. Is there a suitable external wall available for mounting the external evaporator?
2. What is the local external temperature range expected annually?
3. Is there a risk of frost formation to the external refrigerant pipe coils?

The external wall must be capable of supporting the unit and comply with any specific requirements that the manufacturer of the unit may have.

The external ambient air temperature is more difficult to assess for the reasons previously stated. In the UK the temperature range over the heating season varies from −10°C to 15°C with an average of 12°C, but as the external temperature decreases, the heating capacity of the heat pump decreases, therefore the heat pump must be sized for as low a temperature balance as practical for heating. Typically, the surface area of the external evaporator coil is between 50% to 100% larger than the internal condenser coil.

During the heating season the temperature of the evaporating refrigerant is generally 6–12°C lower than the external ambient air temperature, but when the surface temperature of the external evaporator coil is 0°C or less, with a corresponding external air dry-bulb temperature of 2–5°C higher, frost may form on the coil surface and between the convective fins, restricting the air flow over the evaporative pipe coil and consequently reducing the efficiency of the heat pump. For this reason, a means of defrosting the pipe coil must be incorporated within the unit, to maintain the operating efficiency whilst consuming additional imported energy.

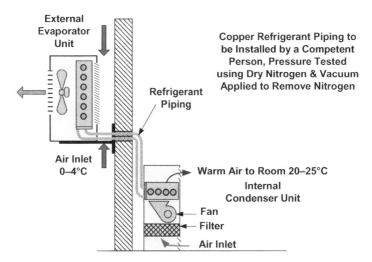

Figure 15.14   Split-unit type air-to-air heat pump

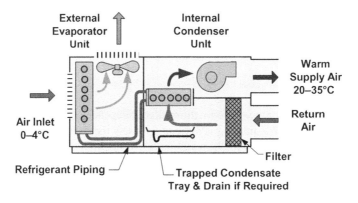

Figure 15.15   Self-contained packaged air-to-air heat pump

AIR-TO-AIR HEAT PUMPS

This type of heat pump is the most commonly employed in either a packaged self-contained unit or a split unit, with heat outputs ranging from 3 kW for domestic units to 44 MW for large commercial units.

Figure 15.14 illustrates the basic arrangement of a split-type air-to-air domestic heat pump unit arranged for heating only. If a dual reversible heating/cooling unit was used, a slightly different configuration of the internal condenser unit would be required, incorporating an internal condensate tray with trapped condensate drain. These domestic versions now incorporate measures to make them more suitable for people who are allergic to various air-borne pollens.

The emission temperature of these air-to-air domestic heat pump units can be higher, depending upon unit type and external ambient air temperature.

Figure 15.15 shows a self-contained air-to-air heat pump unit. This has a larger heat output capacity and is connected to a network of distribution ducting to take the place of a warm air heater unit. This type of unit has the advantage of being supplied as a complete package with the refrigerant pre-charged; the

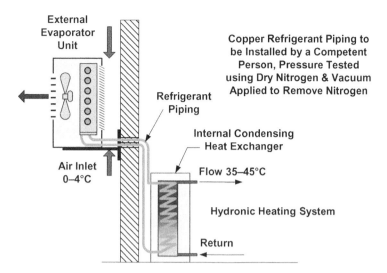

Figure 15.16 Split-unit type air-to-water heat pump

refrigerant unit does not require any work on-site. However, these units are fairly large and require a suitable external wall to obtain the air supply source.

Air-to-air heat pumps are ideal replacements for warm air heaters as they generate an air sink temperature that can be issued directly into the heated space, and they do not require any form of supplementary heating.

AIR-TO-WATER HEAT PUMPS

Sometimes referred to as heat pump water heaters, these differ only from the air-to-air heat pumps by the replacement of the air heat exchanger condenser type heat sink with a water heat exchanger condenser type heat sink. They are normally commercially available as unitary pieces of equipment, or split-type units where the internal condensing heat exchanger takes the place of the traditional fossil-fuelled boiler. The principles of operation are illustrated in Figure 15.16.

The flow temperature achieved leaving the heat pump to the hydronic heating system is around 35–45°C, although under certain conditions a flow temperature as high as 65°C has been claimed. This operating temperature makes this arrangement suitable for providing the primary flow of the heating water for underfloor heating systems without the need to incorporate any supplementary supporting heat source, such as a conventional fossil-fuelled boiler. For a hydronic heating system employing conventional radiators and convectors, the flow temperature will need boosting to raise it up to 82°C: this is accomplished by incorporating an additional heat raising source as illustrated in Figure 15.17.

In this arrangement the heat pump serves as the lead boiler with a supporting boiler fired by a fossil fuel, which is arranged to operate when the flow temperature needs to be raised to satisfy the heating system. The return temperature should be designed to return to the heat pump at the minimum temperature possible for the system concerned: normally a design return temperature of 25–30°C is chosen for this purpose to obtain the maximum efficiency from the heat pump.

The diverting valve shown is controlled to open to the supporting boiler when additional heat is required, thus closing to the heat pump to force the heating water to be circulated from the lead boiler heat pump through the supporting boiler. When the system operating temperature is satisfied, the diverting valve is then arranged to close to the supporting boiler, which should also be interlocked to prevent further firing

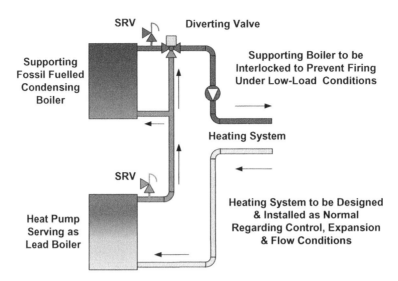

Figure 15.17   Heat pump with supporting conventionally fuelled boiler

under this condition. The heating system is then heated directly by the heat pump until the control system calls for additional heat again.

This arrangement makes use of the maximum amount of heat raised by the heat pump to minimise the fossil fuel used, therefore conserving fuel and reducing the amount of carbon being released into the atmosphere.

Air-to-water heat pumps have been used successfully on larger domestic and commercial heating systems, particularly when swimming pools are involved, and have resulted in more efficient operations – but they have had a poor history for most domestic applications, although developments continue to improve their performance.

These types of heat pumps cannot be used as a direct replacement for conventional boilers on their own; the heating system must be designed to operate at the different flow temperatures that will be experienced.

Air-to-water heat pumps, as with all heat pumps, are physically larger than conventional boilers. This, coupled with high installation and servicing costs, a history of poor reliability and complications when converting an existing heating system, has contributed to them not appearing to be an attractive alternative form of energy.

WATER SOURCE

Water in most cases is a better source for obtaining heat in countries that experience a temperate climate as it generally returns a higher sink temperature than air, but there is a higher installation cost to pay for this improved energy return.

The water can be obtained from natural sources such as rivers, lakes or groundwater, or from man-made structures such as reservoirs, ponds. Heat pump installations from surface water sources such as rivers, lakes and man-made structures are less expensive to install if they are available, but they will also experience a variable temperature range as an energy carrying medium, similar to that of air. The variable temperature of water will also be dependent upon the same seasonal change, altitude, exposure and geographical location factors as experienced by air, with the same difficulties in assessing the heat quantity that can be obtained. Surface sourced water will also be subjected to freezing conditions during certain times of the year.

Water obtained from ground source heat pumps has the advantage of being at a higher temperature than surface water sources and at a reasonably constant temperature, that does not vary significantly throughout the year, making it less difficult to assess the potential energy available.

The methods of harnessing the potential heat from ground sources vary; they can be open loop or closed loop extending down vertically dug wells to depths in the ground beyond the water table, or they can be laid closer to the ground surface and extend horizontally below the ground surface level in a continuous pipe loop. Water taken directly, or indirectly from the water undertakers supply main is neither desirable on cost grounds, nor permitted by the water undertakers.

The Environment Agency or corresponding agencies in other parts of the UK should be consulted during the planning stage before sinking a bore-hole or well, as a water abstraction licence may need to be obtained, even when the water being extracted is returned to the ground, or when it is via a closed loop.

When selecting a ground source water heat pump, the following must be considered:

- Is there a surface water source available?
- Is it practical to sink a bore-hole well to a depth to reach the water table?
- Is there sufficient ground area available to install a closed loop ground coil?

The constraints that have to be overcome for water source heat pumps are more arduous than those required for air source heat pumps, but the increase in performance can make it worthwhile if the above requirements can be met.

WATER-TO-AIR AND WATER-TO-WATER HEAT PUMPS

The heat sink side of the water source heat pump, be it hydronic or warm air, is the same as that for an air source heat pump, which is also the same regardless of what fuel or energy source is utilised.

Figure 15.18 illustrates the two typical surface water heat source heat pump arrangements.

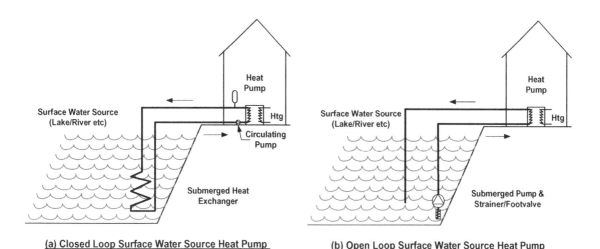

(a) Closed Loop Surface Water Source Heat Pump   (b) Open Loop Surface Water Source Heat Pump

Figure 15.18 Surface water source heat pump arrangements

Figure 15.18(a) shows a closed loop arrangement where a brine solution (water and anti-freeze solution) is circulated as a medium in a closed circuit through a submerged heat exchanger to the evaporator section of the water-to-water heat pump, and is complete with its own sealed expansion vessel and safety relief valve facilities. This method is the one most commonly used for a surface water heat source, as it is less affected by freezing conditions of the surface water source, but precautions must be taken to prevent scale build-up within the circulating piping.

Figure 15.18(b) shows the alternative arrangement of an open loop piping system, where the surface water source is also the medium being circulated directly to the water-to-water heat pump and returned back to the surface water source. This arrangement is more prone to being affected by freezing surface water conditions, with all metallic materials that are in contact with the water being circulated unprotected against corrosion.

Water-to-water heat pumps utilising a surface water heat source are more suitable for industrial applications where the industrial process concerned requires them to have their own storage reservoir of water, although if it is serving as a cooling reservoir, it would be more beneficial to recover the waste heat before the water is returned to the reservoir.

As most domestic residential dwellings do not have a convenient lake, river or other form of surface water source, this method is seldom used for domestic heating applications. Ground source applications are usually chosen in a unitary form as being a more practical arrangement for these dwellings.

GROUND SOURCE HEAT PUMPS

Ground source heat pumps (GSHPs) are classified as a form of water source heat pumps as they use either ground water as the heat source, or they circulate a water/anti-freeze/corrosion inhibitor through a closed loop. They can also be either a vertical deep bore-hole or a shallow continuous pipe loop laid horizontally below the ground surface.

Figure 15.19 illustrates the general arrangement for an open loop ground source heat pump which comprises two bore-holes spaced a minimum of 7 metres apart. The depth of the bore-holes should be enough to extend down below the water table level when it is at its lowest level; normally the maximum considered practical limits the depth to 180 metres. It should be noted that the deeper the bore-hole is, the higher the potential source temperature will be.

A pump, with an extended drive and submersible impeller, is placed down one of the bore-holes, that is also equipped with an inlet strainer to prevent solid particles from being drawn into the pump casing. This pump draws water from the ground water and passes it through the evaporator section of the heat pump before returning back down the other bore-hole to its original source.

The open loop ground source heat pump is expensive to install as two bore-holes have to be drilled to the required depth, but it can return a heat sink temperature of up to 60°C.

Although the water is not not for consumption and is being returned to its source, an abstraction licence is still required before the bore-holes can be drilled.

The bore-holes should also be lined with a casing to prevent ground collapses on to the pump boosted main or return piping.

Figure 15.20 illustrates the general arrangement for a closed loop ground source heat pump, which differs from the open loop arrangement by only requiring a single bore-hole to be drilled. A continuous high-density polyethylene tube with a diameter of between 25 mm to 40 mm is passed down the bore-hole to form a return loop at the bottom of the bore-hole, with the space between the high-density polyethylene tubing and the bore-hole filled with a grout of a type approved by the Environment Agency for this use. The piping is connected to the evaporator section of the heat pump to form a continuous sealed loop. Any joints must be formed by the electro-fusion welding process and be above ground.

The closed circuit is charged with a medium referred to as brine, which is water with a mixture of anti-freeze and corrosion inhibitor added to it. The additives should be selected to have the minimum environmental impact in the event of an accidental leakage.

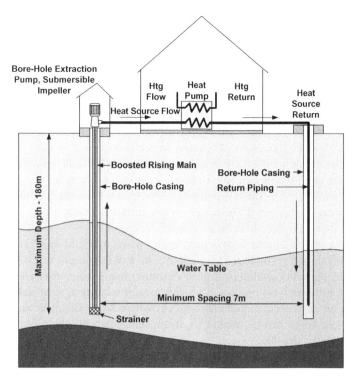

Figure 15.19 Open loop ground source heat pump arrangement

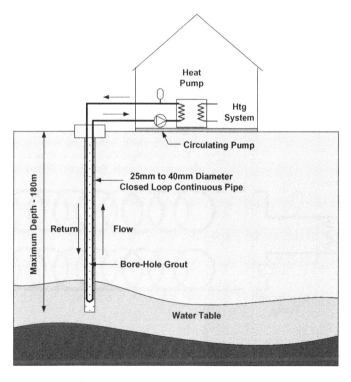

Figure 15.20 Closed loop ground source heat pump arrangement

This solution is continuously circulated around the loop where it absorbs low grade heat from the ground (and water, if applicable: this arrangement does not require the circuit to be submerged in ground water). The medium is then delivered into the evaporator section of the heat pump to be processed into useful higher grade heat for the heat sink, in the form of an air or water heating system.

The open and closed deep bore-hole heat pump arrangements depicted in Figures 15.19 and 15.20 are expensive to install and require a suitable piece of land that has access for a bore-hole drilling rig to enter the site: this is usually more practical for new-build developments than it is for existing properties, where access may be restricted. A less expensive arrangement is the use of a shallow ground source heat pump circuit as illustrated in Figures 15.21 and 15.22; here a trench is excavated to allow the piping to be laid in the trench in a coil form, either horizontally as shown, or arranged vertically in a narrow trench. The recommended depth of the trench varies; in North America the practice is to lay the pipe coil at a depth of between 1 m and 3 m, whilst here in the UK shallower depths have been used, as little as 0.6 m deep. The deeper the trench that can be excavated, the better the performance that can be returned in temperature achieved, but it is not always possible to excavate deeper trenches.

Figure 15.21 depicts a small diameter high density polyethylene tube laid in a continuous coil to form a circuit, where the brine solution is circulated through the coil and evaporator section of the ground source heat pump. Joints in the pipework should be kept to a minimum and where they are required, they should be formed by the electro-fusion weld process to provide greater security against leakage.

Figure 15.22 shows an alternative piping arrangement where larger diameter and slightly less flexible high density polyethylene tubes can be used. This arrangement permits the pipes to be laid in straight continuous lengths at similar depths.

The advantage of using larger pipe diameters is that the frictional resistance encountered by the flow will be less than that for the smaller diameters, and therefore the pipes can be installed in overall total longer lengths for the circuit.

Table 15.4 can be used as a guide to determine maximum pipe circuit lengths.

The shallow ground source heat pump arrangements are less expensive than the deep bore-hole method and the equipment used to excavate the trench can be more accessible in confined spaces, but they do require larger areas of land to lay the pipe circuits. This constraint may not be a problem for existing dwellings that were built a few years ago, but the present trend of building dwellings with very little land around them, prohibits the option of this form of heat pump.

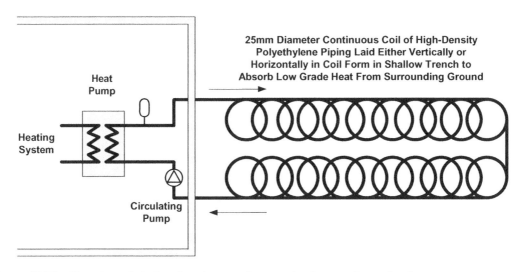

Figure 15.21 Plan view of shallow trench ground source heat pump, looped pattern

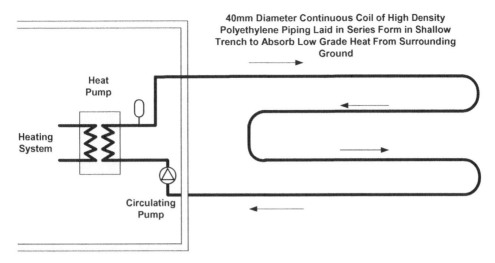

Figure 15.22   Plan view of shallow trench ground source heat pump, series pattern

Table 15.4   Pipe circuit lengths

Pipe Diameter mm	Max Circuit Length m
25	100
32	250
40	450

Some shallow trench heat pumps have been developed using the direct expansion method (DX), where refrigerant replaces the brine in the buried pipe circuit, but care has to be taken to ensure the integrity of the piping system.

SOLAR ENERGY HEAT PUMPS

Solar energy as a heat source for heat pumps has been used in some applications, with the advantage of achieving a higher temperature when solar energy is available. It can be used in combination with the main sources of air or water.

Solar energy heat pumps are either direct or indirect in their mode of operation. In a direct system a refrigerant replaces the brine with the solar collector panels arranged to incorporate refrigerant evaporator tubes within them.

An indirect system differs whereby either water or air is circulated through the solar collector panel, and is then delivered to the evaporator section of the heat pump.

Solar heating as an alternative energy source is discussed in further detail in Chapter 20.

COEFFICIENT OF PERFORMANCE (COP)

The effective performance of heat pumps is expressed as the COP rating, which is an abbreviation for the coefficient of performance obtained under controlled conditions. The coefficient of performance is defined as the ratio of heat transferred at the evaporator section of the heat pump to the energy used to power the compressor element of the heat pump, expressed in watts/watts.

In theoretical terms this is obtained by:

$$COP = T_1/(T_1 - T_2)$$

Where:

T_1 is the condensing temperature of the thermodynamic cycle in degrees Kelvin.
T_2 is the evaporating temperature of the thermodynamic cycle in degrees Kelvin.

In simplistic terms, the coefficient of performance may be stated as being the amount of electrical energy input into the heat pump, compared to the amount of heat energy output into the heating system. A heat pump having a stated COP rating of 3 means that for every one kilowatt of electrical energy input into the heat pump, it will output three kilowatts of heat into the heating system. However, for an accurate coefficient of performance the electrical energy input should include the consumption of electricity by all the appropriate electrical auxiliaries required to operate the heat pump, and not just the compressor unit.

Most domestic sized heat pumps have a stated coefficient of performance ranging from 2 to 4, whereas larger commercial sized heat pumps have a much larger COP range.

HEAT RECOVERY

Heat recovery systems utilise waste heat from buildings or processes by recovering part of the heat for putting back into the building. Heat pumps are not classified as heat recovery systems as the heat that they generate has not been raised previously, therefore it cannot claim to recover the heat.

Typically, heat recovery systems employ some form of heat exchanger which is arranged to recover heat being expelled from the building via extract ventilation systems, or hot exhaust gases that have been produced by some form of combustion process or heat producing machinery.

This is not an alternative form of energy, but it will increase the overall performance efficiency of the building, which will reduce fuel consumption and therefore CO_2 emissions.

16

Combustion, Flues and Chimneys

The subject of combustion has been included in previous chapters on the individual fuels and their method of burning. As a natural extension to this, it is important to have an understanding of the science involved, and the methods of removing the products of combustion safely from the appliance and building.

In the past, particularly for domestic boiler applications, there has been a tendency to look upon the inclusion of the flue as a necessary evil and this matter has not been given the due respect that it deserved. More recently, however, those involved in the design and installation of fuel burning appliances have had to learn about this subject, in order to comply with current requirements for fuel efficiency and safety legislation.

It has been established that all fuels that burn require oxygen in sufficient quantity for the combustion process to be successfully completed and to allow energy, in the form of heat, to be released from the fuel.

The ideal and most efficient solution would be to supply pure oxygen to the burner – which would result in smaller burners and appliances – but this would not be practical or cost effective as it is easier and cheaper to obtain this oxygen from the air which is both freely and readily available.

AIR

This is the invisible and odourless substance that surrounds our planet up to an altitude of 11 000 metres, at which point the troposphere ends and the stratosphere begins. This invisible and odourless gas is in fact a mixture of base gases, sometimes referred to as a cocktail of gases, which usually includes some degree of impurities and contaminants.

Table 16.1 lists the composition of dry air together with the components' proportions to each other. Dry air is free from any contaminants such as smoke, pollen, water vapour etc. It can be observed, from the simplified diagram in Figure 16.1, that oxygen represents only approximately ⅕ of the total unit volume of air available, which equates to approximately $5\,m^3$ of air needing to be supplied to the burner in order

Heating Services in Buildings: Design, Installation, Commissioning & Maintenance, First Edition. David E. Watkins.
© 2011 John Wiley & Sons, Ltd. Published 2011 by John Wiley & Sons, Ltd.
As an aid to lecturers and students, full colour versions of the figures in this chapter may be found at www.wiley.com/go/watkins
These figures are © 2011 by John Wiley & Sons, Ltd.

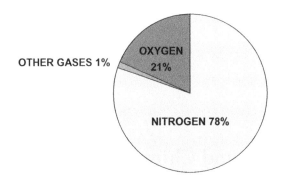

Figure 16.1   Simplified diagram of the composition of air

Table 16.1   Composition of dry air

Constituent	Symbol	Percentage by Volume	Molecular Weight	Boiling Point °C
Nitrogen	N	78.084	28.01	−195.8
Oxygen	O	20.9476	32.00	−183
Argon	A	0.934	39.95	−185.8
Carbon Dioxide	CO_2	0.0314	44.00	−78.5
Neon	Ne	0.00182	20.18	−247
Helium	He	0.00053	4	−269
Methane	CH_4	0.00016	16	−164
Sulphur Dioxide	SO_2	0.00015	64	−10
Hydrogen	H	0.00013	2	−253.1
Krypton	Kr	0.0001	–	−169
Xenon	Xe	0.0001	–	−140
Ozone	O_3	0.00001	–	−112

to provide $1\,m^3$ of oxygen. Therefore, burners have to be constructed to allow for the remaining ⅘ of the air to pass through the burner without contributing to the combustion process.

COMBUSTION AND COMBUSTION REACTION

Combustion is defined as a chemical reaction where an oxidant, usually oxygen, reacts rapidly with a fuel to release thermal energy in the form of high temperature gases. It will also release small amounts of electromagnetic energy in the form of light, mechanical energy in the form of noise and electric energy in free ions and electrons.

When the combustion process takes place at temperatures of between 400°C and 600°C, a reaction occurs as shown in Table 16.2, where the chemical constituent of the combustion input is converted to a different chemical constituent of the output in proportion to their molecular weight. Basically, carbon is converted to carbon dioxide with small amounts of carbon monoxide; hydrogen is converted to form water; methane, propane and butane convert to varying proportions of carbon dioxide and water; and sulphur converts to sulphur dioxide, which may combine with further oxygen to form sulphur trioxide. In addition, there may be small amounts of ash and other impurities present, depending upon the type of fuel.

The nitrogen and other inert gases content of the air just pass through the combustion process and play no part in it; therefore they are not considered to be products of combustion when they are expelled through the flue.

Table 16.2 Stoichiometric combustion reactions of common fuels

Constituent	Symbol	Combustion Reaction	Oxygen/Air Requirements			
			kg/kg Fuel		m³/m³ Fuel	
			O_2	Air	O_2	Air
Carbon	C	$C + O_2 = CO_2$ $(12 + 32 = 44)$	2.7	11.6		
Hydrogen	H_2	$2H_2 + O_2 = 2H_2O$ $(4 + 32 = 36)$	8	34.3	0.5	2.4
Methane	CH_4	$CH_4 + 2O_2 = CO_2 + 2H_2O$ $(16 + 64 = 44 + 36 = 80)$	4	17.3	2	9.7
Ethane	C_2H_6	$C_2H_6 + 3.5O_2 = 2CO_2 + 3H_2O$ $(30 + 112 = 88 + 54 = 142)$	3.8	16	3.5	16.8
Propane	C_3H_8	$C_3H_8 + 5O_2 = 3CO_2 + 4H_2O$ $(44 + 160 = 132 + 72 = 204)$	3.7	15.7	5	24
Butane	C_4H_{10}	$C_4H_{10} + 6.5O_2 = 4CO_2 + 5H_2O$ $(58 + 208 = 176 + 90 = 266)$	3.6	15.5	6.5	31.2
Ethylene	C_2H_4	$C_2H_4 + 3O_2 = 2CO_2 + 2H_2O$ $(28 + 96 = 88 + 36 = 124)$	3.4	14.8	3	14.4
Sulphur	S	$S + O_2 = SO_2$ $(32 + 32 = 64)$	1	4.3		

Stoichiometry – The fixed rational numerical relationship between the relative quantities of substances in a reaction or compound.

The chemical combustion equations given in Table 16.2 are based on the assumption of complete combustion being achieved. The numbers beneath the chemical equations for each fuel are the molecular mass for each compound, multiplied by the number of units concerned where applicable, hence it can be observed that each side of the equation balances out both chemically and numerically.

To explain this further and using natural gas as an example, by referring back to the properties for natural gas given in Table 13.1, we see that natural gas is predominantly methane – with an average value of 92% for North Sea natural gas and the remaining 8% consisting of other hydrocarbon products with some inert gases. We can therefore confidently use the chemical equation given for methane in Figure 16.2.

Also, by referring back to the operating properties in Table 13.2, we can see that one cubic metre of natural gas requires $9.73\,m^3$ of air for complete combustion to occur. Also by referring to Figure 16.1 and Table 16.1, oxygen can be seen to form approximately 21% of that air and equate to two cubic metres.

Using the above information, the chemical combustion equation for methane, is $CH_4 + 2O_2 = CO_2 + 2H_2O$, as demonstrated in Figure 16.2.

Methane comprises one carbon atom bound to four hydrogen atoms. When this is added to two oxygen units, each consisting of two oxygen atoms, making a total of four oxygen atoms, the combustible mixture of $CH_4 + 2O_2$ is achieved. This gives a total of one carbon atom, four hydrogen atoms and four oxygen atoms to this combustible part of the equation.

When this combustible mixture is ignited and the thermal energy liberated, the chemical equation is converted and the single carbon atom becomes bound to two of the oxygen atoms to form carbon dioxide (CO_2), and the four hydrogen atoms become bound to the remaining two oxygen atoms to form two units of water (H_2O), thus converting the equation to $CO_2 + 2H_2O$. It can be seen that there are still one carbon atom, four hydrogen atoms and four oxygen atoms, so they still exist but have been rearranged into a different order. Therefore the ingoing combustible mixture will equal the outgoing products of combustion, both having a total molecular mass of 80.

The result of this equation is that one cubic metre of natural gas combined to two cubic metres of oxygen will be converted, upon complete combustion, to one cubic metre of carbon dioxide together with two

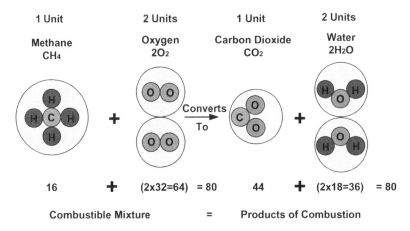

Figure 16.2   Stoichiometric chemical equation for the combustion of methane

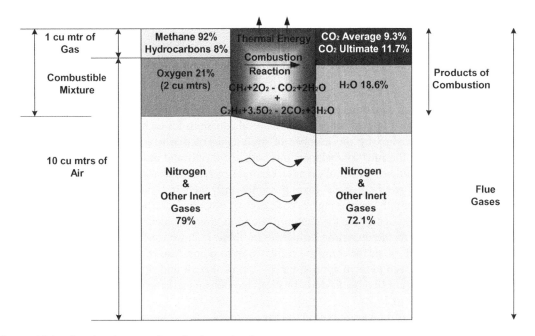

Figure 16.3   Combustion reaction chart – natural gas

cubic metres of water in the form of vapour before it condenses. The remaining $7.73\,\mathrm{m}^3$ of air, consisting primarily of nitrogen and minute amounts of other inert gases, does not form part of the combustion process and just passes through without being altered.

Having arrived at the volumetric quantities required for the products of combustion to be expelled from the flue, it is necessary to convert these figures into a percentage value to assess if complete combustion is being achieved. (The subject of combustion efficiency is discussed in Chapter 17.) Reference can be made to the combustion reaction charts depicted in Figures 16.3 to 16.6, which present the combustion reactions pictorially. It should be noted that these combustion reaction charts are based on typical values for each

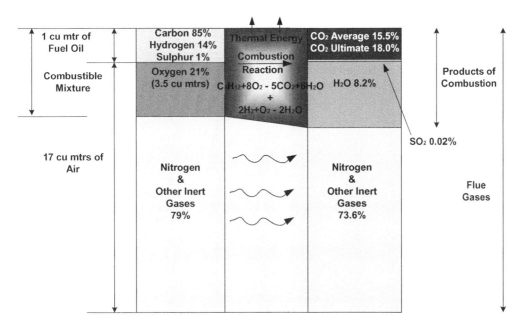

Figure 16.4   Combustion reaction chart – typical fuel oil

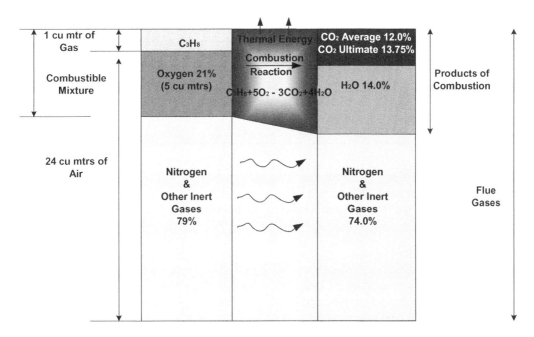

Figure 16.5   Combustion reaction chart – average propane gas

fuel and may vary. The carbon dioxide content indicated in the products of combustion column are stated both as the ultimate value – which is the stoichiometric value assuming complete combustion without any excess air – and the average value, which incorporates a degree of excess air to the burner.

It is extremely difficult, due to a number of reasons including manufacturing tolerances and variances in fuel supply, to provide the exact quantity of air to the burner for combustion purposes, therefore a certain

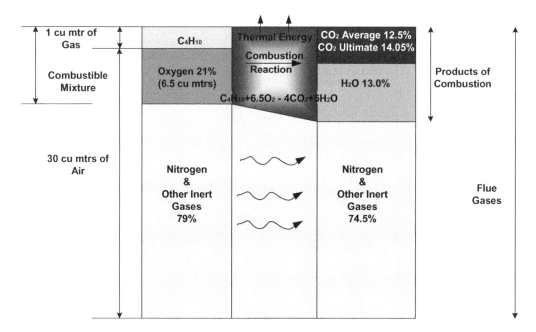

Figure 16.6   Combustion reaction chart – average butane gas

Table 16.3   Air requirements and CO_2 for complete combustion

Fuel	Combustion Air m³	CO_2 %
Fuel Oil	17	18.0
Natural Gas	9.8	11.7
Propane Gas	24	13.75
Butane Gas	30	14.05
Manufactured Gas	4.5	25.0

degree of excess air is provided to ensure that complete combustion will occur, which has the effect of diluting the carbon dioxide content.

Table 16.3 summarises the CO_2 and combustion air requirements for the common fossil fuels.

INCOMPLETE COMBUSTION

So far, this chapter has concentrated on the assumption that complete combustion has taken place. A simple indication of complete combustion is provided by the flame shape and colour (see Figure 16.7) where the flame colour and shape are said to be healthy.

The causes of incomplete combustion are:

- Insufficient air
- Incorrect burner pressures
- Flame impingement

Insufficient or vitiated air – or to be more precise, insufficient oxygen – is a common cause of incomplete combustion, normally as a result of inadequate or blocked combustion air ventilation to the room in which

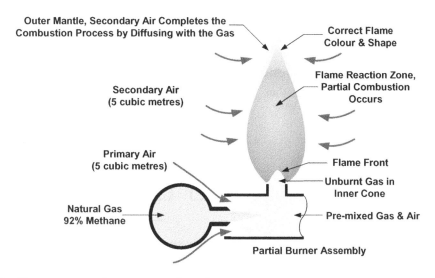

Figure 16.7 Flame structure for pre-aerated gas burner

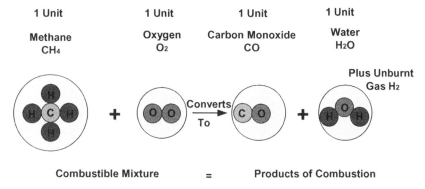

Figure 16.8 Chemical equation for incomplete combustion of methane

the boiler is located, or to the combustion air duct on a room sealed appliance. The result of this can be shown by re-examining the chemical equation for the combustion of methane but changing the oxygen supply by reducing it to 50% of that required.

From the revised chemical equation shown in Figure 16.8, it can be seen that with a reduced supply of oxygen the two units of carbon dioxide within the products of combustion have now been replaced by one unit of carbon monoxide plus water and unburnt fuel.

CARBON MONOXIDE (CO)

Carbon monoxide is a colourless, odourless, tasteless and toxic gas; it is a product of incomplete combustion resulting from all carbon-containing fossil fuels.

Carbon itself is not dangerous; it is a thermal contributing constituent part of all fossil fuels such as oil, natural gas and solid fuels, including wood. On complete combustion it is converted to carbon dioxide as part of the combustion reaction and discharged as a product of combustion through the flue. However,

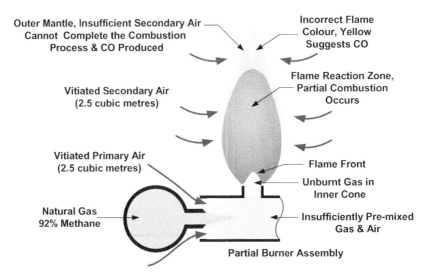

Figure 16.9   Flame structure for pre-aerated gas burner during vitiated air supply conditions

during incomplete combustion, carbon monoxide is produced – which is a useful gas in some industrial processes but a dangerous gas if discharged into a habitable space.

Carbon monoxide was a constituent part of the manufactured coal gas that preceded the supply and use of natural gas (see Table 13.3). It was the component part of this gas that made it toxic.

Having no odour, taste or colour, with today's improved standards for thermal insulation and tight fitting double glazed windows and doors, it has become increasingly important to have adequate living space ventilation, regular maintenance of all carbon fuel burning appliances and for the heated space to be equipped with CO detecting monitors to give both visual and audible alarms.

Figure 16.9 illustrates the expected flame structure during vitiated air conditions.

EFFECTS OF CARBON MONOXIDE POISONING

Mild carbon monoxide poisoning causes headache, nausea, drowsiness and inability to co-ordinate movement. Most people who have been exposed to mild carbon monoxide conditions will recover quickly when moved into the fresh air.

Moderate or severe exposure to carbon monoxide causes serious poisoning resulting in confusion, unconsciousness, palpitations, chest and stomach pains, shortness of breath, diarrhoea and eventual coma. Most victims will not recover from this stage of poisoning and it usually proves fatal. Those that do recover can be permanently affected with symptoms such as memory loss, poor co-ordination and delayed neuropsychiatric symptoms.

Because carbon monoxide does not have any odour, taste or colour it is particularly dangerous as an affected person will not recognise drowsiness as a symptom of poisoning, sometimes confusing it with flu. Consequently, someone developing mild carbon monoxide poisoning can fall asleep and continue to breathe in the contaminated air until severe poisoning or death occurs; hence, it is sometimes referred to as the 'silent killer'.

Carbon monoxide poisoning turns the skin pink and the lips bright red, so despite being asphyxiated, the victim doesn't turn blue as expected.

Table 16.4 lists these conditions both as an air dilution figure and as content in blood.

Table 16.4   Effects of carbon monoxide in air and percent saturation of haemoglobin

Volume of CO in Air %	ppm	Carbon Monoxide Effects on Adults	CO % Saturation in the Bloodstream
0.01	100	No symptoms up to 10%, slight headache develops within 2–3 hours.	**4% in 1½ hrs** **11–20%**
0.02	200	Mild headache, slight dizziness, palpitation, mild nausea and drowsiness after 2–3 hours.	**20–30%**
0.03	300	Flushed skin, breathlessness and increased palpitation on exertion.	**10% in 1½ hrs** **20% in 4 hrs** **30–35%**
0.04	400	Severe headache to front of head, nausea after 1 hour, risk of death after 3 hours.	**35–40%**
0.05	500	Symptoms as above with increased respiration and pulse rates, collapse on exertion.	**40–50%**
0.08	800	Severe headache, dizziness and nausea. Unconsciousness with possible fatal consequences and coma within 2 hours.	**50–65%**
0.16	1600	Severe headache, weakened heart and respiration, dizziness and nausea within 20 minutes Unconsciousness with fatal consequences and coma within 1–2 hours.	**65–70%**
0.32	3200	Severe headache, weakened heart and respiration, dizziness and nausea within 5–10 minutes. Unconsciousness with fatal consequences and coma within 15 minutes.	**70–75%**
0.4	4000	Severe headache, weakened heart and respiration, dizziness and nausea within 3–5 minutes. Unconsciousness with fatal consequences and coma within 15 minutes.	**60% within a few minutes** **75–80%**
0.64	6400	Severe above medical symptoms within 1–2 minutes. Fatal consequences within 15 minutes.	**80–85%**
1.28	12800	Fatal consequences within 1–3 minutes.	**85–90%**

When carbon monoxide is inhaled, it prevents absorption of oxygen into the body and can result in oxygen starvation.

This is because the oxygen-absorbing haemoglobin in human blood, that carries the oxygen from the lungs to the rest of the body, finds it 240 times easier to absorb carbon monoxide than oxygen. Therefore, carbon monoxide attaches itself to the haemoglobin and starves the human body of oxygen.

Any subsequent exertion increases oxygen demand and makes the problem worse, rapidly leading to collapse and potentially death. Children are at greatest risk.

CARBON MONOXIDE DETECTORS

Early CO detectors worked on the principle of detecting a lack of oxygen in the air regardless of what the reason for this was, but today detectors that sense the presence of carbon monoxide have been developed and proved quite reliable. They work on an electrochemical cell that raises an audible and visual alarm when the concentration of CO is sensed at a level of as little as 45 ppm. They can also be arranged to shut down any carbon burning appliance as an added safety measure. These detectors have a working life expectancy of five years.

INCORRECT BURNER PRESSURES

Incorrect fuel pressures, both over-pressure and under-pressure, can have the same result in achieving incomplete combustion.

Under-pressure, regardless of the burner being a natural draught or forced draught, will result in incomplete combustion with the likelihood of carbon monoxide gas being produced.

Over-pressure can result in the burner flame being lost and the production of NOx, which is discussed later in this chapter.

The remedy for this situation is to reset the burner to the correct operating pressure recommended by the burner manufacturer for the heat output required.

FLAME IMPINGEMENT

This normally occurs when the burner has been incorrectly fitted into the appliance. The burner flame will touch on a cold surface, cold being defined as being at a lower temperature compared to the flame, and cause chilling to the burner flame below its ignition temperature of 704°C for natural gas, 480–540°C for propane gas and 410–470°C for butane gas.

The effect of this chilling of the flame will be incomplete combustion, and the production of soot as well as carbon monoxide.

PRODUCT OF COMBUSTION STANDARDS

The minimum standards for the quality of the products of combustion from carbon burning appliances are based on the ratio of the amount of carbon monoxide as a percentage per volume in dilution, compared to the amount of carbon dioxide as a percentage per volume being produced. This is arrived at by using the following formula:

$$CO \div CO_2$$

For example, consider a carbon burning appliance producing 100 ppm of carbon monoxide and 11% carbon dioxide.

First convert the CO 100 ppm to a percentage, $100 \div 10\,000 = 0.01\%$.

Therefore

$$\frac{CO}{CO_2} \frac{0.01}{11} = CO/CO_2 \text{ Ratio } 0.0009$$

This subject and procedure is discussed in further detail in the following chapter on combustion efficiency.

It should be noted that most of the drawings and illustrations depicted in this chapter have been based on burners and combustion processes that employ natural gas (methane), but the information given can be taken for any carbon based fuel unless otherwise stated.

NITROGEN OXIDE

Nitrogen oxide (NOx) is a generic term applied to any binary compound of oxygen and nitrogen, or mixture of such compounds.

The chemical reactions that produce nitrogen oxides can create several different compounds, the proportions of which depend on the specific reaction and conditions in which they are formed.

The nitrogen compounds formed are shown in Table 16.5.

Table 16.5   Common nitrogen oxide compounds

Chemical Formula	Systematic Name	Common Name
NO	Nitrogen Monoxide	Nitric Oxide
NO_2	Nitrogen Dioxide	Nitrogen Peroxide
N_2O	Dinitrogen Monoxide	Nitrous Oxide
		Also known as laughing gas
N_2O_3	Dinitrogen Trioxide	Nitrous Anhydride
N_2O_4	Dinitrogen Tetroxide	–
N_2O_5	Dinitrogen Pentoxide	Nitric Anhydride
HNO_3	–	Nitric Acid
NH_3	–	Ammonia

The largest contributor by far of nitrogen oxides in the atmosphere is the automotive industry in the form of exhaust gases from vehicles powered by the internal combustion engine, but from the list of nitrogen oxide compounds given in Table 16.5, the most important forms that are of interest to the heating industry regarding the reaction of nitrogen in air, are nitrogen monoxide (NO) and nitrogen dioxide (NO_2). These are grouped under the generic term of NOx and are produced during the combustion process, particularly at high temperatures.

NOx

The nitrogen oxide gases of nitrogen monoxide and nitrogen dioxide, commonly referred to as NOx, are products of the combustion procedure of a carbon burning appliance. They contribute to the aggravation of asthmatic conditions, react with oxygen in the air to produce ozone – which is also an irritant – and form nitric acid when dissolved in water. This will eventually form acid rain which will damage trees and entire forest ecosystems. They also contribute to the formation of smog, commonly seen over major cities, although vehicle exhaust gases are the main contributor.

It has previously been established that a certain degree of excess air is designed to be supplied as part of the combustion air supply to carbon fuel burners, to prevent the formation of carbon monoxide. This excess air will contain both oxygen and additional nitrogen; the excess oxygen will not contribute to the combustion process but will pass through the combustion reaction to combine with the large volume of nitrogen also passing through the combustion reaction, and will form NOx gases. This is made possible by the high temperatures involved in the combustion process that make carbon based fuels suitable for providing the thermal energy to heating systems.

Therefore the former practice of providing up to 40–70% excess air to the burner is no longer acceptable and the burner must be set up accurately, to suit both the fuel being used and the heat output required.

FLUES AND CHIMNEYS

Having discussed the science of the combustion reaction that takes place within the combustion chamber of a boiler, together with the chemistry of the products of combustion being emitted, attention is now directed to conveying and discharging the products of combustion, as this has great importance with regard to the successful operation of an appliance.

Before proceeding, it is important to have an understanding of the purpose of flues and chimneys, together with knowledge of the critical factors that affect their performance.

The function of a flue or chimney has a dual purpose:

1. The most obvious purpose is to convey and discharge the products of combustion from the boiler in a safe, environmentally friendly and efficient manner, without causing a nuisance or endangering the safety of anyone, or causing damage or deterioration to the building or any other structure.
2. Equally important, but not always fully appreciated, is to assist in the supply of combustion air to the burner, as the appliance will not operate correctly if this is not achieved.

DESIGN PERFORMANCE FACTORS

There are a number of design factors that contribute to the successful performance of a flue or chimney, each is discussed individually, but they are all inter-related and in practice they must be considered together.

DRAUGHT

In order for a flue or chimney to achieve its two main functions, of expelling the products of combustion and assisting in the supply of combustion air to the burner, a draught must be created. This may be attained by the inclusion of a fan – as in the case of a forced draught burner or fan assisted flue – or by the method of flue construction and operation – as in the case of natural draught appliances – in either situation, the main objective is to create a draught of the correct magnitude in order for the appliance to function correctly and safely.

Natural draught is expressed as a difference in the pressure created by the hot, less dense flue gases within the flue and that of the colder and denser surrounding air. This difference is quite small if created by natural means and is measured in the unit Pa or $N\,m^{-2}$ ($249.1\,Pa$ or $N\,m^{-2} = 1$ inch wg or $2.491\,mb$). This pressure is always termed negative. Flues connected to forced draught or fan assisted draught appliances are not dependent upon creating the draught by natural means, and the fan can be sized to overcome the frictional resistance that will be encountered.

COMBUSTION AND VENTILATION AIR

The cycle of expelling the products of combustion and drawing in air for combustion will be continuous when the burner is operating, but the upward movement will not be at the correct rate unless free passage for the flow of air at the minimum volumes listed in Table 16.6 is maintained to the burner.

Burners are designed to have a steady draught value and it is critical to stabilise this value between the maximum and minimum limits to achieve a safe and efficient performance.

Figure 16.10 indicates rooms or compartments housing boilers that are provided with air inlet vents to furnish the correct amount of combustion air together with air for general room ventilation. If there is more than one appliance within the same room or compartment, then their combined ratings should be used to assess the size of the ventilation openings.

It can be seen from Table 16.6 that for room sealed appliances the requirements are less, as these types of appliances only require ventilation air into the room with no requirements for combustion.

It can also be seen from Table 16.6 that no requirement for combustion air ventilation is required to the room for solid fuel or oil burning appliances having a heat output of less than 5 kW, or gas fired appliances having a heat input of less than 7 kW. This is because it is considered that a sufficient amount of natural

Table 16.6 Ventilation requirements for combustion appliances

Flue Arrangement

Air Vent Locations	Conventional Open-Flue										Room-Sealed									
	Solid Fuel & Oil. Area in cm². Heat Output above 5kW					Natural Gas Area in cm². Heat Input above 7kW					Oil Area in cm². Heat Output above 5kW					Natural Gas Area in cm². Heat Input above 7kW				
	A	B	C	D	E	A	B	C	D	E	A	B	C	D	E	A	B	C	D	E
Appliance in a room or space	5.5	N/A	N/A	N/A	N/A	5	N/A	N/A	N/A	N/A	N/A	N/A	N/A	N/A	N/A	N/A	N/A	N/A	N/A	N/A
Appliance in a compartment ventilated via adjoining room	5.5	11	16.5	N/A	N/A	5	10	20	N/A	N/A	N/A	11	11	N/A	N/A	N/A	10	10	N/A	N/A
Appliance in a compartment ventilated externally	N/A	N/A	N/A	5.5	11	N/A	N/A	N/A	5	10	N/A	N/A	N/A	5.5	11	N/A	N/A	N/A	5	5

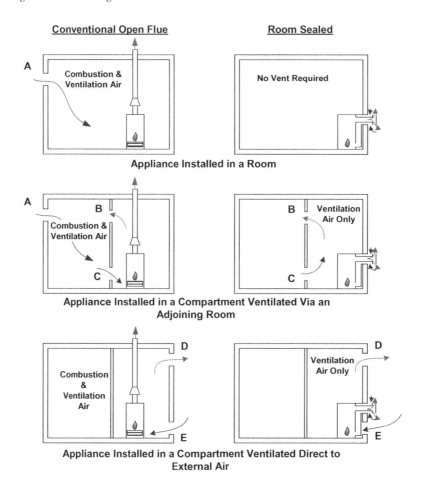

Figure 16.10   Air ventilation entry arrangements as listed in Table 16.6

The above drawing should be read in conjunction with the data given in Table 16.6 indicating the locations for the air vents denoted A, B, C, D and E.

The definition of an appliance compartment is an enclosure specifically constructed or adapted to accommodate one or more gas- or oil-fired appliances. Although not included in the Building Regulations, the information and data listed are equally applicable to solid-fuel burning appliances as they are for gas and oil.

The above illustrations indicate air entry into the room or appliance compartment where the boiler is located, and show the ventilation air entry for combustion purposes only, although for conventional open flue appliances there is an element of air entry that provides room ventilation air as well as combustion air.

The air ventilation requirements for room sealed appliances do not require air for combustion from the room air vents as they receive their combustion air direct from the external environment, whereas conventionally open flued appliances require air direct from the room in which they are located.

ventilation, sometimes termed adventitious (happening by chance) ventilation, through the space around doors and openable windows, even though they may be in good order, will allow appliances below these ratings to operate safely and correctly. However, with the continued improvement in the standards for thermal insulation and weather-proofing around openings etc, this heat output/input rating may have to be revised.

TEMPERATURE DIFFERENTIAL

As a result of the combustion process within the boiler, the flue gases are at a much higher temperature than that of the surrounding air, which contributes to the creation of a draught. The hot combustion gases in the flue are less dense and therefore lighter than the colder, denser air outside the building; draught is created by this heavier colder air pushing the lighter flue gases upwards. Therefore the hotter the gases within the flue, the greater the pressure differential and the greater the draught.

FLUE OR CHIMNEY HEIGHT

As with the temperature differential, the height of the flue or chimney also has an influence on the draught condition: the higher the flues, the greater the pressure differential and therefore the greater the potential draught. However, it should be noted that the higher the flue, the greater the potential is for the flue gases to lose some of their temperature, which will therefore reduce the draught available. For this reason, precautions by way of thermal insulation to the flue should be applied to reduce the heat loss from the walls of the flue pipe.

COMBUSTION STACK EFFECT

From the previous text it can be seen that the draught created for a natural draught burner is dependent upon a combination of both the temperature differential and the height of the flue or chimney, plus the provision for ventilation air to enter the room or compartment by either adventitious means or controlled means, via air vents.

Figure 16.11 serves to illustrate this 'U' tube or 'stack effect' created by the natural conditions prevailing, and also the triangular relationship between temperature differential, height and draught. It should also be noted that the flue or chimney should be as vertical as possible with the number of changes in direction kept to a minimum.

Where bends are required, they should be made using long radius bends so as to restrict the frictional resistance to the flow of the high temperature flue gases to a minimum.

Figure 16.11 indicates how the higher pressure colder external air is forced into the building through any openings to try and fill the void caused by the higher temperature and lighter internal air that is being forced out, this is termed 'stack effect'.

Flues for fan assisted and forced draught appliances are not constrained by the above restrictions and do not rely on creating these conditions.

ADVERSE PERFORMANCE FACTORS

Whilst the previous design performance factors for natural draught appliances may have been adhered to, any interference with the free exit of the flue gases at the flue or chimney terminal may have an adverse effect on the draught available to the burner.

If the flue or chimney is terminated too close to the roof surface, it is probable that the exit of the flue gases will encounter opposition from the effects of wind pressure, as shown in Figure 16.12.

The exact effect, magnitude and shape formed are dependent upon the strength and direction of the wind, together with the angle of pitch for the roof construction, which will determine the depth of the wind pattern.

Where a serious pressure difference exists between the windward and leeward sides of an exposed building, this can increase or even reverse the gas flow in the flue, as any flue terminating in a positive pressure

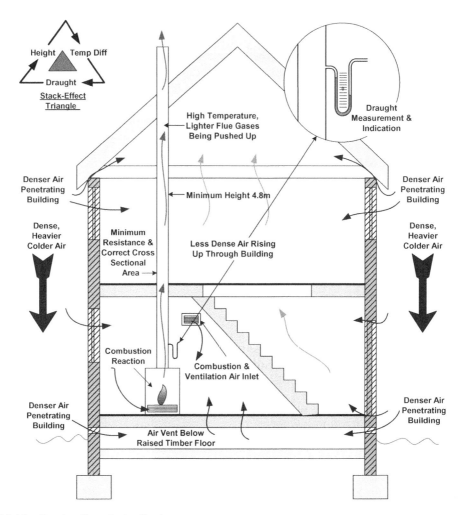

Figure 16.11   Combustion stack-effect

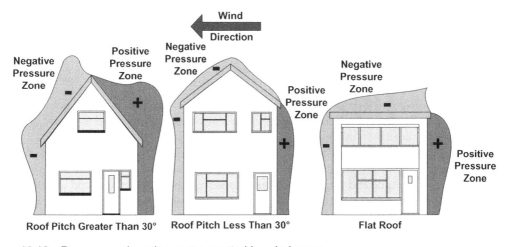

Figure 16.12   Pressure and suction zones created by wind pressure

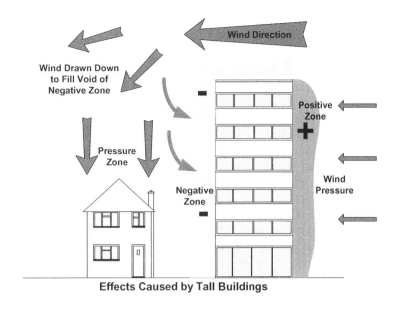

Effects Caused by Tall Buildings

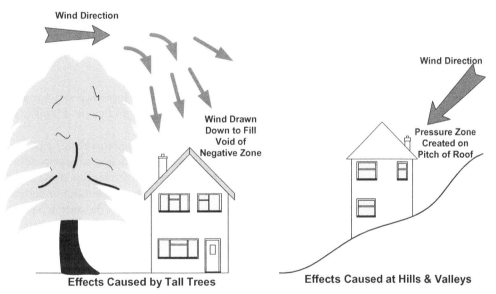

Figure 16.13   Effects of adjacent obstructions on wind

zone would be subjected to downdraught conditions. Likewise, any flue terminating in a negative zone could have the flue gas exit accelerated, causing the burner flame to be pulled off the burner by the increased draught velocity.

As well as this, any adjacent obstacle, such as large trees or buildings that are higher than the flue or chimney, can deflect the wind currents to cause pressure zones as depicted in Figure 16.13. In this figure it is illustrated that the wind pressure can be drawn down onto a flue terminal by the negative pressure zone created as it passes over or around the obstruction.

Whilst termination of the flue within pressure or suction zones can be said to be the main cause for poor draught performance, there are other factors that may contribute towards this condition, and they should be considered during the design phase.

These may include unsuitable flue pipe materials with insufficient insulating properties that allow the products of combustion to cool down rapidly, and so reduce the flue gas temperature difference available.

An oversized flue with a large cross-sectional area would also have the effect of allowing the flue gases to cool down, due to the larger area of the flue wall surface that would be in contact with the flue gases. Both these factors would also cause the flue gases to condense out, this condition is discussed later in this chapter. In certain circumstances, insufficient height of the flue or chimney would contribute to a poor draught condition, therefore for natural draught appliances a minimum height of 4.8 metres should be allowed to obtain a suitable draught.

Figure 16.13 illustrates the effects that the wind can have on the termination of flues from combustion appliances. These effects and the behaviour of wind on buildings has been well documented in the past, based on the experience gained in constructing high-rise buildings in the 1920s in places such as Chicago, New York and other cities in North America.

The failure in town planners to apply this acquired knowledge, particularly during the building boom of the 1960s in London and other cities in the UK, has led to some high-rise buildings being constructed that have resulted in extreme windy conditions occurring around their base, even when there is only a slight breeze blowing.

FLUE AND CHIMNEY TERMINATIONS AND TERMINALS

The previous section described how the wind can seriously impair the performance and termination of flues associated with burners installed in appliances relying on natural draught flue conditions to function correctly.

The Building Regulations, along with other regulations and standards, require that all flue pipes terminate in compliance with minimum siting dimensions to overcome pressure and suction adverse conditions.

This chapter has mainly referred to conventional flues from natural draught appliances, where 'conventional flue' is defined as an appliance that receives its air for combustion from the room or compartment in which it is housed, and discharges its products of combustion externally to the outside air. However, there are a number of specific variations of conventional flues, each having particular constraints or merits regarding their method of termination; these are demonstrated in Figure 16.14, which shows the differences in the conventional flue arrangements.

Figure 16.14(a) indicates the simple flue arrangement of a natural draught created by the stack effect complete with an open flue arrangement, so called because the diverter unit creates an opening in the flue pipe. The draught is produced by natural means causing a small negative pressure within the combustion zone and forcing air to enter by the pressure differential.

Figure 16.14(b) indicates a similar arrangement to that described for Figure 16.14(a), except that a fan has been introduced by the boiler manufacturer, turning it into a fan assisted appliance, to create a negative pressure within the combustion zone and so produce a draught. The only constraint on the flue that this appliance has is to ensure that the frictional resistance produced by the flue is within the fan's capacity to overcome. The boiler manufacturer will give guidance on this matter.

Figure 16.14(c) is a similar arrangement to that described for Figure 16.14(b), except that the flue diverter has been omitted, turning the flue arrangement into a closed flue. This arrangement should only be used when permitted by the boiler manufacturer.

Figure 16.14(d) indicates the natural draught open flue arrangement described for Figure 16.14(a), with a separate booster fan installed within the secondary flue section to overcome the frictional resistance being produced by the flue pipe arrangement. Flue boosting is discussed later in this chapter.

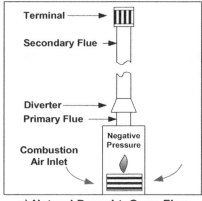

a) Natural Draught, Open Flue

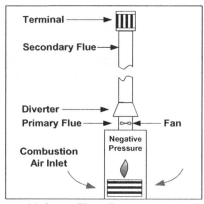

b) Open Flue, Fan Assisted

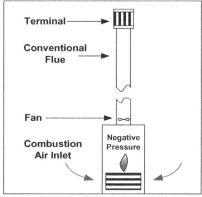

c) Closed Flue, Fan Assisted

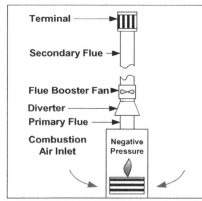

d) Open Flue, Flue Boosted

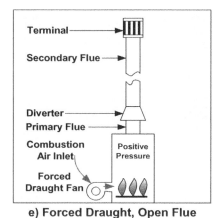

e) Forced Draught, Open Flue

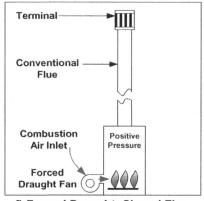

f) Forced Draught, Closed Flue

Figure 16.14 Conventional flue arrangements

Figure 16.14(e) shows a natural draught open flue arrangement, except the appliance has been equipped with a forced draught burner as described in Chapter 13. This will have the same advantages as the fan assisted versions but will create a positive pressure within the combustion zone.

Figure 16.14(f) is a similar arrangement to that described for Figure 16.14(e) except that the flue is a closed arrangement, instead of the more common open flue arrangement.

Figure 16.15, together with the data given in Table 16.7, indicates the termination limitations for the various conventional flue arrangements.

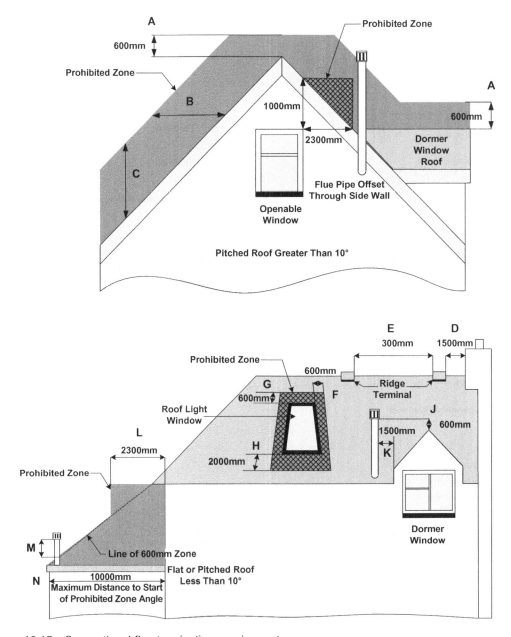

Figure 16.15   Conventional flue termination requirements

Table 16.7 Conventional flue termination distances

Minimum Separation Distances for Terminals mm

Roof Position and/or Location	Gas Fired		Solid Fuel	Oil Fired	
	Natural Draught	Forced or Fan Assisted Draught	Natural Draught	Natural Draught Vaporising Burner	Pressure Jet Burner or Fan Assisted Draught
A Minimum distance above ridge of roof	600	N/A	600	600	N/A
B Distance measured horizontally from roof surface	1500	N/A	2300	2300	N/A
C Distance measured vertically from roof surface	1500	600	1000/1800	1000	600
D Ridge terminal measured horizontally from vertical structure	N/A	1500	N/A	N/A	1500
E Distance measured horizontally between ridge terminals	N/A	300	N/A	N/A	300
F Prohibited zone from side of a roof light	600	600	600	600	600
G Prohibited zone above a roof light	600	600	600	600	600
H Prohibited zone below a roof light	2000	2000	2000	2000	2000
J Termination height above ridge of dormer window if within distance 'K'	600	600	600	600	600
K Distance from flue to dormer	1500	1500	1500	1500	1500
L Horizontal measurement of prohibited zone from a vertical structure above a flat roof	2300	2300	2300	2300	2300
M Vertical height of termination above a flat roof or measured above angle of 600mm prohibited zone	600	250	600	600	250
N Horizontal distance measured between vertical structure and edge of angle of 600mm prohibited zone	10000	10000	10000	10000	10000

The above figures have been extracted from a number of different regulations and standards that are subject to constant review and revision: the reader should check that the latest requirements have been complied with.
The appliance manufacturer's recommendations should also be complied with when designing and installing a flue for that specific appliance.

The above termination requirements are designed to take the flue outlet out of the normally expected pressure and suction zones.

The flue or chimney should be terminated with a correctly fitting and approved terminal, designed to discourage birds from nesting in them and prevent the entry of rain and snow, but also to offer the minimum of resistance to the flow of flue gases exiting through it. The open free discharge area of the terminal should be at least equal to, if not greater than, the internal cross-sectional area of the flue pipe to which it is attached.

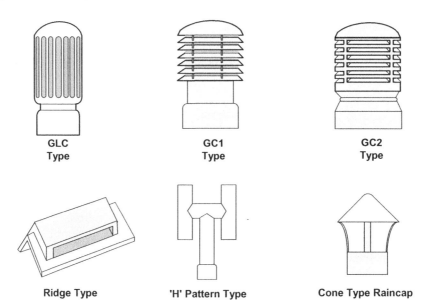

Figure 16.16   Typical common type flue terminals

Common flue terminals are depicted in Figure 16.16, where the GLC, GC1 and GC2 terminal types meet the above requirements, and have been used successfully for both oil and gas installations.

The ridge type terminal has been very popular with some architects due to its aesthetically pleasing appearance, as it can be built into the roof structure at the ridge as part of the roof tiling activity. However, because this type of terminal terminates the flue below the recommended termination heights within the positive and negative pressure zones, it can only be used on force draught or fan assisted flued appliances that incorporate a fan capable of overcoming the positive pressure likely to be created; they should not be used in conjunction with natural draught appliances.

The terminals for flues and chimneys discussed so far have all been applicable to weather situations and conditions that may be described as normal; however, sometimes far more severe conditions can be expected. In these situations the 'H' pattern terminal may be found to go some way to counteracting these conditions and preventing or reducing downdraught or excess pull, but the fumes may still be carried down to ground level. In this case, some form of break or stabiliser should be considered.

DRAUGHT DIVERTERS, BREAKS AND STABILISERS

If the building containing the heating system fuelled by a combustible fuel is located in an exposed geographical area subject to severe weather conditions, or when extreme weather conditions prevail, the type or height of the flue terminal will not be sufficient on its own to prevent downdraught, or excess negative pressure conditions pulling the flame off the burner. In these circumstances one or more of the following items of equipment should be considered for incorporation within the flue system or chimney arrangement.

Draught diverter

Manufacturers of domestic sized conventional open flued boilers normally incorporate draught diverters as an integral component within the casing of the appliance, but larger boilers will usually require this

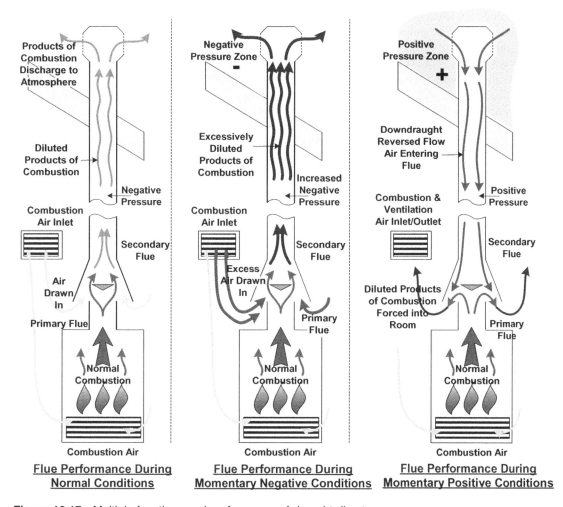

Figure 16.17 Multiple functions and performance of draught diverters

essential piece of equipment to be supplied and installed separately on the appliance flue spigot on top of the boiler casing, forming a demarcation point between the primary flue and the secondary flue.

The function of the draught diverter is to maintain normal combustion conditions within the combustion chamber of the boiler regardless of changing weather airstreams, or atmospheric pressure conditions that cause pressure or suction zones in excess of the termination requirements. It achieves this by performing differently to suit changing weather conditions.

Figure 16.17 illustrates these variable functions during changing conditions, during which the combustion process continues to perform normally, leaving the burner unaffected by these circumstances.

During normal operating conditions where the wind is light, the draught diverter just forms an obstruction to the passage of the products of combustion. Air is drawn in through the opening on the underside of the draught diverter from the room by the stack effect created, which serves to dilute the products of combustion, and so lower the temperature of these flue gases before allowing them to discharge out the top to atmosphere. Care should be taken to ensure that this temperature in the secondary flue is not lowered below its dew point for non-condensing appliances, or condensate will drip out of the bottom of the draught diverter.

During conditions when negative pressures momentarily exist, created by wind pressure over the terminal, the flue gases will be pulled out of the top of the flue to try and fill the partial void created, thus increasing the velocity of the gases within the flue. To prevent the flame being pulled off the burner by this increase in velocity, excess air is drawn in at the draught diverter, thus protecting the combustion zone from these effects and allowing normal combustion to continue. The result is a momentary increase in the dilution of the flue gases, coupled with a further drop in their temperature. The combustion air inlet ventilation grille serves to supply the additional air during this brief period.

When the opposite weather conditions prevail and a positive pressure zone is briefly created, the opening at the top of the flue will serve as a means of relieving this pressure and a downdraught situation will exist, causing a reversal to the flow of the flue gases. To prevent this downdraught from blowing the flame of the burner, this downdraught pressure is allowed to escape out of the flue and into the room by the opening in the draught diverter. Any increase in room pressure is allowed to escape through the combustion air inlet ventilation grille. When this condition exists, diluted products of combustion will be discharged into the room where the appliance is located, and they are also allowed to escape through the combustion air inlet ventilation grille.

It should be appreciated that the purpose of the draught diverter is to function as described during abnormal weather conditions that might occur momentarily; it is not intended for situations where these conditions occur often, or as an alternative to not complying with termination requirements.

Draught break

Draught breaks are no longer employed as, to a certain extent, they have a similar function to that of a draught diverter and the draught diverter, is now the preferred arrangement.

Draught breaks simply consisted of a full size opening at the base of the flue pipe through which air was drawn during normal operation, as shown in Figure 16.18.

During abnormal weather conditions the draught break performed in the same way as the draught diverter by allowing excess air to enter at the base of the flue when negative pressure conditions occurred at the terminal. When positive pressure conditions transpired at the terminal, the downdraught that was created was allowed to bypass the burner by exiting the flue out of the opening at the base of the flue pipe.

Draught stabiliser

A draught stabiliser, unlike draught diverters and draught breaks, is an item of equipment designed exclusively to prevent excessive draught conditions from occurring in the flue when negative pressure zones have

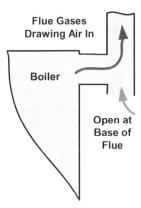

Figure 16.18   Draught break

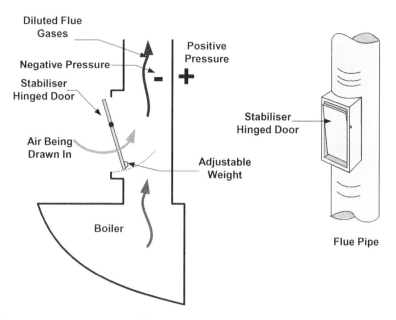

Figure 16.19 Application of draught stabiliser

been created at the flue terminal. It therefore reduces the risk of the flame being pulled off the burner and so stabilises the draught conditions.

Draught stabilisers comprise a top hung hinged and weighted flap or door covering an opening in the flue. The weight can be adjusted by sliding it up and down on the inside of the door so that the stabiliser is able to suit individual burner conditions. The flap will thereafter function automatically by swinging open by an inward movement when the draught exceeds requirements, and allowing the air to bypass the burner by being drawn directly into the flue.

Draught stabilisers have rarely been used on domestic appliances but have been used to protect larger solid fuel burning boilers.

Figure 16.19 indicates the application of the draught stabiliser, showing the hinged door being drawn open by the imbalance in the pressures between the room air and the flue gases, where the velocity has been accelerated by the negative pressure zone that has been created at the flue terminal.

THERMAL INVERSION

In recent years, there has been a necessity for and attention given to improving the thermal performance of new buildings by producing more rigorous standards, and homeowners have been encouraged on environmental and sustainable grounds to improve their existing homes by replacing windows and doors, applying weather stripping and general upgrading of the building and insulation. This has led to a phenomenon known as thermal inversion manifesting itself, under certain operating conditions.

It has also become popular in new housing developments to incorporate an open fireplace in each property. This reintroduction of the open fireplace is viewed as being a character-adding feature of a property that many potential homeowners will find attractive.

Figure 16.20 illustrates the normal operating conditions that exist when a conventional open flue boiler is operating under the heating control system. The combustion air entry required by the boiler is being

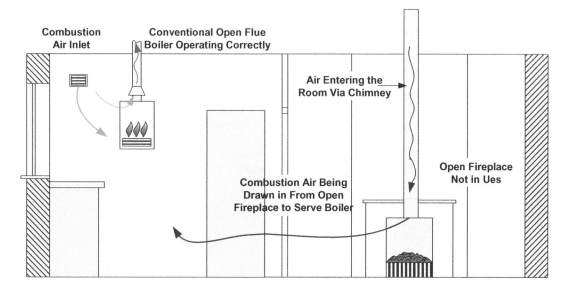

Figure 16.20 Conventional open flue boiler operating normally

supplied via the combustion air inlet grille and by air entering the building via adventitious means – including the open chimney of the fireplace when not in use.

Air will enter the building by any means that it can, but when the building has tight fitting doors and windows the options for air entry are reduced. However, the inclusion of an open fireplace will provide a large ventilation air entry point as well as a heat loss opportunity.

Thermal inversion occurs when the open fire has been ignited and is burning at full rate, together with the boiler operating to satisfy the heating system. During certain conditions, such as the open fire demanding oxygen when it's burning at its maximum rate, together with an inadequate provision of combustion air to satisfy all appliances, a reversal in the normal order of operating conditions occurs when the path that the combustion air normally takes is forced to change, and air plus flue gases from the boiler are drawn out of the draught diverter by the thermal demand from the open fire.

This reversal in normal operation highlights the need to ensure that the calculation for combustion air inlet provision includes all fuel burning appliances, and that it is recalculated following any home improvements that have resulted in a reduction of adventitious air entry. Figure 16.21 illustrates thermal inversion and the conditions that could cause it.

FLUE PRINCIPLES, CONSTRUCTION AND MATERIALS

The method of installing the flue should follow certain general rules, including following the appliance manufacturer's instructions, to enable the appliance to function correctly. It should be installed preferably vertically with the least possible amount of changes in direction, and within the appliance manufacturer's maximum resistance allowance. When bends are required they should be of maximum angle not exceeding 135°.

Figure 16.22 serves to demonstrate these general requirements where the diverter is installed as close to the boiler outlet spigot as possible, or even within the boiler casing, as most domestic appliances are normally arranged. The draught diverter separates the primary flue pipe section from the secondary flue pipe section and as indicated, should be best installed internally within the building if possible to reduce the heat loss from the flue pipe together with reducing the possibility of condensation occurring within the flue.

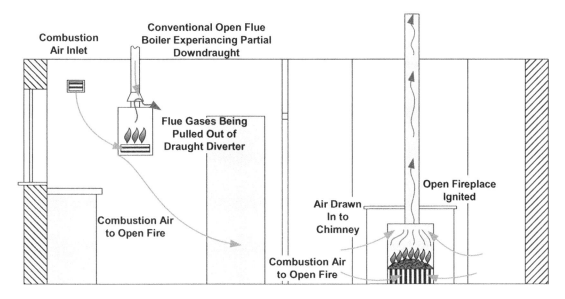

Figure 16.21   Example of a thermal inversion

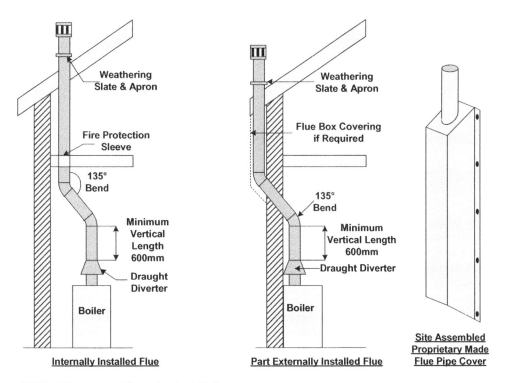

Figure 16.22   Principles of flue pipe installation

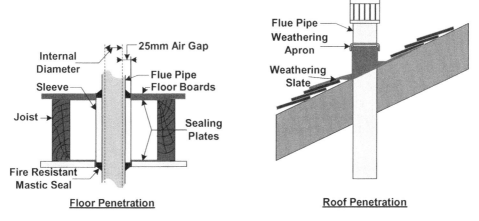

Figure 16.23   Floor and roof penetrations

To assist in the process of removing the products of combustion from the appliance, the flue pipe must rise vertically for a minimum of 600 mm measured from the top of the draught diverter to the first bend on the secondary flue.

Where the flue pipe has to be installed externally and the resultant thermal loss would be such that the draught created would be insufficient for the appliance to operate correctly, it should either be erected using a twin-walled insulated flue pipe material, or the flue pipe should be encased in a flue box cover with the space between filled with vermiculite granules, to reduce the temperature loss from the flue.

Figure 16.23 illustrates a flue pipe passing through a timber floor and roof structure which has been separated from the combustible timber by a non-combustible sleeve that allows a minimum of 25 mm air space between the flue pipe external wall and the sleeve.

USE AND LINING OF EXISTING MASONRY CHIMNEYS

It was common practice for many centuries to build properties with open fireplaces and chimneys constructed from masonry materials. In the 1960s it became fashionable to construct dwellings without fireplaces due to heating systems becoming more common and fireplaces were not built as they were considered sources of large amounts of heat loss.

The chimneys were usually constructed from brick products that were laid as part of the wall construction, with the internal surfaces of the flue passage roughly rendered with a sand/cement lining of unequal thickness. This cement rendering, in countless properties, had over the years become broken down in places, leaving the brick surface exposed to the flue gases from the open fire and leaving the cement mortar between the bricks open to attack from the flue gases and condensate. Properties built before 1965 can be expected to have chimneys in this condition and if they were to be used in this condition for conveying the products of combustion from appliances fired by any of the common fuels discussed, then the acidic nature of the flue gases and condensate would accelerate the breakdown of the cement rendering and mortar between the bricks, eventually showing itself on the internal wall plastering as black tar-like stains.

1965 is significant because that is the year in which the first edition of the Building Regulations was introduced. The building regulations required all chimneys to be internally lined as part of the construction process. This lining requirement was normally performed by using clay tiles or vitrified clay pipes that were built-in as the building height progressed, and that offered a corrosive resistance to the acidic condensate that could occur. This regulation has remained in the various revisions that have been made to the Building

Regulations since their introduction, although the thermal resistance requirement for the flue has been revised.

It was common practice to install flexible flue liners from open flue appliances when it was desired to use the existing chimney, to prevent the flue gases and subsequent condensate from attacking the cement mortar between the brick joints, causing them to disintegrate followed by possible structural failure of the chimney stack.

The gas fired back boiler appliances, that had been popular for many years up until recently, and which incorporated a gas fire front, commonly used existing chimneys that had to be lined. They are rarely used today because of their low SEDBUK rating which precludes them from most applications.

The flexible flue liners are available made from a number of different metals, although stainless steel that has been titanium-stabilised and contains molybdenum has proved to be the most suitable, both in its physical strength and its resistance to corrosion compared to other metals that have been used. They are manufactured from continuous strip metal that is spirally wound together to form an interlocking joint, lightweight flexible hollow flue liner in a number of sizes, and they have an upper temperature limit of 760°C.

Before any installation work commences, the chimney should be thoroughly swept to remove all loose debris and any obstructions, to ensure a clear path is available.

The flue liner should be of at least equal diameter to the flue spigot from the appliance and it should be installed in compliance with the manufacturer's instructions, but in general terms the installation should be as depicted in Figure 16.24, which illustrates the basic components that make up the flue system.

The flexible liner is normally recommended to be inserted into the existing flue passage from the top, although at least one manufacturer recommends that the flue liner is inserted from the bottom and the liner pulled up into place.

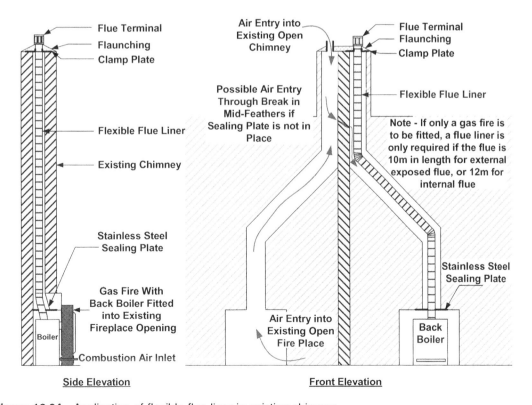

Figure 16.24   Application of flexible flue liner in existing chimney

When the liner has to be inserted from the top, the existing terracotta chimney pot together with the cement flaunching has to be removed. A cord is then passed down the flue passageway to the bottom at the open fireplace. The top of the cord is attached to a plastic plug that will protect the leading edge of the liner and has the same internal diameter as the flue liner into which it is inserted. The liner is then passed down the flue passageway guided by an operator lightly pulling the cord, until it is low enough to connect onto the flue spigot of the boiler, after which a clamp plate can be fixed onto the top to hold the liner in place. The flue terminal can then be fitted into place and the cement flaunching renewed.

A light gauge stainless steel sealing plate should be fitted around the flue liner at the base of the liner to form a seal between the open fireplace and the base of the chimney flue passage. The purpose of this plate is to prevent the entry of any air through cracks or other openings in the chimney wall where there has been a breakdown in the cement mortar joints, and which, if left unchecked, would cause excessive draughts and interfere with the combustion process at the appliance burner. It would also give a passageway for flue gases from a faulty appliance to rise up the space between the liner and structural chimney and enter into a room any (anywhere) where the wall of the chimney breast is not properly sealed.

BALANCED FLUES

The flue systems discussed previously in this chapter have concentrated on the various versions of conventional flues, which are defined as appliances that receive their air for combustion from the room or compartment in which they are housed, and discharge their products of combustion externally to the outside air via a fixed flue pipe, and are suitable for all fuels.

A balanced flue, however, does not suffer from the same constraints as conventional flue arrangements and has some advantages over them. The room sealed version is defined as 'an appliance that takes its air for combustion from external to the room or compartment in which it is located and discharges its products of combustion externally via a flue/air pipe or duct arrangement', with the main advantage being that it is not affected by the wind-created pressures that can cause problems to conventional flue systems.

Figure 16.25 illustrates the working principles of a natural draught balanced flue arrangement where the incoming combustion air pressure is balanced with that of an equal amount of combustion waste gases, is

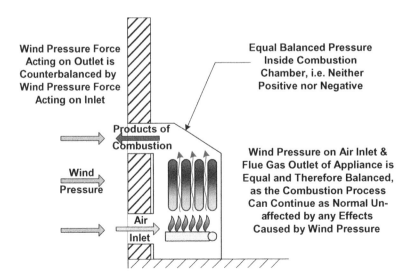

Figure 16.25 Principles of balanced flues

therefore unaffected by changes in wind pressure, and so creates a 'balanced' pressure inside the combustion chamber, hence the term 'balanced flue'.

Balanced flue appliances are suitable for oil and gas burning boilers, but not solid fuel, which should be equipped with a conventional flue.

The majority of domestic boilers work on the principle of balanced flues and are of the 'room sealed' type; however, not all balanced flue arrangements are room sealed as it is sometimes desirable to install larger commercial sized boilers that have provisions for conventional flues in a balanced flue situation. Figure 16.26 illustrates such arrangements.

The two illustrations shown are both for conventional flue appliances installed in a dedicated balanced boiler compartment with a sealed air-tight door to prevent inversion through the building, together with a warning notice on the door regarding its function.

Figure 16.26(a) depicts a twin flue arrangement where the two pipes are installed side-by-side, and Figure 16.26(b) is for a combined terminal arrangement.

ROOM SEALED APPLIANCES

As previously established, a room sealed appliance may be defined as 'an appliance that takes its air for combustion from external to the room or compartment in which it is located, and discharges its products of combustion externally via a flue/air pipe or duct arrangement'. Room sealed appliances are only suitable for domestic sized appliances and are usually the preferred choice of flue arrangement if a suitable wall is available. There are a number of variations of room sealed appliance as illustrated in Figure 16.27.

Figure 16.27(a) illustrates what could be called the traditional domestic heating room sealed boiler, which operates with a balanced flue arrangement creating a natural draught. Because of this, the physical size would be reasonably large and heavy, hence it is required to be floor standing but to still fit below a standard kitchen worktop. The physical size is governed by the need to be able to create the natural draught required.

The combined flue and combustion air intake terminal consists of a duct within a duct that is arranged to be adjustable by being telescopic, so that it will fit a varying thickness of walls without the need for cutting.

If the combined wall mounted terminal terminates at a height of less than 2 metres above the ground in an area that is used as a public access, then the terminal should be fitted with an additional protective guard or cage that should be large enough to give a minimum of 50 mm clearance all round, including the front.

If the terminal is sited 500 mm underneath a painted eaves, or 1000 mm below a plastic rainwater gutter, then a heat deflection shield should be installed to the underside of the eaves or gutter that extends to a minimum length of 1000 mm with a 5 mm air gap between the shield and gutter/eaves.

In order for boilers to be made physically smaller and lighter in weight, the provision of a natural draught has to be sacrificed by the inclusion of an electrically powered fan that will assist in the supply of combustion air and expelling the products of combustion. Figure 16.27(b) and Figure 16.27(d) show the smaller, lighter wall mounted boiler that results from this development, one producing a negative pressure within the combustion zone and the other a positive pressure.

Figure 16.27(c) depicts a twin flue arrangement that consists of a flue within the air supply duct connected via a common terminal similar to those described before. This system permits the wall mounted boiler to be sited away from an external wall, but the total distance must be within the capabilities of the fan that has to overcome the resistance.

An alternative version of this arrangement consists of two separate flue pipes, one air supply and the other exhaust flue.

Figure 16.28 illustrates an alternative flue arrangement that looks like a conventionally flued wall mounted boiler, but incorporates its air supply intake in the roof space that must be ventilated. The fan assisted flue

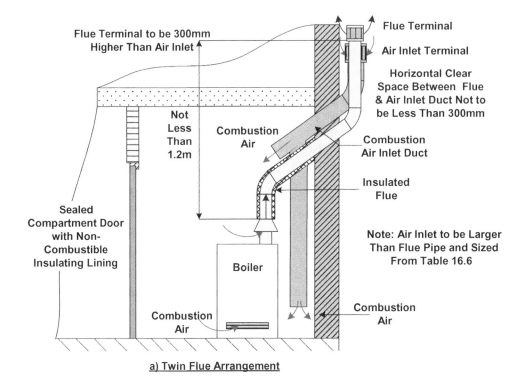

a) Twin Flue Arrangement

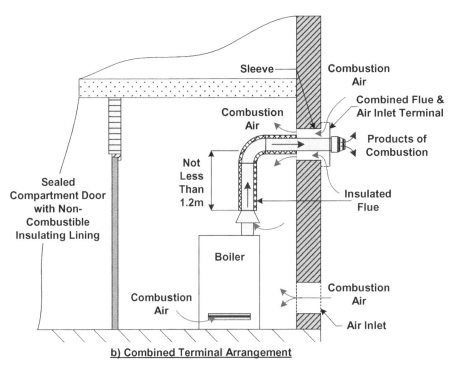

b) Combined Terminal Arrangement

Figure 16.26   Alternative balanced flue arrangements

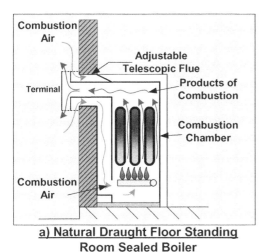

a) Natural Draught Floor Standing Room Sealed Boiler

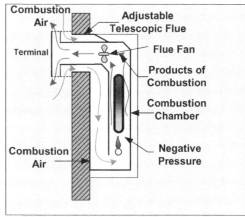

b) Fan Assisted Wall Mounted Room Sealed Boiler (Negative Pressure)

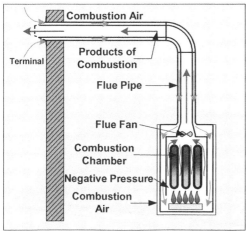

c) Fan Assisted Wall Mounted Twin Flue Room Sealed Boiler (Negative Pressure)

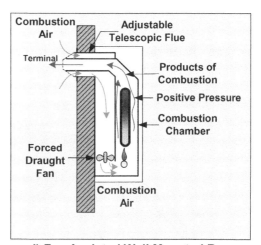

d) Fan Assisted Wall Mounted Room Sealed Boiler (Positive Pressure)

Figure 16.27   Room sealed boiler flue arrangements

within the centre of the twin pipe arrangement discharges the products of combustion vertically through the pitched roof that gives it the appearance of being a conventional flue.

BALANCED FLUE TERMINATIONS

Figure 16.29 illustrates the various positions where balanced flues from room sealed appliances may terminate, subject to the dimensional constraints shown in Table 16.8.

The Building Regulations stipulate differing measurements for natural draught gas-fired boilers based upon the kilowatt rating of the appliance for termination positions 'A', 'M' and 'N', all denoted by an asterisk in Table 16.8. These alternative dimensional requirements are given in Table 16.9.

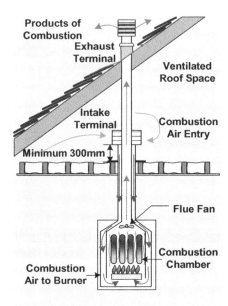

Figure 16.28 Room sealed appliance with twin vertical flue

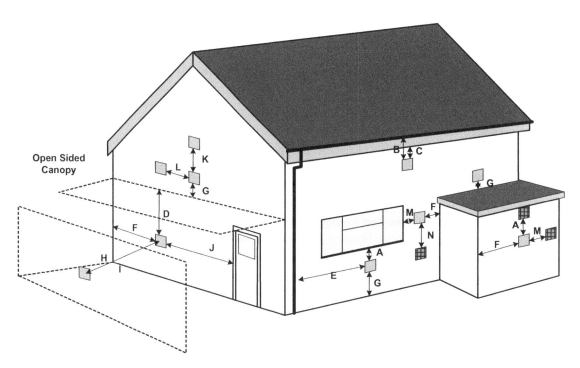

Figure 16.29 Suitable locations for balanced flue mounting (see also Table 16.8)

Table 16.8 Room sealed balanced flue terminations (refer to Figure 16.29 for locations)

Ref	Terminal Location & Definition	Natural & LPG Gas		Oil
		Natural Draught mm	Fan Assisted Draught mm	Pressure Jet Burner mm
A	Directly below an openable window or air brick vent	300*	300	600
B	Below gutters, soil or rainwater pipes	300	75	600
C	Directly below eaves	300	200	600
D	Below balconies or car port roofs/open sided canopies	600	200	Not Recommended
E	Distance from a vertical soil or rainwater pipe	300	150	300
F	Distance from an internal or external corner	600	300	300
G	Above ground, roofs or balcony level	300	300	300
H	From a surface facing a terminal	600	600	600
I	From a terminal facing a terminal	600	1200	1200
J	From an opening in a car port into a dwelling (door, vent or window	1200	1200	Not Recommended
K	Vertically above a terminal on the same wall	1500	1500	1500
L	Horizontally from a terminal on the same wall	300	300	750
M	Horizontally from an air brick/vent on the same wall	600*	300	600
N	Above an openable window or air brick/vent	600*	300	600
P	Terminals sited below the top level of a basement area, light well or retaining wall, which creates an uncovered passageway of at least 1.5m wide. The combustion products should discharge into free open air. Measurement from ground level to terminal **(Not Shown)**	1000	1000	Not Recommended

Table 16.9 Room sealed balanced flue terminations (Building Regulations)

Ref	Terminal Location & Definition	Natural Draught Gas Appliance	
		kW	mm
A	Directly below an openable window or air brick vent	0–7	300
		>7–14	600
		>14–32	1500
		>32	2000
M	Horizontally from an air brick/vent on the same wall	0–7	300
		>7–14	400
		>14	600
N	Above an openable window or air brick/vent	0–32	300
		>32	600

Tables 16.8 and 16.9 stipulate dimensional requirements for the termination of appliance flues, but the manufacturer's recommendations should always be followed if their requirements are greater than those published here.

In general, the terminal should be sited against a suitable wall to allow the products of combustion to disperse into the atmosphere unhindered at all times. If the flue terminal is positioned too close to an obstruction such as the wall of a building, then the combustion products may be deflected back onto the appliance air intake duct, causing the air supply to the appliance burner to become vitiated through oxygen depletion.

The flue should not be terminated in such a way that it will cause a nuisance to a public passageway or an obstruction to the general public, and the Building Regulations do not permit the termination of flues to encroach on neighbouring properties.

It should also be noted that in certain areas local bye-laws and planning permissions require flue terminations to be made at specific minimum heights above ground level, and any projection from the termination wall to be restricted to a maximum distance.

If certain weather conditions prevail, the products of combustion may condense to form water vapour that will give the appearance of steam being generated. Terminal positions where this may be a nuisance should be avoided.

Figure 16.27 indicates some fan-assisted room sealed appliances that operate under either positive or negative pressure conditions within the combustion chamber. If the appliance combustion casing is not sealed totally air-tight, then under positive pressure conditions the products of combustion that may contain carbon monoxide may be forced into the room in which the appliance is located. Therefore, the exercise of checking the combustion chamber seal should form part of any periodic servicing procedure to ensure that the appliance is operating safely.

FLUE BOOSTING

In situations where it is not possible to install a conventional flue arrangement, due to structural constraints, or when a suitable external wall is not available for a room sealed appliance to be installed, then a draught inducing flue boost fan may be selected to overcome the problem.

In this situation where terminating a flue, either conventional or room sealed, in the normal manner is impossible, the choice available is to install either an electrically powered generator which does not require a flue, or a purpose designed and manufactured flue exhausting fan, in conjunction with a conventionally flued natural draught appliance.

The booster fan is installed in the secondary section of the flue pipe, as shown in Figure 16.30. It is rated to overcome the higher resistance created by the flue pipe which has been installed in an unorthodox route, including turning the flue pipe down to avoid obstructions.

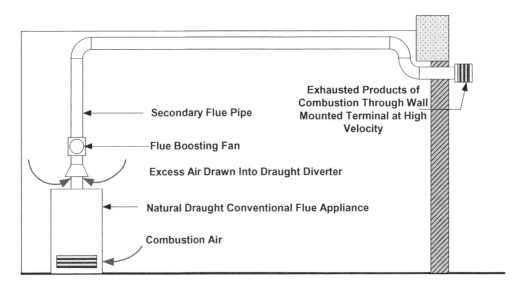

Figure 16.30   Example of draught-inducing flue boosting arrangement

In this application, the performance of the flue is no longer dependent upon the height and temperature difference of the flue to create the draught, as this is achieved by the inclusion of the electrically powered exhaust fan. The fan has also been used to overcome excessive downdraught caused by adjacent high buildings and similar obstructions, as the waste flue gases are exhausted at a higher velocity than would be the case in such situations. Excess air is drawn into the flue via the downdraught diverter to reduce the pull of air over the burner, created by the negative pressure conditions caused by the operation of the booster fan.

The booster fan unit must be made to operate by the dictates of an integral flow switch, which must be electrically wired up between the boiler thermostat and main safety solenoid valve to prevent the boiler from operating if the fan should fail to function.

It is important to note that these fans are purpose made for use in flue applications and are suitable for temperatures up to 200°C and the gases involved.

FLUES FOR HIGH-RISE BUILDINGS

In low-, medium- or high-rise buildings that have been developed or converted into separate dwellings – such as apartments or flats – each having their own individual heating systems, the criteria for terminating flues will be different.

For low-rise buildings no more than three stories high, room sealed appliances may be used in the normal way, providing there is a suitable external wall available. In other applications shared or separate flues would need to be considered, as listed in Table 16.10.

Separate flues

In this arrangement the conventional flue and natural draught appliance are treated in the same manner as a normal conventionally flued arrangement, except that the flue pipe from each appliance is insulated to reduce heat loss and prevent condensation of the flue gases occurring along its entire length.

Figure 16.31 depicts a typical separate flue arrangement for a low-rise building, where each flue pipe is taken individually into a common builders' work shaft, where it rises up the building to terminate at roof level in the normal way.

Due to the long length of each flue pipe, particularly from those situated on the lower floors, the void around the flue pipes in the builders' work shaft should be filled with a good insulating material, such as vermiculite granules mixed in with a weak cement mortar, to keep the products of combustion above their dew point temperature.

Shared flues

The practice of connecting the flue outlets from conventionally flued natural draught appliances into a common single flue shaft or pipe has been generally discontinued, due to operational difficulties: the sizing of the flue shaft to cater for all appliances operating simultaneously differs when fewer appliances

Table 16.10 Applications for separate and shared flues

Flue System		Appliance	Application
Separate Flues		Conventional	Low and Medium Rise
Shared Flues		Conventional	Used with Caution
Shared Flues	Se-Duct	Room Sealed	Medium and High Rise
Shared Flues	'U' Duct	Room Sealed	Medium and High Rise

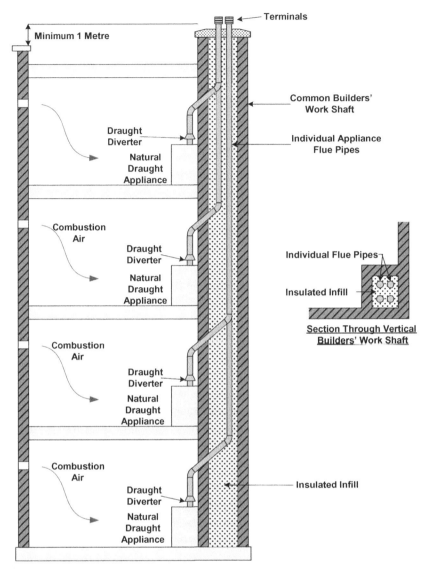

Figure 16.31   Separate flue arrangement for multi-storey building

are operating, resulting in rapid cooling of the products of combustion, and condensation within the flue.

If the flue is undersized to avoid this situation, it will not function when the majority of the appliances are working. Therefore, unless it can be ensured that all appliances will function at the same time, some other arrangement should be considered.

Figures 16.32(a) and 16.32(b) illustrate the single common shared flue arrangement that has been used with partial success; it incorporates a subsidiary flue from each appliance before they connect into the common main flue.

Figure 16.32(a) illustrates an arrangement where two boilers of unequal duty share a common flue; the boiler having the larger heat output is operated for the heating system only during the heating season, and

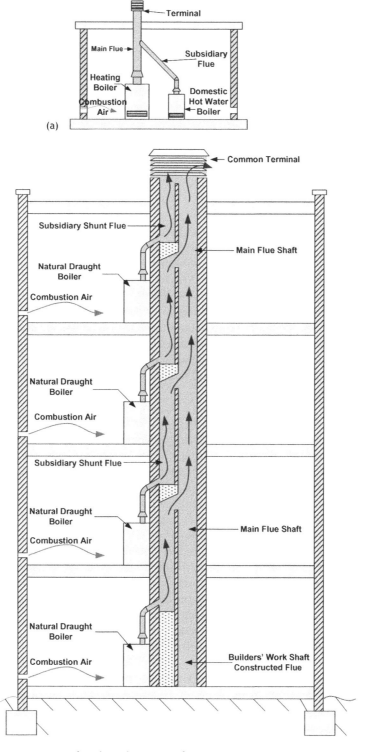

Figure 16.32   Two arrangements for shared common flues

the smaller duty boiler is sized to satisfy the domestic hot water requirements. In this system both boilers operate during the heating season, but during the summer months only the smaller, lesser duty boiler will be operating.

Figure 16.32(b) depicts a shared common flue for multi-storey buildings where each natural draught appliance discharges its products of combustion into its own dedicated subsidiary type shunt flue, which rises up before it joins the main common flue. The common flue then terminates above roof level through a common terminal.

The diagram has been restricted to four floors only for reasons of clarity; taller buildings have been constructed using this flue arrangement.

This flue arrangement is no longer constructed on new multi-storey buildings.

SE-DUCT AND 'U' DUCT FLUE SYSTEMS

The development of a shared flue serving room sealed appliances dates back to around 1948, with the first constructions occurring in the UK in the mid-1950s. By employing room sealed appliances, these flue arrangements have made the need to construct problematical shared flues using natural draught appliances obsolete, and have permitted architects to be less restricted in their approach to building design for multi-storey residential dwellings with individual their own heating systems.

The Se-duct system consists of a vertical builders' work constructed duct, as illustrated in Figure 16.33, which is open at both the top and bottom ends to the atmosphere. As before, the drawing has been restricted to four floors only for reasons of clarity; taller buildings have been constructed using this flue arrangement. Specially adapted room sealed appliances are manufactured which are connected to the duct. They draw their combustion air supply in at the bottom and return their products of combustion back into the same duct via the top outlet duct of the appliance.

The vertical civil constructed flue duct is sized to ensure that the degree of dilution to the combustion air supply to the uppermost appliances is sufficient to restrict the concentration of carbon dioxide entering the combustion air inlet to no greater than 1.5%, to reduce the vitiation of the air quality.

The 'U' duct flue system of construction is similar to the Se-duct arrangement but takes the form of a 'U'. Air for combustion enters the shaft at the top of the building and is brought down the air supply inlet section of the 'U' from roof level to the bottom of the 'U', which then turns to rise up the combined flue and air supply section, where the room sealed appliances are connected into it in the same manner as for the Se-duct. This is depicted in Figure 16.34, where, again, the diagram has been restricted to four floors only for reasons of clarity; taller buildings have been constructed using this flue arrangement. The sizing criteria for the 'U' duct are the same as for the Se-duct system.

The arrangement of the combustion air intake for the Se-duct flue system may vary and would have to be designed to suit the building concerned – it may take the form of a high level air intake for elevated structures, or from ventilated lobbies – in all cases the air intake should be so positioned so as not to become blocked by leaves or snow, or obstructed by parked vehicles.

The base of the flue terminal of a Se-duct system must be at least 250 mm above the roof level or the parapet wall construction if within 1.5 m of the terminal duct to prevent it from also becoming blocked.

The free area of the terminal grille must be twice the cross-sectional area of the shaft duct and uniformly distributed around all the sides or circumference of the terminal, with the mesh of the grille made from a material suitable for its application.

The inlet and outlet grilles on the 'U' duct system should be on the same face of the terminal structure but staggered so that they have a similar exposure, making the terminal balanced under different weather conditions, as with a balanced flue.

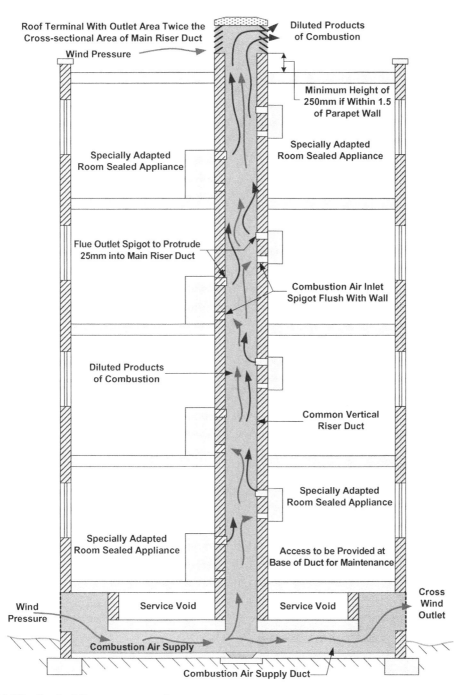

Figure 16.33   Se-duct flue arrangement

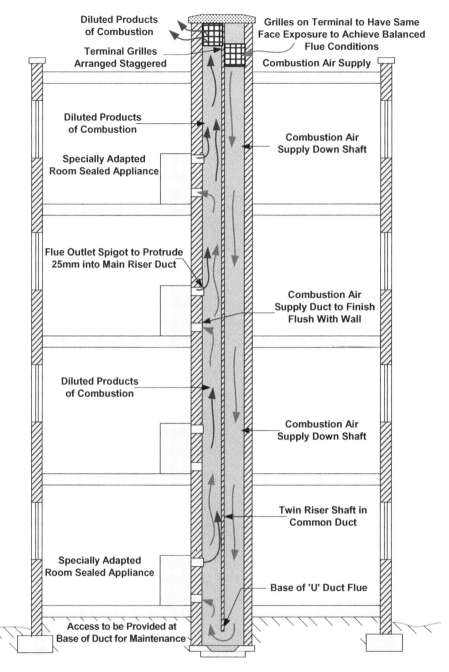

Figure 16.34   'U' duct flue arrangement

The duct shafts for both the Se-duct and 'U' duct arrangements should be constructed using a smooth faced non-combustible and non-porous block, with the combustion air inlet duct finished flush with the inside surface of the duct, and the flue outlet spigot protruding inside the shaft by 25 mm.

FLUE DILUTION

Flue dilution is more suited to the larger domestic or commercial heating installation and is not normally considered economically viable for the majority of domestic heating systems, where diluted flue gases may be safely disposed of at low level without the necessity for a tall and expensive flue system, or having to overcome structural constraints associated with finding a suitable route for them.

The system comprises a vertical short flue pipe from the boiler, complete with a draught diverter which connects into a horizontal looped duct open at both ends, each with a protective grille having a free area equal to that of the air supply/exhaust ducting as illustrated in Figure 16.35.

The system is designed to provide sufficient air into the dilution air supply inlet grille to dilute the carbon dioxide content of the flue gases at the exhaust flue gases outlet grille to less than 1%.

This may be calculated by:

Flow rate in litres per second = 2.69 × rated input of boiler in kW for fan duty.

This means that a boiler having a rated heat input of 35 kW would require a diluting fan capable of producing 94.2 l s^{-1}, or 0.095 m^3 s^{-1}.

The ducting arrangement should be kept as simple as possible and sized on a velocity of 3–5 m s^{-1}, but care should be taken to avoid high discharge velocities at low level if discharging on to public pedestrian walkways.

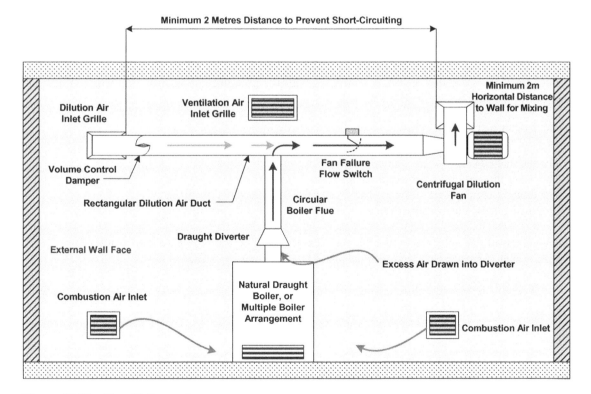

Figure 16.35 Flue dilution system

The dilution air inlet grille and diluted flue gas outlet grille should be arranged to be on the same wall face and spaced at a minimum distance of two metres apart. This is so as to achieve a balanced flue effect, unaffected by wind pressure, and to reduce the risk of short-circuiting the products of combustion back into the air intake grille. It is also recommended that the outlet grille should, if possible, be arranged slightly higher than the inlet grille and be positioned away from any boiler room combustion air or boiler room ventilation inlet grilles.

The exhaust dilution fan should always be located downstream from where the boiler secondary flue connects into the dilution air supply duct, so that it draws the air/combustion products into it. The fan should also be equipped with a safety shut-off air flow switch to prevent the boiler from operating in the event of a fan failure.

It is also considered good practice to arrange the discharge exhaust ducting from the outlet of the dilution exhaust fan to have a minimum horizontal length of two metres before it discharges from the outlet termination grille; this is to ensure that the products of combustion are fully mixed with the diluting air.

ASSESSMENT OF CONVENTIONAL FLUE PIPE SIZES

The calculation of the size and height of a flue or chimney can be a complex matter if done correctly, but for the purposes of domestic and small heating appliances the following simplified method may be used. It is based on commercially available flue materials and limiting variables applicable to such installations.

It should be noted that regardless of the results from any calculation, the size of the flue pipe should not be less than the size of the flue pipe spigot outlet from the appliance together with the minimum height of 4.8 m, taking into account any termination heights above the roof, etc. All the appliance manufacturer's conditions regarding the limitation of bends, as well as the diameter and height, should be complied with.

The objective is to assess a suitable internal diameter of flue required to convey the products of combustion, together with any excess free air at an efficient and realistic velocity, plus a minimum height required to obtain the necessary draught previously discussed.

The information required, before commencing with the procedure of assessing the required internal diameter of the flue, is:

- The type of fuel to be used by the appliance together with its calorific value.
- The type of appliance burner and rated heat input.
- The thermal efficiency of the boiler based on the gross calorific value.
- The boiler flue gas outlet temperature.
- The draught requirement of the appliance burner needed to function correctly.
- Proposed general flue construction.

Having obtained this information, the flue pipe diameter and height may be assessed as follows:

Area of Flue: The internal cross-sectional area of the flue may be calculated by:

$$A = \frac{Fg}{v}$$

Where

A = Cross-sectional area m^2
Fg = Flue gas volume $m^3\,s^{-1}$
v = Flue gas velocity $m\,s^{-1}$

The volume of flue gas for solid fuel, oil and gas, may be taken as approximately $0.0015\,m^3\,kW^{-1}$ of boiler heat input at a flue gas temperature of between 120°C and 260°C.

The flue gas velocity for natural draught appliances varies between $3\,m\,s^{-1}$ to $5\,m\,s^{-1}$.

Therefore, a boiler heat input of 25 kW and a flue velocity of $3.5\,m\,s^{-1}$ would result in a flue pipe internal diameter of:

$$0.0015 \times 25 = 0.038\,m^3$$

Therefore

$$\frac{0.038}{3.5} = 0.011\,m^2$$

which corresponds to a circular equivalent of 125 mm diameter

Circular equivalents are given in Table 16.11.

Having established the internal diameter of the flue pipe, the next stage would be to determine the pressure loss within the flue: this may be done by using the chart depicted in Figure 16.36.

With the example of the 125 mm diameter flue, this can be plotted and found to correspond to a pressure drop of $1.1\,Pa\,m^{-1}$ run of flue pipe.

With the flue diameter and pressure loss having been determined, the required height of the flue can now be established using the chart depicted in Figure 16.37.

Table 16.11 Circular equivalents

Diameter mm	Area m²
100	0.008
125	0.013
150	0.018
180	0.025
200	0.032
250	0.050
300	0.070

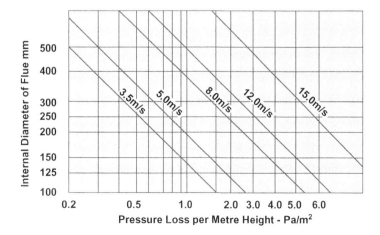

Figure 16.36 Pressure loss in smooth-bore circular flue pipes

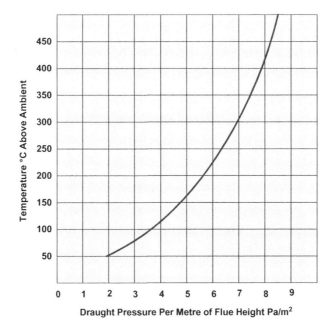

Figure 16.37 Theoretical draught of flue/metre

Continuing with the example, this may be found as follows:

Assume that the burner requires a draught value of $20\,Pa\,m^{-2}$ in order to function correctly and that the flue gas temperature will be approximately 120°C in the secondary flue.

The temperature in the primary flue leaving the boiler will be approximately 180/200°C, but when diluted with air from the draught diverter, the temperature will drop to approximately 120°C in the secondary flue.

Therefore, from Figure 16.37 it can be seen that a draught value of approximately $4\,Pa\,m^{-2}\,m^{-1}$ height will be achieved at a flue gas temperature of 120°C.

As already established, the pressure loss for the example for the 125 mm diameter flue will be $1.1\,Pa\,m^{-2}$, the draught value can be adjusted to suit:

$$4-1.1=2.9\,Pa\,m^{-2}\,m^{-1}\ height.$$

Therefore, with the draught required at the burner being $20\,Pa\,m^{-2}$:

$$20 \div 2.9 = 6.89\ m\ high.$$

Therefore, taking into account any additional pressure loss through flue terminals and bends etc, we would need to install the 125 mm diameter flue approximately 7.5 metres high to enable the boiler to produce an input heat rating of 25 kW with its required draught value of $20\,Pa\,m^{-2}$.

If the available draught is insufficient, either the height may be increased or the velocity reduced by selecting a larger diameter, or a combination of both. If the calculated draught is in excess of requirements, the height or diameter may be reduced, or a damper may be employed to increase the resistance.

The chart shown in Figure 16.38 has been produced using the above method to enable quick estimated selections of heights and internal diameters of flues to be made for given heat inputs.

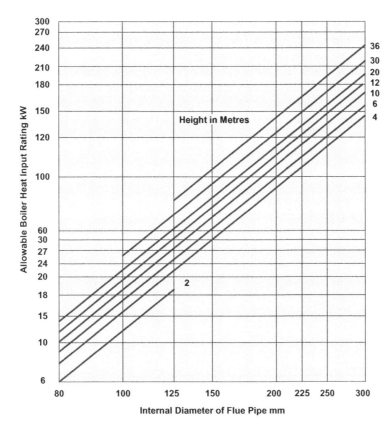

Figure 16.38 Flue selection chart

CONDENSATION WITHIN FLUES

The phenomenon of moisture being released in the form of water vapour from air when cooled down, is well known. Less well known and not always appreciated, are the effects of condensation within conventional flues and chimneys from traditional non-condensing boilers.

It has already been established in the preceding text that all combustible fuels contain a certain amount of moisture, as well as that from any hydrogen content of the fuel being converted into water from the products of combustion, together with moisture being obtained from the combustion air.

The proportion of water vapour produced from the various fossil fuels can be seen by referring to the combustion reaction charts given in Figures 16.3 to 16.6 earlier in this chapter.

Condensation is likely to occur if the waste flue gases are allowed to cool down beyond the dew point temperature of these gases. The dew point temperature is the temperature at which excess water reaches saturation point in the flue gases and can no longer be held in suspension by the flue gases, as water vapour precipitates out as water in liquid form.

In the case of traditional floor standing non-condensing boilers, condensation if it occurs, can disfigure and damage the structure of a chimney or flue and the acidic nature of the sulphur trioxide it contains can, if it is allowed to re-enter the combustion chamber, cause rapid corrosion if the combustion heat exchanger of the appliance is made from cast iron or other vulnerable metal.

Flue pipe materials that have a low thermal resistance will contribute to the cause of condensation by the cold wall surfaces of the flue pipe having a cooling effect on the products of combustion. If the flue is installed externally it will be more susceptible to condensation than if the flue pipe were to be installed internally, where less of a temperature difference exists between the flue gas temperature and the ambient air temperature, particularly during the winter months.

The problem of condensation is reduced somewhat by drier air being drawn into the flue at the draught diverter which will mix with the flue gases: although this has the effect of lowering the flue gas temperature, it also reduces the overall moisture content of the flue gases in the secondary flue.

The exact dew point and moisture content can be determined for any given set of circumstances by the combined use of hygrometric and psychrometric charts, but to avoid or reduce the possibilities of condensation occurring within a flue, the following good practice should be observed.

1. Operate the boiler at its optimum temperature of 80–85°C. On domestic heating boilers this would correspond to the maximum thermostat setting number.
2. Size the flue correctly, not over- or under-sized.
3. Select the flue pipe material and route through the building so as to reduce the heat loss from the flue to a minimum.
4. Ensure that the appliance burner assembly operates at or near its maximum efficiency.
5. Limit the number of changes in direction in the flue pipe route when installing the flue.
6. Check that the flue height is correct to ensure that an adequate draught will be created.

PSYCHROMETRICS

'Psychrometrics' is defined as the science involving the thermodynamic properties of moist air and the effect that atmospheric moisture has on comfort conditions and building fabrics and materials. It is beyond the scope of this work to discuss the subject of psychrometrics in the detail that it deserves or warrants, other than to offer an introduction to the construction of the psychrometric chart and to understand its usefulness regarding the formation of condensation.

The psychrometric chart as depicted in Figure 16.39 may be used to determine a number of factors when both the dry bulb and wet bulb temperatures are known. These are explained as follows:

- Dry-Bulb Temperature – The temperature of air recorded on a standard thermometer.
- Wet-Bulb Temperature – The temperature of air recorded on a thermometer whose bulb is covered by a wetted cloth or wick and subjected to a current of rapidly moving air.
- Both the dry-bulb and wet-bulb temperatures may be recorded by a sling psychrometer.
- Relative Humidity – Expressed as a percentage of saturated air with moisture at the same temperature.
- Dew Point Temperature – The temperature at which condensation of moisture begins when the air temperature is cooled.
- Moisture Content/Specific Humidity – The weight of water vapour expressed in kg of moisture per kg of dry air.
- Specific Volume – The cubic metre of mixture per kg of dry air.
- Specific Enthalpy – The thermal property indicating the quantity of heat in air expressed in kJ per kg of air.
- Sensible Heat Factor – The ratio of sensible heat to total heat.

CONDENSING BOILERS

The preceding text has emphasised the importance of preventing the flue gases from condensing out in the flue together with the subsequent damage that can occur in the combustion chamber of the boiler and the

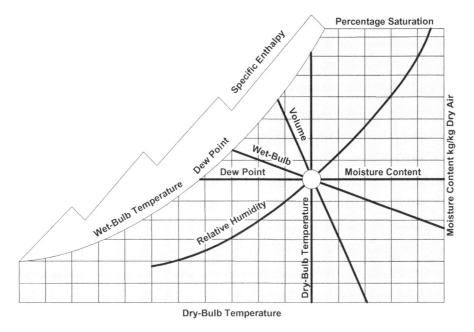

Figure 16.39   Skeleton psychrometric chart

problems that would transpire if this situation were allowed to exist. The consequence of achieving this non-condensing condition and the cost incurred, is a relatively low efficiency rating of the appliance.

All the traditionally flued boilers discussed so far have relied on the temperature difference principle to create the combustion air draught to expel the flue gases, with the result that a certain amount of heat is lost up the flue or chimney. Although the exact efficiency of an appliance varies from one model to another, most of these traditionally flued appliances had an efficiency rating ranging between 70% and 78%, which means that approximately 22% to 30% of the heat raised by the boiler/burner unit is lost up the flue in order for the appliance to function correctly.

These appliances were designed and constructed to operate with these efficiencies, and the efficiency levels were considered acceptable at that time as the 22% to 30% of the heat designed to go up the flue was required to create the draught necessary for the appliance to function correctly.

In recognition of the need to conserve energy and to reduce wastage, the UK Government developed and introduced a Standard Assessment Procedure (SAP) to apply to the construction and alteration of all new and existing dwellings. The SAP uses SEDBUK ratings of heating appliances to calculate the energy needed for heating and domestic hot water systems to assess the energy efficiency of both new and existing dwellings, and is included in the Building Regulations. The SAP is explained further in Chapter 24, Regulations, Standards, Codes and Guides.

SEDBUK

The **S**easonal **E**fficiency of **D**omestic **B**oilers in the **U**nited **K**ingdom (SEDBUK) method was developed under the UK Government's Energy Efficiency Best Practice Programme to provide a fair comparison of the efficiencies for oil- and gas-fired appliances, including LPG. SEDBUK is the average annual efficiency achieved by boilers in typical domestic conditions, making reasonable assumptions about pattern of usage,

Band	SEDBUK Range
A	90% or above
B	86% to 90%
C	82% to 86%
D	78% to 82%
E	74% to 78%
F	70% to 74%
G	Below 70%

Figure 16.40   SEDBUK efficiency bands

climate, control and other influences. It is calculated from the results of standard laboratory tests together with other factors, such as boiler type, ignition arrangement, internal store size and fuel used.

It is expressed as a percentage indicating how much of the heat obtained from burning the fuel is transferred to the heating and/or domestic hot water system. These percentage figures are grouped into bands on an 'A' to 'G' scale as shown in Figure 16.40, with band 'A' being the highest efficiency rating.

The system of applying a SEDBUK rating to heating appliances was introduced by the UK Government as an initiative to provide a temporary means of assigning an efficiency rating pending a European Directive on boiler efficiencies.

Boiler manufacturers are not compelled to display the SEDBUK rating band on their appliances, but in practice many do.

Domestic heating boilers are required to have a SEDBUK rating of 'A' or 'B' to comply with current Building Regulations.

DEVELOPMENT OF CONDENSING BOILERS

In order to improve on the efficiency of heating boilers and achieve the higher band ratings listed in the SEDBUK efficiency table, a radical redesign of heating boilers was required that would extract more heat from the combustion process and therefore tolerate flue gases condensing within the boiler combustion chamber and heat exchanger, without causing corrosion or premature failure of the appliance.

Figure 16.41 illustrates the heat balance for a typical non-condensing floor standing natural draught boiler with a cast iron heat exchanger and conventional flue, although it could equally apply to a low water content fan assisted room sealed appliance with finned tube heat exchanger operating on standard non-condensing principles.

The appliance has been arranged to operate on a standard flow temperature of 82°C to the heating system and a return water temperature of 71°C back to the boiler. The temperature of the flue gases will be approximately 200°C to ensure that they do not condense within the flue or combustion chamber.

Of the 100% of the energy supplied to the appliance burner as fuel, only approximately 75% of this energy is transferred to the heating system as useful heat, with about 22% of the energy being lost up the flue, as products of combustion and being required to create the natural draught for the combustion process to continue correctly.

These appliances often had poor thermal insulation, leading to relatively high heat losses through the boiler casing of 3% or more of the rated heat input.

The heat to water efficiency depicted in Figure 16.41 is shown as 75% based on the calculated consumption expected over the heating season, although the efficiency rating of appliances complying with this

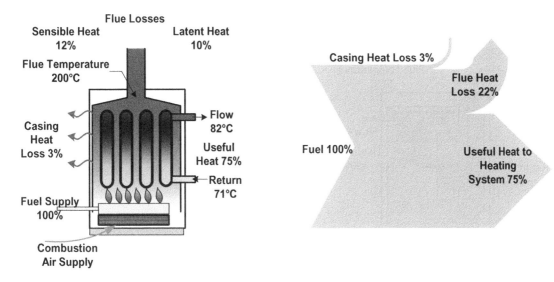

Figure 16.41 Typical non-condensing conventional boiler (SEDBUK 'E', 'F' and 'G')

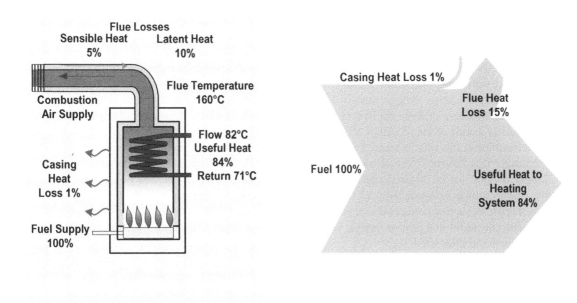

Figure 16.42 Typical non-condensing high-efficiency boiler (SEDBUK 'C' and 'D')

traditional boiler design could vary from a low of 60% to a high of 76%, depending upon model type and design.

This type of appliance would have a SEDBUK band rating of 'E', 'F' or 'G'.

Figure 16.42 illustrates a more efficient heat exchanger within the boiler, together with a higher level of casing thermal insulation enabling a greater level of useful heat output to the heating system to be achieved without cooling the flue gases beyond the dew point temperature.

The sensible heat flue losses are reduced to around 5%, but the latent heat flue losses remain at 10%.

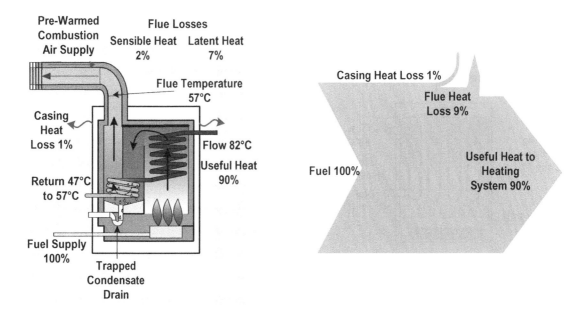

Figure 16.43   Typical single-coil condensing boiler (SEDBUK 'A' and 'B')

The combustion air is supplied through the outer section of a dual supply twin wall flue pipe, which has the effect of pre-heating the combustion air supply as it is passed around the outer part of the combustion chamber, which combines with other design features to improve the overall efficiency.

The flow and return water temperatures from the heating system remain at 82°C and 71°C respectively.

The seasonal efficiency of these appliances is around 84%, which places these high-efficiency appliances in the 'C' and 'D' bands of the SEDBUK efficiency rating method.

Figure 16.43 illustrates what is termed a single-coil condensing boiler, which incorporates an additional heat exchanger that has the effect of cooling the flue gases below its dew point temperature of 57°C for natural gas, or 47°C for an oil-fired appliance. This cooling effect of below the dew point temperature of the flue gases results in releasing some of the latent heat contained in the water vapour.

More of the sensible heat contained in the flue gases is also transferred to the heating system water, which together with the additional heat exchanger achieves an overall higher efficiency than the previous appliances described and illustrated.

The seasonal efficiency of these appliances is around 90%, which places these single condensing high-efficiency appliances in the 'A' and 'B' bands of the SEDBUK efficiency rating method.

Figure 16.44 illustrates a double-coil condensing boiler where a further heat exchanger has been added. This provides a second stage condensing process and can be incorporated as an extension to the second heat exchanger as shown, or can be a separate heat exchanger having its own flow and return from an independent heating circuit operating at a lower temperature, such as that required for an underfloor heating circuit.

The additional heat recovered from the flue gases raises the overall appliance efficiency to around 94% by extracting more of the heat from both the sensible and latent heat content.

CONDENSING HEATING SYSTEMS

Figures 16.41 to 16.44 serve to illustrate the development from the traditional non-condensing boiler arrangement operating with system flow and return temperatures of 82°C and 71°C respectively, and

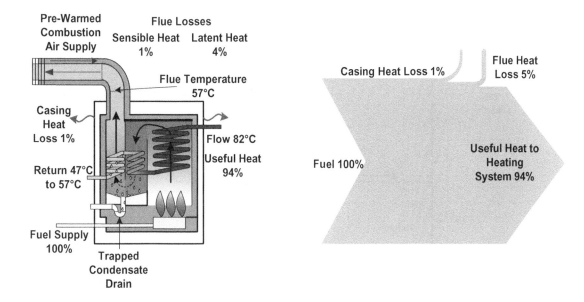

Figure 16.44 Typical double-coil condensing boiler (SEDBUK 'A')

achieving an overall efficiency in the low 70 percents. The efficiency of these appliances can be increased to the low 80 percents by improving the heat exchanger and casing insulation, whilst still retaining the system operating conditions that have become standard to UK heating practice, but without designing in the problem of condensation within the flue and combustion chamber.

As explained earlier in this chapter, increasing the efficiencies of fuel burning appliances by extracting more heat from the flue gases, which results in the temperature of the flue gases dropping below the dew point temperature of 57° for natural gas or 47° for hydrocarbon oil, would require a radical redesign of the boiler combustion process and the selection materials for the heat exchanger and flue passageways that would not be attacked by the aggressive nature of the condensed flue gases that have resulted.

The appliances illustrated in Figures 16.43 and 16.44 exemplify the condensing process within the combustion chamber by substituting the cast iron heat exchanger with a coil or drum type heat exchanger manufactured from a high grade stainless steel that is more resistant to the acidic and aggressive condensate that will occur. In order for this condensation within the flue gases to occur and in turn release a portion of the latent heat content, the temperature of the heating system return water entering the boiler must be low enough to allow the dew point temperature of the flue gases to be reached and condensation to occur.

Therefore, just replacing the traditional non-condensing boiler with a higher efficiency condensing appliance, without changing the heating system operating flow and return temperatures, will not achieve a continuous condensing operation, or the high efficiencies that have been claimed.

If the work just involves a boiler change, where a condensing boiler is to replace an old low-efficiency non-condensing boiler, the new boiler will be more efficient than the old one but will only be in the condensing mode when the heating system first fires up, where the return water temperature is still below the dew point temperature of the flue gases before it has reached full operational temperatures; therefore it will not achieve the efficiency rating of which it is capable.

For the condensing boiler to realise its full efficiency potential, the heating system must be designed to produce the operating conditions that it requires; this would involve selecting a greater temperature difference across the heat emitters, so that a lower return temperature back to the boiler is accomplished, as portrayed in the two-pipe direct return heating system shown in Figure 16.45.

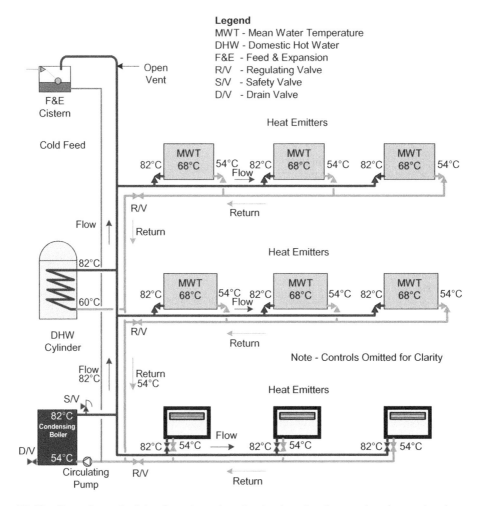

Figure 16.45 Operating principles for a two-pipe direct-return heating system in condensing mode

The system shown is the two-pipe direct return heating system shown in Figure 2.10 in Chapter 2, but modified in its operating conditions by designing the system to have a mean water temperature of 68°C across the heat emitters as explained in Chapter 8 and illustrated in Figure 8.4 and Table 8.2. This modification to the design of the heating system attains a return temperature of 54°C, which is below the dew point temperature of the flue gases for a heating appliance fuelled by natural gas, allowing it to operate continuously in the condensing mode. This would also result in larger heat emitters and smaller pipe sizes being required when compared to the conventional system depicted in its original form in Figure 2.10.

This arrangement is easy to accomplish when designing the system as new, but would involve extensive work in altering an existing heating system to achieve a continuous condensing function when just changing the boiler on the system.

This example uses the two-pipe direct return piping arrangement to demonstrate the differences between a condensing heating system and a non-condensing heating system, but the criteria and approach used are equally applicable to all other heating piping arrangements discussed earlier in this work.

It can be observed from Figure 16.45 that the return temperature from the domestic hot water cylinder is above the dew point temperature of the flue gases, but this will only occur when the temperature of the domestic hot water is about to be reached: most of the time the return temperature will be lower than 60°C as it will still be transferring its heat to the secondary water. This situation will only occur during the non-heating season of the summer months, when the heating system is off and the boiler is directing its heat to satisfy the domestic hot water requirements only. This situation would have to be tolerated as the temperature of the domestic hot water must be heated to 60°C.

The example chosen has kept the system operating flow temperature of 82°C, which when coupled with a system return temperature of 54°C, achieves a mean water temperature in the heat emitters of 68°C and a room temperature difference of 47°C when the room design temperature is 21°C. As previously explained, this results in larger heat emitters and smaller pipe sizes being required, and if a lower system flow temperature were to be selected, the heat emitters required would be even larger. However, there is merit in selecting lower system flow temperatures when the heating system concerned consists extensively of under-floor heating.

Condensing boilers do not pretend to be the answer in the search for alternative energies or appliances, but they are a step in the right direction in the conservation of fossil fuels by converting more of the fuel into useful heat. However, the predicted life expectancy of a condensing boiler is about a third of the life expectancy of the traditional non-condensing appliances that they are replacing, which means that we are manufacturing more boilers than we were before together, which requires additional manufacturing energy.

PLUMING

'Pluming' is the name given to the visible plume of water vapour contained in the products of combustion issuing from the appliance flue terminal, sometimes mistaken as smoke.

As established earlier in this chapter, water in the form of vapour is present in the flue gases from all appliances burning fossil fuels, regardless of whether they are condensing or non-condensing.

The flue gases leaving the appliance flue terminal from a non-condensing boiler will be at a temperature of 120°C or higher. Water will be present in the form of vapour, but invisible most of the time at that temperature, the only exceptions being during extreme cold weather periods, or when the boiler initially fires up from cold and the flue gases have not yet reached full operating temperature. Here the water vapour is cooled down beyond its dew point temperature when it encounters the cold air temperature, but this situation does not normally last long before the water vapour becomes a gas due to the higher temperature, and disperses unnoticed in the air.

As the flue gases from a condensing appliance will be leaving the flue terminal at a lower temperature than that experienced with non-condensing appliances – normally in the region of 55°C, but possibly as low as 30°C – the water vapour contained in the flue gases will start to condense out into a liquid as it is below its dew point temperature. It will be inevitable that some of this water vapour will be carried through the flue of the appliance and will be visibly noticeable in the form of a plume of vapour that is in the process of condensing out when it comes into contact with the cold air. However, if excessive amounts of water vapour are being experienced in the form of a plume issuing from the appliance flue terminal, then this is an indication that the flue gases are not condensing completely in the appliance flue where they are supposed to, and are not being removed by the condensate drain, but are still at a temperature above their dew point until they are discharged into the atmosphere.

The termination of flues from condensing appliances have additional requirements over and above those specified for flues in general: these include the preclusion of terminating them where the pluming effect would cause a nuisance to neighbouring properties or in areas that form a public access way. In these situations the flue should be terminated at a height that is sufficient not to cause a nuisance.

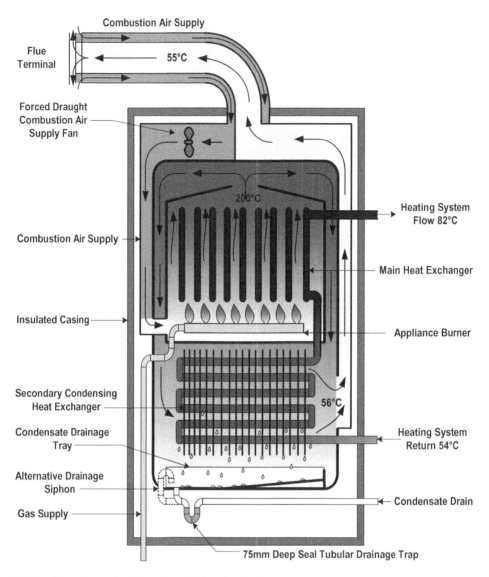

Figure 16.46   Operating principles of a condensing boiler

CONDENSATE DRAIN

One of the most noticeable features of condensing boilers is that they require an additional pipe connection, which is the drain pipe for conveying the condensed water from the appliance to drain.

The correct removal of the condensate from the appliance is an essential part of the installation and must comply with the requirements of the Building Regulations Approved Document Part H.

The condensing appliance incorporates an acidic resistant drip tray as indicated in Figure 16.46, which collects the condensed water from the flue gases in contact with the condensing heat exchanger coil and channels it towards the drain outlet which drains the condensate out by gravity to drain. Some appliances discharge the condensate by siphonic action, also shown in Figure 16.46, as an alternative method; this method discharges the condensate in full bore plugs of water at periodic intermittent intervals rather than by a slow drip of water, which is more prone to freezing if the condensate drain is located externally.

Regardless of which method of draining the condensate liquid is preferred by the appliance manufacturer, all condensate drains should incorporate a 75 mm deep seal trap as either an integral component provided by the boiler manufacturer, or installed as part of the installation works. If the water seal of any integral condensate drainage trap is less than 75 mm deep, then an additional compliant trap should be installed that has the correct depth of water seal with the two pipework sections separated by a tundish fitting.

The water that has been condensed out from the products of combustion can be quite acidic, having a pH value ranging between 3 and 5; therefore pipes manufactured from all the common metallic materials such as copper, low carbon mild steel or even standard AISI 304 stainless steel, are not suitable and should not be used. Pipe materials considered suitable for the expected low pH value and the discharge temperature of 54°C or lower are the regular thermoplastic waste pipe materials such as PVCU or ABS, both with HDPE traps, all of which are commonly available.

Figure 16.47 illustrates acceptable methods of disposing of the condensate waste from the condensing boiler. Figure 16.47(a) illustrates the termination via an externally located trapped gully, where the condensate drain should be extended to terminate below the gully grating, but above the water level within the trap. When the condensate is to be disposed of by this method, the appliance trap is permitted to have a shallow seal of 38 mm deep for a gravity drip discharge, but if the appliance disposes the condensate by discharging it by siphonic intermittent plugs of water, then the trap seal should be of the 75 mm deep seal type to prevent the seal of water contained within the trap from self siphoning out when subjected to the intermittent full bore flow being discharged periodically.

The minimum diameter for the condensate drain pipe should not be less than 22 mm and should preferably be installed internally. If any section of the condensate drain pipe has to be installed externally, then the diameter should be increased to 32 mm if its exposed length is in excess of 3 metres, with all externally installed pipes insulated against frost with a weatherproof covering.

Figure 16.47(b) illustrates the preferred method of condensate disposal by connecting the condensate waste directly into a soil, waste or vent pipe where it can then be conveyed as part of the domestic effluent from the building.

The condensate waste can also be connected into a rainwater pipe or hopper, if the surface water is legally connected directly to a combined sewer, but not if it is connected to a separate system of foul and surface water drainage, or if it is disposed of by a system of surface water soakaways, both of which can kill vegetation and plants and contaminate drinking water catchment areas, which can kill fish and other water life if it enters into a water course.

This method also indicates the arrangement for incorporating a second trap in the condensate drainage system when the appliance trap does not have a 75 mm deep seal. A visible air break, provided by a tundish drain, should be installed in the piping between the two traps.

The height of the condensate waste pipe connecting into the soil pipe should be at least 450 mm above the horizontal section of the drain as shown, to reduce the possibility of compression at the base of the vertical soil pipe being relieved through the condensate waste pipe back into the condensing appliance.

Figure 16.47(c) illustrates the most commonly employed method of condensate disposal for domestic dwellings, due to its simplicity. The condensate waste pipe is simply connected into an existing waste pipe downstream of the trap from a sink or washing machine from the appliance's 75 mm deep seal trap as shown. The depth of this trap seal is critical with this method so as to prevent either self- or induced-siphonage from occurring.

Figure 16.47(c) also illustrates an alternative route sometimes recommended by some sources for the condensate waste pipe, where it connects into the existing waste pipe upstream of the trap connected to the sink or washing machine. The author warns against this arrangement as there is a high risk of the water

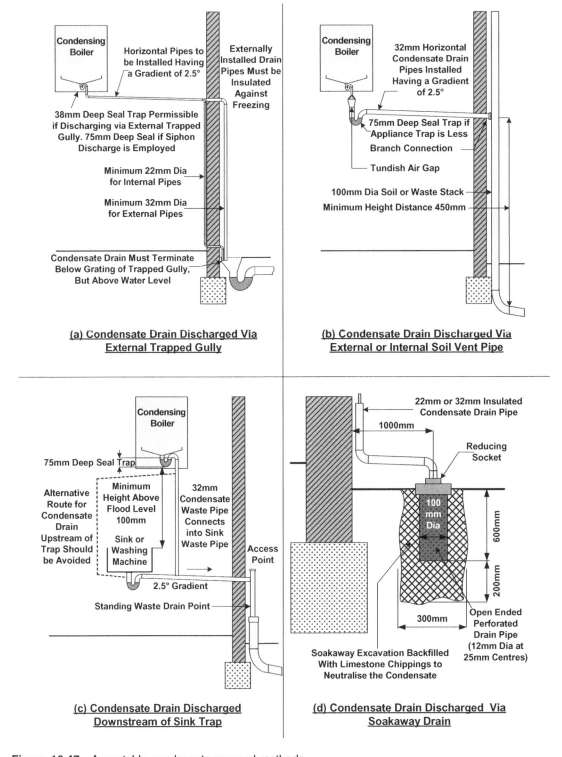

Figure 16.47   Acceptable condensate removal methods

seal in the condensing appliance trap being subjected to induced siphonage when a large quantity of water is being emptied from the sink bowl, or discharged from the washing machine. This arrangement also creates an unvented section of waste pipe between the two trap seals, which will inevitably be vented through the boiler when the appliance trap seal has been lost by induced siphonage.

Figure 16.47(d) illustrates the method of disposing of the condensate by a purpose constructed soakaway. If all the previous options are not possible then a purpose built soakaway may be used as a last resort. The soakaway should be excavated as shown to a depth of approximately 800 mm and a diameter of approximately 300 mm. The soakaway should be clear of any foundation by a distance of around 1000 mm and not be located close to any other services such as water and gas mains, or electrical cables.

A short length of 100 mm diameter thermoplastic perforated drain pipe should be placed in the excavated hole for the soakaway, with the space between the perforated pipe and excavation walls backfilled with limestone chippings – these chippings will serve to neutralise the acidic nature of the condensate making it less aggressive to plant life.

The soakaway illustrated is minute when compared to those constructed for the disposal of surface water, but is considered sufficient for the slow trickle or dripping of the condensate from a domestic condensing appliance, as the volume expected will be quite small and soak into the ground without causing any local flooding.

BOILER EFFICIENCY

Boiler manufacturers have traditionally stated the efficiency of their appliances when operating correctly by expressing it as a percentage which was calculated as follows:

$$\frac{\text{Heat Output}}{\text{Heat Input}} \times 100$$

The difference between the potential 'heat input' of the fuel to be consumed and the actual 'heat output' achieved to the heating system water is the designed loss in efficiency, which is mainly accounted for by the amount of heat being passed up the flue, as previously discussed, plus a small amount that is lost through the boiler casing via radiation and convection. As the heat lost through the casing is small, it still enters the heated space and contributes to the heat input and therefore can be discounted.

The present day method of expressing the efficiency of appliances under the SEDBUK system is less simplistic as it calculates the seasonal efficiency over the heating period. This takes into account periods of time when the external ambient temperature is quite mild and not at the standard base design temperature.

This procedure is based on empirical data making assumptions about seasonal temperatures and therefore leads to some boiler manufacturers claiming appliance efficiencies in excess of 100%.

17

Combustion Efficiency Testing

Having studied the subjects of the combustible fuels and combustion, together with the theory of the combustion reaction during the combustion process, which has been explained in the preceding chapter, it is important to understand the need for complete combustion to occur in order to achieve combustion efficiency of the appliance.

Boiler or appliance efficiency should not be confused with combustion efficiency. It has been established that manufacturers express their appliance efficiency as a percentage of heat input to that of heat output, with the remainder of the heat passing up through the flue. For example, an appliance stated to have an efficiency of 90% means that 90% of the energy released from the fuel will be converted into heat transferred into the heating system water, with the remaining 10% of the energy converted into heat passing up through the flue, in order to create or assist with the appliance's required draught and therefore operate at its maximum efficiency.

The subject of combustion efficiency is concerned with the effectiveness of converting fuel to heat by the combustion process and is expressed as 100 less stack losses as a percentage. Therefore if the boiler manufacturer states that their appliance has an efficiency of 80%, then the appliance should be commissioned and maintained to achieve this figure for the appliance to be totally efficient.

BURNER EFFICIENCY

In order to test for the combustion efficiency of an appliance, it is essential to know what the efficiency should be and what readings should be expected. Therefore, use should be made of the combustion reaction charts, together with their supporting text in the preceding chapter to determine the expected values for such constituents as CO_2, CO, SO_2, H_2O, Nitrogen and NOx, together with the flue temperature and general condition.

It is also important to appreciate that an efficient burner is one that operates with the minimum of excess air, coupled with the non-production of smoke.

Heating Services in Buildings: Design, Installation, Commissioning & Maintenance, First Edition. David E. Watkins.
© 2011 John Wiley & Sons, Ltd. Published 2011 by John Wiley & Sons, Ltd.
As an aid to lecturers and students, full colour versions of the figures in this chapter may be found at www.wiley.com/go/watkins
These figures are © 2011 by John Wiley & Sons, Ltd.

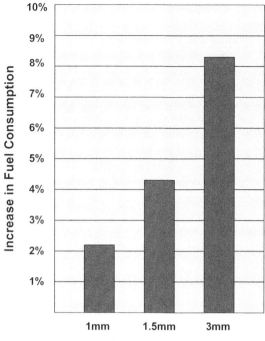

Figure 17.1   Increase of fuel consumption by the effects of soot

The effects of incomplete combustion will produce excessive smoke, resulting in the loss of usable heat – as less heat is being liberated – and causing soot to accumulate on the heating surfaces – which serves to insulate the heat exchanger, reducing the transfer of heat to the heating system water or air, and in the worst situation, retarding the velocity of the flue gases to a point where pressurisation of the combustion zone occurs.

Soot not only insulates these surfaces, but also increases the flue gas temperatures, with the overall result of increasing fuel consumption to achieve the required heat output; Figure 17.1 shows this effect as a percentage of increased fuel consumption.

Incomplete combustion will also produce a formation of carbon in the combustion chamber which will interfere with the combustion process.

Also, if more air is supplied than is required to achieve complete combustion, then heat will be lost due to the large volume of excess air carrying heat up the flue and overloading the gas passages of the appliance burner and heat exchanger.

COMBUSTION EFFICIENCY TESTING

The combustion efficiency test comprises a number of test procedures, all of which have to be completed and recorded before any final conclusions may be arrived at regarding the operating condition of the appliance, although individual test results may give an indication of any problems or deficiencies in combustion that may exist.

The method of determining the efficiency of a fuel burning appliance has developed over the last one hundred years, from a fairly inexact method to a more sophisticated and accurate set of procedures, to determine the amounts of the various constituents present in a flue gas.

Some of the test procedures previously used, but now considered obsolete, have been included in this text to help explain how this important subject has developed to today's recommended methods.

The basic test procedures explained use a traditional combustion efficiency testing kit, consisting of a CO_2 indicator, smoke tester, flue thermometer and draught gauge. The current preferred method uses an electronic flue gas analyser, which is required for domestic boilers under the requirements of CPA1 (Combustion Performance Assessment).

As the electronic flue gas analysers vary from one make to another and one model to another, the manufacturer's specific instructions for the test instrument should be adhered to.

The test results for the concentration of pollutant in the flue gases are stated as being either a percentage (%) of the total volume, or expressed as parts per million (ppm), which is basically the same as a percentage except that it expresses the concentration as being divided into a million parts instead of one hundred parts. Table 17.1 may be used as a quick conversion between the two units of measurements.

Alternatively, the concentration is sometimes expressed in the variable unit of milligrams per cubic metre ($mg\,m^{-3}$). This unit of measurement is dependent upon temperature and pressure, although Table 17.2 may be used to convert it to a fuel-specific unit.

The following test procedures are explained in no particular order.

FLUE GAS TEMPERATURE

The flue gas temperature may be ascertained by using a probe type flue gas thermometer, as depicted in Figure 17.2, or by an electronic digital flue gas analyser, as shown in Figure 17.3.

Regardless of which method is used, the temperature of the incoming combustion air to the burner must also be obtained so that the temperature difference between the flue gases being expelled and the incoming combustion air can be determined to establish the true flue gas temperature.

The probe of the thermometer or flue gas analyser is inserted into the test hole in the flue pipe, which should be as close as possible to the flue pipe outlet spigot from the boiler and before any draught

Table 17.1   Conversion of percentages to ppm

Parts per Million ppm	Percentage %
100,000	10
10,000	1
1,000	0.1
100	0.01
10	0.001
1	0.0001

Table 17.2   Conversion factors for energy related units

Pollutant	Natual Gas	Light Fuel Oil
CO	1 ppm = 1.074 mg/kWh 0.931 ppm = 1 mg/kWh 1 mg/m³ = 0.859 mg/kWh 1.164 mg/m³ = 1 mg/kWh	1 ppm = 1.110 mg/kWh 0.900 ppm = 1 mg/kWh 1 mg/m³ = 0.889 mg/kWh 1.125 mg/m³ = 1 mg/kWh
NOx	1 ppm = 1.759 mg/kWh 0.569 ppm = 1 mg/kWh 1 mg/m³ = 0.859 mg/kWh 1.164 mg/m³ = 1 mg/kWh	1 ppm = 1.822 mg/kWh 0.549 ppm = 1 mg/kWh 1 mg/m³ = 0.889 mg/kWh 1.125 mg/m³ = 1 mg/kWh

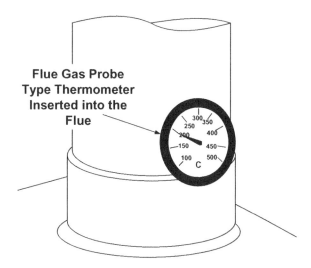

Figure 17.2   Application of flue gas probe type thermometer

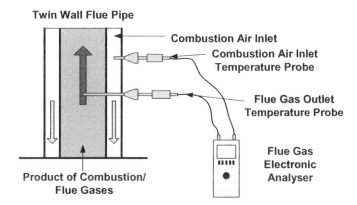

Figure 17.3   Application of electronic flue gas analyser and flue gas temperature

regulator, this being termed the 'hot spot point' and is considered the point where the highest temperature will exist, as well as the highest concentration of CO_2. If the flue pipe has not been provided with a test point, such as in older type installations, then a hole must be drilled using a 6 mm diameter drill bit. The hole should be sealed after the efficiency test procedures have been completed.

To obtain a true reading of the flue gas temperature the appliance must have been operated normally for at least fifteen minutes with the probe inserted at an angle of 90° to the flue pipe wall. The probe must also be located so that the tip of the probe is centrally positioned within the flue passage to avoid a false reading of the flue gases.

The temperature obtained from this test procedure will not determine the efficiency of the appliance on its own, but it will give an indication of the following:

- Excessive or inadequate draught conditions where excess or insufficient oxygen to the burner is occurring.
- Dirty and sooted-up surface of the appliance heat exchanger acting as an insulating material.
- Low or over-firing of the appliance burner.

Figure 17.3 indicates the application of an electronic flue gas analyser recording both the inlet combustion air temperature and the flue gas outlet temperature, complete with a read-out of the temperature difference recorded. The combustion air inlet temperature can be recorded as shown by inserting the probe into the outer flue pipe wall, into the air supply passage with the flue gas temperature recorded as normal. If a twin wall/passage flue pipe has not been used, then the combustion air inlet temperature must be recorded at the air inlet point to the appliance burner.

The flue gas temperature recorded on condensing boilers will give an indication about whether the appliance is operating in the condensing mode continuously or just intermittently, as this will affect the efficiency rating of the boiler. The condensing temperatures for different fuels are:

Natural Gas	57°C
Kerosene	47°C
LPG	54°C
Light Fuel Oil	50°C

DEW POINT TEMPERATURE

The dew point of the flue gas is the temperature at which the water vapour contained in the gas changes to the liquid state. The dew point is condensation, and the liquid formed is condensate.

Below the dew point temperature curve illustrated in Figure 17.4, moisture exists as a liquid; above this curve the moisture is contained as a gas, whereby this gas has not yet reached saturation point.

The dew point temperature is determined by the moisture content of the individual combustible fuel and therefore each fuel has a different dew point temperature, which is an important aspect when designing condensing heating systems. Condensation occurs when the flue gas temperature is cooled to a point where the flue gas becomes saturated and cannot hold any further moisture in suspension; this is the dew point temperature and below this temperature condensation will start to appear.

CARBON DIOXIDE (CO$_2$) CONTENT

The level of carbon dioxide in the flue gas provides an indication of the efficiency of the appliance burner. If the carbon dioxide content is as high as possible for the fuel being burnt, then complete combustion is being achieved with the appliance operating efficiently.

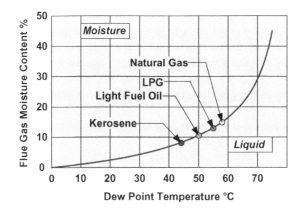

Figure 17.4 Dew point temperatures for combustible fuels at atmospheric pressure

Table 17.3   Summary of maximum achievable CO_2 content

Fuel	Maximum CO_2
Natural Gas	11.7
Fuel Oil	18.0
Propane	13.75
Butane	14.05
Coke	17.5
Wood	20.0
Manufactured Solid Fuel Briquette	19.3
Anthracite	18.0
Manufactured Town Gas	25.0

A low CO_2 content will indicate excess air being provided to the burner and therefore an inefficient appliance.

The previous chapters have discussed the importance of understanding the production of carbon dioxide in the combustion process and the maximum content that can be achieved for each fuel, but in practice the maximum is seldom possible to achieve as there is nearly always a small degree of excess air being provided to the burner. Table 17.3 summarises the maximum carbon dioxide content in the flue gases that is achievable for the fuels in common use.

The carbon dioxide content of the flue gases may be obtained by the use of either an electronic flue gas analyser or a chemical CO_2 absorption indicator as illustrated in Figure 17.5.

Before commencing the test with this type of apparatus, the plunger valve must first be depressed whilst the indicator is in the upright position to nullify any previous tests. Then the adjustable scale on the side must be reset so that the reagent liquid level corresponds to the zero CO_2 mark.

The CO_2 test is conducted with the boiler operating for over fifteen minutes by inserting the probe into the same test hole employed for the flue gas temperature test. The aspirator valve is depressed at least six times to warm the sampling tube and expel any air that would otherwise dilute the flue gas sample as it is taken into the indicator.

A sample of the flue gas may then be taken by holding the sample tube connector on top of the indicator plunger valve causing it to be depressed, then operating the aspirator valve twenty times in succession, ensuring that the plunger valve remains depressed. On the twentieth operation, the depressed plunger valve is released by removing the connecting plug, allowing the plunger valve to reseal before releasing the aspirator valve.

The CO_2 indicator is then inverted twice to ensure that the CO_2 in the flue gas sample is thoroughly absorbed by the reagent liquid: this will create a partial vacuum by the gas left in the indicator, so that the fluid level is partially raised up into the centre of the instrument tube by an amount equal to the CO_2 absorbed.

The indicator is then placed on a flat, cool, level surface and the CO_2 content in the sample is read off the scale against the height of the fluid column. The instrument should then be inverted twice more and the level read again until a constant reading is obtained: if this cannot be obtained in less than six additional operations, the reagent fluid should be replaced.

The reagent fluid chemical used is a strong alkali: care should be taken to avoid contact with the skin or clothing as it will cause a burning effect, and any contact should be immediately washed off with liberal amounts of water to render it harmless, or neutralised by applying vinegar. Figure 17.6 demonstrates the test procedure.

The electronic flue gas analyser is much simpler, and will draw the sample into the analyser by its integral diaphragm pump where both the oxygen concentration and carbon dioxide concentration are measured and analysed as an air ratio λ.

The Shandon Fyrite CO₂ Indicator

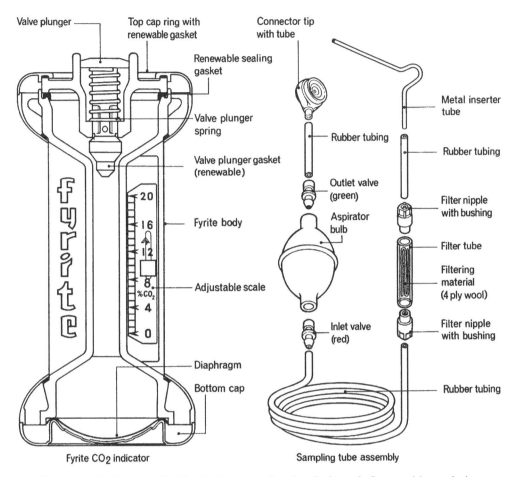

The reagent fluid is normally effective for two or three hundred tests before requiring replacing.

Figure 17.5   Chemical carbon dioxide absorption indicator

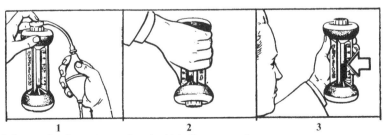

1) Depress the plunger twenty times to obtain flue gas sample.
2) Invert the indicator twice for the reagent to absorb the CO_2 sample.
3) Read the CO_2 content from the vertical scale on the side of the instrument.

Figure 17.6   Carbon dioxide test procedure

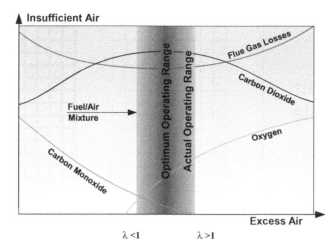

Figure 17.7 Combustion chart illustrating air ratio, λ (Lambda – the eleventh letter in the Greek alphabet used as the mathematical symbol for wave length)

The oxygen required for combustion is supplied to the burner via the ambient air from either the room in which it is located, or the external air in the case of room sealed appliances. To achieve complete combustion, surplus air, other than the theoretical amount required, has to be available for the combustion process. The ratio of this excess combustion air to the theoretical air requirement is termed the air ratio λ (lambda).

This air ratio is determined from the concentration of CO, CO_2 and O_2. These relationships are shown in the combustion diagram shown in Figure 17.7. During combustion each CO_2 level is related to a specific CO level during insufficient air conditions ($\lambda < 1$), and O_2 levels during excess air conditions ($\lambda > 1$).

SMOKE TEST

Excessive smoke is evidence of incomplete combustion occurring and loss of heat through the flue; this waste increases as the smoke density intensifies. Excessive smoke will also lead to rapid deposits of soot within the flue and combustion chamber of the appliance, which will contribute to an even greater loss of heat.

The smoke test is applicable to all combustible fuels, except it does not normally form part of the combustion efficiency test procedures for domestic gas burning appliances.

The method of assessing the amount of smoke issuing from a flue or chimney has changed over the years since the first attempt to measure this was devised by Professor Ringelmann in France, and this is one of the few tests that the electronic gas flue analysers do not incorporate.

The Ringelmann method, which is now obsolete, was based on visibility by comparing the colour or shade of the smoke leaving the top of the flue with a printed set of cards with the shading scale that has been reproduced in Figure 17.8. This method has been included in this work for historical reasons and takes no account of solids, but it is important for the student reader to understand how the subject of combustion efficiency has developed from these basic and inaccurate methods to those used today.

The Ringelmann smoke grading test

Each card shown in Figure 17.8 from 0 to 5 is 20% darker in shading than the preceding card, Thus card number 0 = 0% and card number 5 = 100% black smoke. If in a 5-minute period, 2 minutes resemble card

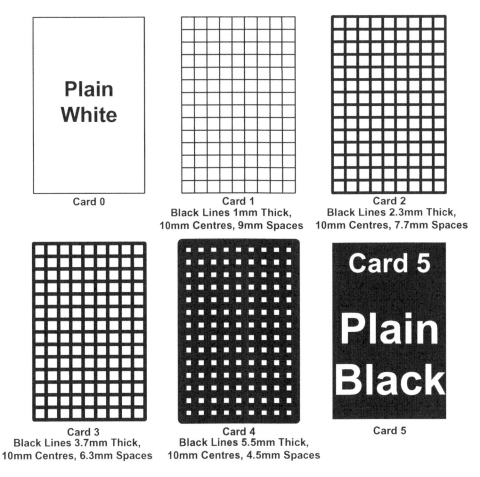

Figure 17.8   Ringelmann smoke density scale grading cards

shading number 4, 1 minute compares to card shading number 2 and 2 minutes compare to card number 0 smoke, the average smoke density for 5 minutes would be $(2 \times 80 + 1 \times 40 + 2 \times 0) \div 5$ equals 40% black on the Ringelmann chart.

Chart number 2, or 40%, was generally considered to be the maximum acceptable density.

The Ringelmann cards were used by placing them vertically in a line side-by-side, at a point of approximately 15 m from the observer and in the sight line of the flue or chimney.

At this distance, the lines become merged and the cards appear to be of different shades of grey, ranging from light grey to black. The observer glances from the smoke issuing from the flue or chimney to the series of grading cards and finds the card that most corresponds with the colour of the smoke, and does this continuously over a one minute period, recording the estimated average density during that minute. The tests should be repeated for every minute for a period of 5 to 15 minutes.

Although a system that compares the density of smoke by sight has obvious failings, it can give an instant indication as to the condition of an appliance.

A more accurate method is to take a sample of the flue gases from within the flue and measure the amount of smoke present.

A smoke test pump, similar to a bicycle pump complete with filter paper and smoke scale chart as shown in Figure 17.9, may be used for obtaining a smoke sample that can be used to produce the result.

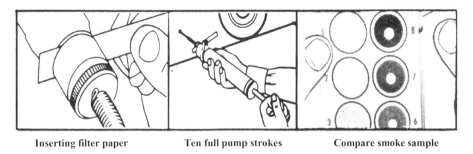

| Inserting filter paper | Ten full pump strokes | Compare smoke sample |

Figure 17.9 Smoke pump test procedure

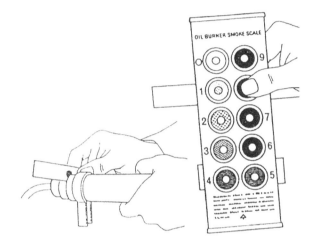

Figure 17.10 Smoke scale comparison chart

The smoke sample is taken from the same test hole used for the temperature and CO_2 tests previously described, ensuring that the boiler has been operating for a minimum of fifteen minutes. The probe from the test pump tube should be placed inside the flue so that it will obtain the flue gas sample from the central area of the flue passage and care should be taken not to contaminate it with soot that may have been deposited on the flue walls.

The hand pump should be operated a few times to warm the pump body, thus preventing condensation from occurring in the pump and expelling any air within the sample tube. A clean strip of Whatman Number 4 filter paper can then be inserted into the slot provided on the pump sample tube and clamped in position. A sample of the flue gas is then drawn over the filter paper by operating ten full strokes of the test pump, pausing a few seconds between strokes. It is important that the strokes are full and not varied, so as to obtain the correct volume of $0.00018\,\mathrm{m^3}$ of smoke laden flue products from ten full strokes.

The filter paper is then released from the pump and withdrawn from the holding slot and the 6mm diameter smoke stain colour/shading compared with the smoke scale based on BS 799 Part 2. This smoke scale card consists of ten circular discs each with a 6mm diameter hole in the middle, with each disc colour getting progressively darker in shading from pure white (0) to deep black (9), as can be seen in Figure 17.10. The filter paper with the smoke deposit on it can then be placed behind the holes so that the stain shows through the hole to allow the shade to be compared.

Table 17.4   Interpretation of smoke scale reading

Smoke Scale No	Rating	Condition
1	Excellent	Extremely Light
2	Good	Slight Sooting
3	Fair	May be Some Sooting, Requires Annual Cleaning
4	Poor	Borderline Condition, Requires Instant Cleaning
5	Very Poor	Heavy Sooting

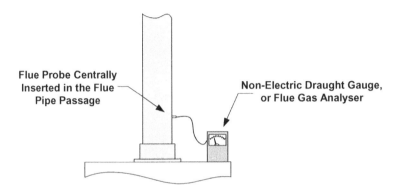

Figure 17.11   Flue draught test

The smoke scale illustrated in Table 17.4 interprets the readings in terms of sooting produced. Some boilers can tolerate a higher smoke scale number than others, but manufacturers' specific recommendations should be referred to. Generally, smoke scale reading number 2 should be aimed for, and scale reading number 3 should not be exceeded.

Smoky combustion recorded on the smoke scale chart may be attributed to:

- Insufficient draught or air supply
- Poor fuel supply or low quality fuel
- Defective or incorrectly set burner
- Excessively high CO_2 content

FLUE DRAUGHT TEST

In order to determine the flue draught necessary to convey the products of combustion away in boilers equipped with atmospheric/natural draught burners, the flue draught must be established.

Because the natural draught created is fairly low, the draught can be measured by either an inclined manometer, non-electrical draught gauge or an electronic flue gas analyser, as shown in Figure 17.11.

Regardless of which instrument is being used, the probe from it is inserted into the same test hole used for all the previous tests described and – having first zeroed the instrument scale before commencing – the probe and instrument can now be activated to record the draught being created.

NITROGEN OXIDES (NOx)

The electronic flue gas analysers have a number of advantages over the other methods described, including the ability of some models to analyse nitrogen monoxide (NO) and nitrogen dioxide (NO_2). High values of NOx are an indication of excess air above that needed to provide efficient combustion.

ELECTRONIC FLUE GAS ANALYSERS

Although the merits of the electronic flue gas analysers have been discussed, they do have two main disadvantages: one is the initial purchase cost of the instrument, that can vary considerably between models; the other is that the instrument needs to be checked and recalibrated on an annual basis, an exercise that is also expensive.

18

Circulating Pumps

It would be understandable if students believed that pumps are a modern day invention, but they have been around in one form or another for a very long time. The best example is that of the Archimedean screw pump, invented by the Greek mathematician and physicist Archimedes (287–212 BC), which is still used today and – apart from the motive power source – has changed very little in its design. This form of pump is of no interest in the use of hydronic heating systems, but since ancient times many different forms of pumps have been developed.

In general, pumps are classified by their action, but the heating industry uses the rotodynamic form of pump, which involves the transfer of mechanical energy to a fluid by means of a rotating impeller, causing an increase in pressure due to centrifugal force. The centrifugal type of pump is by far the most commonly used pump in building services, although it comes in many different forms, of which only a few are of interest to the heating industry.

Before considering the theory and detail of these pumps, mention should be made of the fact that no other single piece of equipment or device has had such a dramatic impact as the development of the gland-less inline centrifugal circulating pump, which is comparatively noiseless in its operation and has made hydronic heating systems both possible and affordable in all domestic and other applications. Although this pump has been improved upon many times since its introduction, it can justifiably claim to be the main significant component that gave birth to the domestic hydronic heating industry.

Prior to the introduction of the inline domestic glandless circulating pump in the 1950s, hydronic heating systems installed in domestic dwellings worked either by thermo-siphon or gravity circulation – which required larger diameter pipes arranged with slight gradients to assist circulation – or large and noisy pumps incorporating a seal type gland to prevent leakage from the pump casing around the physical drive shaft that connected the pump and motor and penetrated the pump casing. Such pumps required their own designated plant room so that they were housed out of sight and sound. The glandless circulating pump does not require a physical connection such as a drive shaft between the pump and motor to function, and therefore there is no risk of water leaking back into the electric drive motor – making it ideal for domestic applications. In either case, for most dwellings these arrangements were not practical or affordable and

Heating Services in Buildings: Design, Installation, Commissioning & Maintenance, First Edition. David E. Watkins.
© 2011 John Wiley & Sons, Ltd. Published 2011 by John Wiley & Sons, Ltd.
As an aid to lecturers and students, full colour versions of the figures in this chapter may be found at www.wiley.com/go/watkins
These figures are © 2011 by John Wiley & Sons, Ltd.

could only be installed in larger dwellings, where the high cost could be justified. For this reason, the development of this type of circulating pump, which also coincided with the development of the pressed steel panel radiator, revolutionised the domestic heating industry.

CENTRIFUGAL ACTION

The centrifugal type of rotodynaminc pump is essentially a high speed machine containing a rotating element of one or more impellers, arranged off-centre inside a volute shaped casing with a central inlet and outer edge branch discharge outlet.

The motive force driver converts part of the output torque velocity energy into pressure energy by centrifugal force, which is described as an outward energy movement from a rotating central point that is also a function of the impeller vane peripheral velocity. The high speed rotation of the impeller adds energy to the liquid after it enters the central inlet eye of the impeller, whereby the pump casing serves to collect the liquid as it leaves the impeller and guides it out the discharge branch. The impeller has vanes or blades shaped in such a way as to guide the liquid to the outer edge of the casing and to impart energy during this guidance.

Figure 18.1 illustrates the principal components of the centrifugal pump together with the action of the centrifugal force.

This centrifugal action results in a continuous, non-pulsating steady state flow rate and discharge pressure, the extent of which is dependent upon the speed of the impeller, size of the impeller, shape and spacing of the impeller vanes, size of the volute casing and the relationship between the volute space separating the impeller from the pump casing. Any variation or adjustment of one or more of these factors will result in a change in the pump performance, all of which makes this pump type ideal for most building services and other liquid movement applications.

An example impeller is shown in Figure 18.2.

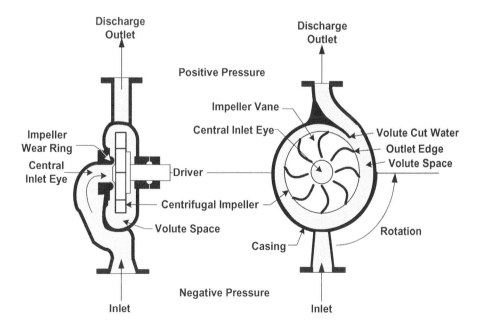

Figure 18.1   Principles of centrifugal force

Figure 18.2   Centrifugal pump impeller

CENTRIFUGAL PUMPS

As previously established, centrifugal pumps are widely used for all building services applications and are therefore available in a variety of different forms and perform different duties. These include high pressure booster pumps for water distribution and fire protection, submersible forms for drainage applications, as well as many variations of circulating pumps.

It is not the intention to provide a detailed text on all pumping systems and pump types as a large volume would be required to do the subject justice. This chapter is concerned with circulating pumps associated with hydronic heating systems only, together with their details and constraints.

CIRCULATING PUMPS

Circulators are sometimes referred to as accelerators because they accelerate the natural gravity or thermo-siphon circulation that exists – or would exist, if there were no such thing as frictional resistance within the piping system to prevent or hinder it. They circulate the heat carrying medium around the pipework system from the heat raising boiler plant to the heat emitters and back again. Because of this function, they are known as relatively low head pumps compared to other building services pumps such as cold water booster pumps, or fire protection pumps that have to raise water from a low level to a higher level, plus providing a working pressure at the outlet point. Circulating pumps are only required to generate sufficient pressure or head to overcome the frictional resistance encountered by the piping circuit. Figure 18.3 serves to illustrate this matter, disregarding any physical height of the heating system.

Most circulating pumps for heating systems are classified as low head pumps, except those pumps required for some district heating schemes and similar installations. But the volume – or flow rate – of water required to be pumped can be quite large for circulating to all circuits and not just the index circuit that the pump head is concerned with. Therefore, heating systems require low head, large flow rate centrifugal circulating pumps to meet the requirements of the heating system.

The form of pump normally required for these conditions is either an inline pump or, for larger duties, an end suction pump as shown in Figure 18.4.

Inline pumps have the advantages of requiring less physical space, being neater in appearance and less costly and – in the case of smaller systems such as domestic installations – can be pipeline mounted. They are limited to the lower duty range and this can make maintenance more difficult if the pump has to be removed completely from the pipeline at its connections.

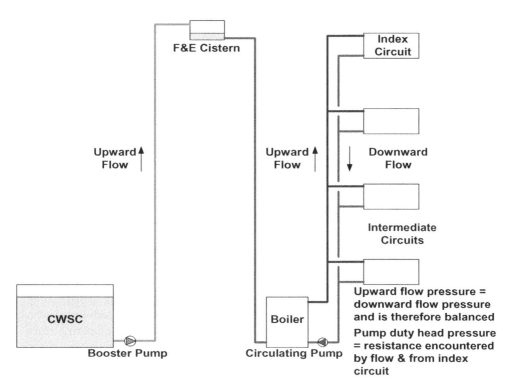

Figure 18.3 Circulating pump head requirements

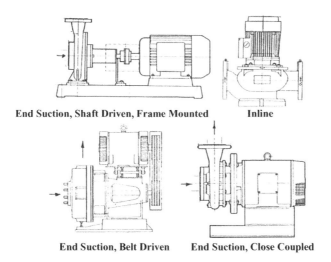

Figure 18.4 Variations of circulating pump forms

End suction pumps are more versatile and can be obtained in a number of forms. The end suction, shaft driven version has the advantage of allowing the motor to be removed without disturbing the pipework connections, but is usually more expensive and requires larger plant space to accommodate it.

The close coupled version is more compact, but maintenance is more difficult and greater protection, by way of seals, has to be provided to prevent leakage from the pump casing to the motor via the drive shaft.

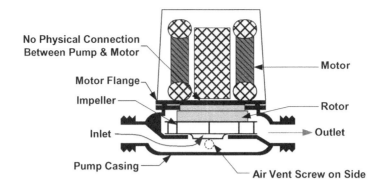

Figure 18.5   Glandless circulating pump

The belt driven type is a compromise between the two previous versions, where the motor can be mounted vertically on top of the pump as shown, or by the side of it. The belt drive gives the advantage of enabling fine adjustments to be made to the pump performance duty by either tightening or slackening the belt tension, or even changing the belt or belt pulleys. However, it does have the disadvantage of requiring constant checking of the belt drive for tension and wear in order to maintain its optimum performance.

DOMESTIC INLINE CIRCULATORS

Circulating pumps used for domestic heating applications are nearly always of an inline type known as a glandless or canned rotor pump. It is this development, where there is a physical barrier between the electrical workings and the water passages of the pump, that makes it ideally suited for domestic applications, as the risk of water coming into contact with the electrical components is greatly reduced.

Figure 18.5 illustrates the basic principles of a glandless circulating pump. The motor windings induce the rotor, that is attached to the impeller, to rotate at a speed of between 1700 to 3000 r.p.m. by the changing magnetic fields in the windings through a stainless steel rotor sealing plate separating the pump body from the motor.

The rotor is constructed from steel with short-circulated copper bars embedded in its top plate surface, and it is these windings that induce a current in these bars, causing it to rotate and forcing the impeller to turn with it as the two are connected. The exact detail will vary from one manufacturer to another, but the basic principle will remain the same.

As the name of the pump implies, it does not require any physical drive shaft or sealing glands between the pump and motor that could eventually leak due to wear and deterioration, as there is a non-moving sealing plate between the two, which makes it an ideal and safe piece of machinery for domestic applications.

Figure 18.6 illustrates a variation of this type of pump favoured by some manufacturers. It is termed a canned rotor, as the rotor takes a different form and resembles a can arranged over the rotor to form the seal between the pump and motor.

PUMP SELECTION

In order for the heating system to function correctly and operate as the design is intended it to, it is important to make a correct pump selection for the duty that is required. The method of establishing the

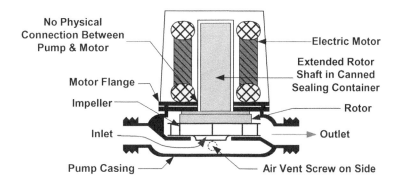

Figure 18.6   Canned rotor version of glandless circulating pump

performance duty of the pump is described in Chapter 9, forming part of the results obtained from the pipe sizing exercise, where the procedure for calculating the mass flow rate and pump head discharge pressure is fully explained.

The mass flow rate (MFR) is the total volume of water required to be circulated by the circulating pump around the heating system to convey the heat that is required to maintain the designed room temperatures.

The mass flow rate is expressed as kg/sec (litres per second) and should include an additional allowance of 10% to cater for the balancing procedure of the heating system.

The pump discharge pressure head is the total resistance for the index circuit only of the heating system, including the frictional resistance for all pipework changes in direction and pipe fittings that the pump duty has to be capable of overcoming, in order to circulate the mass flow rate required. This is normally expressed in the unit Pascal, as $Pa\,m^{-1}$ run of pipe, or as kN/m^2 on older publications. For larger installations the pump head could be expressed in the more practical unit of metres head or bar.

For conversion:

Head of water in metres	$= 9810\,Pa$, or $9.81\,kPa$ ($9810\ kN\,m^{-2}$, or $9.81\ kN\,m^{-2}$)
One bar pressure	$= 100\,000\,Pa$, or $100\,kPa$, or 10.2 metres
1000 mbar	$= 1\,bar$
One millibar (mbar)	$= 100\,Pa$
Pressure in Pa	$= 0.102$ metres head

It is usual to include an additional 10% to the calculated pump head to cater for unforeseen frictional resistance.

Having arrived at the duty that is required for the pump to perform, it can now be selected from manufacturers' performance graphs.

Using the pump duty of $0.42\,kg\,s^{-1}$ @ $20\,kPa$ calculated in the example in Chapter 9, it can now be plotted on the performance graph depicted in Figure 18.7.

The performance graph illustrated has three duty curves printed on it. By plotting the calculated duty point required onto it, it can be observed that the duty is at the top part of curve 3, but below the performance curve line, which means that with a resistance encountered of $20\,kPa$ ($2.04\,m$), the pump will actually circulate $0.66\,kg\,s^{-1}$. This will allow the heating system to be commissioned and balanced to obtain the original calculated duty of $0.42\,kg\,s^{-1}$ at $20\,kPa$.

The performance curve represents a range of pump heads and flow rates obtainable by either adjusting the pump speed setting on a variable performance pump, or in the case of larger duties, selecting the manufacturer's model number for the pump type from the range available.

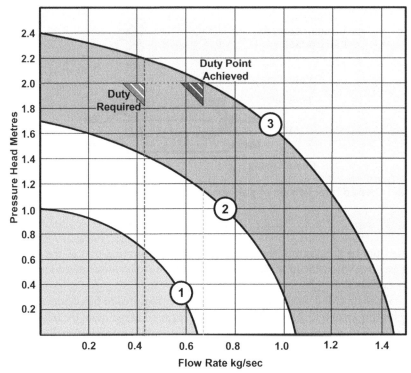

Duty required 0.42 kg/s at 20 kPa (2.04m)
Duty achieved 0.66 kg/s at 20 kPa (2.04m)

Figure 18.7   Typical pump performance graph

FIXED OR VARIABLE PERFORMANCE PUMPS

It is quite common for pump manufacturers to claim that their pumps are capable of performing certain duties. For example, using the performance graph depicted in Figure 18.7, it could be claimed that this pump is capable of achieving flow rates up to 1.425 kg s^{-1} and pressure heads of up to 2.4 metres, but by studying the performance curve it can be seen that the pump cannot achieve both these figures together. It can achieve a pressure of 2.4 metres with virtually no flow rate, or a flow rate of 1.425 kg s^{-1} with virtually no pressure, therefore the pump must be selected within the duty range of its performance curve, as in the example given.

The example performance duty graph also shows three duty curves that represent three settings for a variable speed pump, or individual duties of three separate pumps contained within a manufacturer's model range.

Variable speed pumps are common to most domestic and small commercial circulating pumps; they mean that the pump manufacturer only has to produce a single pump that has the facility of achieving a wider range of duties. In the example given, the pump speed has to be adjusted by selecting from three settings where speed setting number 3 was selected to achieve the duty required. This adjustment is normally done by arranging the pump to operate at different speed settings, controlled electrically by an adjustment screw on the side of the pump regulator.

Not quite so common, but used extensively in the early days of the production of these domestic glandless circulating pumps, the variable duty adjustment was sometimes achieved by a lever located on the back of the pump body, which serves to adjust the distance gap between the pump impeller and the pump casing whilst the pump motor remains at a constant speed. This method of adjustment relied on altering the efficiency of the pump by varying the volute space of the pump impeller within the casing.

This method of varying the pump performance, although electrically inefficient and no longer favoured, did have the advantage of being able to free-up any seized impeller that had become stuck by a build-up of sludge if the pump had been dormant for a period of time. This was achieved by operating the adjustment lever a few times, which would have the effect of physically moving the impeller within the casing, allowing it to become free to move.

Fixed duty pumps differ in so far as they normally apply to circulating pumps having larger duties, and the range of duty is too great for any adjustment to be made. This means that the pump manufacturer has to produce a number of pumps, albeit of the same design type, to cover the range of duties that may be required. Each pump would therefore have just one fixed duty curve, and using the example given in Figure 18.7, each of the three curves would represent a separate pump within a model range.

A detrimental effect resulting from the practice of relying on adjusting the pump speed instead of calculating the duty correctly, is that if the system resistance does not match the pump discharge pressure, the current being drawn by the motor will increase, with the consequential effect of increased inefficiency and early failure of the pump motor.

Even if the pump duty has been correctly calculated, it will only function in this mode when all system zones are fully open, allowing the water to be circulated throughout the entire system. If part of the heating system has been isolated, or certain zones closed by the control scheme, including thermostatic radiator valves closing, then the system resistance will be altered, affecting the efficiency of the pump and the current being drawn.

A more recent development in the manufacture of domestic circulating pumps, which has been pioneered in larger commercial sized circulating pumps, is the pump's ability to sense this change in system resistance and automatically reduce the pump/motor speed to compensate and retain its original efficiency. The practice of employing inverter pumps in larger systems has been common practice for a number of years and is now being applied to domestic heating systems.

PUMP EFFECTS

A question often posed by students concerns the best location for the circulating pump within the heating system. Although it can be argued that the return will permit the pump to operate at a lower temperature and therefore prolong the life of the pump, there is no ideal common position. Each system should be considered on an individual basis.

Before considering the factors that determine the position of the circulating pump, it is important to have an understanding of the effects that the circulating pump will have on the heating system and how it works within the system.

The pump by its nature will have both a positive pressure (push) on its discharge outlet, and a negative pressure (pull) on its suction inlet, the magnitude of which will be determined by its total head pressure generated and its relationship with the system entry position of the cold feed.

Figure 18.8 serves to illustrate this matter where the pump is located on the heating flow, which has the effect of creating a predominantly positive pressure on the system by using most of the head pressure generated to push the water around the heating circuit. This pressure is at its greatest at the discharge outlet of the pump, then gradually diminishes as it travels around the heating system, where it encounters the frictional resistance of the heating system until it ceases altogether when it reaches the point of entry for the cold feed. This point is termed the 'neutral point'; at this point neither positive nor negative pressure exists. From this neutral point back to the pump the pressure converts to negative, the amount increasing gradually until it reaches the pump where it changes back again to positive pressure for the process to be

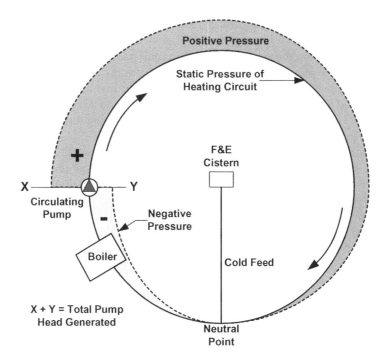

Figure 18.8   Predominantly positive pressure (pump pushes water around system)

repeated. The reason why this pressure changes from positive to negative at the entry of the cold feed has never been fully understood, but if a series of pressure gauges were to be fitted around the system this phenomenon could be observed.

In Figure 18.9, the opposite occurs. Here the pump is positioned on the return with the cold feed entry into the heating system positioned forward of it where the majority of the pressure created is negative – that is, pressure that is less than existed when static with no pump running, so that the circulating pump has created a partial vacuum. This has the effect of pulling the water around the heating system.

As with the predominantly positive system, the same effect is repeated but in reverse, where the negative pressure gradually decreases until it reaches the cold feed neutral point position and reverts back to a positive pressure between the neutral point and circulating pump.

It can be appreciated that by changing the position of the circulating pump in the heating system, or the neutral point position, the system characteristics can be altered from positive to negative or vice versa to any degree required. If the circulating pump was to be relocated halfway round the heating system, it would have the result of creating a system of half positive and half negative, each using approximately half the total pump head pressure available.

Having established that the heating system can be both positive and negative, this will result in any open end, such as a vent pipe, likely to be subjected to either a pressure greater than atmospheric and in danger of pumping out, in the case of a positive system, or pressure less than atmospheric and drawing air in, as in the case of a negative system.

It should be noted that any slight weakness in the pipework joints which proved to be sound when subjected to static pressure, could leak when a greater positive pressure is applied, or draw air into the heating circuit when subjected to a negative pressure. This latter situation could be quite a problem to find and may go undetected for some time, as there are no apparent external signs and the only indication is a recurring air lock in one of the heat emitters that requires frequent air venting.

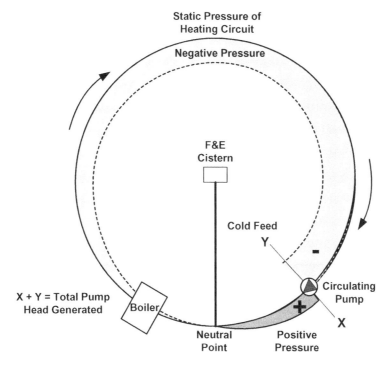

Figure 18.9   Predominantly negative pressure (pump pulls water around system)

This situation emphasises the importance of hydraulically pressure testing all the pipework on completion and not just relying on static pressure alone to test all pipework joints.

PUMP POSITION

Having discussed and understood the effects that the circulating pump will have on the heating system, its position can now be determined.

To demonstrate the pressure effects on a heating system, the area around the boiler on Figures 18.8 and 18.9 has been enlarged and detailed in schematic form in Figure 18.10, with alternative open vent positions added.

Figure 18.10(a) indicates the heating system with the pump arranged to operate under a positive pressure by its effects. The circulating pump has been located on the flow pipe immediately by the boiler, which will have the effect of creating a negative pressure on one of the open vent pipe positions, with a positive pressure being produced on the alternative open vent pipe position.

This pressure, be it positive or negative, may be calculated back from the pump position denoted as 'Y' on the drawing as follows:

Equivalent length of pipe from 'Y' back to position 'W' at the neutral point cold feed entry multiplied by the pressure drop per metre for the pipe size concerned, will equal the portion of the pump head that is negative.

Equivalent length of pipe from 'Y' forward to position 'W' multiplied by the pressure drop per metre for the pipe size concerned, equals the portion of pump head that is positive.

For open vent position 'A':

Total negative pump head – Equivalent pipe length × pressure drop from 'Y' back to 'X' equals negative pump head at P1.

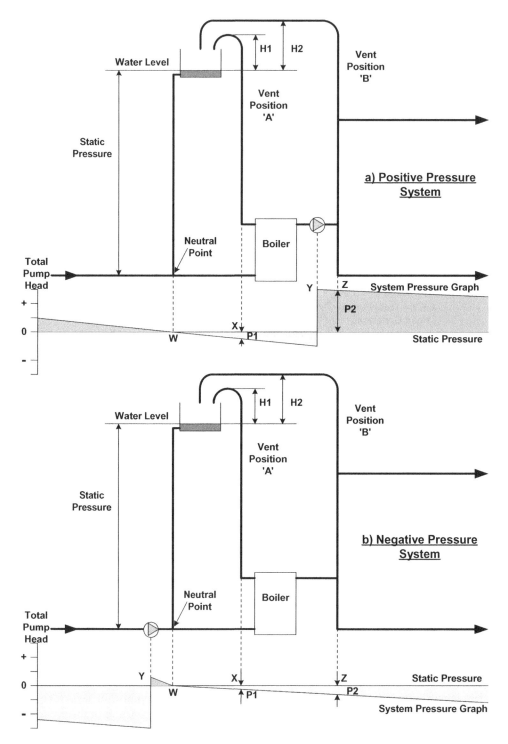

Figure 18.10   Heating system pressure graphs

P1 negative pump head expressed in metres will be the distance the level of the water will be lowered in the vent pipe at H1.

If a branch connection is made within this distance then air will be drawn into the system.

For open vent position 'B':

Total positive pump head – Equivalent pipe length × pressure drop from 'Y' forward to 'Z' equals positive pump head at P2.

If the height measured at H2 is less than P2 measured in metres, water will be pumped out of the open vent pipe.

In Figure 18.10(b) the same heating system has been arranged to operate as a predominantly negative pressure system, although if the circulating pump were to be repositioned between the cold feed entry point and the boiler, the heating system would operate as a predominantly positive pressure system.

The procedure for calculating the pressure at P1 and P2 is the same as described for the positive arrangement in Figure 18.10(a).

If the results are such that the possibility exists that either air can be drawn into the heating system, or water pumped out over the open vent pipe, then consideration should be given to raising the open vent pipe loop if it is practical without being too high, or the pump position in relation to the neutral point and open vent pipe should be reconsidered, without compromising the operational and safety aspects of these pipes, as explained in Chapter 2.

CIRCULATING PUMPS FOR SEALED HEATING SYSTEMS

The pump positions and their effects considered so far have concentrated on open vented heating systems operating at atmospheric pressure. With sealed heating systems, the procedure for calculating the circulating pump duty together with its operational effects – i.e. positive or negative pressure – are exactly the same as for the open vented system, as in Figure 18.11.

The following points should be understood in order for the heating system to operate correctly and safely.

- Due to the fact that the heating system is sealed, it is designed to operate slightly pressurised, at a pressure above its normal charge static pressure. Therefore the effect of the circulating pump will be to either increase this pressure by an amount equal to its duty pump head and operate as a positive system, or to lower this pressure by an amount equal to its duty head, which may or may not be sufficient to reduce the pressure below its initial charge static pressure and can be considered as operating as a negative pressure system.
- The sealed expansion vessel will take the place of the cold feed entry point, therefore the neutral point or position of system pressure change is now found to be at the point of connection of the sealed expansion vessel to the heating system.
- Due to the absence of an open vent pipe, the problem of pumping out the vent pipe or drawing air into the heating system will not exist.
- As previously established, the heating system is designed to operate at a pressure above the initial charge static pressure, therefore it is desirable to position the pump in relation to the expansion vessel so as to retain a pressurised system; see Figure 18.11(a).
- If the pump is positioned so as to create a negative situation as depicted in Figure 18.11(b), then cavitation may occur, causing damage to the system, especially if higher operating temperatures have been selected.
- For the reasons previously mentioned, it is desirable to arrange the pump and sealed expansion vessel to operate with a positive pressure pump head, but it should be noted that the piping system and components should be capable of withstanding the combined pressure of the expansion vessel operating pressure and the maximum pump head.

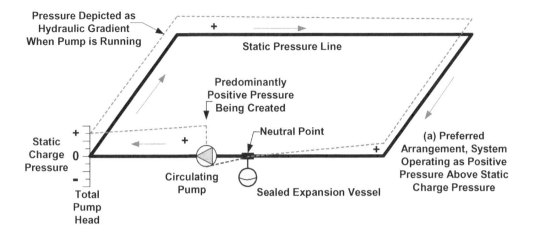

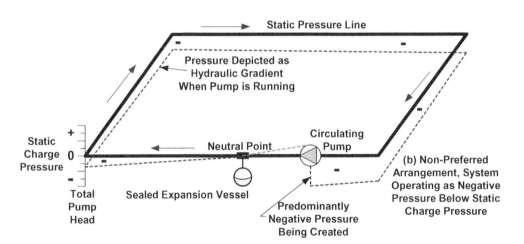

Figure 18.11   Pump effects on sealed heating system

A compromise position for the sealed expansion vessel, which may be suitable for some applications, is to locate the sealed expansion vessel approximately halfway round the heating system to create half the system operating at positive pressure conditions, and the remaining half operating under negative pressure conditions.

CAVITATION

The problem of cavitation can occur in most pumping applications where there is insufficient net positive suction head (NPSH) available, i.e. too low a suction pressure leads to cavitation.

The term 'cavitation' is derived from the word cavity, meaning an empty or hollow space, and the empty spaces found in a liquid when cavitation occurs are bubbles of water vapour, similar in nature to the bubbles formed when water boils. This boiling can be induced without heating the liquid, but by lowering the absolute pressure on it.

Water has a boiling point temperature of 100°C at atmospheric pressure measured at sea level: any change in this pressure will have a corresponding change in its boiling temperature, and so if the pressure is increased above this level the boiling point will also be raised above 100°C. It also follows that if the pressure is reduced below atmospheric pressure, the boiling point of the water will be less than 100°C.

Applying this to a hydronic heating system, it has already been established that the circulating pump will create both positive and negative pressures, the magnitude of which is dependent on its relationship to the entry of the cold feed, termed the neutral point. Therefore it is reasonable to suppose that a situation could easily exist whereby the circulating pump could cause sufficient reduction in suction pressure to lower the boiling point to below that of the operating temperature of the heating system (see Figure 18.12).

Although the example depicted in Figure 18.12 can be said to be extreme, similar conditions could easily exist where the boiling temperature of the circulating water will be dangerously close to that of the system operating temperature.

In the example illustrated, the local pressure in the pump suction pipework and boiler heat exchanger has been reduced to below that of the liquid vapour pressure, with the inevitable result of vapour bubbles

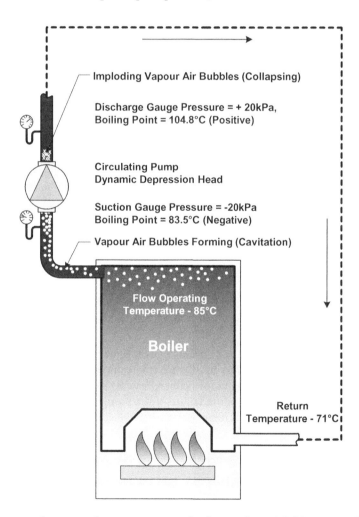

Zero gauge pressure equals atmospheric pressure which equals 101.3 kPa at sea level (1013mb)

Figure 18.12 Example of cavitation and vapour pressure

forming, accompanied by a boiling sound sometimes referred to as 'kettling', as it can be likened to the sound of a kettle coming to the boil.

At this point, although the sound can be concerning, no damage to the heating system has occurred as the velocity of the liquid flowing through the piping is swiftly moved on into the circulating pump where a change to higher pressure almost instantly transpires. At the same time, because the vapour pressure is now higher, the vapour bubbles will collapse by imploding with an audible cracking or banging sound similar to that experienced with the water hammer surge condition, causing the pipework to violently shake as the vapour/liquid mixture reverts back to a total liquid form.

This will occur in all cases where the system pressure is higher than the vapour pressure as the vapour bubbles cannot exist even if the system pressure is only marginally higher than the vapour pressure.

EFFECTS OF CAVITATION

It is not the cavities themselves that are necessarily destructive, but the residual effects left after the cavities have imploded and disappeared.

When the air bubbles violently implode – a great proportion of them imploding in contact with the pump casing, impeller or pipe walls – the result of the liquid rushing in to fill the vacuous spaces causes the generation of a constant series of minute water hammer surge-like blows onto the metallic surfaces concentrated on a small area. This means that extremely high pressures are momentarily developed with every implosion and when repeated often enough over a sufficiently long period of time, will quickly damage the metal and eventually result in a failure.

Severe cases of cavitation can be a frightening experience, when the whole pipework system vibrates alarmingly, pulling itself off the wall and in some instances fracturing the pipe with the accompanying knocking sound. On examining the components that have been in the affected zone, the pump impeller will be seen to have had large chunks of metal taken out of it, usually around the edges of the impeller blades, the result sometimes likened to something that resembles the after effects of being gnawed by rodents.

The pipework and fittings connected immediately to the pump discharge will show evidence of severe erosion to the internal pipe wall surface. The rate at which the erosion occurs will depend upon the severity of the cavitation, which itself will be resultant upon the difference between the operating pressure and the vapour pressure. This erosion rate may also be accentuated if the liquid itself already has corrosive tendencies, for instance if it contains large amounts of dissolved oxygen of a slightly acidic nature.

PUMP ARRANGEMENTS

Sometimes it is required to install pumps in multiple arrangements for either control methods, practical reasons or economics of the installation. In all cases the pumps should be arranged to operate for a minimum period of time to prevent the motor being repeatedly stopped and started, a condition referred to as 'hunting', which will result in premature failure of the pump unit.

Pumps can be arranged to operate in either 'series' or 'parallel'.

Pumps in series

Figure 18.13 illustrates the situation where two identical duty pumps have been installed in series. In this situation where more than one pump is installed on the same pipeline, the mass flow rate will remain unaffected but the pressure head will be increased accordingly. This is due to the fact that the discharge from the first pump becomes the suction conditions for the second pump, therefore the mass flow rate

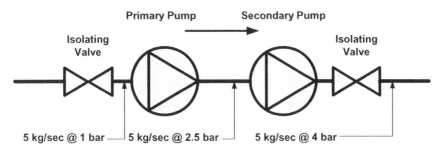

Figure 18.13 Pump in series (single mass flow rate, double pressure head)

remains the same but the pressure head leaving the discharge outlet from the first pump will be the suction inlet pressure for the second pump, and so on for any additional pumps.

In the example illustrated in Figure 18.13, the discharge pressure from the second pump has been doubled from that developed by the first pump plus the initial inlet suction pressure for the first pump, but the mass flow rate remains constant throughout the process.

Pumps chosen for this situation should have the same mass flow rate duty as each other, but the pressure head being generated by each pump may vary according to requirements in order for this arrangement to function correctly. If the mass flow rates differ from each pump, the pump having the largest duty could have an adverse effect on the pump having the lesser mass flow rate duty.

This arrangement of pumps in series is a fairly uncommon practice, unless the pressure output duty from the pump needs to be upgraded due to changes to the heating system. However, this arrangement is the basis of the multistage pump as illustrated in Figure 18.14.

The multistage centrifugal pump can be arranged vertically, as depicted in Figure 18.14, or horizontally, similar to an end suction type pump.

The multistage pump illustrated has four stages and is simplified here to show the principles of its operation, although numerous stages may be incorporated for this type of pump. Each stage has its own impeller fitted to a common drive shaft that extends the length of the pump and incorporates its own suction inlet and discharge outlet. With each stage arranged so the discharge outlet from one stage delivers into the suction inlet of the next stage.

This type of pump provides a neat compact unit that can produce high discharge pressures when required.

Pumps in parallel

A more common arrangement than pumps in series is to install them in parallel, as this composition can be used in shared duty applications. With this assembly, when both pumps are operating the mass flow rate is increased equal to that of the combined duty of both pumps, but the pressure head generated remains unaltered. Pumps used in this particular arrangement may be selected with different mass flow rate duties, but the pressure head produced by each pump must be equal to the other, or a situation could exist where the pump having the greater pressure head could act against the pump with the lesser pressure head.

Figure 18.15 illustrates the procedure and principles of arranging two pumps to operate in a parallel situation.

Figure 18.16 illustrates some basic common pump arrangements which may be found in industry. Variations of these arrangements also exist.

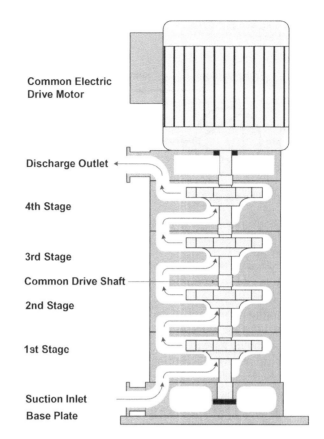

Figure 18.14   Simplified arrangement of a vertical four-stage multistage centrifugal pump

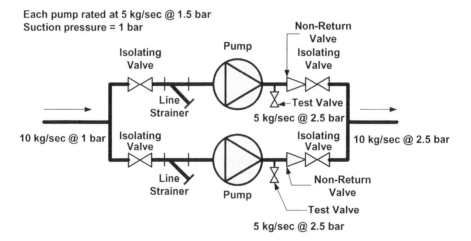

Figure 18.15   Pumps in parallel (double mass flow rate, single pressure head)

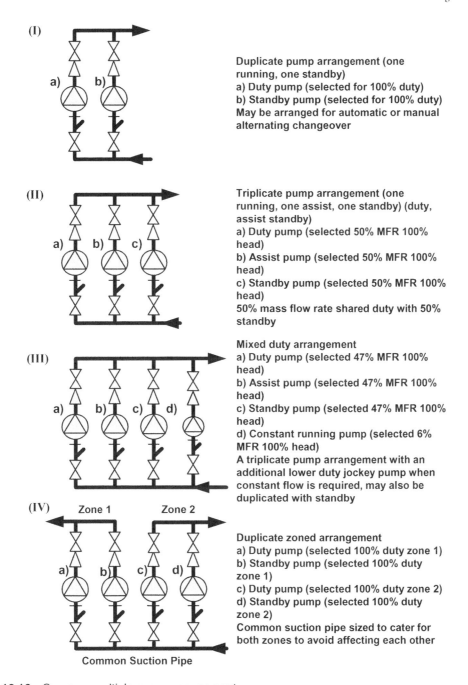

Figure 18.16 Common multiple pump arrangements

1. The duplicate pump array is the most common form of pumping arrangement used, where one pump is arranged to operate providing 100% of the duty required, with a second pump being provided to give a 100% standby of the duty in the event of the duty pump failing. The pump assembly can be arranged to automatically change over from standby to duty in the event of a main pump failure, or changed over manually by a selector switch. In either case it is considered good practice to allow the pumps to change

designation from duty to standby and vice versa, on a frequent basis to ensure that even wear of each pump is achieved and also to provide a regular check that the standby pump is still operational if required to be called upon. This alternating of duty/standby pump can be arranged to function automatically so that every time the duty pump operates it immediately becomes the standby pump when it stops, with the standby pump then becoming the next duty pump when called upon.

2. The triplicate pump assembly can sometimes be a more economical method of providing the duty when larger duties are required than the duplicate arrangement. With this method, two pumps of equal duty are selected to operate as duty and assist duty pump each providing 50% of the flow rate required, with a third pump of equal duty provided as the standby pump to either one of the operating pumps. This arrangement is said to give 50% standby duty facilities.

3. Although the pump system indicated in Figure 18.16 shows three pumps, there are instances where four or more pumps may be used that are arranged for three pumps running providing 33.3% flow each, with the fourth pump providing 33.3% of the duty as standby. These configurations should also be arranged to provide an alternating duty/standby facility as previously described.

4. The mixed duty pump arrangement is a variation of the triplicate pump arrangement, but with an additional constant running pump added, sometimes referred to as a jockey pump, when it is required to maintain a minimum flow at all times. This constant running pump may also be provided with a standby pump if considered desirable to the system.

5. The zoned multiple pump arrangement is used on larger pumping systems when the heating scheme has been divided up into a number of zones, each with their own control system. Each zone is served independently of the others by its own duplicate pump arrangement and organised as a duty and standby system. The sizing of the common pump suction line is critical with this arrangement to ensure that operation of each zone pump does not affect the operation of another zone.

Pump connections

In order for a pump to be maintained and operate satisfactorily, it is important that the valve arrangement indicated in Figure 18.17 is adhered to. This arrangement is common to both singular and multiple pump assemblies, with the exception of the non-return valve and test valve, unless there is a reason for preventing a reversal of flow back through the pump.

ELECTRIC MOTORS

Pump manufacturers supply their pumps as a complete pump/motor assembly unit, with the exception of very large pumps where the motors may be supplied separately.

Electric motors for centrifugal pumps can be either vertically or horizontally mounted, with many of the centrifugal pumps for hydronic circulating systems being close coupled with the pump impeller mounted on a motor drive shaft, or canned rotor, as previously described.

When selecting a pump/motor unit, the manufacturer will need to be informed of the following information:

- Electrical supply available, i.e. 230 or 400 volts, single or three phase and number of cycles (Hertz).
- Preferred motor running speed, which would normally be either 1450 rpm or 2900 rpm. Domestic circulating pumps would normally have a variable speed motor revolving at a speed of somewhere between the two.
- Method of starting.
- Ambient operating temperature and conditions of installation.

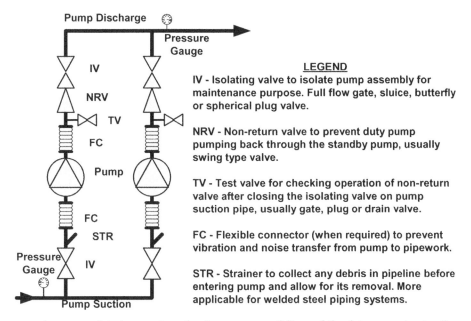

Figure 18.17 Pump valve arrangements and connections

Pumps may be arranged to have automatic changeover on failure of the duty pump to standby, or manual changeover. In either case the duty and standby pump designation should be frequently alternated to obtain even wear of each pump and check that the standby pump is operational.

Table 18.1 Classification of motor insulation material

Class	Maximum Temperature °C
Y	90
A	105
E	120
B	130
F	155
H	180
C	Above 180

- Degree of electrical protection against ingress (IP rating)
- Class of electrical motor insulation.

Table 18.1 shows the temperature limitation of the electrical insulation of motors, with class 'F' being chosen for standard pumping applications.

INDEX OF PROTECTION (IP RATING)

IP stands for index of protection for enclosures of electrical equipment. It was originally referred to as 'Ingress Protection' but is now standardised as an index, as shown in Figure 18.18.

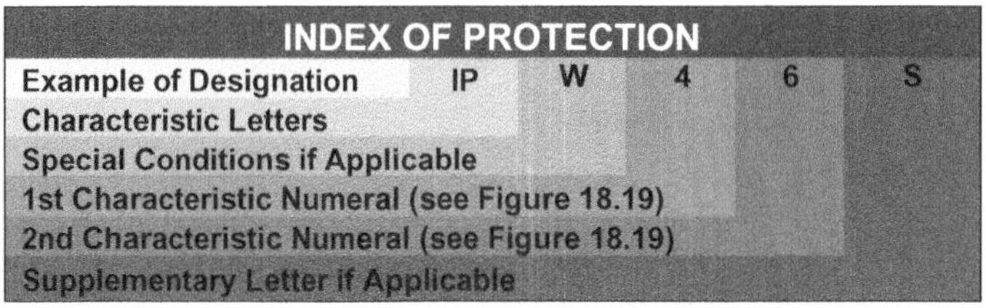

INDEX OF PROTECTION					
Example of Designation	IP	W	4	6	S
Characteristic Letters					
Special Conditions if Applicable					
1st Characteristic Numeral (see Figure 18.19)					
2nd Characteristic Numeral (see Figure 18.19)					
Supplementary Letter if Applicable					

Figure 18.18   Index of Protection classification

The designation to indicate the degree of protection consists of the characteristic letters IP followed by two numerals indicating the conformity with the conditions stated in the table given in Figure 18.19.

A third numeral is sometimes used in parts of Europe to indicate the degree of mechanical protection.

- Single Characteristic Numeral – When only one class of protection is required to be indicated, the omitted numeral is replaced by the letter X, for example, IP5X.
- Supplementary Letters – Additional information may be indicated by a supplementary letter following the numerals. This additional procedure should be clearly stated prior to motor manufacture. The absence of these letters S or M implies that the intended degree of protection will be provided under all normal conditions of use, both running and not running.
- S – Tested against harmful ingress of water when equipment is not running (stationary).
- M – Tested against harmful ingress of water when equipment is running (mechanical operation).
- W – (Placed immediately after the letters IP.) Tested for use under specified weather conditions that are stated and agreed between manufacturer and user.

Some European specifications use a third characteristic numeral to define a degree of mechanical protection.

METHODS OF STARTING ELECTRIC MOTORS

Electric motors may be started by either a direct-on-line method, or by a reduced voltage method. The choice of starter is normally dictated by:

- The electrical load rating of the motor.
- The frequency of starting; i.e. number of starts required per hour.
- The local electricity supply company's restrictions with regard to the permitted starting currents and voltage drop limitations.

Direct-on-line (DOL) – This is the simplest and least costly method of starting electric motors. The expression 'direct-on-line' means that full supply voltage is directly connected to the starter of the motor by means of a manually operated or automatic switch, which is coupled with a condenser (capacitor), connected in series with the starter windings to give a phase difference of nearly 90° between currents in the windings, enabling the motor to be turned. All domestic electric motors incorporate direct-on-line starters.

The electrical suppliers normally limit this method of starting motors to under 10 kW in load because of the initial electrical surge current that occurs during the starting procedure.

Index of Protection

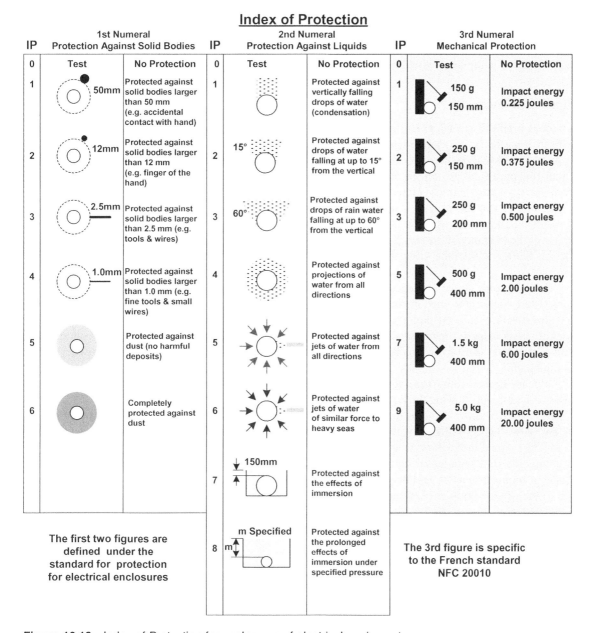

Figure 18.19   Index of Protection for enclosures of electrical equipment

The star-delta method is the most common form of reduced voltage starting method used for three-phase pump motors. As with all reduced voltage starting methods, it involves the reduction of voltage applied to the stator windings of the motor at the start.

Star-delta is a two-position method of direct-on-line starting three-phase motors, in which the windings are connected firstly in star for acceleration of the rotor from standstill, then secondly in delta for normal running.

ANCILLARY EQUIPMENT FOR PUMPS

This chapter has concentrated on circulating pumps, but to enable the pump to function correctly, additional items of equipment should be considered as part of the installation.

Pipeline strainers

Although briefly discussed under the subject of oil supply, pipeline strainers are covered in more detail here with their particular association with circulating pumps.

A strainer should be incorporated on the pipeline inlet suction side of all rotating machinery, including circulating pumps, to protect the wetted moving parts of the machinery – with the possible exception of domestic circulating pumps connected to smooth bore pipework.

The function of the strainer is to collect any loose or free moving scale or debris and retain it in the perforated basket, from where it can be removed at periodic intervals. Removal of debris is often a condition of the warranty of the pump manufacturers to protect their equipment from being damaged by this foreign matter within the circulatory system, by preventing it from entering the pump casing and impeller.

Circulating pipe systems, particularly those constructed from steel pipe with welded joints, could be subject to loose scale formed during the welding process becoming detached from the pipe and entering the flow stream, even if the piping system has been flushed through. Also, particles of jointing material etc, could cause damage to the impeller if allowed to enter into the pump, resulting in premature failure of the pump, or at least scoring the impeller and casing's internal surfaces, resulting in a reduced efficiency of the pumping unit.

A pipeline strainer comprises a single or multiple perforated baskets designed to retain any foreign matter inside it whilst still allowing the liquid to flow through it reasonably unhindered.

The 'Y'-type strainer depicted in Figure 18.20 is commonly used for pipelines of 80 mm diameter and below, where there is sufficient withdrawal space below to remove and replace the strainer basket.

In larger piping installations where the pipework is at low level, which is usually the case for pump connections, adequate withdrawal space for the basket of the 'Y'-type strainer may not be available; here a single or mono strainer should be installed which allows the basket to be removed for cleaning from the top of the strainer body, as illustrated in Figure 18.21.

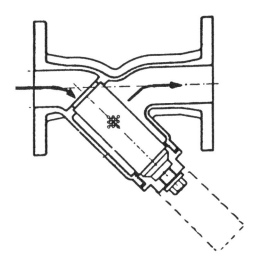

Figure 18.20   Detail of 'Y'-type pipeline strainer showing withdrawal space

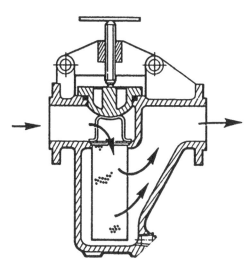

Figure 18.21   Single inline mono strainer with top withdrawal space

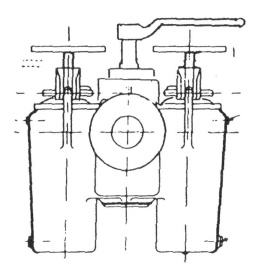

Figure 18.22   Duplex strainer

Both these types of strainer designs would require the circulating piping system to be shut down so that the strainer can then be isolated by valves installed either side of the strainer, to permit the body to be drained and the basket removed for cleaning.

In situations where it would not be acceptable to shut down the circulating system for the time required to inspect and clean the strainer basket, then a duplex model may be installed, as indicated in Figure 18.22.

A duplex model strainer incorporates two strainer baskets within a single common body where the liquid flow is directed into one half of the strainer, leaving the remaining half isolated and available for cleaning.

Although the duplex strainer is more expensive, it is less costly than installing two mono strainers in parallel and requires less space.

Strainers, regardless of their type, consist of a perforated cylindrical basket and are arranged to allow the flow to enter the container where the debris is safely retained. The particle size of the debris retained will depend upon the size of the perforations in the strainer basket. Most strainer manufacturers can provide baskets with varying degrees of filtration, from coarse to fine, which would need to be specified at the time of ordering.

The degree of filtration can be expressed in either mesh or micron rating, which relates to the aperture size and number of holes or perforations per square area. Appendix 3, Mesh/Micron Rating, shows this relationship, which can be explained as follows.

The term 'mesh', which is commonly used, is derived from the old imperial unit of measurement of mesh per inch, which is the square root of the number of perforations per square inch; hence a mesh rating of 10 would be $10 \times 10 = 100$ holes per square inch of filtration area.

The term 'micron' is used in the metric system of measurement to denote one millionth of the main unit, hence micrometre (μm) is used to denote one millionth of a metre. In filtration terms, the micron rating relates directly to the aperture in millimetres, which is 1000 times larger than the hole size, hence an aperture of 0.1 mm will have a micron rating of 100.

In the absence of any specific information from the manufacturers of pumps and rotating machinery, it would be common to install a strainer having a mesh rating of 100 on the inlet side of the pump.

The decision of when to clean out the basket of a strainer for a circulating water system is usually undertaken during routine servicing and maintenance periods, which is not a precise way of determining when the strainer requires cleaning.

Where this is considered not an acceptable means of determining when the strainer needs cleaning, a more accurate method of determining this would to be install pressure gauges on the inlet and outlet sides of the strainer, or a single differential pressure gauge across the unit: both methods will indicate an increasing pressure drop through the strainer, which indicate that the strainer basket needs to be cleaned.

Pressure gauges

All heating systems, in common with other piping systems, are subject to operating pressures, the degree of which is dependent upon the design. Within the heating system there may be a need to observe the pressure at given points around equipment such as boilers, pumps and strainers etc, in order to be satisfied that it is operating correctly.

The most common instrument used to give this indication is the Bourdon tube pressure gauge, mounted directly onto the piping adjacent to the equipment where the pressure needs to be determined.

The Bourdon tube pressure gauge in its simplest form is a piece of thin walled tube rolled or drawn to an elliptical or similar section, where it is coiled to form approximately 270° of a circle as shown in Figure 18.23.

One end of the Bourdon tube is mounted so that it is held rigid but can admit pressure, and the other end is closed, but free to move. When pressure acts on the inside of the tube it tries to straighten, causing the free end to move out. The free end is connected by a quadrant and pinion which moves the gauge pointer to indicate the pressure being applied on the calibrated face of the gauge. Most pressure gauges are within ±2% accuracy.

These pressure gauges can be obtained calibrated in most forms of pressure measurement, including metres head, although gauges with this form of measurement are often referred to as altitude gauges.

Pressure gauges should be calibrated to a scale approximately twice that of the working pressure so that any increase in pressure may be accurately indicated, therefore a heating system having a working pressure of 4 bar should have a pressure gauge calibrated up to 8 bar, with a second adjustable pointer that is set to indicate the working pressure of 4 bar.

The only exception to this is on high-pressure systems in excess of 20 bar, where they can be arranged to have a calibrated scale of 1½ times the working pressure.

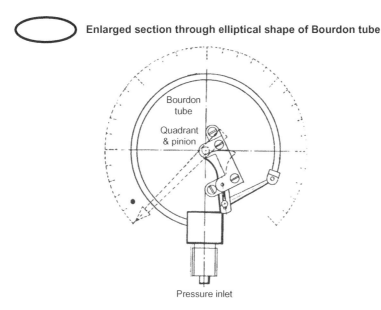

Enlarged section through elliptical shape of Bourdon tube

Bourdon tube

Quadrant & pinion

Pressure inlet

Figure 18.23   Bourdon tube pressure gauge

Where pressure gauges are to be fitted close to rotating machinery, such as circulating pumps, vibration may be experienced that could cause the gauge needle to shake or jump, making it difficult to read and damaging the mechanism. In this situation the face of the pressure gauge should be filled with oil or glycerine that will have the effect of dampening down any oscillations of the needle, but in severe cases of pressure pulsations a diaphragm gauge with internal hydraulic damper should be fitted.

The Bourdon tube pressure gauge is named after the watchmaker and engineer Eugene Bourdon who first patented it in France in 1849. It was said to be able to register pressures of 10 000 psi (689 bar).

As previously mentioned in the section on strainers, pressure gauges are sometimes used to register any pressure drop across equipment such as strainers – and although two pressure gauges can be used, located on both the inlet and outlet, it is more effective to install a single pressure differential gauge across the strainer as shown in Figure 18.24. Here the pressure will increase on the inlet as the strainer basket becomes dirtier and offers greater resistance to the flow.

In critical situations where it is required to know about a pressure drop across the strainer as soon as it happens, then pressure switches can be arranged to sound an alarm when they reach a set point.

Pressure gauges should be mounted directly onto the equipment or adjacent piping, or piped away to a remote wall mounted board and be complete with a quarter turn gauge cock to isolate the gauge for maintenance purposes.

Gauge siphons need only to be used for steam or high-temperature water systems to prevent the Bourdon tube from expanding from the heat, when it would give a false pressure reading on the gauge.

Thermometers

Thermometers are sometimes required to indicate that the operating temperature is being achieved. They are mounted directly onto the pipework and may be either a glass filled vertical type, or gauge type with bi-metallic or mercury in a steel form. These work on similar principles to those of the Bourdon tube pressure gauge, where the tube is filled with mercury and when the bulb is heated, the mercury expands causing the Bourdon tube to move, thus giving an indication on the calibrated scale.

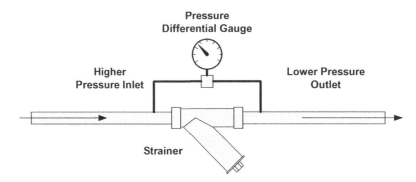

Figure 18.24   Differential pressure gauge fitted to pipeline strainer

The calibration of the scale should reflect the operating temperature of the heating system, plus an additional 25% over-temperature capacity.

NOISE AND VIBRATION

All rotating machinery when operating will produce a degree of vibration and in some cases this will be accompanied by noise. The extent of noise and vibration generated would require a whole volume to explain, but it is not the intention of this work to examine the subject in detail but to make the reader aware of the subject in relation to circulating pumps.

In order to prevent the vibration, that may have been generated from being transmitted both through the building structure and connecting piping, certain precautions should be considered.

Figure 18.25 illustrates methods in common use, or that have been in common use, to isolate or absorb vibration and noise being transmitted through the building structure.

The piping connections made to the circulating pump are shown using flexible synthetic rubber or braided stainless steel couplings on both the inlet and discharge sides of the pump, and a motor unit that will permit the rotating unit to move and vibrate in any direction without transmitting this vibration along the piping system.

The pipework either side of the flexible pump couplings should be securely fixed by rigid pipe supports to prevent any movement being transmitted from the flexible couplings.

These types of flexible couplings will also allow the connecting pipework to remain firm and avoid any damage that may occur to pipe joints that would be subjected to any transmitted vibration over a prolonged period of time.

Regarding the structure, it is common practice to support the pumps on some form of firm base. This serves the purpose of affording a degree of protection for the pumps from mechanical damage from plant and equipment that may be moved around the plant room.

There are a number of suitable methods used to achieve this; three of the most commonly employed methods are indicated in Figure 18.25.

Two of these methods show a concrete support base that is cast *in situ*; the third is seldom used today and employs a 50 mm thick cork pad located halfway up the concrete base, thus isolating the top section from the lower section. This method may be used successfully when small amounts of oscillation are expected as shown in Figure 18.25(a).

The method shown in Figure 18.25(b) has a solid concrete support base with the pump supported on anti-vibration mountings located between the pump base frame and the concrete support base. These anti-

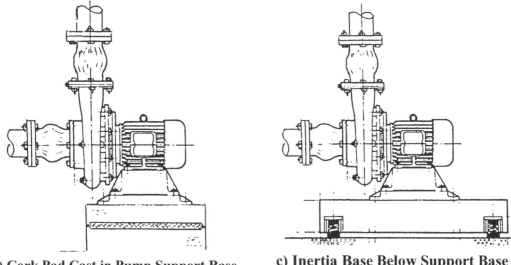

a) Cork Pad Cast in Pump Support Base

c) Inertia Base Below Support Base

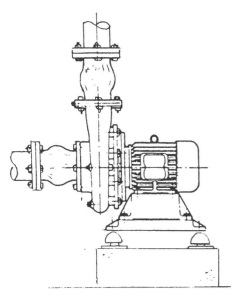

b) Anti-Vibration Mountings Between Pump & Base

Figure 18.25   Pump vibration isolation

vibration mountings consist of synthetic rubber bushes or spring supports that absorb the vibration from the pump, thus isolating it from the building structure.

Both these methods are constructed on-site, but the method shown in Figure 18.25(c) is manufactured off-site by specialist manufacturers who will construct the inertia bases from steel sections complete with isolating damper mountings, which only require fixing to the floor slab when supplied to site. This can be an advantage in situations where it would be impractical to pour concrete on-site or wait for it to cure before the pump can be installed.

19

Domestic Hot Water

The subject of domestic hot water is a separate plumbing matter which, in certain situations interfaces with heating. It is not the intention in this volume to cover the topic of domestic hot water in complete detail.

It is the intention to limit this topic to that element of providing the primary source of energy in the form of heat to raise the temperature of the domestic hot water as this interfaces with the heating system.

It should be appreciated that if a hydronic heating system is being planned, then the incorporation of the domestic hot water as the means of providing the primary heat source should be considered on economic running costs, as this is often the least costly means of raising the temperature of the domestic hot water when fossil fuels are being used.

Domestic hot water may be provided by either of two means:

1. Central storage system or systems – employing a central storage vessel, or number of storage vessels, located in a central area of the building complex with distribution piping extending from the storage vessel to serve all domestic hot water draw-off points.
2. Local water heaters – gas-fired or electric proprietary made water heaters located adjacent to the fixtures that they are required to serve.

Alternatively, an amalgamation of both central storage vessels with selective local water heaters may be considered to be the best solution for some building applications.

CHOICE OF DOMESTIC HOT WATER SYSTEM

The system to be selected will be dependent upon a number of factors concerning the type and intended use of the building in question.

Heating Services in Buildings: Design, Installation, Commissioning & Maintenance, First Edition. David E. Watkins.
© 2011 John Wiley & Sons, Ltd. Published 2011 by John Wiley & Sons, Ltd.
As an aid to lecturers and students, full colour versions of the figures in this chapter may be found at www.wiley.com/go/watkins
These figures are © 2011 by John Wiley & Sons, Ltd.

- Type of Building – Flats or apartments, together with office blocks, are equally served by either local water heaters or a central storage vessel, with the choice being influenced by other factors. Hotels, hospitals and factories with large demands for domestic hot water are almost exclusively served by a central storage system, which is better suited to provide large draw-offs and for peak demands for domestic hot water.
- Boiler Plant – If it is planned to have a source of primary heat available in the form of hot water supplied from a central boiler plant, then it could be more economically viable to consider utilising this readily obtainable source of heat for a central domestic hot water storage vessel or vessels.
- Occupancy of Building – Buildings that are normally sparsely occupied cannot justify the cost of providing a central domestic hot water storage system, or the operating costs associated with it. Local water heaters will be more practical, as well as more economical, to provide small quantities of domestic hot water. Buildings that are more densely occupied and require large quantities of domestic hot water, particularly places of manual work that operate a shift system, would be better suited to a central storage arrangement.
- Building Complexity – The physical size and architectural layout of the building should also be considered. A building development that is spread out over a large area with only few fixtures that require domestic hot water would be more suited for local water heaters. A building complex with fixtures grouped closely together, or a multi-storey building with washroom areas repeated directly above each other, would be more suited to a system employing a central storage of domestic hot water. An exception to this would be in the case of a high-rise block of residential apartments, where the tendency is to install either local water heaters or individual smaller central storage vessels that enable the occupants to have greater control over the use and cost of the domestic hot water.
- Mixture of Systems – Circumstances often occur whereby it would be advantageous to install a mixture of local water heaters together with a central storage system, for example in a building development where there may be fixtures requiring domestic hot water but which are remotely located from the main areas served by the central domestic hot water storage vessel.
- Multiple Storage Vessels – As in the case for mixing the domestic hot water systems, there are sometimes situations where the storage would be better provided by a number of smaller domestic hot water storage vessels, each located in close proximity to their point of use, but still being heated from the same central boiler plant.

ASSESSMENT OF DOMESTIC HOT WATER STORAGE

The importance of correctly assessing the amount of domestic hot water that should be stored should not be underestimated. If the storage vessel has been under-sized, the occupants will be inconvenienced by continually running out of domestic hot water and then having to absorb the costs to correct the problem. Because of this, there is a temptation to allow a safety factor, which could result in the over-sizing of the domestic hot water storage vessel and in turn incur higher installation expenditure together with high operating costs.

There are a number of recognised engineering publications that give guidance on domestic hot water storage quantities for given applications and different types of buildings. Table 19.1 collates some of the most common building types and may be used as a guide for the minimum quantity of domestic hot water that should be stored.

RECOVERY TIME

The recovery time is the term given to the time taken to heat the contents of the domestic hot water storage vessel from cold to its desired temperature.

Table 19.1   Assessment of domestic hot water storage

Type of Building	Storage per Person Litres	Minimum Storage Litres
DWELLINGS		
1 Bedroom	114	114*
2 Bedroom	75	114*
3 Bedroom +	55	114*
HOTELS		
First Class (5 Star)	45	136
Average (3 to 4 Star)	36	114
Budget	35	114
Motel	32	90
Hostel	30	90
OFFICES		
With Canteen	5	15
Without Canteen	5	10
FACTORIES		
Heavy Engineering	10	15
Light Engineering	5	10
RESTAURANTS	6	Per Guest
COLLEGES & SCHOOLS		
Day (Secondary & Primary)	5	15
Boarding	25	114
SPORTS & LEISURE CENTRES		
Sports Hall	20	Per Participant
Field Sports	35	Per Participant
SHOPS & STORES	15	Per Staff
MUSEUMS, THEATRES, CINEMAS & GALLERIES	1	Per Visitor
BARS & NIGHTCLUBS	2	Per Customer
HOSPITALS		
General	200	Per Patient
Maternity	225	Per Patient
Mental	90	Per Patient
Nursing Homes	45	Per Resident
Nurses' Homes	20	136
LAUNDRIES Sized on requirements of washing machines		

*Minimum storage of 114 litres per dwelling with 2 hour recovery.

The recommended maximum recovery time for a domestic hot water vessel remained at 4 hours for many years but was considered unacceptably long. It was eventually replaced by the present recommendation of 2 hours, which had become the custom many years before being adopted officially.

BS 1566 for copper cylinders for domestic purposes require that any heat exchanger fitted inside them must be capable of achieving a reheat time of 25 minutes or less. This must not be confused with the design recovery time, as a very large energy heat capacity from the boiler would be required to achieve this. However, some heating control systems that are arranged to give priority to recovering the domestic hot water temperature may achieve this quicker heat-up time, or come close to it.

DOMESTIC HOT WATER TEMPERATURE

Domestic hot water should be heated and stored to a temperature of between 60°C and 65°C, which is the recommended temperature to store water. This is considered a safe temperature for able bodied and able minded people as well as one that will not support legionella.

Legionnaires' disease is a fatal form of pneumonia and is caused by the bacterium *Legionella pneumophila*. It exists in all water supplies and although everybody is susceptible, men are more susceptible than women, people over the age of 45 are considered more at risk, and those who are smokers, diabetics, alcoholics, or suffer from respiratory disease are at even greater risk.

Research has gone into understanding legionnaires' disease since it was first recognised in 1976. One of its aspects is its dependency on and relationship with water temperature, although research continues in this field. See Figure 19.1 for an illustration of this relationship.

Legionella pneumophila is present but inactive in water at temperatures below 20°C, above this it starts to multiply at an increasing rate with the rise in the temperature up to 45°C, when it starts to be killed off. *Legionella pneumophila* bacterium is said not to exist in water temperatures above 60°C.

For the above reasons it is recommended to store domestic hot water at temperatures of 60°C to reduce the risk of *Legionella pneumophila* bacteria being present, but not to exceed 65°C, to reduce the risk of people being scalded when using the water.

This temperature is considered safe for use by people who are not afflicted by either physical or mental problems, as they have both the physical and mental ability to remove their hands from the hot water before being scalded or burnt.

Although domestic hot water is stored at these temperatures, it is normally used at lower temperatures by mixing with cold water at point of use, which also has the effect of increasing the useful storage capacity of domestic hot water. Table 19.2 lists the guidance temperatures for domestic hot water use.

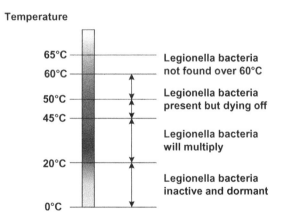

Figure 19.1   Relationship of temperature to Legionella bacteria in water

Table 19.2   Domestic hot water usable temperatures

Sanitary Appliance	Average Temperature °C
Wash Hand Basin	41–43
Bath	44*
Shower	40
Bidet	38

*Building Regulations Approved Document G requires a maximum temperature of 48°.

In situations where it would be considered unsafe to use domestic hot water at the storage temperature of 60°C to 65°C, such as for use by children or the elderly and infirm, then a thermostatic mixing valve should be incorporated on the domestic hot water supply outlet point.

DOMESTIC HOT WATER STORAGE VESSELS

Vessels for the storage of domestic hot water have traditionally been manufactured from copper conforming to BS 1566, although other materials have been and still are used. These are discussed later in this chapter.

Copper cylinders conforming to BS 1566 Part 1 are available in three construction forms, namely types 'D', 'G' and 'P'.

- Type 'D' Cylinders – Directly heated cylinders heated by either a direct heating system, or more commonly by a thermostatically controlled electric immersion heater.
- Type 'G' Cylinders – Double feed indirectly heated cylinders suitable for both a gravity or forced primary circulation, although more commonly used for gravity primary circulation.
- Type 'P' Cylinders – Double feed indirectly heated cylinders suitable for forced primary circulation only.

Domestic hot water systems are either of the directly heated type, employing a type 'D' direct cylinder, or more commonly, the indirect cylinder, utilising a type 'G' or type 'P' cylinder incorporating an internal heat exchanger.

All domestic cylinders are now pre-insulated with polyurethane foam, a section of which must be neatly removed if a contact thermostat is to be used, so that it fits snugly.

BS 1566 for open vented copper cylinders also categorises the three types of cylinders into three pressure grades, grades 1, 2 and 3, which refer to the maximum working head pressure that should be exerted on the base of the cylinder. Prior to the year 2002 a fourth grade was included that was suitable for a lower working head pressure. Figure 19.2 illustrates the pressure grading of cylinders.

Table 19.3 lists the dimensions and grades of copper cylinders.

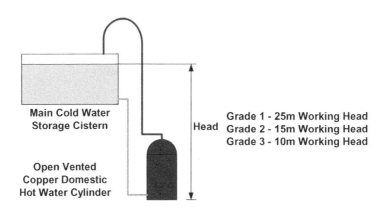

Figure 19.2   Application of copper cylinder pressure grades

Table 19.3 Dimensions and grades of copper cylinders

BS Ref	Dia mm	Height mm	Capacity Litres	Metal Thickness mm							
				Grade 1 Max Working Head 25 m		Grade 2 Max Working Head 15 m		Grade 3 Max Working Head 10 m		Grade 4 Max Working Head 6 m	
				Base	Top & Shell	Base	Top & Shell	Base	Top & Shell	Base	Top & Shell
0	300	1600	96	1.6	1.2	1.6	0.9	1.6	0.7	0.9	0.55
1	350	900	72	1.6	1.2	1.6	0.9	1.6	0.7	0.9	0.55
2	400	900	96	1.8	1.2	1.6	0.9	1.6	0.7	0.9	0.55
3	400	1050	114	1.8	1.2	1.6	0.9	1.6	0.7	0.9	0.55
4	450	675	84	2	1.6	1.6	1	1.6	0.7	0.9	0.55
5	450	750	95	2	1.6	1.6	1	1.6	0.7	0.9	0.55
6	450	825	106	2	1.6	1.6	1	1.6	0.7	0.9	0.55
7	450	900	117	2	1.6	1.6	1	1.6	0.7	0.9	0.55
8	450	1050	140	2	1.6	1.6	1	1.6	0.7	0.9	0.55
9	450	1200	162	2	1.6	1.6	1	1.6	0.7	0.9	0.7
9E	450	1500	206	2	1.6	1.6	1	1.6	0.7	0.9	0.7
10	500	1200	190	2.5	1.8	1.8	1.2	1.6	0.9	1.4	0.8
11	500	1500	245	2.5	1.8	1.8	1.2	1.6	0.9	1.4	0.8
12	600	1200	280	2.8	2	2.5	1.4	2	1.2	1.8	1.1
13	600	1500	360	2.8	2	2.5	1.4	2	1.2	1.8	1.1
14	600	1800	440	2.8	2	2.5	1.4	2	1.2	1.8	1.1

Note – Grade 4 cylinders are no longer included in BS 1566.
Other grades and sizes are available to special order.

DIRECT DOMESTIC HOT WATER SYSTEMS

A direct type domestic hot water cylinder does not contain an internal heat exchanger, and with a direct type domestic hot water system the water that is used for domestic purposes also passes through the boiler heat exchanger, therefore the secondary and primary water is the same.

Figure 19.3 illustrates a typical vertical pattern direct type 'D' copper cylinder; horizontal patterns are also available. The illustration shows a secondary return connection together with a primary flow and primary return connection – these connections would have to be specified to the supplier as they are not provided as standard. Figure 19.3 also indicates alternative positions for the 2¼ inch electric immersion heater boss; again, the required position would have to be specified.

The direct type domestic hot water system employing a boiler is rarely used today. This system had the disadvantage of suffering from scale formation build-up in the boiler heat exchanger and connecting piping, as the water being circulated is continually replaced at the rate at which the domestic hot water is drawn off.

Figure 19.4 illustrates a direct domestic hot water supply system employing a type 'D' direct domestic hot water cylinder heated by an electric immersion heater, as is in common use today.

This system does not interface with the heating system and is an independent stand-alone system.

INDIRECT DOMESTIC HOT WATER SYSTEMS

An indirect system of domestic hot water supply employs either a type 'G' or type 'P' cylinder that contains an internal heat exchanger through which the primary water from the heating boiler circulates without

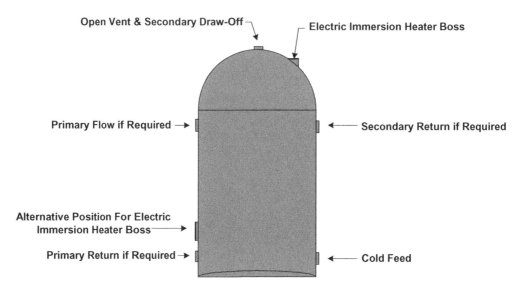

Figure 19.3   Vertical pattern direct type 'D' domestic hot water cylinder

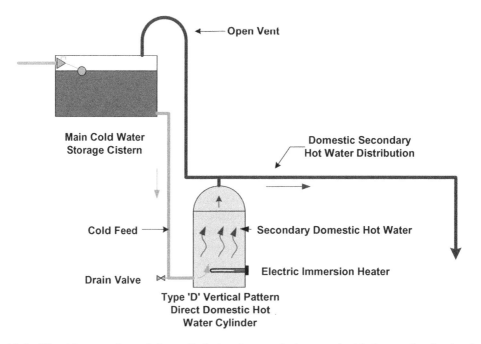

Figure 19.4   Direct type system of domestic hot water employing an electric immersion heater (no connection to the heating system)

mixing with the secondary water and transfers part of its heat to the secondary or consumable domestic water.

Because the primary water and the secondary water are separate and do not mix, the problem of scale formation and corrosion that existed with the direct type system is much reduced. Because the primary and secondary waters do not mix, type 'G' and type 'P' indirect cylinders are referred to as 'double feed

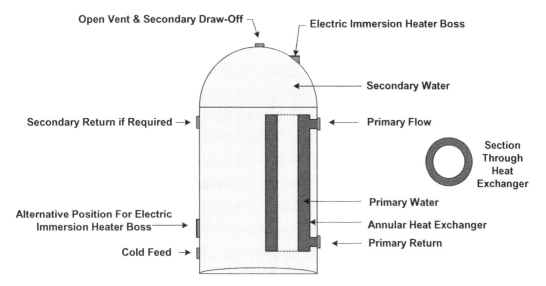

Figure 19.5 Vertical pattern indirect type 'G' double feed domestic hot water cylinder with annular pattern heat exchanger

open vented indirect domestic hot water cylinders, as the primary water is fed from the feed and expansion cistern and the secondary water is fed separately from the main cold water storage cistern.

The domestic hot water cylinder conforming to BS 1566 type 'G' is ideal for functioning with a thermo-siphon or gravity circulation of the primary flow and return from the boiler to the heat exchanger, inside the double feed indirect domestic hot water cylinder.

Double feed indirect cylinders have been available with a choice of heat exchangers. BS 1566 specifies a coil type primary heater that should be capable of passing a steel ball of a given diameter through the coil, but a better choice is an annular type heat exchanger in which the pressure drop or resistance to the flow through it is less than a coil, which is critical for a gravity primary flow and return. Figure 19.5 illustrates a typical double feed indirect copper cylinder equipped with an annular heat exchanger, which is best described as a cylinder within a cylinder.

The design of the annular pattern heat exchanger offers the least resistance to the primary flow water through it, as the gravity circulation's slow velocity permits the flow of water to spread out around the internal section of the heat exchanger, thus utilising the entire heating surface area of the annular heat exchanger to transfer the heat. It is not suitable for a forced primary circulation, as the higher velocity primary water would just short-circuit from the inlet connection to the outlet connection, and so would use only a small percentage of the heat exchanger, and thereby not achieve a full transfer of heat.

The principles of the thermo-siphon gravity circulation are discussed and explained in Chapters 2 and 4, with the application illustrated further in Figure 19.6. The double feed part of the indirect cylinder's name is shown by the two cold feeds, one from the main cold water storage cistern feeding the secondary domestic hot water, and the other from the feed and expansion cistern feeding the heating system.

It can be seen that the secondary and primary waters are completely separate from each other, with the only interface being in the cylinder, whereby the primary water transfers part of its heat through the heat exchanger from the higher temperature primary water to the lower temperature secondary water.

The primary water is circulated at the boiler operating temperature of 82°C, but the secondary domestic hot water is limited to 60/65°C by either a thermostatically controlled motorised two-port zone valve, or a non-electric temperature sensing valve.

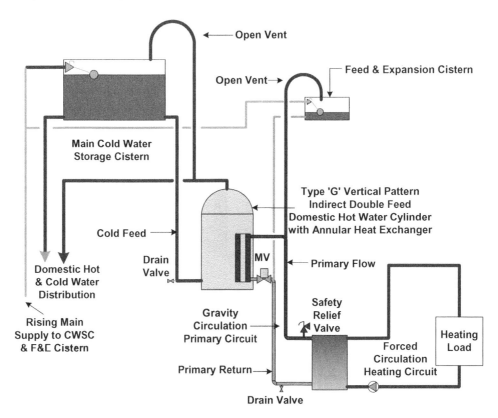

Figure 19.6   Indirect system of domestic hot water with type 'G' cylinder and gravity primary circulation

The use of gravity circulation to provide domestic hot water for residential dwellings is no longer permitted by the Building Regulations.

The domestic hot water cylinder conforming to BS 1566 type 'P' is designed to function with a forced primary circulation employing a pump to circulate the primary water through a coil type heat exchanger. This is now the required method of satisfying the provision for domestic hot water specified by the Building Regulations for domestic dwellings. It is also the easiest and most versatile method compared to gravity circulation, but there are occasionally applications when gravity circulation would be a better choice.

Figure 19.7 illustrates a double feed vertical pattern indirect type 'P' domestic hot water cylinder complete with an internal coil type primary heater, which is required by BS 1566 to be able to recover the cylinder contents from 10°C to 60°C in 25 minutes. The coil heat exchanger should not have a pressure drop greater than 0.5 bar at a flow rate of $0.251 s^{-1}$.

This form of heat exchanger is more suitable for fully pumped primary flow and returns than the annular type of heat exchanger as the water cannot short-circuit through the heat exchanger because it has to follow the passage through the length of the coil. If it was used for a gravity primary circulation, the frictional resistance encountered would be high, reducing the circulation pressure available.

The fully pumped forced circulation method of providing domestic hot water employing an indirect type 'P' cylinder, is the procedure required by the Building Regulations for domestic dwellings. It is also the most commonly used arrangement for all types of buildings due to its versatility, as it frees the designer of the physical constraints imposed by gravity circulation of having to work within a limited frictional resistance that can be allowed for the flow through the primary flow and return.

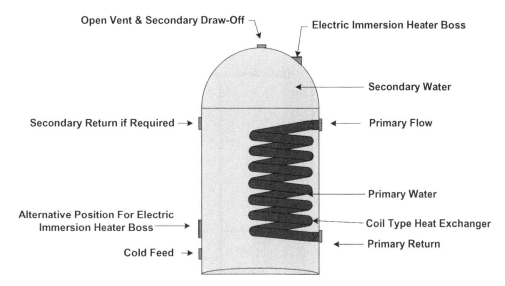

Figure 19.7 Vertical pattern indirect type 'P' double feed domestic hot water cylinder with coil pattern heat exchanger

This versatility enables the design engineer to locate the domestic hot water storage vessel anywhere within the building, including locating it below the level of the boiler, providing the circulating pump has been sized correctly to overcome the increased frictional resistance that will be encountered by the more tortuous and maybe lengthy path for the primary flow and return.

The Building Regulations did not need to regulate against the use of the thermo-siphon gravity circulation, as the fully pumped forced primary circulation had become the most favoured method of providing the primary heat source for raising the temperature of domestic hot water long before this regulation was made.

In almost all situations the provision of a forced circulation arrangement for the domestic hot water is the most efficient method; however, there are occasions when a gravity circulation would be more efficient if it was designed and installed correctly, but the current Building Regulations prohibit this.

Figure 19.8 illustrates the versatility of the fully pumped forced primary circulation arrangement, employing an indirect domestic hot water cylinder equipped with a coil type primary heater, whereby the domestic hot water temperature is regulated by a thermostatically controlled motorised two-port zone valve.

SINGLE FEED INDIRECT DOMESTIC HOT WATER SYSTEMS

The indirect domestic hot water cylinders previously discussed and illustrated, conforming to BS 1566 Part 1, have been described as 'double feed', the reason for which has been demonstrated in both Figures 19.6 and 19.8. BS 1566 Part 2 covers 'single feed' indirect domestic hot water cylinders that are used in an indirect system that has all the advantages of a direct domestic hot water system – such as not requiring a separate feed and expansion cistern, open vent and cold feed to the heating system – without having the disadvantages of scale formation and corrosion, as the primary and secondary waters do not mix because the indirect principle is still maintained.

The working principles of this system are based on creating an air lock within the upper and lower domes of the internal heat exchanger of the single feed indirect domestic hot water cylinder, as depicted in Figure 19.9. The shape of the heat exchanger is designed to capture this air lock in order to keep the secondary and primary waters separated.

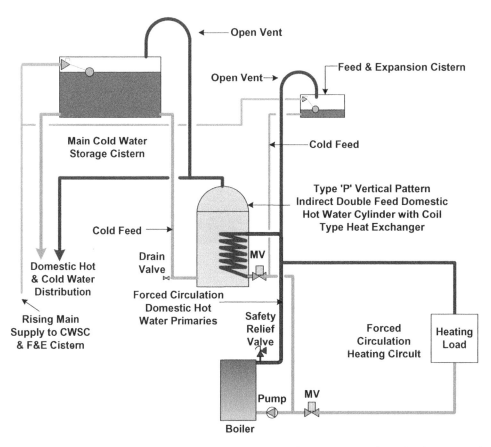

Figure 19.8   Indirect system of domestic hot water with type 'P' cylinder and forced primary circulation

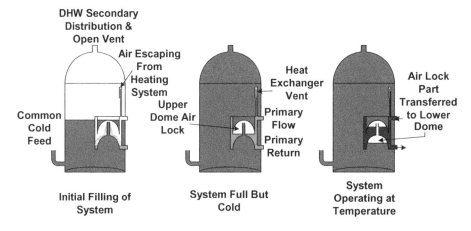

Figure 19.9   Single feed indirect domestic hot water copper cylinder

The double feed indirect domestic hot water cylinders previously described can be obtained in almost unlimited sizes and capacities, suitable for both small residential dwellings and large commercial buildings, but the single feed indirect domestic hot water cylinders are only available in a limited number of sizes, suitable for residential dwellings only.

The air lock indicated in Figure 19.9 is required to form a separation barrier of air between the primary and secondary waters within the cylinder, and has to be created when initially filling both the heating system and domestic hot water system. This is achieved by allowing water from the main cold water storage cistern to slowly fill the cylinder via the common cold feed; if this is done too quickly, the turbulence created inside the cylinder may cause the volume of air required for the air lock to be reduced below that required for an effective separation barrier. As the level of the water within the cylinder begins to rise to just below the top of the heat exchanger, the water begins to pass into the heating system through the vertical tube located at the top of the lower heat exchanger dome. As the heating system fills, air is expelled from the heat exchanger vent and subsequently through the main vent on top of the cylinder.

Air from the heating system must be vented off through radiator vents and manual air valves, which should be fitted at any high points on heating systems that are not self venting.

When the heating system is full but still cold, an air lock would have been created within the upper dome of the heat exchanger as shown in Figure 19.9. During normal operation when the primary water has been heated, the increased volume of the primary heating water caused by expansion would force the water level within the heat exchanger below the air lock to rise, and in turn force the air pocket to partially transfer back down the connecting tube into the lower dome, as well as rise up the heat exchanger vent tube. When the heating system cools down and the primary water contracts, the water levels return to their original levels with the air lock separation barrier returning to the upper dome.

Figure 19.10 illustrates the working principles of the single feed indirect domestic hot water cylinder within the heating system.

There have been a number of variations to the design of the single feed indirect heat exchanger made by several manufacturers since its initial conception in the late 1950s. The system has been successful as a method of providing domestic hot water, providing that it has been designed correctly, whereby the air lock being created is equivalent to a minimum of $\frac{1}{20}$ of the total water content of the heating system. If this air volume is inadequate the air lock will be gradually lost, resulting in the primary and secondary waters being mixed.

UNVENTED DOMESTIC HOT WATER

The domestic hot water systems described and illustrated so far in this chapter have been of the traditional type, which are open vented and gravity fed from the main cold water storage cistern, that have been in common use in the UK for many decades.

In the UK, since their introduction in the 1985 Building Regulations Part G and subsequent Model Water Byelaws 1986 that preceded the present Water Regulations, unvented domestic hot water storage systems (UVDHWSS) have been permitted to be fed direct from the local water supply undertaker's water main, offering an alternative method of providing domestic hot water that presents many advantages.

These regulations apply to domestic hot water storage vessels with a capacity greater than 15 litres, which do not incorporate an open vent pipe to atmosphere. These are required to be installed by a competent person who has been specifically trained on such systems and holds the necessary certification of competency.

For many decades before their introduction to the UK, unvented domestic hot water storage systems were in common use in numerous countries, including North America, Australia and many parts of Europe. The slow introduction of this domestic hot water system into Great Britain was mainly to do

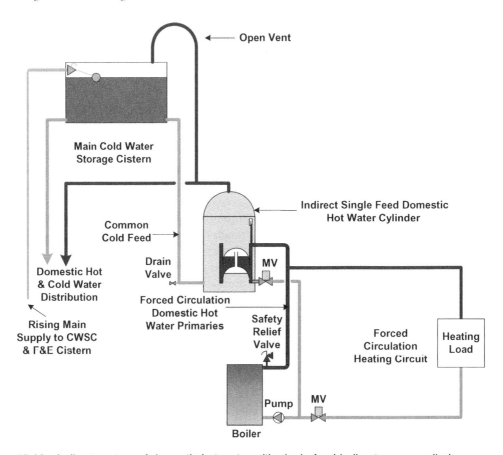

Figure 19.10   Indirect system of domestic hot water with single feed indirect copper cylinder

with historic reasons, because of the way that British water supply system has developed. The quality of the water supply in the UK has to be of a drinkable standard, which means that all precautions necessary have to be taken to prevent that quality from being contaminated by cross-connections to water services that are not of a drinkable standard, such as domestic hot water systems. Traditionally this protection was provided by the installation of the main cold water storage cistern, serving as a protective barrier by incorporating an air gap between the stored water level and the float valve on the incoming water supply.

The change in the Building Regulations, together with the Model Water Byelaws and subsequently the Water Regulations that permitted the installation of unvented domestic hot water storage systems, where the temperature of the water did not exceed 100°C, was made following the need to bring British practice for building water services in line with current European practice. This was made possible by the improved quality in the manufacture of the control components and the requirement for all installers of these systems to be fully trained and certified competent.

Table 19.4 summarises the merits for both the unvented and fully open vented systems of domestic hot water.

The development of the unvented domestic hot water storage system in the UK has gone hand-in-hand with the development of the sealed heating system, as the reasons for selecting these methods are similar.

Table 19.4 Domestic hot water central storage selection

UVDHWSS	Open Vented Storage System

a) System does not require a cold water storage cistern and therefore the risk of freezing damaging roof space services has been eliminated.
 	
   This also vacates the roof space for other purposes and permits greater flexibility in the location of the domestic hot water storage vessel.
	a) The cold water storage cistern serves as a reserve of water in the event of a water mains failure.
	
	   This limited reserve can be used for both sanitation and fire fighting purposes.

b) Achieves higher operating pressures that permit the use of high resistance shower heads.	b) Because both the domestic hot and cold water services are fed from the cold water storage cistern, pressures will be balanced and remain constant without fluctuations.
c) There is no risk of contamination via the cold water storage cistern due to the omission of this item.	c) There is no risk of backflow contamination from the domestic hot water system back into the water mains due to the integral air break.
d) System may be subjected to fluctuating water mains pressures caused during peak demands throughout the day.	d) System pressures will be dictated by the physical height of the cold water storage cistern.
e) The flow of water will be limited by the capacity of the incoming cold water main, which in most dwellings is 15mm diameter.	e) Pipe diameters can be sized to accommodate the low working pressures which can be compensated to provide a greater volume.

PRINCIPLES OF UNVENTED DOMESTIC HOT WATER STORAGE SYSTEMS

Figure 19.11 illustrates a schematic layout of a combined sealed heating system producing the primary heat for an unvented domestic hot water storage system. The domestic hot water system can be used equally well in conjunction with an open vented heating system, although it is more commonly used as shown.

The storage cylinder is similar to a standard copper or stainless steel indirect cylinder with a coil type heat exchanger to provide the primary heat, although the cylinder must be rated for the higher operating pressure that it could be subjected to; that includes the maximum pressure that the water main may exert on the system if there is a failure of the pressure reducing valve.

The difference between the unvented storage vessel and the open vented storage vessel occurs in the method by which the cold feed to the cylinder is delivered, and the way that the increased volume of water created by the expansion of the heated secondary water is catered for, plus the additional control devices that are required for safety reasons.

The cold feed to the unvented domestic hot water storage vessel is arranged to be supplied direct from the incoming water supply main, as the traditional cold water storage cistern used for the open vented system is omitted.

The cold feed being supplied direct from the incoming water main requires measures to be taken to prevent backflow contamination to the water main from the domestic hot water system, and to maintain a constant working supply pressure.

The arrangement shown in Figure 19.11 illustrates an unvented domestic hot water storage system incorporated into a sealed system of heating, which is as described and illustrated in Chapter 2. The sealed system of heating transfers its primary heat via the coil heat exchanger within the storage cylinder to the secondary domestic hot water in the normal way, but has an additional control device in the form of a high-limit energy cut-out thermostat set at 90°C. This supplementary thermostat is arranged as a safety mechanism to close the motorised valve on the primary return in the event of the main control thermostat failing to operate at its setting of 60/65°C, and has to be manually reset after the fault has been identified and rectified.

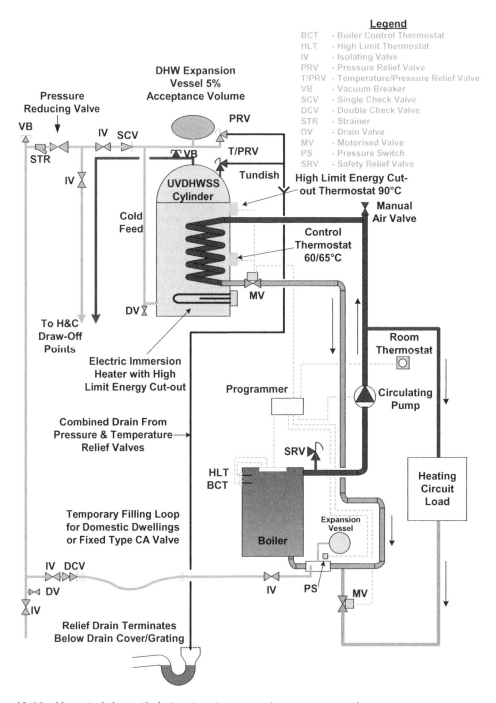

Figure 19.11   Unvented domestic hot water storage system arrangement

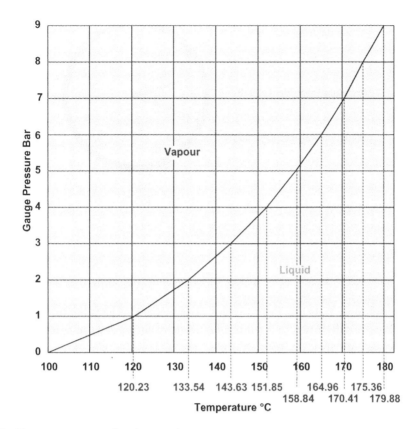

Figure 19.12 Vapour pressure of water graph

Due to the omission of the traditional cold feed, open vent pipe and cold water storage cistern, the increased volume of water created by the rise in water temperature has to be accommodated in a sealed expansion vessel similar to that described for the sealed heating system, except that it must be suitable for domestic hot water use and not one that is manufactured for hydronic heating systems only.

A pressure relief valve similar to the safety relief valve discussed in Chapter 11, is also fitted by the manufacturer to relieve any excess pressure above its normal working pressure, to protect the storage vessel and sealed expansion vessel from splitting. This increase in pressure can result from a failure of the pressure reducing valve on the incoming water main, or membrane inside the expansion vessel, or malfunction of the control thermostat. Figure 19.12 gives an indication of the relationship between the vapour pressure and gauge pressure.

The incoming mains water supply to the unvented domestic hot water storage vessel is required by the Water Regulations to be protected from backflow by a single check valve, as illustrated in Figure 19.11. This valve should be of a type that will also prevent backflow occurring when both the upstream and downstream pressures are equal – hence the term 'check valve' and not 'non-return valve'.

Figure 19.11 also illustrates a pressure reducing valve on the incoming mains water supply to the unvented domestic hot water storage vessel, which controls the pressure to below the maximum pressure required by the system. This valve is sometimes fitted by the manufacturer of the vessel if the pressure is critical, but may be required to be fitted separately if the incoming water pressure fluctuates over a wide pressure range.

For this reason the pressure of the water supply should be monitored over a period of at least one week, to record both the maximum and minimum pressures that occurred over that period. This should be

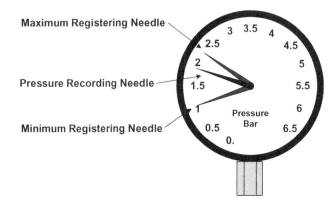

Figure 19.13   Pressure monitoring gauge for recording water supply pressures

undertaken using a pressure gauge that incorporates two record indicating needles, as shown in Figure 19.13, that is temporarily fitted to the water supply.

Another component shown in Figure 19.11, but not a requirement of the Water Regulations, is the inclusion of a vacuum breaker valve fitted to the incoming water main supply: this is often a requirement in other countries that use unvented domestic hot water storage systems. The purpose of this device is to break any siphon that may occur in the event of a failure in the water supply, by allowing air to enter the pipeline and thus preventing any partial vacuum being created at the inlet of the storage vessel. When the water supply is restored, the valve will automatically close again and prevent any loss of water.

A device that is used by some manufacturers of unvented domestic hot water storage vessels is the vacuum breaker valve, shown on the secondary domestic hot water distribution. This is provided to prevent a partial vacuum occurring when the system is being drained down at any draw-off point – this could cause a collapse of the cylinder by the external pressure on the walls of the vessel being higher than the internal pressure.

Some manufacturers of unvented domestic hot water storage vessels fit vacuum breakers direct to the storage vessel shell instead of the distribution piping as shown.

The Water Regulations and Building Regulations also require the domestic hot water to be prevented from exceeding 100°C by the inclusion of an additional safety component, such as a temperature relief valve as shown in Figure 19.11. This valve is fitted directly into the storage vessel and set to open when the water temperature reaches 95°C. It is often provided in a combined pressure and temperature form, but the fact that it also has pressure relieving properties does not mean that the separate pressure relief valve does not have to be fitted.

This valve protects the system should the thermostatic control and energy cut-out thermostat fail to maintain the water temperature within safe limits.

DISCHARGE PIPES FROM SAFETY DEVICES

The Water Regulations and Building Regulations Part G specify performance requirements for the safe discharge and termination of discharge pipes from expansion valves (pressure relief valves), temperature relief valves and combined temperature and pressure relief valves which are illustrated in Figure 19.14.

These regulations require that both the pressurised section and gravity flow section of all discharge pipes are installed in a non-ferrous metallic pipe, which means either copper or stainless steel tube should be used with all pipe joints suitable for the fluid flow having a temperature of 100°C. This would prohibit

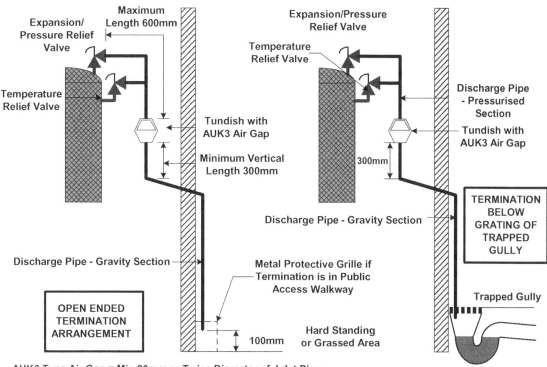

AUK3 Type Air Gap = Min 20mm or Twice Diameter of Inlet Pipe

Discharge Pipes to be of a Metallic Material and to have a Continuous Fall

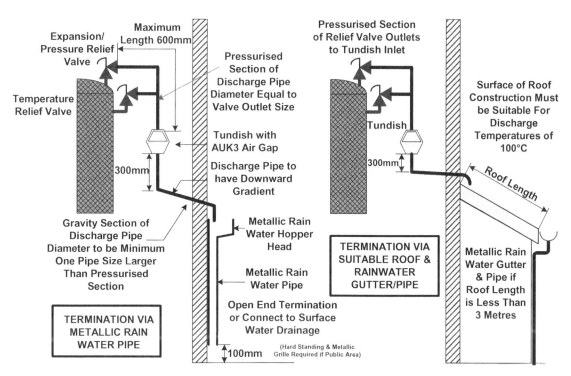

Figure 19.14   Relief valve discharge pipe and terminations from UVDHWSS

Table 19.5   Simplified sizing of copper/stainless steel discharge pipes

Relief Valve Outlet Size (Inches)	Minimum Size of Pressurised Section of Discharge Pipe to Tundish (mm)	Minimum Size of Gravity Section of Discharge Pipe From Tundish (mm)	Maximum Pipe Length Permitted Expressed as Straight Pipe (No Elbows or Bends) (Metres)	Resistance Expressed as Equivalent Length of Pipe for Elbows & Bends (Metres)
½	15	22	Up to 9	0.8
		28	Up to 18	1.0
		35	Up to 27	1.4
¾	22	28	Up to 9	1.0
		35	Up to 18	1.4
		42	Up to 27	1.7
1	28	35	Up to 9	1.4
		42	Up to 18	1.7
		54	Up to 27	2.3

It should be noted that the term 'bends' includes all labour pulled bends at any angle.

the use of pipe joints using some types of synthetic rubber 'O' rings that are not rated at these temperatures.

The pressurised section of the discharge pipe should have a diameter of not less than the valve outlet diameter and should be connected to a tundish having a type AUK3 air gap.

The total length of the pressurised section of discharge pipe, from expansion/pressure relief valves, temperature relief valves and combined pressure/temperature relief valves to the tundish inlet, should not exceed 600 mm. The tundish should be located in a visible position with the outlet of the gravity flow section from the tundish having a minimum diameter of at least one commercial pipe size larger than the inlet pipe to the tundish. If the tundish is not visible, then an additional air break such as that illustrated terminating over a rainwater hopper head should be provided where it can be easily observed.

The Building Regulations give simple guidance on the sizing of the gravity section of the discharge pipe from the outlet of the tundish, which is summarised in Table 19.5, but it should be appreciated that this section of the discharged flow behaves in a similar manner to that of waste water drainage flow due to the entry of air from the tundish, and should be sized accordingly.

It should be emphasised that the guidance note stating that the gravity section of the discharge pipe from the tundish should be a minimum of one commercial pipe size larger than the tundish inlet pipe is only a guidance note and the size must be calculated using the information given in Table 19.5.

Example – Calculate the gravity section of the discharge pipe from the tundish which has a 15 mm diameter inlet from the relief valve whereby the discharge pipe length is 6 m and has 5 elbows.

From the table, for a 15 mm diameter relief valve discharge into the tundish, a 22 mm diameter pipe is the minimum size tube that is permitted, which can be up to 9 m in length.

The 5 elbows required on the discharge pipe would each equate to a resistance of 0.8 m, which when multiplied by 5 ($5 \times 0.8 = 4.0$ m) equals 4.0 m of resistance encountered. Subtract the 4.0 m from the maximum permitted length of 9 m = 5 m, which is one metre short of our required length.

Therefore selecting the next tube size up, 28 mm can have a maximum permitted length of 18 m, but each elbow now has a resistance equal to 1 m of equivalent pipe diameter.

Therefore, 5 elbows at 1 m each equals a total of 5 m.

Therefore, 18 m minus 5 m for the elbows = 13 m total length of pipe permitted. As the example has 6 m of pipe run, the result of the calculation is 7 m longer than our requirement and should be used.

This example shows that quite often two or more pipe diameters may be required above the inlet pipe size of the tundish.

The gravity section of the discharge pipe from the outlet of the tundish should extend down vertically for a minimum length of 300 mm before any bend is fitted, to reduce the risk of high temperature water splashing out of the tundish caused by the back-up from the resistance encountered. It should then have a continuous fall to its termination point.

SUMMARY OF CONTROLS FOR UVDHWSS

Because the system becomes pressurised when heated, and there is an absence of the traditional open vented controls from the cold water storage cistern via the open vent and cold feed, additional controls are required above those required for the open vented system.

These may be summarised as follows:

- Main primary heat control make or break thermostat set at 60/65°C operating a motorised valve.
- High limit energy cut-out thermostat set at 90°C that does not automatically reset itself when the temperature decreases. This thermostat also operates the primary heat source control valve as a secondary safety control component.
- If an electric immersion heater is provided as a secondary heat energy source it must also be fitted with a high limit energy cut-out thermostat set at 90°C.
- Expansion/pressure relief valve to relieve any excess pressure within the system and protect all components to within their pressure limitations. The set pressure should be as recommended by the unvented domestic hot water storage vessel manufacturer.
- Temperature relief valve arranged to relieve the system and prevent the temperature of the domestic water exceeding 100°C. This safety valve should be set to operate at a temperature of 95°C.
- Single check valve required to prevent any secondary domestic hot water returning back into the incoming water supply main.
- Pressure reducing valve set to maintain the downstream operating water supply pressure to within the manufacturer's recommended maximum pressure.
- Sealed expansion vessel suitable for domestic hot water and sized to be able to accept at least 5% of the total domestic hot water content of the system at a temperature of 100°C.
- Vacuum breaking valve fitted if required to break any siphoning of the domestic hot water back into the water supply main and possible implosion of the unvented domestic hot water storage vessel.

UNVENTED DOMESTIC HOT WATER STORAGE VESSEL WITH INTERNAL EXPANSION CHAMBER

As an alternative unvented domestic hot water storage vessel arrangement to the one shown in Figure 19.11, Figure 19.15 indicates an arrangement where the expansion of the secondary domestic hot water is contained within an air chamber formed inside the storage vessel.

The only difference between this arrangement and the one illustrated in Figure 19.11 is that a dome has been formed on the end of the secondary domestic hot water distribution pipe. This creates an air chamber above it to accommodate the expansion of the domestic hot water when it has been heated. This air chamber must be large enough to accept 5% of the domestic hot water system for the reasons discussed previously; if it is under-capacity an additional separate sealed expansion vessel must be fitted to the system to supplement the internal air chamber. This arrangement can, in certain turbulent conditions, partially lose the air lock in a similar way to that described for the single feed indirect cylinder: this will be noticed by the regular activation of the pressure relief/expansion valve, but the air chamber can be restored fully by draining down the system and slowly refilling it.

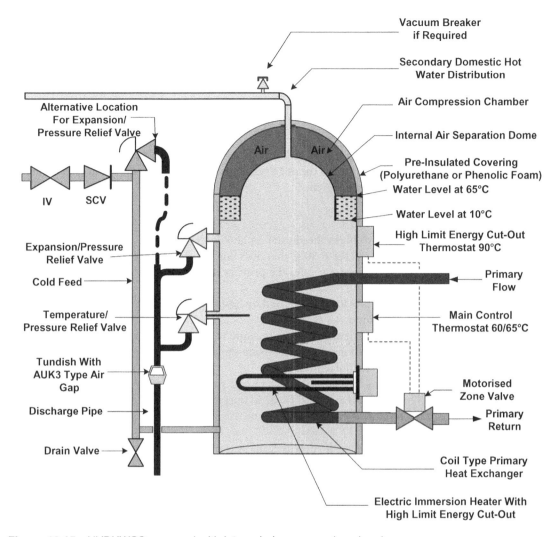

Figure 19.15   UVDHWSS arranged with internal air compression chamber

NON-STORAGE UNVENTED DOMESTIC HOT WATER SYSTEMS

The unvented domestic hot water systems illustrated and described earlier in this chapter have all been based on maintaining a storage of hot water at a temperature ready for use.

Figure 19.16 depicts a pressurised system being fed direct from the water supply mains that does not require a reserve of domestic hot water, but incorporates a purpose-made thermal heat store type cylinder, complete with a coil type heat exchanger that works in reverse to that of the standard indirect domestic hot water cylinder.

The primary water circulates through the cylinder and is maintained as a heat store at a temperature of 60°C by a standard control thermostat and protected by an additional high limit non-automatically resetting thermostat.

The cold feed to the domestic hot water is supplied from the water supply mains into the bottom of the coil of the heat exchanger in the cylinder, which transfers part of its heat from the heat store as it passes through the coil and exits the top of the coil heat exchanger as the secondary domestic hot water distribution.

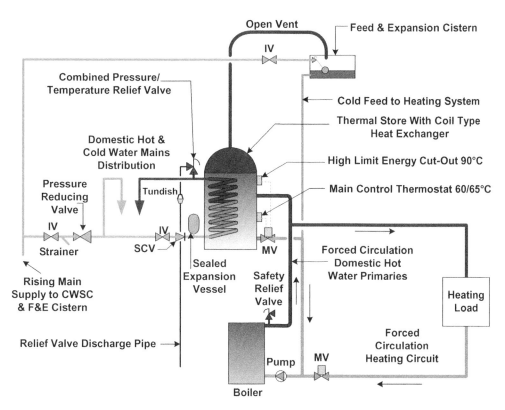

Figure 19.16 Thermal store domestic hot water system

It should be noted that this is not a standard indirect domestic hot water cylinder used in reverse, but is purpose-made for this application. It is designed to operate as an instantaneous heater from the heat store to deliver approximately 100 litres before the temperature of the domestic hot water falls to below 40°C. The system is equipped with the same control and safety devices as other unvented domestic hot water storage systems, except the sealed expansion vessel will be smaller due to the lower volume of domestic hot water content.

The heating system may be either the open vented system as illustrated, or a pressurised sealed system.

DOMESTIC HOT WATER CYLINDER WITH TWIN HEAT EXCHANGERS

There are both merits and disadvantages associated with unvented domestic hot water storage systems as well as sealed heating systems; these have been discussed earlier in this chapter as well as in Chapter 2.

One of the reasons often given for preferring unvented domestic hot water systems is the higher operating pressures, especially for showers. Figure 19.17 illustrates a system that combines both open vented and unvented domestic hot water systems supplied from a common indirect cylinder.

This is a purpose-made vessel whereby the primary water is supplied from the heating system to the main coil type heat exchanger in the normal way; it then transfers part of its heat to the secondary water stored within the cylinder to provide domestic hot water to all draw-off points. The secondary water then transfers part of its heat to a second smaller coil type heat exchanger contained within the cylinder, which has been supplied direct from the water supply main, to provide mains pressurised domestic hot and cold water to a fixed shower.

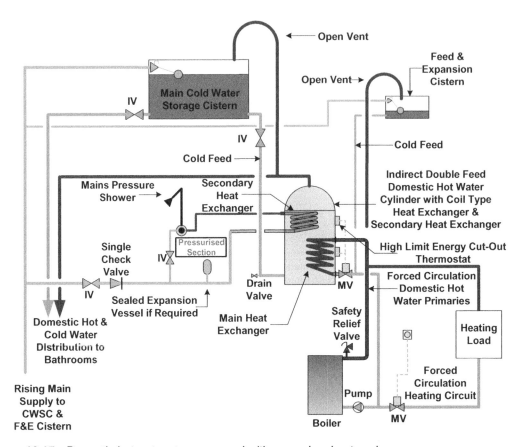

Figure 19.17   Domestic hot water storage vessel with secondary heat exchanger

STRATIFICATION

Stratification is the term applied to the phenomenon that exists within any storage vessel where the contents are raised in temperature. A steady temperature gradient as depicted in Figure 19.18 will result, with the hottest temperature existing at the top and a gradual lower temperature occurring at the base.

As previously established, the recommended position for the cylinder control thermostat is approximately one third up the height of the cylinder and set to a temperature of 60°C. Assuming that the thermostat is functioning correctly it will maintain that temperature at that particular point. If it was possible to record the temperatures it would be observed that there is a steady rise in temperature above the thermostat, which could be anything between 5 to 10°C higher than the thermostat setting, depending upon the physical height of the cylinder. Also, it would be observed that the temperature of the water below the thermostat is somewhat lower than the set temperature.

If a significant portion of the cylinder contents were to be drawn off, it would simply be replaced by incoming cold water at ambient temperature, which would then be heated over a period of time to restore the design temperature of 60°C.

However, if only a small quantity of water is drawn off, the replacement colder water would mix with the remaining hot water in the cylinder, having the effect of lowering the overall temperature of the remaining water, which would then have to be reheated to restore the desired storage temperature, using energy to do so.

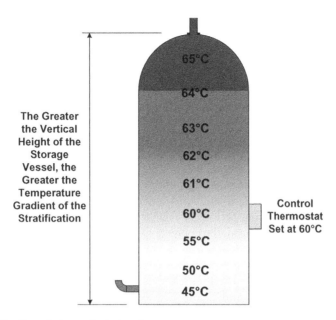

Figure 19.18   Stratification of domestic hot water

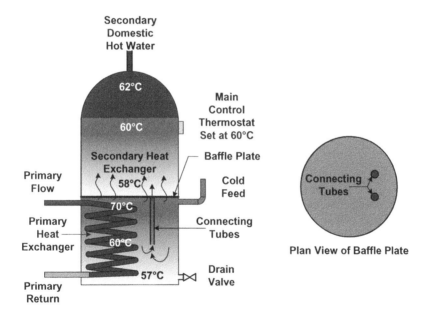

Figure 19.19   Stratification reducing domestic hot water cylinder

The phenomenon of stratification has mainly been accepted as something that exists, but there have been a number of innovative attempts to reduce it's effects and in so doing, increase efficiency. These include the practice of fitting a deflection plate over the inlet of the cold feed to reduce the turbulence caused by the velocity of the incoming cold water and thus reduce the cooling effect.

Figure 19.19 illustrates an inventive attempt at reducing the stratification problem in residential domestic dwellings, mentioned here so that the reader will be able to identify it if they come across it.

It takes the external form of a standard domestic hot water cylinder, apart from the odd positions of the entry of the cold feed connection and the low primary flow and return to the heat exchanger on the side of the cylinder.

Internally, the cylinder is divided into two halves by the introduction of an internal baffle plate, which acts as a secondary heat exchanger and incorporates two inverted tubes to join the two sections up, with the lower section housing a coil type primary heat exchanger. This creates a situation whereby water stored in the upper portion of the lower section can reach a temperature of 70–73°C when the main control thermostat is set at 60°C, and transfers part of that heat to the water stored in the upper section.

When a small quantity of domestic hot water is drawn off, the incoming cold feed water mixes with this higher temperature water to cool it to approximately 60°C before it is transferred to the upper section via the connecting tubes. This is said to achieve a larger effective capacity of usable domestic hot water for the storage capacity of an equivalent standard indirect cylinder.

This arrangement is effective during periods of regular usage of domestic hot water, but during long periods of no domestic hot water demand, the contents of the upper section of the cylinder could exceed 65°C.

Figure 19.20 illustrates another method that is frequently used for domestic hot water storage systems in excess of 500 litres capacity. Here, a circulating pump is added as shown, and arranged to operate for a minimum period of time when the stratification thermostat mounted on the top of the storage vessel reaches a temperature in excess of 66°C. When the circulating pump is activated it circulates the water to mix the temperature of the water and obtain a fairly even temperature of 60°C.

The circulating pump has to be arranged to operate for a minimum period of time to protect the starters of the pump from burning out: this time period will depend upon the pump manufacturer's recommendations, but is normally in the region of 10 to 12 minutes, with a similar period of time between starts.

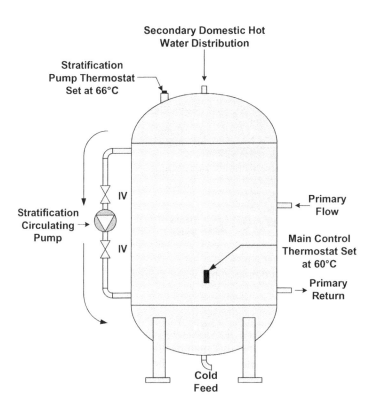

Figure 19.20   Stratification reducing circulating pump

Alternatively, the pump can be arranged to run for approximately 20 to 30 minutes every two hours during the time that the primary heat is being supplied.

BOILER LOADING/ELECTRICAL POWER FOR DOMESTIC HOT WATER

During the heat loss calculation exercise conducted in Chapter 7, Heat Requirements of Buildings, an arbitrary figure expressed in kilowatts was added to the heat loss total of the building to arrive at a boiler rating for the example building used.

The amount of power required to raise the temperature of a known volume of water may be calculated using the following formula:

$$Q = \frac{M \times \Delta t \times SpHt}{Rt \times 3600}$$

Where

Q   = Quantity of heat required in kW
Δt = Temperature difference (initial temperature to final temperature)
SpHt = Specific heat of water in kJ
Rt   = Recovery time expressed as a decimal in hours
M   = Mass quantity of water expressed in kg

For example – Calculate the amount of heat required to raise 140 litres of water to 60°C in two hours.

One litre of water weighs 1 kg at 4.4°C; therefore 140 litres will weigh 140 kg, or approximately 140 kg at its initial incoming temperature which is normally taken as being 10°C.

The specific heat of water as previously established is a constant and for the purposes of the calculation is 4.2 kJ.

$$\frac{140 \times (60-10) \times 4.2}{2 \times 3600} = \frac{29400}{7200} = 4.08 \text{ kW}$$

Therefore, in this example 4.08 kW of heat is required to raise 140 litres of water from 10°C to 60°C in two hours and should be added to the total heat loss calculated for the dwelling in order to arrive at a suitable rating for the boiler.

However, if it is desired to heat the water by a 3 kW electric immersion heater, the recovery time may be calculated by transposing the formula as follows:

$$Rt = \frac{M \times \Delta t \times SpHt}{Q \times 3600}$$

Therefore:

$$\frac{140 \times (60-10) \times 4.2}{3 \times 3600} = \frac{29400}{10800} = 2.72 \text{ hours}$$

Therefore, a 3 kW electric immersion heater fitted in the cylinder having a capacity of 140 litres will raise the temperature from 10°C to 60°C in 2.72 hours, or approximately 2¾ hours.

As previously discussed in this chapter, BS 1566 for type 'P' indirect copper cylinders specifies that the coil type heat exchanger must be capable of recovering the quantity of the cylinder in 25 minutes at the same temperature difference used in the example calculation. For this rapid recovery time to be achieved the kilowatt power requirement must be calculated.

(25 minutes = 0.416 as a decimal of an hour).

$$\frac{140 \times (60-10) \times 4.2}{0.416 \times 3600} = \frac{29400}{1497} = 19.64 \text{ kW}$$

Therefore, to achieve a 25 minute recovery time, we would need to put nearly 20 kW of heat into the domestic hot water to heat 140 litres of water from 10°C to 60°C.

Adding 20 kW to the heat loss of the building to arrive at the boiler loading would not be an economic or efficient action. However, if the control system was one that permitted the domestic hot water to be arranged as a priority by ignoring the heating requirements during this period, then a 25 minute recovery time could be achieved, if the boiler had a heat output rating of 20 kW or more.

ELECTRIC IMMERSION HEATERS

The merits and uses of electric immersion heaters have been mentioned briefly throughout this text, and Chapter 10 explains their electrical requirements.

Electric immersion heaters can be used as a supplementary or standby means of heating power, fitted into an indirect domestic hot water cylinder with its main primary heat arranged to be furnished from the hydronic heating system, or as the solitary source of primary heat installed in a direct type domestic hot water system.

Immersion heaters consist of an electrical resistance element sheathed in copper, tinned copper, or a corrosion-resistant element – such as titanium or nickel alloy – suitable for hard or aggressive waters and have the provision for a thermostat to be installed, which usually has to be purchased separately.

The function of the thermostat is to control the power input to the resistant element by switching the electrical supply on or off when it has reached its set temperature, normally 60°C, but since April 2004 all new electric immersion heaters or replacement electric immersion heaters must be fitted with a resettable double thermostat (RDT).

A resettable double thermostat incorporates an over-temperature cut-out device independent of the main control thermostat that has to be reset manually by a qualified person after the fault has been investigated and rectified. The function of this over-temperature cut-out device is to prevent the domestic hot water from boiling in the event of the main control thermostat failing, or the primary water overheating due to a boiler fault and should therefore be set to cut out at 90°C. It should be noted that if the main thermostat fails and the cut-out device trips the electrical supply, then the whole RDT will need replacing.

Domestic versions have a unique 2¼ inch BSP male thread as standard and are available in 1, 2, 3 and 4 kW ratings at a 230 volt single phase electrical supply, with the most commonly used being 3 kW. Higher kilowatt ratings up to 12 kW can be obtained still with a 230 volt single phase electrical supply and 2¼ inch BSP male thread.

Figure 19.21 illustrates the different types, mounting positions and arrangements for electric immersion heaters, including the selection of the correct proportional length for the domestic hot water cylinders that they are to be fitted into. Top mounted vertical applications should extend approximately two-thirds the depth of the cylinder with side mounted horizontal applications extending around three-quarters of the cylinder diameter, but the immersion element end should not be closer than 50 mm to the far wall of the cylinder.

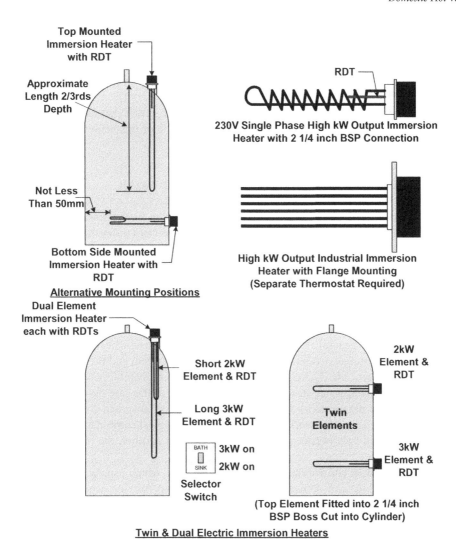

Figure 19.21 Types, mountings and arrangements for electric immersion heaters

Figure 19.21 also illustrates arrangements for a more economical use of electrical power whereby only a proportion of the cylinder contents can be heated in situations where it is not desirable to heat the entire cylinder contents.

This can be achieved by fitting two separate thermostatically controlled electric immersion heaters controlled by a selector switch, so that only one of them can operate at any one time, with the top immersion element having a lower kilowatt rating than the lower immersion element. This permits the top section of the cylinder to be heated when only small quantities of domestic hot water are required, or the whole of the cylinder to be heated by the lower immersion element when larger quantities of domestic hot water are needed.

Domestic hot water cylinders are manufactured with a single 2¼ inch BSP immersion heater boss if specified, but in some situations a second boss must be cut and inserted into the wall or dome of the cylinder if twin immersion heater elements are required.

Alternatively, a single entry duel immersion heater element may be installed that functions in the same way as two separate elements. This comprises two electric immersion elements arranged for a vertical application from a single boss connection, one short element for partial cylinder heating and a longer element for heating the complete contents of the cylinder. Both immersion elements are controlled by their own dedicated RDT thermostats independent from each other.

The duel immersion heater unit is also controlled by a selector switch that can be obtained as part of the electric immersion heater package and which only permits one of the elements to be operated at any one time.

COMBINATION BOILERS FOR HEATING AND DOMESTIC HOT WATER

Units that combine space heating and domestic hot water provision have been available for many years, albeit they consisted of a traditional boiler with a traditional domestic hot water storage cylinder, sometimes with a cold water storage cistern together with a feed and expansion cistern, all pre-plumbed and mounted on a support frame. This made for a large and heavy unit that just required to be placed on site, requiring only the distribution pipework connections to be made along with flues, gas connections and electrics.

Today, the term 'combination boiler' is applied to a more compact unit that still performs the same functions but achieves them differently. Combination boilers comprise a low water content heat exchanger together with a second instantaneous type heat exchanger to furnish domestic hot water on demand.

Figure 19.22 illustrates the general arrangement of heating and domestic hot water supply provided by a combination boiler. The heating is arranged as a sealed pressurised system and the domestic hot water operates on an instantaneous principle. Therefore it can be seen that the advantages of these compact units are the absence of any hot or cold water storage provision – similar to the case described for sealed heating systems – and pressurised water services – similar to those provided by unvented domestic hot water systems.

Figure 19.23 illustrates the general working arrangement of a typical combination boiler unit; although the exact configuration will vary between one manufacturer and another, the basic operating principles will be similar.

The boiler is activated by its control thermostat calling for heat, which usually begins by going through a flue fan proving procedure and sometimes a circulating pump proving sequence, whereby pressure

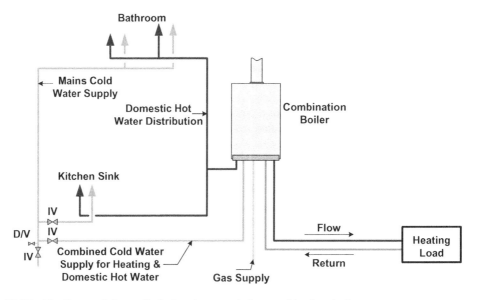

Figure 19.22 Heating and domestic hot water supply by combination boiler

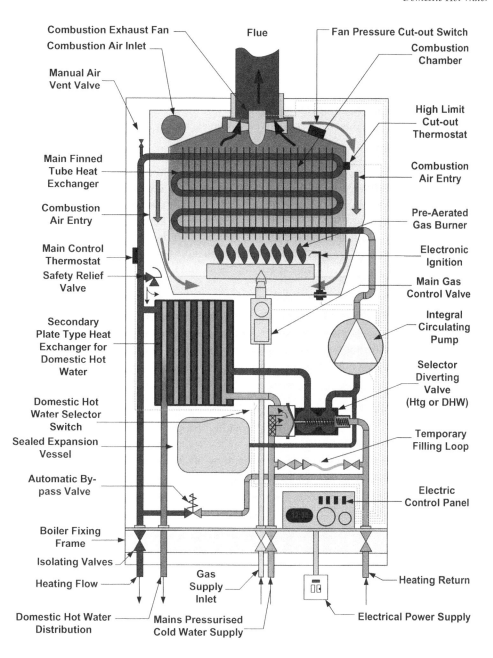

Combustion Exhaust Fan
Combustion Air Inlet
Manual Air Vent Valve
Main Finned Tube Heat Exchanger
Combustion Air Entry
Main Control Thermostat
Safety Relief Valve
Secondary Plate Type Heat Exchanger for Domestic Hot Water
Domestic Hot Water Selector Switch
Sealed Expansion Vessel
Automatic By-pass Valve
Boiler Fixing Frame
Isolating Valves
Heating Flow
Domestic Hot Water Distribution
Mains Pressurised Cold Water Supply
Gas Supply Inlet
Flue
Fan Pressure Cut-out Switch
Combustion Chamber
High Limit Cut-out Thermostat
Combustion Air Entry
Pre-Aerated Gas Burner
Electronic Ignition
Main Gas Control Valve
Integral Circulating Pump
Selector Diverting Valve (Htg or DHW)
Temporary Filling Loop
Electric Control Panel
Heating Return
Electrical Power Supply

This is not a working arrangement drawing but shows operating principles only.

Figure 19.23   Basic working principles of combination boiler (shown open to DHW)

switches have to be satisfied before ignition and flame monitoring systems energise the main gas control valve and establish the gas burner at a low flame ignition rate. After a short period of time, and when the flame monitoring system is satisfied that the burner has been successfully established, the gas valve will open fully, or if incorporated, modulate under its own thermostatic control.

The water is then circulated through the main finned tube heat exchanger whereby it is rapidly heated before passing on to the heating system to transfer its heat and return to the boiler. When the heating system has achieved its design temperature, controlled by one of the control systems described in earlier chapters, the boiler will shut down its burner through a timer delay cycle before being allowed to start up again.

When domestic hot water is required, the boiler will be activated by one of the domestic hot water draw-off valves being opened, thus allowing the mains pressurised cold water to flow through the appliance and in turn operate the diaphragm within the diverter valve. This will have the effect of closing the diverter valve to the heating return and opening it to the domestic hot water primary return, allowing the domestic secondary hot water to be heated by the secondary plate heat exchanger almost instantly as it flows through it.

The temperature of the domestic hot water is controlled separately by its own thermostat.

When all the domestic hot water draw-off valves have been closed, the flow of domestic hot water through it will cease, causing the diverter valve to revert to satisfying the heating requirements.

Combination boilers have a number of attributes, namely that they are amongst the high-efficiency group of appliances, are compact in size, which saves on space that would be occupied by other equipment in a traditional heating system, and will provide domestic hot water on demand when required.

However, combination boilers are not the best suited appliances for all applications as they have several shortcomings that must be considered before specifying them, and they are suitable for domestic residential dwellings only.

Combination boilers will perform just as well as any other type of boiler in satisfying the heating requirements of the building, but are dependent upon the water supply pressure and flow available for satisfying the domestic hot water requirements. The water pressure must be checked for its suitability before deciding to install one, as the domestic hot water flow will only be as good as the incoming water supply main can provide and in some cases may need to be upgraded to a larger incoming water main size.

Combination boilers have proved to be suitable for single bathroom applications where the water supply available can satisfy the expected domestic hot water demand, but in dwellings where the expected simultaneous domestic hot water demand is higher, combination boilers could struggle to meet the flow required. In situations such as these, some other form of providing heating and domestic hot water should be considered.

Combination boilers should be located close to the draw-off points to avoid long dead-leg sections of distribution piping for the reasons discussed earlier.

If combination boilers are correctly specified, they can prove to be a very efficient method of providing heating and domestic hot water from a single compact appliance.

20

Solar Energy for Water Heating

Solar energy for water heating is often mistakenly referred to as a new form of alternative energy, but the truth is that harnessing the energy from the Sun to provide heating has existed since the start of time. The first documented science being applied to it can be traced back to work undertaken in northern Europe in the late 1700s, but more recently there are many documented examples of solar water heating installations dating back to the early 1900s.

What is new is the worldwide interest and the political emphasis on installing alternative green forms of sustainable energy as described in Chapter 15, Alternative Fuels and Energy, including the harnessing of solar energy.

The Sun radiates considerable energy upon the Earth, and this can in part be captured and utilised to obtain different forms of energy as illustrated in Figure 20.1.

Although the subjects of photovoltaic and photosynthesis are interesting and important as potential sustainable forms of future energy, this work is only concerned with the utilisation of the thermal energy being radiated from the sun as it affects the plumbing and heating industry. It is a promising form of free energy.

SOLAR ENERGY

For any engineered system incorporating the thermal energy obtained from the Sun, we must have an understanding of our planet's relationship with the Sun together with the seasonal weather changes for different geographical areas, plus the many variables associated with this subject.

The Earth rotates around the Sun in a slightly elliptical path as shown in Figure 20.2. The Earth's distance from the Sun will vary at different times of the year along with its rotational angle facing the Sun, which will have the effect of increasing or decreasing the amount of daylight being received.

It can also be observed from Figure 20.2 that the position or angle of the Sun in relation to a fixed point location on Earth changes every minute of every day on an annual cycle, which will affect the potential effectiveness of any solar collector panel.

Heating Services in Buildings: Design, Installation, Commissioning & Maintenance, First Edition. David E. Watkins.
© 2011 John Wiley & Sons, Ltd. Published 2011 by John Wiley & Sons, Ltd.
As an aid to lecturers and students, full colour versions of the figures in this chapter may be found at www.wiley.com/go/watkins
These figures are © 2011 by John Wiley & Sons, Ltd.

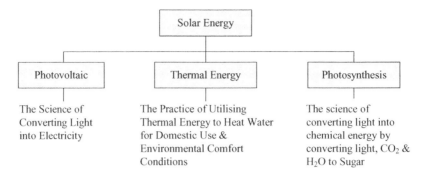

Figure 20.1   Solar energy forms

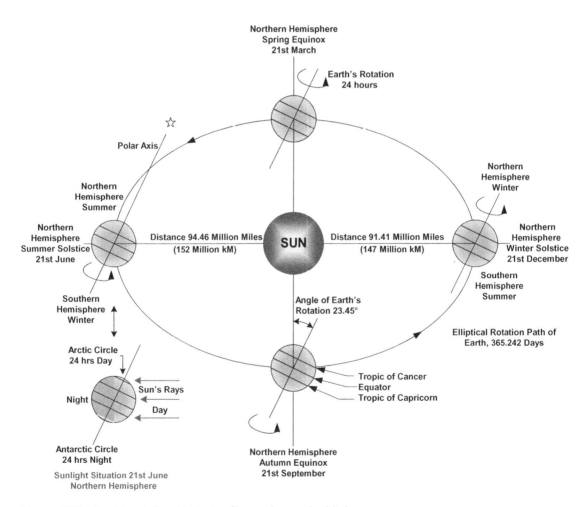

Figure 20.2   Earth's relationship to the Sun and annual orbital passage

Azimuth – The angular distance from a north or south point on the horizon to the intersection with the horizon of a vertical circle passing through a given celestial body.

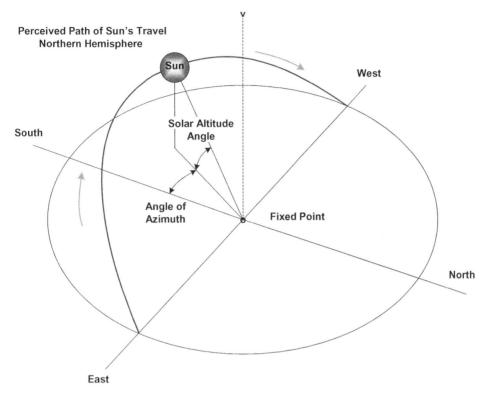

Figure 20.3   Perceived daily path of Sun indicating solar altitude and solar azimuth

Figure 20.3 illustrates this situation viewing the changing position of the Sun from a fixed-point ground location where the perceived orbital path of the Sun travelling from east to west across the horizon, together with its solar altitude angle and azimuth angle can be seen.

Figure 20.3 shows the two angles used to define the angular position of the Sun, which are:

- Solar altitude – the angular elevation of the centre of the Sun above the horizontal plane.
- Solar azimuth – the horizontal angle between the vertical plane containing the centre of the Sun and the vertical plane running in a true north–south direction.

Solar azimuth is measured clockwise from due south in the northern hemisphere and anti-clockwise, measured from due north in the southern hemisphere. Values are negative before solar noon and positive after solar noon.

Figure 20.3 illustrates the Sun's apparent daily arc of travel across the sky in relation to a fixed northern hemisphere point on Earth, showing how it will vary in the solar altitude angle and solar azimuth angle, both of which will be constantly changing throughout the daylight period of the day. Also, as established in Figure 20.2, these angles will also vary each day, as the Earth's position in relation to the Sun changes due to the Earth's annual orbit around the Sun.

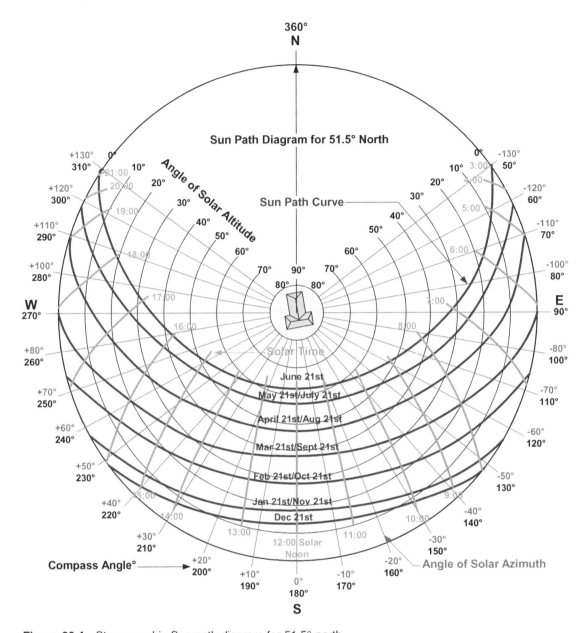

Figure 20.4   Stereographic Sun path diagram for 51.5° north

This constantly changing position can be plotted on a stereographic sun path diagram as illustrated in Figure 20.4, where the path of the Sun can be tracked for any given date, time or location provided that the degree of latitude is known.

This diagram presents the solar altitude and azimuth data in an alternative form. The diagram represents a plan view presented as a hemispherical sky in stereographic projection, with the building plan geographically orientated in the centre at the latitude coordinate of 51.5° North for the example used, with the horizon around the circumference.

Concentric circles and radial lines form a co-ordinate grid with the circles representing contours of equal solar altitude and the radial lines representing solar azimuths. The vertical and horizontal diameters correspond to the N–S and E–W lines respectively.

Arcs are superimposed on the grid to represent the Sun's path of travel across the sky, indicating the situation for different months of the year. Corresponding solar times for the different Sun paths are shown by other radial lines intersecting the Sun path arcs.

The Sun's position at any time of the day and month of the year can be read off in terms of solar altitude and solar azimuth against the co-ordinates of the grid.

It should be noted that the times given on the diagram are solar times, where 12:00 noon indicates the solar midday. Any national times of daylight savings or geographic variances must be converted if actual clock times are required, but normally this is not necessary as solar times are used for assessing potential solar thermal energy.

For latitudes other than 51.5 North, these may be obtained from the tabulated data published in CIBSE guide book A, 'Environmental Design', or the ASHRAE book, 'Fundamentals.'

SOLAR RADIATION INTENSITY

Solar radiation passes through the Earth's atmosphere at an angle perpendicular to the Sun's rays and has an annual mean irradiance of approximately $1370\,\mathrm{W\,m^{-2}}$, (known as the solar constant), at the Earth's surface. In passing through the atmosphere, part of this radiation is diffused by atmospheric water vapour, dust, ozone and aerosol contaminants, which has the effect of making the sky appearing blue on clear days: this is termed 'diffused radiation'; the remainder, termed 'direct radiation', is transmitted largely unhindered.

Cloudy skies will have an effect on both direct and diffused radiation reaching the Earth's surface, and there will a reduction of radiation caused by shading from geological features and adjacent buildings. Even shading from such items as telephone and electricity poles will cause a reduction in the radiation received at particular points.

Figure 20.5 illustrates in simplistic form the average amount of solar radiation received annually at ground level in $\mathrm{W\,m^{-2}}$ for different parts of the British Isles, showing that the south-west receives the most and the north-east the least, although there are local discrepancies and variances to these figures.

It can be appreciated from the previous text that there are many factors affecting the irradiation at any particular location, but the graph depicted in Figure 20.6 may be used to determine the average daily amount of solar irradiation received in the UK and Ireland over the course of a year. It should be emphasised that this graph is for the average amount of radiation received on a daily basis for each calendar month of the year, although nature quite often departs from the average and it is not uncommon for the daily irradiation energy to achieve $7\,\mathrm{kW\,m^{-2}}$. This radiation intensity will vary from as little as $60\,\mathrm{W\,m^{-2}}$ at the winter solstice, falling horizontally on the ground, to a peak of $1200\,\mathrm{W\,m^{-2}}$ during the peak of mid-summer. This is often intermittent during the day due to cloud conditions and will also vary from zero at sunrise, rising to the daily peak at solar noon, before falling back to zero at sunset.

SOLAR THERMAL SYSTEMS

Thermal energy received from the Sun may be harnessed to provide:

- Solar energy for space heating
- Solar energy for heating swimming pools
- Solar energy as a heat source for heat pumps (see Chapter 15, Alternative Fuels and Energies)
- Solar energy for raising domestic hot water temperature

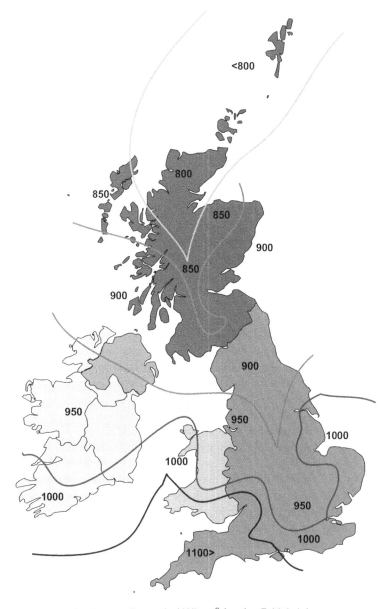

Figure 20.5 Average annual solar irradiance in kWh m^{-2} for the British Isles

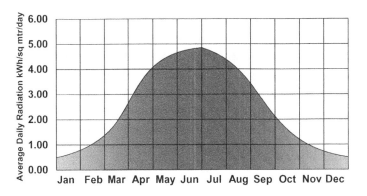

Figure 20.6 Average daily solar radiation received in the British Isles

Solar thermal energy can be utilised for a number of different applications, but the performance of each arrangement will be dependent upon the amount of useful solar irradiation available.

SOLAR THERMAL ENERGY FOR SPACE HEATING

Solar space heating has not been commonly employed in the UK and other northern hemisphere countries with a similar latitude, mainly due to the low amount of solar irradiation received during the normal heating season months when the heat is required. This fact, together with the low operating temperatures that can be achieved and the reduced amount of winter daylight time available for harnessing this energy, generally prevents solar energy from being selected for direct space heating applications.

However, as the length of the solar time available for harnessing this energy is longer either side of peak winter, at the start and end of the heating season, as depicted in the Sun path diagram in Figure 20.4, sufficient energy can be obtained during this period to supplement a conventional fuelled heating appliance and so reduce the amount of fossil fuel energy required to provide space heating.

Figure 20.7 illustrates the operating principles of a solar heated warm air space heating system, utilising limited solar thermal energy available during the winter season to pre-heat the return warm air, or fresh

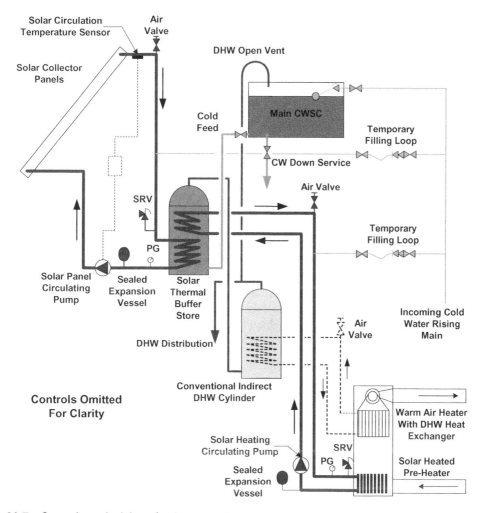

Figure 20.7   Operating principles of solar space heating system

air make-up portion, before the air passes on to the main, conventionally fuelled heat exchanger for the final operating temperature to be achieved, thus saving on fossil fuel energy. Due to only low grade heat being available during this time of the year, the utilisation of thermal energy for hydronic heating applications is considered impractical, but may be economic and save fuel for warm air heating applications, as well as hydronic underfloor heating systems, both of which can use this low grade thermal energy as a means of reducing the consumption of the fossil fuel normally used for producing the desired space heated temperatures.

The system illustrated employs a twin coil type heat exchanger thermal buffer store type cylinder, that is used to serve both as a means of pre-heating the cold feed to the conventional indirect domestic hot water cylinder, and as a means of utilising this thermal store of low grade heat by circulating it through a pre-heat heat exchanger to raise the return/fresh air temperature into the warm air heater, which then raises the temperature further to its final design operating temperature.

The practice of installing solar space heating systems is far more frequently used in countries that enjoy longer daylight time than the UK during the winter season, as they are able to harness more thermal energy, making the system economically viable. In the British Isles it is easy to make a case for installing a solar space heating system on environmental grounds as it will achieve a degree of fuel conservation, together with the reduced amount of the resultant carbon that would have been released into the atmosphere. However, due to the low amount of thermal energy that is obtainable during the peak of the winter season in the UK, it is not possible to justify the installation of a solar space heating system for economic reasons, as the life expectancy of the system is far less than that required to achieve fiscal savings. However, if the costs of the solar collector panels, thermal buffer storage cylinder and circulatory system are ignored – as they form part of the solar domestic hot water system that is economically viable – then the cost of the additional materials and equipment required to provide the limited amount of space heating that can be achieved is far less, thus making the system more attractive on economic grounds.

SOLAR THERMAL ENERGY FOR HEATING SWIMMING POOLS

Swimming pools are ideal applications for being heated by solar thermal energy if the practical requirements for locating solar collector panels – i.e., sufficient space and suitable southerly orientation – can be met. This is due to the relatively low temperatures used for swimming pools in comparison with other heating applications, as they range from 27°C for competitive pool facilities to 30°C for infant and disabled swimming pools (see Table 20.1). The pool heater size should be restricted to raising the temperature of the swimming pool water to between 0.25°C and 0.5°C per hour so as to reduce the possible thermal shock to the pool construction materials, which would damage the structure and waterproof finish.

Figure 20.8 illustrates a typical swimming pool treatment and circulating system arranged to be heated by solar energy, where the heat is transferred via a dedicated heat exchanger. If the solar heating system has been designed correctly then it should be capable of heating the pool totally by solar energy during the summer months as well as the end of the spring season and start of the autumn season. During the winter

Table 20.1   Recommended swimming pool temperatures

Swimming Pool Use	Recommended Maximum Water Temperature °C
Competitive Swimming and Diving Pools	27
Public Recreational/Leisure Pools	28
Private/Domestic Leisure Pools	29
Infant/Children and Disabled Use Pools	30

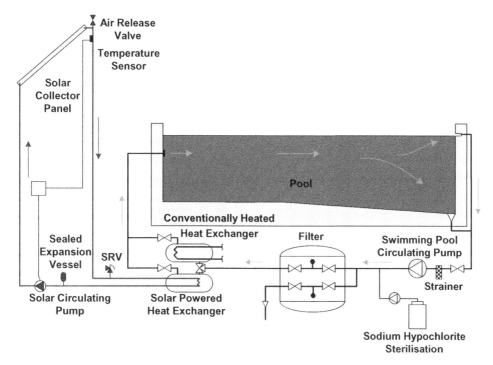

Figure 20.8   Solar energy arranged for heating swimming pool

season, solar energy will be capable of furnishing part of the heat required throughout periods of clear skies, but will need to be supplemented by a conventionally heated heat exchanger energised by either electricity or fossil fuel.

As with all solar heating systems, the effectiveness is dependent upon having a sufficient solar collector panel area arranged at the correct solar altitude angle and free from any shading at all times.

SOLAR ENERGY HEAT SOURCE FOR HEAT PUMPS

The principles and operation of heat pumps are explained in Chapter 15, Alternative Fuels and Energy. The use of solar energy as a heat source for heat pumps has been employed successfully in some applications as it has the advantage of achieving a higher grade heat sink, or usable temperature, than that achieved by a standard air-to-water or air-to-air heat pump when solar energy is available; it can also be used in combination with the main heat sources of air or water.

Solar energy heat pumps are classified as either direct or indirect in their mode of operation. In a direct system as illustrated in Figure 20.9, a refrigerant replaces the brine that is circulated as the refrigerant cycle between the heat pump condenser and the solar collector evaporator, with the solar collector panels arranged to incorporate refrigerant evaporator tubes within them. As this type of unit is supplied and installed as a packaged unit, the distance or length of refrigerant piping is limited to the capabilities of the unit's ability to function under the restrictions of the refrigerant cycle. This restriction also results in the heat obtainable being limited as the collector panel is relatively small, making it only suitable for smaller applications. It can also produce a wider variance in operating heat sink temperatures to the heating system, which can be either hydronic or warm air.

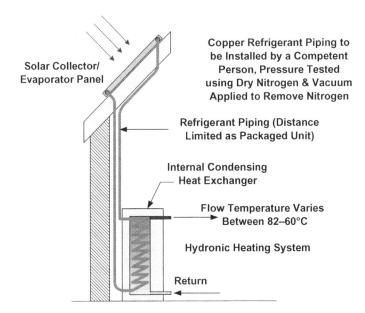

Figure 20.9   Split-unit, packaged type solar powered direct heat pump

An indirect system differs because either water or air is circulated through the solar collector panel as normal, which is then delivered to the evaporator section of the heat pump to function conventionally and provide the heat sink to a hydronic or warm air heating system.

Figure 20.10 illustrates a typical solar energy to water heat pump arrangement.

With this arrangement, the heat pump is arranged for the heat source to be provided by a solar heating system employing a solar collector panel or panels to circulate the heated brine through the evaporator section of the heat pump, whereby a high temperature and a large amount of heat can be provided to the heat sink, which may be either hydronic or warm air.

Provided that sufficient space with the correct angled and orientated roof area is available, this system may be engineered to impart large amounts of high temperature heat for conversion into space heating and any domestic hot water requirements.

SOLAR THERMAL ENERGY FOR RAISING DOMESTIC HOT WATER TEMPERATURE

The utilisation of solar energy for raising the temperature of domestic hot water is far more commonly employed in the UK than for any other application previously described, exploiting this freely available energy source. This is mainly due to a less costly installation compared to the other systems described, and if the conditions are favourable, it can provide up to 100% of the normal domestic dwelling's hot water requirements during the summer season and a reasonable proportion of the requirements during the other seasons. This will also equate to an annual reduction in conventional fossil fuel used for providing space heating and domestic hot water of between 15 and 25%, making it easier to justify on economic grounds. The figures given for fuel savings are average; these can vary considerably for each system and they should be calculated for each specific application to ascertain the expected annual fuel saving.

The principles of harnessing this energy remain similar, although the arrangement of the system can vary depending upon its size and capacity, but as with other forms of domestic hot water systems the primary heat source can be categorised as being either direct or indirect.

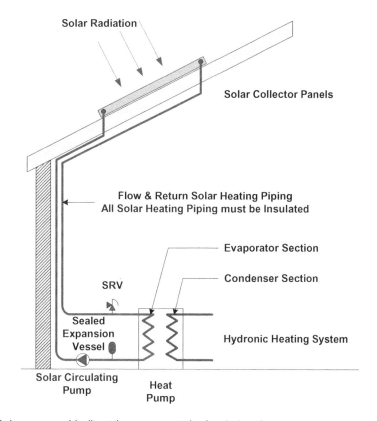

Figure 20.10 Solar-powered indirect heat pump to hydronic heating system

DIRECT SOLAR DOMESTIC HOT WATER SYSTEMS

Within a direct solar domestic hot water system, the domestic hot water is circulated directly through the solar collector panels, as indicated in Figure 20.11.

In the direct system of solar heating, the domestic hot water is circulated through the solar collector panel by a circulating pump, with the cold feed from the main cold water storage cistern connected into the return from the cylinder back to the solar collector panel, usually by the inclusion of a reversed mounted injector or diverting tee. With the direct solar heated domestic hot water system it is important that the solar collector panel is mounted at a height below the normal water level in the main cold water storage cistern, ideally closer to the bottom of the storage cistern, so that the water level within the solar collector panel always remains flooded during peak domestic hot and cold water draw-off demands.

The direct solar hot water system depicted in Figure 20.11 shows that there is a possibility of the temperature of domestic hot water exceeding 65°C; in fact, during the summer months it can reach 100°C. To protect the occupants of the building from scalding hot water, a thermostatic mixing valve must be fitted on the domestic hot water distribution piping to control the temperature of this water and restrict it to 65°C by mixing it with cold water as shown.

The direct type solar heated domestic hot water system has a number of major disadvantages when compared to the indirect system, which make it an undesirable system in the UK and countries with similar climates.

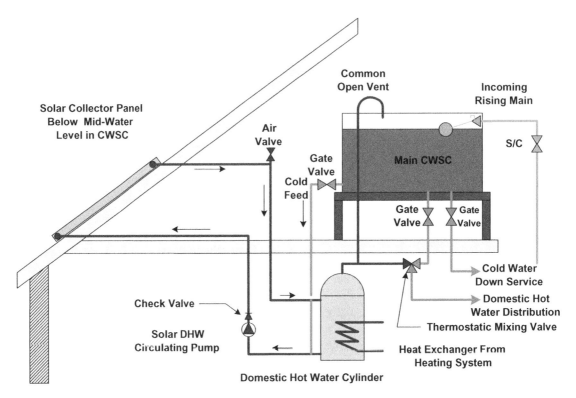

Figure 20.11   Basic principles of direct type solar-heated domestic hot water system

These disadvantages are listed below and should be carefully considered before selecting this type of system:

- As the water being circulated through the solar collector panels is the same water that is being drawn off at the point-of-use at the sanitary fixture units, pillar valves or showers, etc, there is a high risk of build-up of scale similar to that which occurs on a standard direct type domestic hot water system. This scale will reduce the bore of the circulating tubes and solar collector panels, thus reducing the system efficiency.
- As the solar heating section of the system forms part of the direct domestic hot and cold water service, it must comply with the Water Regulations, and all materials and component parts must comply with the Water Supply (Water Fittings) Regulations.
- There is an increased risk of excessively high domestic hot water temperatures up to 100°C, or even steam, being achieved during the summer months, therefore thermostatic temperature control in the form of a mixing or blending valve should be installed, arranged to mix a portion of cold water with the domestic hot water to maintain the blended outlet water at 65°C maximum.
- During the winter months, there is a high risk of the water within the solar collector panels and circulating pipework becoming frozen, causing blockage to the system and damage to its components. Hence, this system should normally only used in countries where freezing conditions do not normally exist.
- During the winter months there is a possibility that the solar collector system will work in reverse and emit heat from the collector panel, causing the domestic hot water storage cylinder to cool down, thus wasting energy rather than conserving energy.
- There is also a high risk of creating the conditions for *Legionella pneumophila* bacteria to multiply during reduced solar irradiation periods, if the standby heating system is not operational.

INDIRECT SOLAR DOMESTIC HOT WATER SYSTEMS

The indirect form of solar heated domestic hot water systems is the predominant system employed in the UK. It differs from the direct system by keeping the primary solar water separate from the domestic hot water. This is achieved in a similar manner to that of a standard hydronic space heating and domestic hot water system, by separating the primary and secondary waters by the inclusion of a heat exchanger incorporated within the domestic hot water storage cylinder.

The reason that the indirect system of solar heated domestic hot water is so prevalent in the UK, together with other countries having weather systems similar to or more severe than the British Isles, is that it alleviates most of the disadvantages associated with the direct solar heated domestic hot water system, such as:

• As the primary solar heated water is separate from the secondary domestic hot water, the problem of scale formation and corrosion to the solar collector panel and primary circulation tubing is almost completely removed.
• Another advantage of the primary and secondary waters being separate is that the primary circulation water can be treated with a properly chemically formulated glycol based antifreeze/corrosion inhibitor, as the two waters will not mix and the risk of the chemical being drawn off at its point of use has been prevented.

Figure 20.12 illustrates a typical indirect solar heated domestic hot water system arranged as a non-pressurised scheme with a feed and expansion cistern, open vent pipe and cold feed. The primary water is circulated by a circulating pump through the solar collector panel to be heated, and then to the coil type heat exchanger incorporated in the indirect domestic hot water cylinder. The feed and expansion cistern, open vent and cold feed pipes serve the same function as they do for an open vented space heating system, described in Chapter 2, Wet Heating Systems.

This system also permits greater temperature control by being arranged to only operate when there is sufficient solar heat available and when the domestic hot water storage cylinder calls for heat. A non-return type check valve inserted into the primary circuit prevents the system operating with a reverse gravity circulation, preventing the solar panel becoming a radiator, emitting heat rather than absorbing it.

Figure 20.13 illustrates the same system but arranged as a pressurised sealed scheme, whereby the expansion of the primary water is contained in the sealed expansion vessel with the water supply connected to the incoming water main via a temporary filling loop for initial charging. This system has the additional advantage of permitting the solar collector panels to be positioned without the constraints of being governed by a water level in the header/feed and expansion cistern. This also presents the opportunity to install further supplementary solar collector panels and possibly accomplish a closer optimum position to harness the solar irradiation.

The scheme indicated in Figure 20.13 incorporates a fully open vented system of domestic hot water being supplied from the main cold water storage cistern, but it can be equally suitable for an unvented pressurised system of domestic hot water; however, caution should be adopted when planning to use a thermostatic mixing/blending valve to control the temperature of the domestic hot water distribution.

SOLAR HEATED THERMAL STORE

The indirect solar heated domestic hot water systems illustrated previously in this chapter employ a single domestic hot water storage cylinder equipped with two coil type heat exchangers. The heat exchanger located in the lowest section of the cylinder is the primary heat received from the solar collectors, whilst the second heat exchanger located in the upper section of the cylinder is the back-up heat from the standby conventionally fuelled boiler serving the space heating system. This arrangement suffers the disadvantage

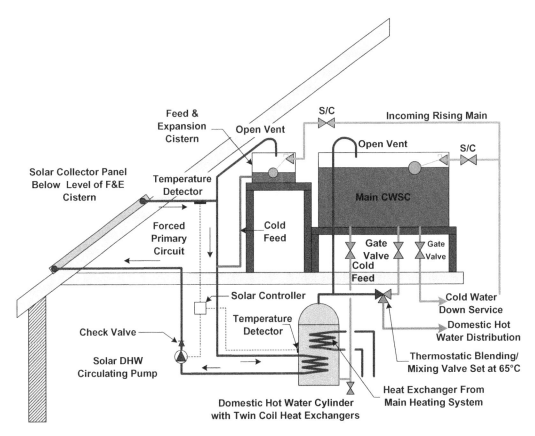

Figure 20.12   Indirect open vented solar-heated domestic hot water system

of solar heat being wasted when it is not hot enough to raise the temperature of the domestic hot water to its usable temperature of 65°C. In this situation the solar heating is switched off and the standby heat from the heating system is activated to raise the temperature. There is also the possibility that the solar heating system could operate in reverse and emit heat, if precautions are not taken.

Figure 20.14 illustrates an arrangement that overcomes this problem by incorporating a second storage cylinder, which serves as a thermal buffer store for the cold feed to the main domestic hot water storage cylinder.

The advantage of the additional indirect domestic hot water storage cylinder is that it serves as a pre-heat thermal store to the cold feed before it enters the main domestic hot water storage cylinder. The pre-heat is achieved by the solar heating system, which can be operational during daylight hours all year round. It will be capable of providing a high percentage of the domestic hot water required during the summer months without any further heat being applied by the standby main boiler, with a reduced amount of heat realised during the winter months.

During the winter months there is insufficient solar irradiation to provide 100% of the domestic hot water requirements, but there will be enough to raise the temperature of the domestic hot water by a few degrees even on cloudy days, where it will then pass on to the main storage cylinder in a pre-heated mode for the standby main boiler to raise the temperature by the remaining degrees required, thus saving on fossil fuel used. In the previous systems employing a single twin coil heat exchanger, the solar primary heating system is deactivated and over-ridden by the cylinder thermostat when there is insufficient solar heat available to provide the total heat required for the domestic hot water system.

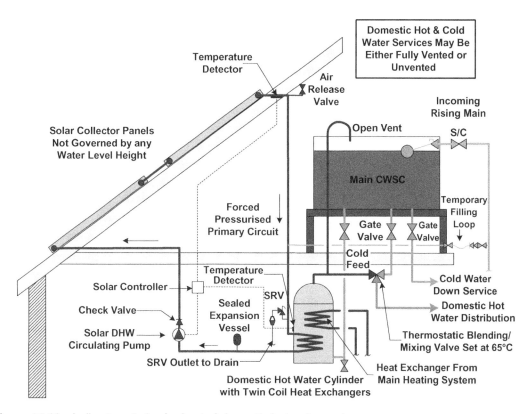

Figure 20.13   Indirect sealed solar-heated domestic hot water system

One of the main disadvantages of this arrangement, apart from the increased installation cost, is one that is common to all alternative forms of energy, and that is the large amount of space required to house the equipment. This essential design prerequisite needs to be incorporated at an early design stage.

FROST PROTECTION

As mentioned earlier in this chapter, the water contained in the indirect circulatory piping system between the solar collectors and the domestic hot water storage cylinder, or thermal storage buffer cylinder, can be protected against freezing by the addition of an antifreeze solution to the water. The most vulnerable part of the system will be the water contained in the exposed solar collector panels mounted on the roof, or in similar exposed situations. The external ambient temperature experienced at this location will be determined by the geographical location of the building and the prevailing weather conditions that occur there: even in sheltered southern areas of the UK where the external ambient temperature rarely descends below –5°C, the wind chill factor that can occur could result in the surface temperature of the solar collector being much lower. Therefore, it is imperative that the correct antifreeze solution at its required concentrated strength is applied to the circulatory part of the solar heating system.

The most commonly used antifreeze solutions employed for protecting solar heating systems are water/ethylene glycol, or water/propylene glycol fluids, both of which are toxic. The antifreeze solution should be added to the circulatory water at a generous concentration to prevent the fluid freezing and becoming solid during extreme cold weather spells. Refer to the manufacturer's recommendations for guidance,

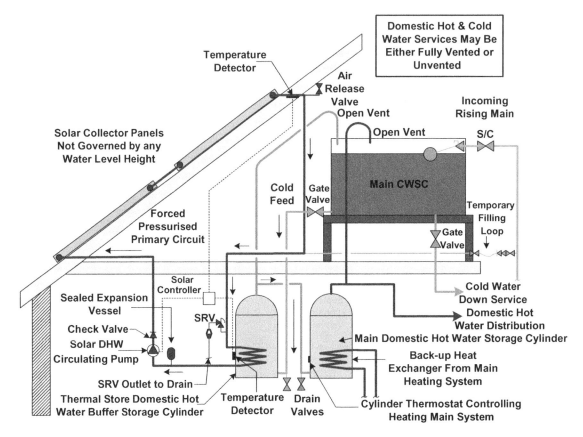

Figure 20.14   Indirect sealed solar-heated domestic hot water system with thermal buffer storage cylinder

but as a general rule the chart indicated in Figure 20.15 gives the freezing points of various water/glycol solutions at different concentrations. Local knowledge should be sought regarding the lowest temperature that can be expected at any particular location, plus an additional allowance should be made for the gradual breakdown and decay of the antifreeze solution which will become less effective in a relatively short period of time.

As a general rule, the concentration of antifreeze solution should be such that it will be equal to the expected lowest ambient temperature multiplied by two and a half. Therefore, for an expected ambient temperature of −5° (−5 × 2.5 = −12.5°) select a minimum concentration of 35% glycol solution.

Figure 20.15 gives guidance on selecting the concentration of glycol antifreeze solution required to protect the solar heating system for various external ambient temperatures, but it should be emphasised that this is for guidance only and if there is any doubt about the temperature then the concentration should be increased. It should also be appreciated that the effectiveness of the glycol solution will deteriorate with age, and decay when subjected to heat, oxygen and bacteria. For this reason, the effectiveness and concentration strength of any antifreeze/corrosion inhibitor should be checked as part of any annual servicing procedure as it may well require renewing every five years. This concentration check may be carried out using a refractometer or pH litmus paper strips obtained from antifreeze solution suppliers.

The addition of a glycol-based antifreeze solution to the solar heating primary circulating water will also have the effect of reducing the specific heat carrying capacity of the water/glycol solution. This means that more fluid will have to be circulated to carry the same amount of heat as water requiring larger pipes with

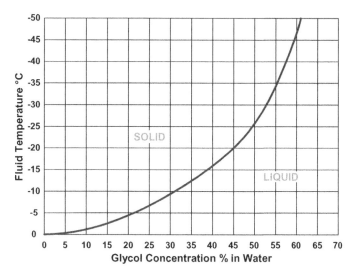

Figure 20.15　Freezing points of aqueous glycol concentrations

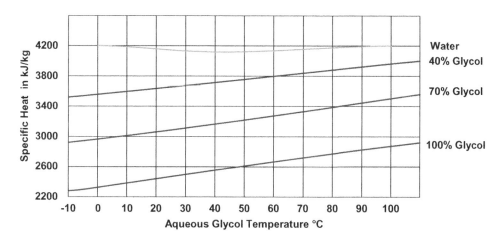

Figure 20.16　Specific heat capacities of aqueous glycol solutions

circulating pumps having larger duties. Fig 20.16 indicates the effect of various concentrations of glycol antifreeze on the specific heat carrying capacity of the circulating fluid at different circulating temperatures, where it can be observed that a fluid containing a 100% solution of glycol antifreeze solution has a very much reduced specific heat carrying capacity.

The addition of a glycol antifreeze solution into the circulating heat carrying water medium from the solar collector panels will also alter the boiling point of the solution, with the vaporisation point increasing with both temperature rise and pressurisation of the system – as depicted in Figure 20.17.

As an alternative, the circulating fluid may be replaced by a silicon or hydrocarbon based thermal fluid as discussed in Chapter 1. However, these fluids are more expensive and have a greater level of toxicity, and require twin-walled tube or annular heat exchangers to ensure mixing with domestic hot water does not occur. Thus, they are not considered acceptable or viable for domestic or commercial installations.

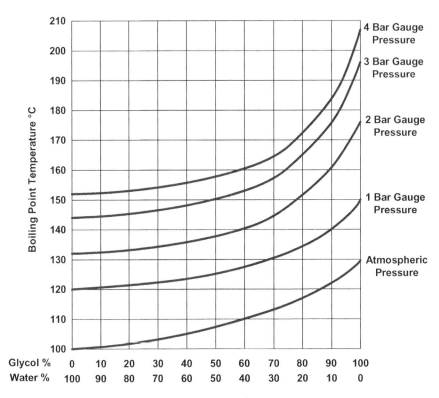

Figure 20.17 Boiling points of aqueous glycol concentrations

DRAIN-BACK SOLAR HEATING SYSTEMS

The drain-back or drain-down solar heating system is an alternative arrangement for systems that may be subjected to freezing conditions. Drain-back systems are circulation water heating schemes in which water that is untreated with any antifreeze solution is forced from the domestic hot water storage cylinder to the closed solar collector panels, where it is heated. Circulation continues until usable solar heat is no longer available. When freezing temperatures are anticipated, or power outage occurs, the pump is stopped and the system drains back automatically by gravity to a sealed storage collection vessel, as shown in Figure 20.18. This sealed drain-back vessel also serves as an expansion vessel that incorporates a safety relief valve to protect against excessive system pressure.

By arranging for the fluid in the system to drain back automatically into a storage vessel and leaving the solar collector panels vacated of water, there is no reason for adding an antifreeze solution to the water for frost protection and therefore no need to accept the limitations and disadvantages associated with glycol solutions. Nevertheless, the circulating water should still be treated with a suitable corrosion inhibitor to protect the materials of construction, although the excess air in the system can be a cause of nuisance or corrosion due to the oxygen content; for this reason the materials used for the circulatory system must be corrosion resistant.

The solar primary heating circuit is only partially filled with water, leaving a permanent air pocket in the upper section of the circulatory system, including partly in the drain-back vessel. When the circulating pump is activated it draws the fluid out of the drain-back vessel and thereby lowers the fluid level in the vessel and delivers it into the solar collector panels. At the same time, it pushes the air out of the collector panels and into the upper portion of the drain-back vessel, as illustrated in Figure 20.19.

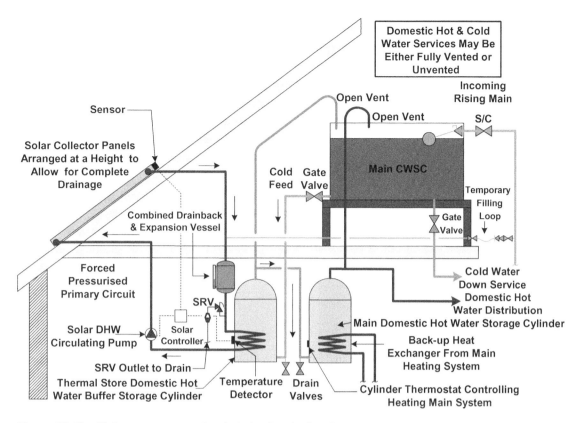

Figure 20.18 Piping arrangement for drain-back solar heating system

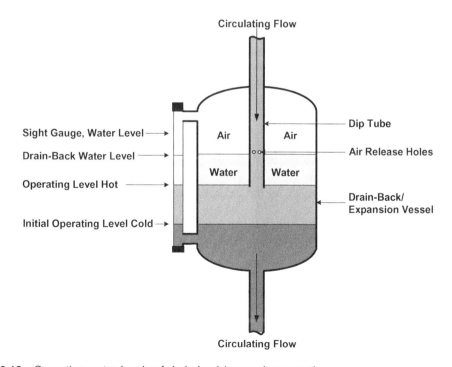

Figure 20.19 Operating water levels of drain-back/expansion vessel

The drain-back solar heating system is discouraged, or even banned, in some countries, due to the difficulty in correctly sizing and selecting the drain-back/expansion vessel and ensuring that the circulating fluid drains back on cessation of the pump running. During drain-back the fluid in the solar collector panel falls briefly to below atmospheric pressure, which can cause the fluid in the panel to vaporise momentarily, accompanied by a kettling sound. For this reason the piping connections to the solar collector panel should be arranged to have a continuous fall back towards the drain-back vessel to assist the drainage procedure.

The drain-back vessel has an internal dip tube which serves to reduce the noise that would be generated by the water cascading through the vessel. The dip tube should have air release holes as shown, to allow the air from becoming completely trapped in the upper dome section of the vessel when the fluid is in a drain-back mode.

The drain-back vessel should be equipped with a means of viewing the water levels in the vessel during the various operational sequences to ensure that it has been correctly filled with water, as over-filling or under-filling can have both a safety effect as well as an operational effect, as the correct water level is needed to create the air pocket inside the vessel.

SOLAR COLLECTOR PANELS

The solar collector panel or panels are located outside the building, facing towards the Sun's path where they can absorb solar irradiation and convert it into thermal energy. The part of the solar collector panel that receives the solar energy is termed the 'absorber', and has been manufactured in many design forms over the years, including plain uncovered flat plate and tube type collector panels made from various materials ranging from steel to copper. These have evolved into today's two most common types.

Flat plate solar collectors

There are many variations in the design of flat plate solar collector panels as illustrated in Figure 20.20. They comprise a single or double glazed front panel, made usually from a low iron content glass of varying physical strengths and capable of receiving a relatively high transmittance of solar radiation. Plastic translucent materials such as polycarbonate have been used as the front solar receiving material. They have the advantage of being more resistant to being hit by stones, or hailstorms, etc, but if the circulating water stagnates within the collector tubes for any period of time the absorber can reach temperatures as high as 300°C, therefore plastic has limited use as a material for solar collector covers.

Below the glazed top an absorber plate similar to a radiant panel is located, which sits above an insulated base, all housed in a steel box or tray. The absorber plate is available in a number of designs made from either copper, aluminium or in some cases, steel. The surface of the absorber plate is coated to achieve a matt black coloured plate to aid the absorption of the radiation and to reduce the amount reflected back.

The absorber plate can be fabricated using a flat plate with vertical tubes bonded to the underside, each connecting to a flow and return header, or tubes spaced between the flat plates, or welded corrugated sheets.

The flat plate solar collector may be considered as resembling a radiator in reverse, whereby it absorbs radiated heat from the Sun rather than emits it, the plate is heated and then transfers the heat into the water contained inside the tubes that is being circulated around the solar heating system.

Evacuated tube solar collectors

Similarly, evacuated tube solar collector panels are also available in a number of variations.

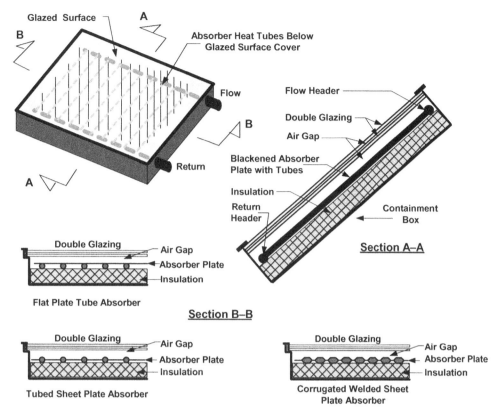

Figure 20.20   Flat plate solar collector panel arrangement

Figure 20.21 shows the general arrangement, which comprises a series of sealed glass tubes that have been partially evacuated of air and that contain a solar collector tube connected to a main common flow and return manifold, contained within a header box. The evacuated tubes are arranged over a parabolic shaped concentrator trough that serves to deflect solar radiation back and concentrate the irradiation onto the collector tubes.

As with the flat plate solar collector panel, the evacuated tubes are contained within a steel or aluminium box, with or without a transparent cover that helps to keep the evacuated tubes assembly clean.

Figure 20.22 illustrates a form of evacuated tube solar collector known as a 'heat-pipe'. This arrangement comprises a series of sealed glass evacuated tubes, each enclosing a second smaller diameter sealed glass tube centrally located in an aluminium heat absorbing fin. This smaller internal glass tube is partially filled with an alcohol based liquid, which when subjected to solar radiation, causes the alcohol liquid to heat up, whereby it vaporises and rises up the tube and into the slightly larger diameter condensing head that has been immersed into the primary flow of an enlarged header pipe.

The purpose of enlarging the header manifold is to permit the physical entry of the heat-pipe into the stream of the primary flow and to reduce the main primary flow from its design velocity of approximately $0.7\,\text{m s}^{-1}$ to around $0.1\,\text{m s}^{-1}$. This slower velocity will aid the transfer of heat from the heat-pipe condensing heads to the circulated primary flow, causing the alcohol vapour to condense back into a liquid and gravitate down the heat-pipe, where it will be vaporised again for the process to be repeated. For this reason, heat-pipe types of evacuated tube solar collector panels will not function if laid horizontally on a flat surface, as they require a recommended minimum gradient of 20° to permit the alcohol condensate to drain back into the heat-pipes.

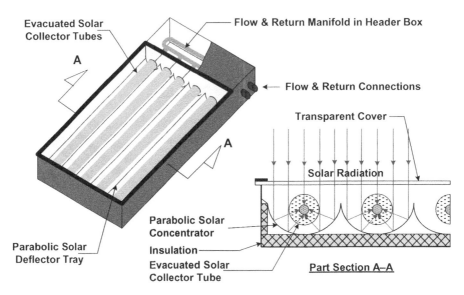

Figure 20.21   Evacuated tube solar collector panel arrangement

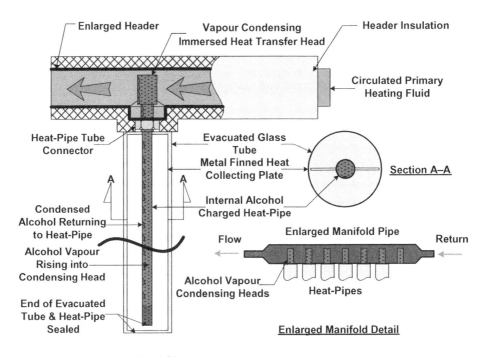

Figure 20.22   Detail of heat-pipe evacuated tube solar collector

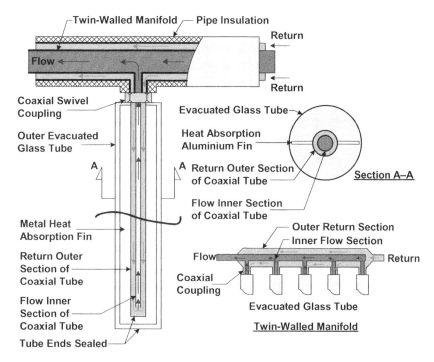

Figure 20.23   Detail of twin-walled coaxial evacuated tube solar collector

Heat-pipe evacuated tube solar collectors have an advantage in situations where the solar collector panels are fixed to a surface that does not face due south (due north in the southern hemisphere), as they can be adjusted on the swivel coupling to the manifold to angle the tubes' metal absorption fins to the optimum angle to receive the Sun's radiation.

Figure 20.23 illustrates an alternative and more common form of evacuated tube solar collector that uses the circulated primary flow fluid to circulate through the inner section of the evacuated tube. This arrangement employs a series of evacuated tubes complete with heat-absorbing aluminium fins as described for the heat-pipe system, but the inner glass tube with this array is assembled as a twin-wall tube with the outer section sealed at the end and the inner tube section left open. The primary water is circulated into the solar collector panel's outer section of the twin-walled manifold, arranged to permit the evacuated tubes to connect into this header, keeping the two water sections separate by use of a coaxial type coupling. The return water is circulated down the outer section of each evacuated tube, where it turns into the inner section of the tube at the bottom to travel back up the tube, having been heated on its journey, finally entering the inner section of the twin-walled manifold where it is then circulated to the heating requirement.

This form of evacuated tube can also be rotated by the plug-in type coaxial swivel coupling to angle it towards the optimum position to receive the solar radiation.

The water is kept apart in the two sections by a coaxial coupling that relies on synthetic rubber 'O'-rings that will require replacing periodically due to their deterioration from the high temperatures that can exist.

This arrangement is not suitable for a drain-back system unless it is installed with the evacuated tubes facing up, allowing them to drain out by gravity.

Figure 20.24 illustrates a variation of the twin-walled evacuated tube solar collector, where the twin-walled tube is replaced with a twin 'U'-tube collector arrangement with each end of the 'U'-tube connected to a separate flow and return manifold contained in the header box. The return liquid from the solar heating system or domestic hot water storage cylinder enters the return manifold enclosed within the header

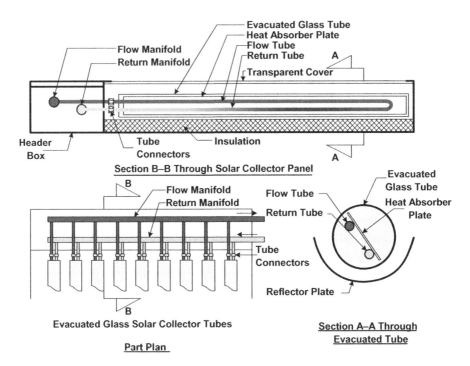

Figure 20.24   Detail of twin-tube evacuated tube solar collector

box, where it is then directed into the return tube section of the evacuated tubes and onto the flow section, where it is heated by the absorbed heat by the attached metal absorber plate, before entering into the flow manifold to be returned to the solar heating system.

The 'U' tube solar collector tubes are bonded to a radiated heat aluminium solar absorber plate angled as shown, to allow the tube ends to be connected to the manifolds by the flow tube passing over the return manifold. The individual ends of the 'U' tube are connected separately to the manifolds and therefore cannot be rotated, as has been the case for the previous versions of evacuated solar collector tubes shown.

SOLAR COLLECTOR PANEL LOCATION

All solar collector panels are at their most efficient when they are fixed with the collector panel arranged facing the Sun at an angle of 90° to the direct radiation path (see Figure 20.25).

It has already been established and demonstrated in Figures 20.2, 20.3 and 20.4 that the perceived position of the Sun in relation to any fixed point will be constantly changing. In fact, with the exception of both the summer and winter solstice, the apparent position of the Sun will only be at the same solar altitude angle and angle of solar azimuth for a few seconds twice each year.

To maintain maximum efficiency, the solar collector panel would have to track the perceived path of the Sun across the sky, changing every few seconds to adjust its solar altitude angle and angle of solar azimuth for every day of the year. Although this tracking arrangement has been successfully utilised for large photovoltaic power generating stations, it would not be considered either economically viable or practical for individual solar heating systems; therefore a more practical solution has to be adopted.

The solar collector panels are normally fitted to a static structure that has a southerly aspect (northerly aspect in the southern hemisphere) which will determine both the angle of solar azimuth and angle of solar altitude.

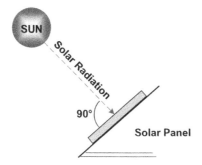

Figure 20.25 Ideal angle for solar collector panel efficiency

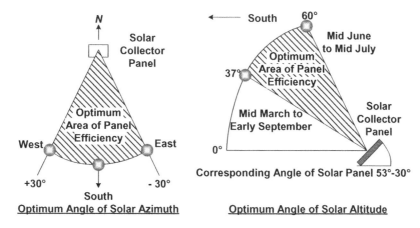

Figure 20.26 Optimum solar angles for 51.5° North

The optimum angle of solar azimuth would be due south (due north in the southern hemisphere), but a variance of 30° to the west or east would be considered acceptable without too much of a decrease in solar collector panel efficiency, providing that the angle of solar altitude is also suitable (see Figure 20.26).

Further north the angle of solar altitude will require to be lowered to 30° from the horizontal to obtain panel efficiency.

The optimum angle of solar altitude varies on the monthly height of the Sun, but to achieve the best results the solar collector panel should be angled as shown in Figure 20.26 for the location depicted in the example used for the Sun path diagram shown in Figure 20.4. If the solar collector panel is orientated facing due south, the higher angle of solar altitude of 60° would be the most favourable, but if the angle of solar azimuth varies by 30° either east or west, the lower angle of solar altitude of 37° would be the optimum.

For every one degree deviation from the optimum angle, either solar altitude or solar azimuth, there will be a reduction in the performance efficiency of the solar collector panel; the exact decrease will depend upon the panel manufacturer's specific design.

SOLAR COLLECTOR PANEL FIXING

Fixing solar collector panels to either vertical or horizontal surfaces requires a support frame to angle the panel to the correct orientation to receive the radiation from the sun. This support frame should be purpose-made to suit site conditions and the angle required.

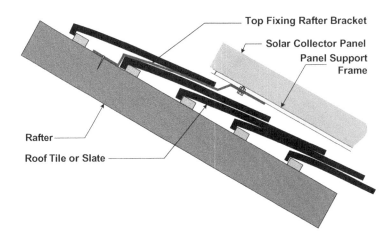

Figure 20.27   Pitched-roof solar panel fixing arrangement

It is quite common to fit solar collector panels to pitched roof surfaces if the structure permits, as these are considered the most suitable locations, as they provide a readily available inclined surface at high level, together with a degree of protection from vandalism and accidental human contact with the high temperature surfaces. However, the angle of the roof pitch and its orientation may not be the best for receiving the Sun's radiation, but the position will have to be accepted together with the reduction in solar panel efficiency.

One method of fixing the solar collector panel to the pitched roof is to integrate it into the roof tile or slates surface – similar to the method of fixing a roof light – where the solar panel is weathered into the roof finish by use of a front apron, back gutter and side flashing. This gives a neat appearance and allows the flow and return piping, together with any electric control cables, to be accessed from the roof void beneath, and permits connection through the underside of the panel. It also eliminates any need for these services to be weathered as they do not penetrate the roof surface.

Another method is to fix the solar collector panel proud of the roof using a stainless steel support frame bolted to rafter brackets, as shown in Figure 20.27, where a series of brackets are fitted to the rafters under the tiles or slates and arranged to slip between the tile design as illustrated, to retain the integrity of the roof.

An alternative and neat fixing arrangement when using some types of contoured roof tiles is to attach moulded fixing brackets to a thermo-plastic version of the contoured roof tile.

Solar collector panels that sit proud of the roof surface will require the piping and cables to be weatherproofed where they pass through the roof. This can be achieved by using conventional rigid metal slates, similar to those employed for weather soil pipes etc, and that are large enough to permit the pipe insulation to pass through. The top of the weathering slate should have a weathering apron above it. Alternatively, flexible synthetic rubber domical weathering cones have been used, but they will degrade over a period of time due to the heat that they are in contact with, and will require replacing periodically.

Control cables that penetrate through the roof from temperature or light sensors have been sealed using silicone sealant packed around them, but this is poor practice and leaks will soon develop. A superior practice is to install the cables in a weatherproofed conduit, weathered in a similar manner to that of the pipes, or if the cables are suitable, they can be weatherproofed using of a flexible synthetic rubber grommet.

SOLAR COLLECTOR PANEL SIZING

The sizing of the solar collector panels is not a precise science due to the large number of variables involved, which include the following:

- Degree latitude, north or south of the equator.
- Orientation of the available solar collector surface facing towards the equator, or the angle of azimuth.
- Angle of pitched slope available or solar altitude angle of solar collector surface available.
- Physical space of solar collector surface area available.
- Extent of any solar shading obstruction that will impact on the solar collector panel positioning.
- Actual amount of solar radiation received annually.
- Actual amount of solar intensity received during daylight periods.

The degree of latitude will be dependent upon where the building is to be built, but the orientation of the building, together with the angle of any pitched roof, will be determined by the architects' and their client's brief during the design phase of the project. The design of the building will also determine the shape and size of any solar collector panel space available and it is therefore prudent for the heating design engineer to be involved in the building design from an early stage.

The extent of any solar shading will, to a certain degree, be out of the control of the design engineer and will need to be considered when designing any solar heating system.

The biggest variable of them all, which will be totally out of anyone's control, will be the amount of solar radiation that will be received from direct sunlight annually, as well as the intensity of the sunlight during that period.

BS EN12975 requires manufacturers of solar collector panels in the UK to publish, amongst other information, performance and efficiency data based on certain assumed conditions. Any deviation from these conditions will result in a reduction in these efficiency figures.

The amount of heat absorbed per square metre by the solar collector panels will vary slightly from one manufacturer to another; it will be rated at around 800 watts per square metre of effective solar panel collector area for a clear sunny summer's day, to about 200 watts per square metre of effective solar panel collector area on overcast cloudy days, falling to about 60 watts per square metre of effective solar panel collector area during the winter months.

Using the stated solar collector panel performance and applying an efficiency percentage reduction for any deviation from the optimum absorption angle, this could range from about 70% for being on the edge of the optimum absorption area, to around 85% for an angle somewhere in the middle.

Therefore, for a solar collector panel having a performance rating of 800 watts per square metre and an efficiency factor reducing the performance to 85%, what would be the expected heat collected for an area suitable for installing 6 square metres of effective solar collector panels?

800 watts × 85% = 680 watts × 6 m² of solar collector panel = 4.08 kW of heat raised.

Alternatively, as a rule-of-thumb method for raising the domestic hot water temperature, apply a figure of 1.5 m² of solar collector panel for every person resident in the building.

SOLAR HEATING CONTROL SCHEMES

A solar heating control system does not differ from any other form of heating and domestic hot water control system as the requirements and objectives are the same. The heating system must comply with the requirements of the Building Regulations Approved Document Part L for the conservation of energy and must be able to operate safely within the pressures and temperatures that could be realised.

In essence, to meet the above requirements the solar heating control system should include the following strategy:

- Solar thermal energy is said to be freely available if we exclude the system installation cost. Although this free thermal energy can be harnessed, it cannot be readily conserved, but as it is used as a primary source of thermal energy its function is to conserve any conventional fossil fuel that is being used as a back-up fuel. Consequently, the control system should be arranged to operate the solar primary heating source as efficiently as the solar energy available will permit, thus only permitting the fossil fuel back-up system to function when the solar heated water temperature needs to be raised to its final working temperature: in this way the conventional fossil back-up fuel is being conserved by the fact that less is being consumed.
- As solar energy cannot be turned on or off, the system's safety controls must include the provision to both prevent and protect against over-pressure of the system by the selection of materials and relief components. It must also incorporate measures to prevent excessively high temperatures occurring and prevent human or animal contact with any high-temperature section of the system.

Figure 20.28 illustrates the basic control philosophy for a solar primary energy scheme to heat the domestic secondary hot water supply which has a conventional fossil fuelled back-up boiler that provides the main heating to the building.

A pump circulates the brine fluid through the solar heating collector panel where its temperature is raised by solar thermal energy and is passed onto the lower coiled tube heat exchanger within the domestic hot water cylinder. If the temperature of the secondary water in the upper portion of the cylinder is below 60°C, then the cylinder thermostat will activate the motorised valve on the primary return from the conventional fossil fuelled boiler to supplement the thermal energy further to reach its desired operating

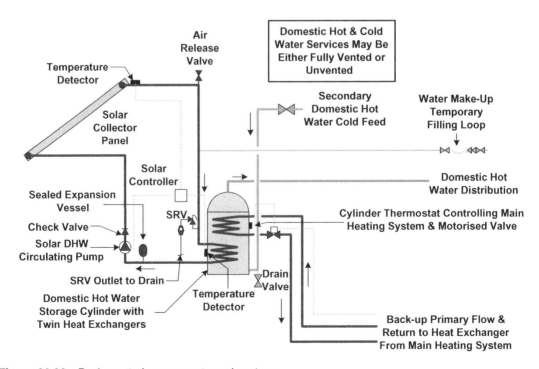

Figure 20.28   Basic control components and system

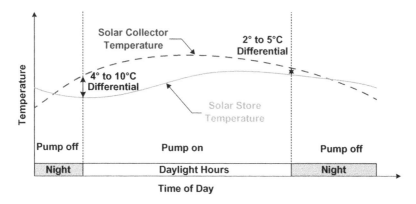

Figure 20.29 Temperature differential between solar collector and solar store

temperature of 60–65°C. To prevent the secondary domestic hot water exceeding 65°C, a three-port thermostatic mixing valve, as indicated in Figures 20.11, 20.12 and 20.13, should be employed to prevent dangerous excessively high domestic water temperatures being reached. Alternatively, the cylinder temperature detector can have a high limit setting to switch the solar heating circulating pump off when a storage temperature of 60°C is reached.

The cylinder temperature detector's main function will be to inform the solar controller of the temperature of the secondary domestic hot water being stored. This can be compared to the temperature exiting the solar collector panel; when this temperature is lower than the stored water temperature, then the solar heating circulating pump will be switched off to prevent the solar heated water from cooling the stored domestic hot water.

This temperature differential is depicted in Figure 20.29, which shows the pump running and pump stopped durations dependent upon the difference in the temperature sensed in the solar store and the solar absorber panel. This temperature difference would be between 4° and 10°C before switching the circulating pump on, and 2° and 5°C before switching the pump off.

Additionally, a light sensing photovoltaic cell can be fitted to the solar collector panel that will only permit the solar heating circulating pump to operate during daylight hours, thus preventing the system from functioning in reverse and losing heat.

21

Water Treatment

It has already been established in Chapter 1 that water is the most commonly employed medium for conveying heat from its point of generation to its points of heat transfer for space heating, used in conjunction with hydronic heating systems – and in the case of domestic hydronic heating systems, it is almost exclusively used.

The reasons for its common use have also been established, the main ones being its abundance and ready availability at a low cost. However, it has also been established that water has a number of limitations, namely high freezing point, low boiling point, restricted heat carrying capacity and high frictional resistance to flow. All of these disadvantages are addressed in the design of the heating system.

The main disadvantages of water as a fluid medium are its ability to corrode metallic materials and – when combined with certain temperatures – its ability to cause deterioration of some non-metallic materials.

WATER SUPPLY

In the UK, the water supply undertakers are required by law to supply a 'wholesome' supply of water, conforming to the World Health Organization (WHO) standards for drinking water. The water undertakers are not required to treat the water to make it less aggressive to any particular materials, unless that aggressiveness in the water is itself hazardous to health, although it is not uncommon for some water undertakers to treat their water supply to make it less aggressive to their own water supply mains.

It is therefore essential that an understanding of the quality of the water being supplied is obtained, together with a detailed comprehension of the potential vulnerability of the materials used and their compatibility with the water. In almost all cases, some form of preventative water treatment will be required to protect the hydronic heating system and extend the working life of the heating system.

Heating Services in Buildings: Design, Installation, Commissioning & Maintenance, First Edition. David E. Watkins.
© 2011 John Wiley & Sons, Ltd. Published 2011 by John Wiley & Sons, Ltd.
As an aid to lecturers and students, full colour versions of the figures in this chapter may be found at www.wiley.com/go/watkins
These figures are © 2011 by John Wiley & Sons, Ltd.

WATER TREATMENT

The British Standard for the treatment of water for domestic heating systems, BS 7593, lists the following objectives that are required to be met when applying any form of water treatment.

- To minimise corrosion of the heating system metals
- To inhibit the formation of scale and sludge
- To inhibit the growth of microbiological organisms
- To maintain the design engineering specifications of the heating system
- To minimise the chemical action and chemical change that takes place in the system primary water and system components.

HARDNESS OF WATER

Hardness of water is caused by the presence of such substances as bicarbonates, sulphates and chlorides of calcium and magnesium and small amounts of nitrates. These substances find their way into the water by rainwater absorbing any free carbon dioxide gas in the atmosphere as it falls to form carbonic acid.

$$\underset{\text{(Water)}}{H_2O} + \underset{\text{(Carbon Dioxide)}}{CO_2} = \underset{\text{(Carbonic Acid)}}{H_2CO_3}$$

As the rainwater soaks into the ground, the carbonic acid passes through calcium carbonate in the form of chalk or limestone and converts it into calcium bicarbonate.

$$\underset{\text{(Carbonic Acid)}}{H_2CO_3} + \underset{\text{(Calcium Carbonate)}}{CaCO_3} = \underset{\text{(Calcium Bicarbonate)}}{Ca(HCO_3)_2}$$

If the water containing carbonic acid and calcium bicarbonate in solution were to be used in a hot water or heating system, or any heat applied application, we would find that when the heat is applied, the carbon dioxide gas has been driven out of the solution and the calcium carbonate would come out of the solution in the form of a scale deposit.

$$\underset{\text{(Calcium Bicarbonate)}}{Ca(HCO_3)_2} + Heat = \underset{\substack{\text{(Calcium Carbonate)}\\\text{(Scale Deposit)}}}{CaCO_3} + \underset{\substack{\text{(Carbon Dioxide)}\\\text{(Released)}}}{CO_2} + \underset{\substack{\text{(Water)}\\\text{(Remaining)}}}{H_2O}$$

From this we would find that hardness due to calcium carbonate had been removed by heating the water and for this reason it is termed 'temporary hardness'. If the water also contained calcium sulphate ($CaSO_4$), it would be observed that it still remained in solution after heating and therefore cannot be removed by the application of heat. Hardness of this kind is called 'permanent hardness' and plays no part in forming scale deposits on pipework or heat exchangers.

Most waters contain a degree of both permanent and temporary hardness and to summarise, bicarbonate of calcium and magnesium give the water its temporary hardness, whilst sulphates of calcium and magnesium give the water its permanent hardness.

CLASSIFICATION OF HARDNESS

The units of hardness may be expressed in a number of different forms of measurement, the most commonly used internationally is to express the parts of calcium carbonate ($CaCO_3$) by weight per million parts of water, usually abbreviated to ppm.

Table 21.1   Units of hardness in water

Unit Name	Parts Per Million (PPM) CaCO₃
1 Milligram per Litre	1 Part per Million
1 Russian Degree Ca	2.57 Part per Million
1 Part per 100 000	10 Part per Million
1 French Degree CaCO₃	10 Part per Million
1 Clark Degree CaCO₃	14.29 Part per Million
1 Grain per UK Gallon CaCO₃	14.29 Part per Million
1 Grain per US Gallon CaCO₃	17.15 Part per Million
1 German Degree CaO	17.86 Part per Million
1 Gram per Litre	1000 Part per Million

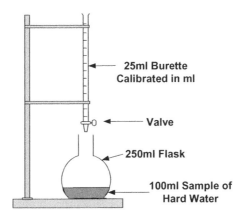

Figure 21.1   Clark's apparatus for determining hardness in water

Table 21.1 may be used to convert from one unit of hardness to another, or alternatively, it has been common practice in the UK to express the degree of hardness in water as degrees Clark.

The simplest method of determining the amount of hardness in water and the easiest to understand, is Clark's method. Here, a calibrated burette is filled with a standard pure soap solution such as Clark's, Wanklyn's or Boy's, which is allowed to drip 1 ml at a time into a flask containing a 100 ml sample of hard water. After each ml of soap solution is added to the sample of water, shake the flask vigorously until a lather is formed that lasts for 5 minutes, count the number of ml of soap solution that were required to form the lather, and deduct one ml from the total (as one ml of soap solution is required to produce a lather when there are no calcium or magnesium salts present) to obtain the degree of hardness in degrees Clark. The apparatus for Clark's method is shown in Figure 21.1.

Example – 20ml of soap solution used = 19 degrees of hardness present.

The above experiment will determine total hardness. If both the temporary and permanent hardness need to be known, take another sample of water from the same source, heat the water to boiling point for 5 minutes and allow to cool, then repeat the above experiment after the water has been filtered: as the temporary hardness has been removed by the application of heat, the result left would be for permanent hardness only.

$$\begin{array}{ccc} \text{Total Hardness} & - \text{ Permanent Hardness} = & \text{Temporary Hardness} \\ 19 & 8 & 11° \text{ Clark} \end{array}$$

Therefore:

Permanent hardness	8° Clark × 14.29 = 114.3 ppm
Temporary hardness	11° Clark × 14.29 = 157.2 ppm
Total hardness	19° Clark × 14.29 = 271.5 ppm

From Table 21.2, a total hardness of 271.5 ppm would be classified as hard water.

Table 21.2 Classification of water hardness

Designation	Parts Per Million CaCO₃	Degrees Clark
Soft	0–50	0–3.5
Moderately Soft	50–100	3.5–7
Slightly Hard	100–150	7–10.5
Moderately Hard	150–200	10.5–14
Hard	200–300	14–21
Very Hard	Over 300	Over 21

EFFECT OF HARD WATER

Hard water has little to no effect on a modern heating system as it is a closed system circulating the same water continuously, and after the initial heat-up of the water, when the calcium carbonate has been driven out and formed a minute deposit of scale, the water is then free of any temporary scale-forming hard water.

However, in a secondary domestic hot water system where the water is constantly being replaced by fresh supplies as it is being used, there will be a gradual build-up of scale if there is any temporary hardness present in the water supply.

The rate of scale build-up in the system will depend upon the amount of temporary hardness present, but when the water temperature is raised to 50°C and above, the calcium bicarbonate will start to be converted to calcium carbonate and deposit itself in the form of scale on the surfaces of pipes and heat exchangers. The higher the temperature of the water above 50°C, the faster the rate of scale formation. This scale build-up will create an insulating barrier that will eventually cause the system to overheat, as thermostatic controls will no longer be able to sense the true temperature of the water accurately, leading to a failure of the system. If this situation continues and is ignored, a dangerous overheating condition could occur resulting in a premature failure of heat exchangers and temperature sensing equipment.

In situations involving domestic hot water, because the water is being continually drawn off for domestic and ablutionary use, chemical treatment would not be suitable as the chemicals would not remain in the system and the water would not be suitable for ablutionary use.

BASE EXCHANGE SOFTENING

The most common method of softening the water is by the application of the base exchange process, whereby a base exchange water softener correctly specified and installed will soften both temporary and permanent hardness.

This method makes use of naturally occurring minerals called 'zeolites' such as siliceous green sand, which possess the property of softening hard water. Zeolites may also be obtained in a synthetic form made from sodium silicate and aluminium sulphate, or from the action of sulphuric acid on coal. Synthetic zeolites have the ability of removing approximately three times more hardness than naturally occurring zeolites, but naturally occurring zeolites have the advantage of being more durable than the artificial variety.

In this process, the hard water containing calcium bicarbonate or magnesium sulphate passes over the zeolite (Z) where the sodium base of the zeolite is exchanged for the calcium and magnesium base salts in the hard water. This may be shown as follows.

Water containing:

$$\underset{Ca(HCO_3)_2}{Calcium\ bicarbonate} + \underset{Na_2Z}{Sodium\ zeolite} = \underset{CaZ}{Calcium\ zeolite} + \underset{2NaHCO_3}{Sodium\ bicarbonate}$$

or

$$\underset{MgSO_4}{Magnesium\ sulphate} + \underset{Na_2Z}{Sodium\ zeolite} = \underset{MgZ}{Magnesium\ zeolite} + \underset{Na_2SO_4}{Sodium\ sulphate}$$

After a period of use, the zeolites will become exhausted of their water softening capacity and they will have completely converted to calcium or magnesium zeolites. However, they may be changed back or regenerated into their original chemical structure by flushing the zeolite bed with a brine solution of sodium chloride, or in more everyday terms, water and common salt. If no zeolite crystals are lost in this regeneration backwashing process, the bed can be used indefinitely provided that it is regenerated at suitable intervals.

The regeneration action may be shown as:

$$\underset{2NaCl}{Sodium\ Chloride} + \underset{CaZ}{Calcium\ Zeolite} = \underset{Na_2Z}{Sodium\ Zeolite} = \underset{CaCl_2}{Calcium\ Chloride\ (To\ drain)}$$

or

$$\underset{2NaCl}{Sodium\ Chloride} + \underset{MgZ}{Magnesium\ Zeolite} = \underset{Na_2Z}{Sodium\ Zeolite} = \underset{MgCl_2}{Magnesium\ Chloride\ (To\ drain)}$$

INSTALLATION OF BASE EXCHANGE WATER SOFTENER

Figure 21.2 illustrates a typical installation of a base exchange water softener for a domestic application, although the principles are similar for larger commercial installations.

The unit is installed on the incoming cold water rising main, but after the wholesome cold water supply branch connection to the kitchen sink. This provides an unsoftened drinking water supply point within the dwelling as it is believed that there is a relationship between drinking soft water and people developing cardiac problems, and also provides a softened supply of water to the building for all other ablutionary purposes.

The method of initiating the regeneration process may be either manual or automatic, with a water meter registering the amount of soft water that has been processed, so that the exact capacity of the zeolites can be measured before it becomes exhausted. Alternatively, it may be activated by a time clock set to regenerate in the early hours of the morning.

The waste water from the regeneration backwash process should be taken to drain via a tundish outlet and trapped gully or waste pipe. For this reason, base exchange water softeners have been criticised for wasting water during the regeneration backwash period and this should be taken into account when considering their use.

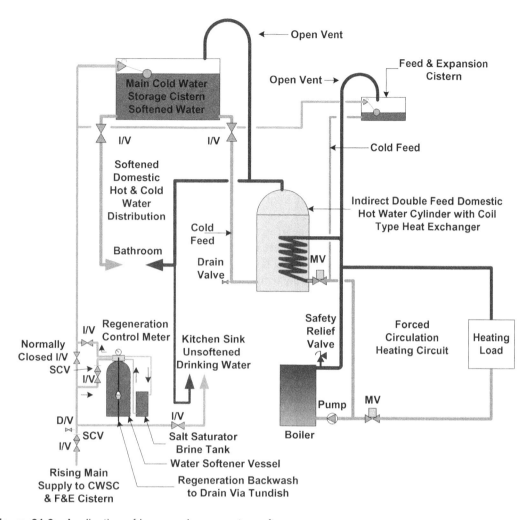

Figure 21.2 Application of base exchange water softener

Soft water does occur naturally in some water supply areas and this is indicated by how easy it is to obtain a lather from soap. In these areas it has been found that the water is aggressive to lead, meaning that it is plumbo-solvent whereby the lead is dissolved into the water, causing a risk of lead poisoning. For this reason lead as a piping material was limited in its use in water supply areas.

ELECTROLYTIC ACTION

Every student of plumbing or public health engineering is aware of the problems that can exist when different metals are in contact with each other in the presence of water, therefore care should taken when selecting materials for plumbing services.

However, when it comes to heating systems it appears that as many different metallic materials are mixed together as possible. It is not uncommon to have pipe materials made from copper, stainless steel or low carbon mild steel, heat emitters manufactured from pressed steel, cast iron, cast steel or aluminium, heat exchangers made from cast iron, stainless steel, aluminium or copper, plus a variety of alloy materials containing zinc, lead, nickel, tin, and chromium as well as the base element metals mentioned above.

The destruction of steel by electrolytic means is inevitable when it is in contact with copper or any other more noble metal in the presence of a conductive medium such as water. Similar conditions exist in a 12 volt car battery, where the more noble lead plates will cause the less noble zinc plates to be sacrificed by corroding at a rate proportional to the amount of electricity taken from the cell.

Almost identical conditions prevail in a hydronic heating system whereby copper, brass and solder perform as the noble elements and steel, cast iron and aluminium serve as the sacrificial metals, with the circulating water serving as the conductive electrolyte.

Electrolytic action may be described as the corrosion between two dissimilar metals in the presence of a conductive liquid, whereby one of the metals is sacrificed to preserve the remaining more noble metal – see Table 21.3 for the relationship between different metals in the electro-motive series table, sometimes referred to as the electro-chemical series table.

The greater the distance the two metals are apart in the table, the greater the potential electric current.

One of the most serious conditions is found where copper is in intimate contact with the steel or aluminium of a radiator, which is not a manufacturing fault of the heat emitter, as has sometimes been said. This condition is accelerated with higher operating temperatures, which is why heating systems suffer more corrosion during the heating season than they do during the dormant summer months, even though the colder water would dissolve and carry more oxygen.

Table 21.3   Electro-motive series table

Noble End or Cathode – (Protected End)

Element	Chemical Symbol	Electrode Potential Volts
Gold	Au	+1.42
Platinum	Pt	+1.2
Titanium	Ti	+1.03
Silver	Ag	+0.8
Silver Solder	–	Varies
Mercury	Hg	+0.8
Copper	Cu	+0.35
Brass	–	Varies
Lead	Pb	−0.125
Tin	Sn	−0.135
Lead Tin Solders	–	Varies
Nickel	Ni	−0.25
Cobalt	Co	−0.3
Cadmium	Cd	−0.4
Stainless Steel	–	Varies
Iron	Fe	−0.44
Steel	–	−0.62
Chromium	Cr	−0.71
Zinc	Zn	−0.76
Manganese	Mn	−0.86
Aluminium	Al	−1.67
Magnesium	Mg	−2.34
Sodium	Na	−2.71
Potassium	K	−2.92
Lithium	Li	−3.02

Base End or Anode – (Sacrificial or Corroded End)

Metal cutting burrs and excess soldering flux will also accelerate the formation of an electrical current if not removed. Soldering flux is particularly aggressive as the flux will remove the copper oxide and convert it into copper chloride, which in turn will plate out its copper content onto a steel surface before the system is flushed through. It will also have a similar effect on stainless steel as the flux used with this metal is even more aggressive.

The resultant product of electrolytic corrosion is magnetite, or black rust as it is more commonly known, whereby black iron oxide sludge accumulates usually along the bottom of radiators and destroys the steel in its immediate vicinity.

DEZINCIFICATION

The electro-motive series Table 21.3 includes some alloys as well as base elements, including brass, which is susceptible to dezincification. In this particular form of corrosion of brasses, the zinc content of the alloy is dissolved preferentially from the brass, leaving a porous mass of copper in the shape of the original component thereby causing it to leak. This has been a particular problem in some soft water areas and is more prevalent at higher water temperatures, where the water can be fairly aggressive to zinc. Today most, but not all, pipe fittings and valves are manufactured from dezincification resistant (DZR) brass. Some imported items may still be manufactured from other brasses.

BLACK IRON OXIDE

Magnetite is black iron oxide and consists of just under 30% oxygen and just over 70% non-metallic iron with a specific gravity of 5.18.

The chemical name of this compound is 'ferrous oxide', Fe_3O_4, and it is commonly mistaken for dirt in the heating system. It can be easily identified as the black iron oxide is magnetisable, and passing a magnet over the top of any dried spillage will pick up the black iron oxide in powder form.

An added problem caused by this product of corrosion is the effect it can have on the circulating pump, whereby the motor of the pump creates a magnetic field with the result that the black sludge is attracted to the pump rotor where it will eventually cause clogging of the impeller and the inevitable motor seizure.

The vast majority of pump failures are caused by this abrasive black oxide sludge building up over a period of time and leading to a pump seizure.

HYDROGEN GAS

As well as ferrous oxide being formed as a result of electrolytic action, another by-product of this corrosion is the generation of hydrogen gas. This is shown in Figure 21.3.

During the process of electrolytic corrosion, iron will go into solution in the water and almost immediately precipitate out as iron hydro-oxide, which is a compound of iron, hydrogen and oxygen. Most of the oxygen will remain bound to the iron for the subsequent formation of ferrous oxide, while the hydrogen will be released in the form of a gas.

The released hydrogen gas will accumulate in the top of radiators which will cause the water level in the heat emitter to be lowered, rendering the radiator ineffective.

Hydrogen gas can be identified as it has a distinctive odour when released via the radiator air vent, and should not be mistaken for air which may have entered the heating system through minute leaks on the negative side of the pump that are too small to manifest themselves during heating off periods.

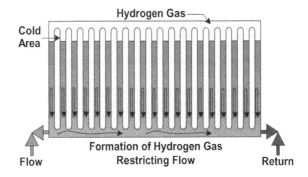

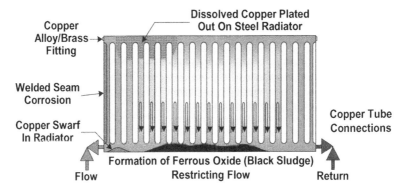

Figure 21.3　Effects of electrolytic action

Hydrogen gas is lighter than air and will burn with a yellow flame when ignited, but this should not be attempted as hydrogen gas is highly flammable and can be explosive.

BACTERIAL ACTIVITY

Water in the feed and expansion cistern will become stagnant for the best part of its lifetime, but stagnant water is rarely lifeless, as on the contrary, it provides an ideal breeding ground for bacteria and fungi, even if the cistern is equipped with a correctly fitted approved cover.

Organic impurities will enter into the feed and expansion cistern via the cistern vent, open vent pipe and overflow/warning pipe, being only restricted by the gauze type filters, which will not stop all air-borne substances from entering. This bacterial activity will eventually find its way into the circulating water, adding to the corrosion actions taking place.

An indication of bacterial activity occurs when venting off what appears to be an air lock in the top of a radiator, and a foul smelling odour reminiscent of rotten eggs may be noticed. This gas is hydrogen sulphide which has been produced by the decomposing organic impurities and bacteria and is an ignitable and toxic gas.

The occurrence of a slime-like jellified mass that occasionally materialises in the feed and expansion cistern is not always the result of micro-organisms multiplying, but may be the consequence of poorly applied chemical corrosion inhibitor that has been added to the heating system via the feed and expansion cistern, but has not been drawn into the system, allowing the heavily concentrated inhibitor in the cistern water to congeal into this jellified mass.

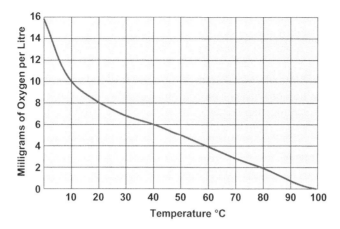

Figure 21.4 The solubility of free oxygen in pure water at atmospheric pressure

OXIDIC CORROSION

It has been established that the black sludge found in a heating system is a compound of black iron oxide, comprising iron and oxygen.

The fact that black iron oxide sludge accumulates in all untreated heating installations proves conclusively that free oxygen must be available for its formation.

Oxygen is soluble in water; the solubility of the gas varies with the temperature and pressure of the water. The solubility of oxygen in water is greatest in cold water, reducing to zero in water at boiling point at atmospheric pressure; however, this solubility will increase if the pressure is increased (see Figure 21.4).

Oxidic corrosion begins after the initially filling the heating system with water and decreases proportionally with the reducing free oxygen content, until, theoretically at least, there should be no free oxygen remaining within a few days after the filling, when all available free oxygen has been used up by the formation of oxides and escaped through the open vent, or automatic air release valves.

It also follows that no further oxidic corrosion can take place within the heating system unless fresh supplies of oxygen are admitted, this would usually occur via a design fault such as pumping over the open vent, or any micro leaks on the negative side of the circulating pump.

OTHER FORMS OF CORROSION

Although electrolytic action is the main form of corrosion within a heating system, there are several other corrosion processes that regularly occur. One such process involves the fact that water will cause iron to dissolve. This is a continuous process, as the water never becomes fully saturated because it precipitates from the solution in the form of sludge. This solubility of iron in water is very much greater in hot water than in cold.

It should also be mentioned that many pressed steel radiator manufacturers have in the past supplied radiators internally coated with a thin film of oil that serves to protect the steel against corrosion during the transit and storage of the radiators, before they are commissioned into the heating system. See Chapter 5, Heat Emitters.

This film of oil will also partially protect the steel when the heating system is initially filled with cold water, but will eventually combine with the iron oxide sludge, forming a sticky tar-like substance if the water is not adequately treated. This oil and iron oxide substance will contribute to causing an early failure of the circulating pump.

Oily deposits occasionally have the advantageous but non-lasting effect of concealing micro leaks, due to the surface tension preventing the water from penetrating the film. However, air can be admitted into the system at places where negative pressure exists, providing fresh supplies of oxygen to accelerate the corrosion process.

Corrosion around the vicinity of welded seams of pressed steel radiators can be observed. This is most likely caused by decomposition products of oil combined with welding flake scale and acidic compounds and particles of carbon, which together create a short-circuited electrolytic current.

ALUMINIUM HEAT EMITTERS

It is sometimes mistakenly believed that by installing heat emitters manufactured from aluminium in place of pressed steel radiators, the problem of corrosion will not be so great. However, the opposite is the case. By referring to the electro-motive series in Table 21.3, it can be seen that the distance between copper and aluminium is greater than it is between copper and steel, therefore the electrolytic action between copper and aluminium will be more severe.

If aluminium heat emitters are specified, then a correctly formulated corrosion inhibitor suitable for use with aluminium must be used. See Chapter 5, Heat Emitters, for further details of aluminium radiators.

pH VALUE (POTENTIAL HYDROGEN)

The pH value is the recognised method of expressing the intensity of the acidity or alkalinity of a solution, and is the logarithm of the reciprocal of the hydrogen ion concentration.

The pH scale is numbered from 0, which is extremely acid, through to 14, which is extremely alkaline, with 7 being neutral; i.e. neither acidic nor alkaline, see Figure 21.5.

Water contains hydrogen ions and hydroxyl ions. In one litre of chemically pure water there is only one ten-millionth of a gram of hydrogen ions. This is expressed as:

$$\frac{1}{10\ 000\ 000} \text{ gram of hydrogen ions}$$

Expressed in terms of indices this equals 10^{-7} $(10^7 = 10\ 000\ 000)$

$$\text{And } 10^{-7} = \frac{1}{10\ 000\ 000}$$

Rather than express the pH value in this way it is more convenient to state that the hydrogen ion potential is 7, with the term hydrogen ion potential being abbreviated to pH.

As previously stated, the pH value is defined as the logarithm of the reciprocal of the hydrogen ion concentration.

The term pOH refers to the hydroxyl ion concentration, which is related to the pH value. The sum of the pH value and the pOH value always equals 14; therefore for a pH value of 9, the pOH value must be $(14 - 9) = 5$.

Acids such as hydrochloric, carbonic, nitric and phosphoric when added to water, will increase the concentration of hydrogen ions, thus reducing the pH value. Alkali such as caustic soda or lime reduces the concentration of hydrogen ions, which increases the pH value.

Water with a high concentration of hydrogen ions and a low pH value will increase the corrosion to metallic materials such as cast iron, steel and copper, hence the condensate waste pipe from condensing

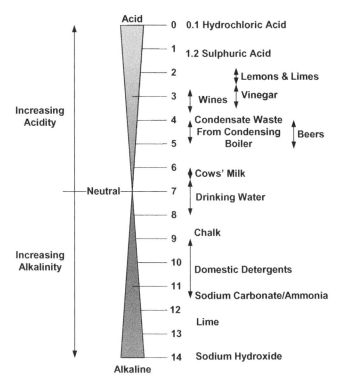

Figure 21.5   The pH scale

boilers should be installed in a non-metallic material. Water with a low concentration of hydrogen ions and a high pH value can be corrosive to aluminium.

Base exchange softening of water will raise the pH value of the water, which is unacceptable in any heating system that contains aluminium. If a base exchange water softener is to be installed and the heating system incorporates aluminium products, then the water supply to the heating system must be taken off separately before the water main passes through the softener.

CORROSION INHIBITION

Having read the preceding text one might be forgiven for thinking that the heating system will disintegrate shortly after being filled with water, but by chemically treating the water with a properly formulated corrosion inhibitor, the water will be rendered less corrosive.

It is now a requirement of the Building Regulations Part L that all new heating systems and systems involving boiler changes should have a suitable corrosion inhibitor added to the circulating water at a correct concentration for the metallic materials involved.

Inhibitive chemicals have been known for many years, in fact the inhibitive properties of sodium nitrate were originally reported in 1899. There are many basic chemicals that have corrosion inhibitive properties, these include ordinary household soda, egg preservative and caustic soda.

Chemicals used for corrosion inhibition include phosphates, silicates, benzenoids, molybdates, vanadates and chromates, the latter being particularly toxic and a skin hazard, as well as being restricted in use.

Heating system water can be treated with a concentration of 0.01% benzotriazole to reduce the attack and formation of ferrous oxide, whereas for a general all-round corrosion inhibition, a mixture of 1.0% sodium nitrite has been successfully used in the past.

It is not recommended that any of the above chemical solutions should be used in a modern hydronic heating system as chemical companies have developed commercially available inhibitors – formulated as a cocktail of different chemicals, and classified as being either film forming or one that will deactivate the medium, usually by the removal of dissolved oxygen.

Most commercially available inhibitors apply an invisible, near-monomolecular, adherent and impervious film uniformly over all components made of ferrous material. Some claim to provide full and lasting protection and being capable of arresting any corrosion that may have started (as in the case of an existing and previously untreated system), whilst others will merely slow down the corrosion rate without suppressing it completely.

The corrosion inhibitor that is eventually specified should be suitable for and capable of preventing corrosion. It should have the following properties:

- It must be compatible with all the materials used in the construction of the heating system that are in contact with the circulating water. Particular attention should be given to the material used to manufacture the boiler heat exchanger and the use of aluminium heat emitters, as not all commercial corrosion inhibitors are suitable for aluminium materials.
- It must remain effective through the operating temperature range, including maximum and minimum temperatures.
- It must be suitable for mixing with the quality of the water supplied for the circulating water as the source may be a private water supply and may contain chlorine or lime.
- It must be harmless to health for the environmental purpose that it is intended for with any application restriction complied with, such as its suitability for use in combined heating and domestic hot water systems employing single feed indirect cylinders, or any other application where it may be possible for the heating and domestic hot water services to mix. Also, any manufacturer's stated handling precautions for the chemical should be rigorously followed.
- It should contain a disinfectant and biocide to prevent any bacterial or fungal growths, particularly in the feed and expansion cistern. If the corrosion inhibitor does not contain a biocide within its compound formation, then a separate one that is compatible with the inhibitor should be applied.
- It should also be capable of rendering any soldering flux or jointing compound residue non-aggressive, but should not be searching so that it will break down any compound and cause leaks to occur at pipe joints or valve glands.
- It must be able, at the manufacturer's recommended concentrations, to arrest and prevent with long-lasting effects all forms of corrosion including ferrous oxide, electrolytic corrosion and oxidic corrosion.

APPLICATION OF THE INHIBITOR (NEW SYSTEMS)

The majority of chemical corrosion inhibitors that are commercially available are obtained in a liquid form, although they have also been produced as a non-dusting type powder that requires premixing before they are added to the heating system water. Examples are given in Table 21.4. In all cases the inhibitor supplier's instructions regarding their application, including the level of concentration, should be adhered to, which consists of the following procedure.

All systems

On completion of the heating installation, the system should be filled with cold water and pressure tested for soundness to ensure that there are no leaks.

Table 21.4 Guide to chemical inhibitors (water at neutral pH value)

Chemical Inhibitor	Metallic Material					
	Copper	Zinc	Steel	Cast Iron	Aluminium	Soft Solder
Chromates	E	E	E	E	E	0
Nitrites	P	I	E	E	P	A
Benzoates	P	I	E	I	P	E
Borates	E	E	E	V	V	0
Phosphates	E	0	E	E	V	0
Silicates	P	P	P	P	P	P
Tannins	P	P	P	P	P	P

Legend
E Effective
I Ineffective
P Partially Effective
A Aggressive
V Varies
0 Not Available

After a successful pressure test of the heating system and before adding the inhibitor, the system should be thoroughly flushed through with clean water to remove all debris, scale or excess flux and jointing compound residue.

Open systems

Isolate the water supply to the feed and expansion cistern, drain off a portion of the circulating water so that the water level in the system empties the feed and expansion cistern and drops to a level just below the top of the highest heat emitters. Pour the chemical inhibiting compound in liquid form into the feed and expansion cistern at the concentrated strength required, and then top up the system with cold water by opening the isolating valve on the water supply to the feed and expansion cistern. When all air has been expelled, pour a small amount of inhibitor, or separate biocide if the chemical inhibitor does not already contain this property, so that the water in the feed and expansion cistern is protected from bacterial and organic growths.

Sealed systems

This includes systems employing single feed indirect domestic hot water cylinders.

Isolate any automatic water make-up equipment, or cold feed to the single feed indirect cylinder if applicable. Drain down as described before to a level just below the top of the highest heat emitters; remove an air valve from one of the highest radiators, or some other convenient union type fitting located at a high point on the heating system. Apply the corrosion inhibitor by the application of a tube form inhibitor by pumping it into the radiator, or by a flexible tube inserted into the radiator and with a funnel at the other end to enable the inhibitor to be poured into the system. Replace the air valve or plug that has been removed, fill the system back up and expel any trapped air.

Initial start-up

After the addition of the inhibitor, start the boiler up as part of the commissioning procedure and allow the circulating pump to circulate the water/inhibitor mixture to allow it to be evenly distributed to all parts of the heating system. This will permit the inhibitor to apply an evenly formed protective film to the internal parts of the heating system and allow the protective procedure to commence.

Most inhibitor suppliers provide a permanent information/warning label that can be affixed to the drain-off valve or feed and expansion cistern informing any future maintenance operatives that the heating system is protected by an inhibitor and that it must be replaced if the system is to be drained down for any reason.

APPLICATION OF THE INHIBITOR (EXISTING SYSTEMS)

Heating systems requiring a boiler change, or existing systems that have never been protected by any form of inhibitor, should still have a chemical corrosion inhibitor added to the circulating water.

The heating system should be isolated from the water supply and drained down completely to allow any remedial work or boiler change to be undertaken. Before refilling with water, the heating system should again be thoroughly flushed through to remove all swarf and jointing residue from the remedial work and to remove any sludge that may have formed since the heating system was installed. If, after two flushing procedures have been conducted the water being drained off is still not clear, a flushing chemical or power flush may be required to remove any stubborn sludge/scale. The flushing procedure should be repeated until the water being drained off is clear.

When the heating system is known to be over 12 years old, caution should be applied when using chemical flushing agents or high-pressure power flushes as they may cause leaks to occur at weak points of the system, but it can be argued that it would be better to find these potential weak points under controlled conditions rather than let them occur at a later date when untold damage may result.

NON-DOMESTIC HEATING SYSTEMS

All heating systems are treated in the same way regardless of size: they all use similar materials with the same operating conditions, the only variation is the amount of inhibitive chemical added to the system.

On heating systems installed in some large buildings or building complexes, it is sometimes desirable to monitor the strength of the inhibiting chemical additive to ensure that it is still performing correctly and continuing to give full protection. Here the circulating water can be continuously monitored by a probe inserted into the circulating piping system and arranged to measure the conductivity of the water, or any change in its pH value that will indicate that the heating water has become diluted of its protective chemical inhibitor and needs topping up.

It would be normal for the monitoring equipment to be arranged to activate some form of pre-mixed chemical treatment plant that will automatically dose the circulating water, via an injection nozzle, with the inhibitive chemical to return the strength of the solution to its required concentration.

SACRIFICIAL ANODES

The effects caused by electrolytic corrosion have been discussed earlier in this chapter, where it was explained that heating circulating water has the advantage of being continuously circulated, enabling it to be chemically treated by the application of a corrosion inhibitor.

However, with regard to domestic hot water services, the addition of any chemical additive would not be suitable, as unlike heating system water – which is a medium to convey the heat and is therefore being continually circulated – domestic hot water is a consumable commodity requiring frequent replenishment as it is being used. For this reason, any chemical would also require frequent replacement at the same rate that the domestic hot water is being used. Also, any chemical additive would have to be suitable for human contact if used for ablutionary purposes.

British Standard 1566 for copper domestic hot water storage cylinders, both direct and indirect, requires them to be tested for their resistance to corrosion by the manufacturer, but they may also be specified to be fitted with an aluminium sacrificial anode as standard by the cylinder manufacturer.

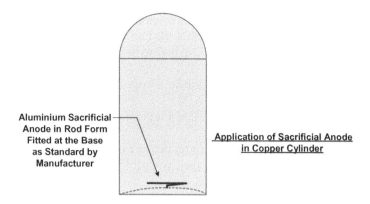

Figure 21.6 Application of sacrificial anodes

The aluminium sacrificial anode protects the copper cylinder in the same way as the magnesium sacrificial anode protects the galvanised steel domestic hot water tank. It allows time for a layer of scale to build up on the internal surfaces of the copper cylinder by being sacrificed by electrolytic action, as can be seen in Figure 21.6.

SCALE REDUCERS/WATER CONDITIONERS

There have been a few innovative devices developed to reduce the formation of scale within piping systems as an alternative means to the traditional base exchange water softener. Amongst the more successful methods is the use of magnetic/electromagnetic water conditioning units.

Magnetic scale reducers take the form of an enlarged in-line pipeline unit that does not require any electrical supply for it to operate. The detailed mechanism of how a magnetic scale reducer works is not fully understood, but it is currently believed that the magnetic field that the flow of water passes through causes an increase in the particle size and structure of the crystal form of scale. This produces a loose powdery deposit that remains in suspension, as opposed to a hard plate-like deposit that adheres to the surface of the pipes and heat exchangers etc, with this loose powdery deposit being washed through with the flow of water.

The restrictions of these devices are that their effects are time-limited and if the water is stored, or allowed to stagnate for any period of no demand, the condition of the water would revert back to its pre-magnetic treatment state. Consequently, water treated in this way should be used immediately with the scale reducing unit located in close proximity to its point of use.

It has also been found that many of the magnetic scale reducing devices available are flow sensitive and will only perform correctly during a narrow flow band; any flow rate outside this flow band will reduce its efficiency, particularly at low flow velocities, where treatment is totally non-effective.

Electromagnetic water conditioning devices work on an electro-chemical principle, where the flow of water is passed through a zinc–copper combination to produce an electrolytic cell. This provides an electrical potential which discourages the particles of scaling salts from attracting one another and prevents the formation of a layer of scale.

Some manufacturers of water conditioning devices claim that they are also suitable for protecting heating systems by reducing the electrolytic action potential of the water and thus the formation of black sludge. It is easier to use the heating system model to arrive at the correct size as the flow rate is constant, because it is dictated by the circulating pump.

Consumable crystal scale reducers are another in-line method of reducing scale formation. This is similar to a filter whereby a cartridge or compartment within the unit contains consumable crystals of sodium hexametaphosphates or natural polyphosphates, which have to be periodically replaced at the rate that they are consumed.

These crystals have the ability of holding the carbonate in suspension, thus preventing it from converting into scale.

Some types of these scale reducers work by forming a thin microscopic protective film on the internal surfaces of the piping and components.

Note that most chemical crystals available under proprietary names are suitable for drinking water supplies, but in all cases they should be checked for their compatibility as well as their suitability for their proposed application.

It should also be noted that scale reducers are not water softeners and although they may reduce the formation of scale within a system, they do not soften the water or give the advantages of soft water.

Figure 21.7 illustrates possible locations for water conditioning and/or scale reducing devices that do not rely on water being stored.

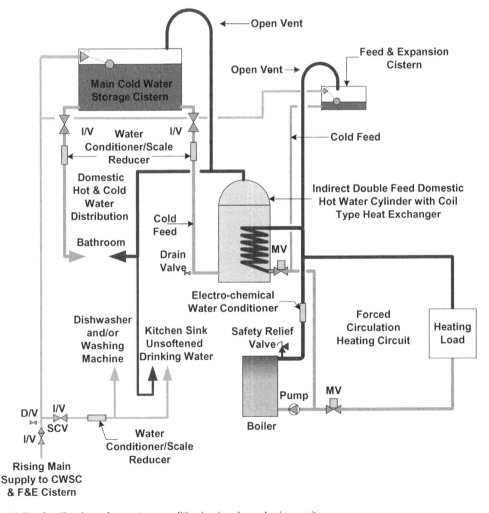

Figure 21.7 Applications for water conditioning/scale reducing units

22

District Heating

District heating is the term used to describe the method of supplying heat from a central heat-generating source via a network of piping to a geographic district such as a village, town or city suburb, with each individual dwelling and building connected to the district heating system, which forms part of the infra-structure services with water, drainage, gas and electricity.

The earliest record of a district heating scheme dates back to 1873 in Germany, but was pioneered mainly in Denmark and other parts of Scandinavia as well as parts of North America, around the turn of the twentieth century.

There have been a number of less ambitious district heating schemes in the UK, such as local authority housing estates and other small-scale developments. One of the more notable schemes is the district heating scheme for a residential estate in Pimlico on the north side of the Thames, supplied from a co-generation plant at Battersea power station on the south bank of the river. The reason why district heating has never become commonly used in the UK is the subject of conjecture, but as the emphasis is now correctly on increasing the efficiency of heating systems and the use of sustainable energies, the merits of district heating schemes become more evident than they have been in the past. These include:

- The maintenance of a single large central heat-generating plant should be better controlled and more reliable than the maintenance of hundreds of individual smaller domestic boilers, and accomplish an overall higher degree of continued system efficiency.
- The heat-generating plant can be arranged to operate using more than one fuel by the use of dual burners taking advantage of fluctuating and constantly changing fuel costs. The plant can also utilise renewable energies on a more economic scale than that achieved by numerous small domestic boilers. There is also a limited opportunity in some instances to recover heat by the use of a combined heat and power system from power stations, waste disposal incineration plants and other heat-producing process plants; again, this would not be practical with many small domestic boilers.
- Due to a more professional approach to system installation and regular maintenance, the heat distribu-tion network may be arranged to operate at higher temperatures, making the distribution piping smaller.

Heating Services in Buildings: Design, Installation, Commissioning & Maintenance, First Edition. David E. Watkins.
© 2011 John Wiley & Sons, Ltd. Published 2011 by John Wiley & Sons, Ltd.
As an aid to lecturers and students, full colour versions of the figures in this chapter may be found at www.wiley.com/go/watkins
These figures are © 2011 by John Wiley & Sons, Ltd.

These temperatures would not be possible or safe for the individual heating systems in domestic dwellings, where the control over the installation and maintenance cannot be guaranteed.

• To the home occupier or owner, a district heating scheme eliminates the need to locate a boiler in each property, but still requires the space to house a heat exchanger but without the problems associated with installing a flue system. It will also eliminate any on-site fuel storage requirements and the associated risks related to incomplete or incorrect combustion.

With the need to conserve more energy and become more efficient, the above advantages associated with district heating schemes increase in their importance, and make the idea of heating many dwellings from a central point become more attractive; however, there are a number of disadvantages that should also be considered:

• Land must be allocated for constructing the heat-generation station, including provision for any fuel storage and/or space for incorporating any forms of renewable energies.
• Heat losses from the distribution mains must be kept to a minimum or the fuel and energy efficiency will be compromised.
• Operating and heat unit charge costs must be competitive to make the system financially viable and attractive to each consumer.
• Higher installation costs, due to the need to construct buildings to house the heat-generation plant and lay distribution piping mains, will need to be justified against the operating costs and the environmental advantages.

The first three objections can be designed out with careful planning and engineering; however, the installation cost will always be higher compared to each individual dwelling having its own boiler plant and these would no doubt be a mixture of different makes, models and sizes; therefore, it will always be difficult to justify on economic grounds alone.

With the emphasis on environmental reasons and the conservation of energy, district heating schemes that employ a degree of sustainable energies will make the system more interesting and feasible and can therefore be justified by comparing the long-term environmental savings to the initial installation costs.

THE DISTRICT HEATING SYSTEM

The district heating system consists of three main sections of work, namely:

1. The heat-generating station complex.
2. The heat distribution network.
3. The individual dwelling end use heating installations.

THE HEAT-GENERATING STATION

The design and location of the heat-generating station housing the boilers and ancillary equipment will be dependent upon the site area available, site access, physical size of the heat raising and distribution equipment required, the choice of fuel or fuels available, and provision of fuel storage if required.

As previously mentioned, there are merits in having dual fossil fuels available such as gas and oil, whereby the burners fitted to the boilers have dual fuel-burning capabilities that can take advantage of changing fuel costs and their availability. There is also merit in incorporating a degree of renewable energies selected to suit the site in question. These can be in the form of combined heat and power from adjacent power stations, desalination plants or heat recovery from any nearby heat-producing process plant or waste incineration station, etc. However, these facilities are not going to be available for most potential district heating systems.

More practical sustainable energy means would be to consider incorporating biomass boilers, but these will require a greater amount of manual attendance compared with what would normally be mainly an unmanned automated plant, monitored remotely at some control centre overseeing and monitoring a number of district heating schemes. A more useful means would be to include a ground source heat pump arrangement if the site is suitable, or solar photovoltaic and wind power turbines to generate and supplement the electrical power required to operate the heat-generating plant. Solar thermal energy would be more useful, incorporated in each individual dwelling, so reducing, the heat required to raise the temperature of the domestic hot water. The inclusion of any form of renewable energy if selected and designed correctly, would increase the construction cost of the installation, but would have the effect of reducing the consumption of fossil fuels at a greater synergy than the combined use of many small individual plants installed in each dwelling.

The number of boilers selected to generate the heat required will be dependent upon the total calculated heat loading for the district heating scheme proposed, together with its expected annual diversity. Normally this would result in a number of boilers capable of modulating to suit the changing heat demand from the total number of dwellings, with the exact number of boilers selected combining to provide the full designed heating load. The boilers should be sized so that during the summer months only one single boiler is required to satisfy the domestic hot water requirements for the development; this will also permit planned maintenance to be carried out on the boilers during this period without interrupting the heat supply provided to the district.

The location and architectural layout of the heat-generating station will be determined by the physical size required to accommodate the heat-producing plant together with all of the ancillary equipment, plus the provision to extend this installation if considered necessary. One failing of this scheme is its inability to cope with the increase in heat requirements caused by additional properties being built at a later date together with extensions to the original properties.

There should be adequate access for fuel deliveries and plant replacement/maintenance activities, and the flues should be designed in a sympathetic manner for the development concerned.

The location of the heat-generating station should also take into account the associated piping network required for distributing the heat to avoid unnecessarily long distribution mains to some areas of the district heating area.

All electrical switchgear and control panels should be housed in a room separate from the main heat-generating plant and associated water conveying piping and mechanical equipment, both for safety reasons and to permit maintenance to be carried out.

On some very large district heating schemes it may be advantageous to assimilate more than one heat-generating station interconnected by the distribution heating mains. In this situation, the total number of boilers and heat-producing equipment should be sized to provide the whole heat load for the development, and be dispersed around the network area within the heat-generating stations.

DISTRIBUTION HEATING MAINS

These convey the heating medium from the heat-generating station to each individual dwelling; they should follow the most direct route possible and should be designed to keep the heat losses from the network to a minimum.

Smaller district heating schemes are normally designed to operate at low pressures and low flow temperatures of 82°C with a return of 71°C, or lower if condensing boilers are being used. Larger district heating systems may be arranged to operate with higher flow temperatures, usually in excess of 95°C, or sometimes – if considered desirable – within the medium temperature/pressure classification range. The higher operating temperatures enable smaller pipe sizes to be selected, but should only be adopted if there is no risk of people coming into contact with the higher temperature piping.

On very large district heating schemes, systems that can be truly described as district heating, even higher operating temperatures and pressures may be used. In these schemes high-pressure heating mains are used to distribute the high temperature water to localised heating substations, where the high temperature is transferred via a heat exchanger to a lower temperature and pressure for safe local distribution to each dwelling. Alternatively, the high-temperature water can be taken into each individual dwelling with their own heat exchanger, which serves as a substitute boiler for supplying domestic heat to the dwelling.

The variations in the heat-distribution systems are indicated in Figure 22.1, which shows the basic concept of their operating principles. The radial networks are suitable for small developments and comprise a flow and return piping network supplying each property direct with a flow temperature of 82°C at low pressure. In this system, the water that passes through the boiler plant will also pass through the dwelling's heating system, as there is no heat exchanger separating the primary and secondary water.

The interconnected networks are suitable for larger district heating systems where two or more heat-generating stations are incorporated, designed so that the heat from one station is returned to another station. The piping network has to be balanced correctly at the commissioning stage to avoid links being short-circuited. Figure 22.1 indicates this system operating at low pressure and temperature, but it may be beneficial to operate at higher temperatures and pressures if the application suits.

A ring network has the heating pipes arranged to form a ring as shown, and may be set at either low, medium or high pressure and temperature. The flow around the ring is directed in one direction with branch connections into local substations. This has the advantage of allowing shut-downs for maintenance on supply mains that can be supplied from either direction, therefore not causing disruptions to any of the dwellings supplied. The temperature may be transformed at these substations to provide a lower, more suitable and safer operating temperature to each dwelling.

The basic layout of the low temperature/low pressure direct radial district heating scheme is depicted in Figure 22.2. In this small-scale system the heat is generated at a central boiler house and distributed via a series of flow and return piping networks, with branch connections taken into each dwelling. Individual control for each dwelling is performed by a time-controlled thermostat operating a three-way bypass valve so that only the heat required will pass through the heat meter and be returned back to the heat-generating station. The flow temperature of the heating medium is restricted to 82°C as it enters into each dwelling, where people can come into contact with it, but the return temperature may be as low as 47°C for oil-fired condensing boilers, or higher, depending upon the design of the boiler plant and choice of fuel used.

Figure 22.3 illustrates a variation of the low temperature/low pressure scheme where the system is arranged as an indirect system by employing a heat exchanger for every dwelling. This separates the water in the main district heating piping network from the water in each individual dwelling's self-contained heating system. This will enable a higher flow temperature to be used for the district heating radial or interconnected piping network, provided that the section entering into each property is adequately insulated and protected from possible human contact.

The district heating scheme indicated in Figure 22.4 depicts the principles employed for a larger system, or a district heating arrangement that is better suited for distributing heat over a much larger geographic area.

The heat is raised to a high temperature and pressure at the heat-generating station. It is then distributed at that temperature through a network of high-pressure heating mains to a series of heat-distributing substations strategically located around the distribution area. The heat is then transferred through a series of heat exchangers so that it can be distributed to smaller localised areas at lower temperatures that are safer for buildings where people could come into contact with them. The arrangement shown in Figure 22.4 has just a single heat exchanger together with a single pipe circuit providing heat to individual dwellings, but

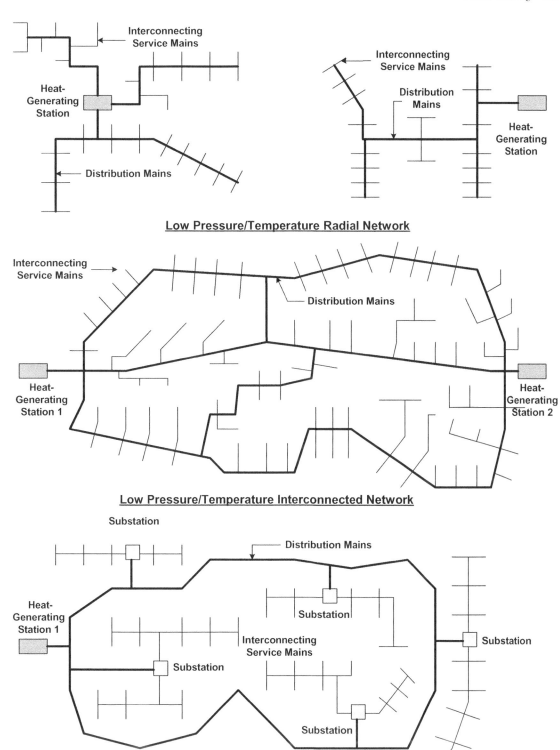

Figure 22.1   Heat-distribution networks

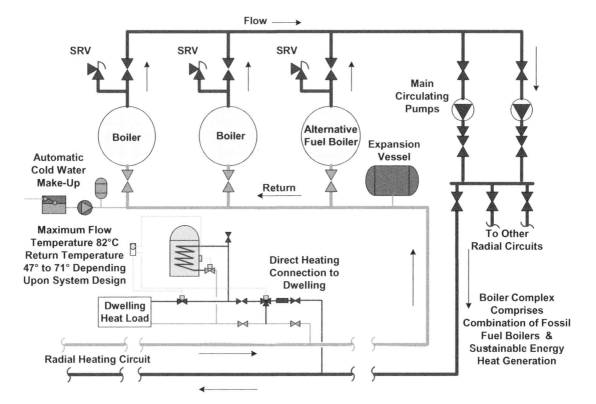

Figure 22.2 Basic low temperature/pressure direct district heating scheme

normally there would be a number of heat exchangers, and an even greater number of pipe circuits, arranged as a radial system from the substation to distribute heat to the localised area.

DISTRIBUTION PIPE MATERIALS AND INSTALLATION

Most distribution heating mains are required to be installed below ground for ease of delivery; they are either directly buried or installed in pre-formed masonry constructed ducts. They must be effectively insulated to reduce pipeline heat losses to a minimum and protected against accidental physical damage and both external and internal corrosion. If possible they should be installed above ground if the building development permits.

Traditional metallic piping materials, such as plain black low carbon mild steel tube, have been used due to their high pressure and high temperature rating and their resistance to physical damage, but they must be thermally insulated after installation and protected against corrosion. This was originally achieved by applying a sectional fibreglass insulation and wrapping in a bitumastic covering, which is a time-consuming and not always effective exercise.

More suitable materials have been developed over a number of years, such as any one of a number of proprietary made pre-insulated plain black low carbon lightweight steel tubes, or alternatively a similar composition but employing lightweight stainless steel or copper tube, although these two materials add significantly to the installation material cost and are restricted to the smaller pipe sizes.

These proprietary piping systems are available with a range of pre-insulated fittings and have been developed and extensively used in North America and Scandinavia: for this reason their compatibility with

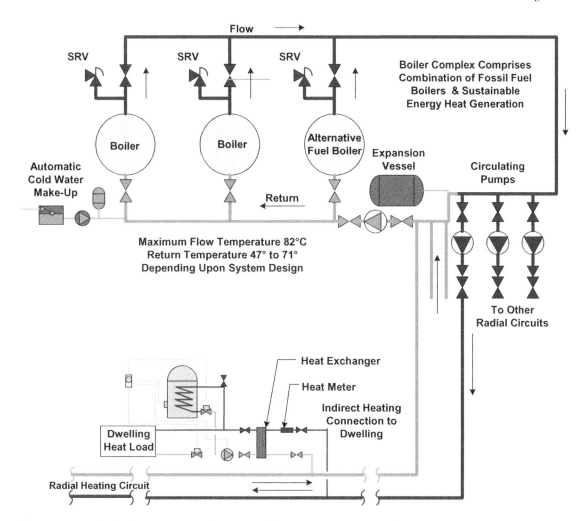

Figure 22.3   Basic low temperature/pressure indirect district heating scheme

British Standard tube used in heat-generating plant rooms and dwellings etc must be checked at the points of tube transition.

These proprietary made pipes are pre-insulated usually with polyurethane or fibreglass insulation, complete with a weatherproof protective covering of high density polyethylene with the pipe ends left exposed to enable the jointing procedure to be made. The completed joint can then be covered with a proprietary made section to maintain the integrity of the weatherproofed protective covering.

Figure 22.5 illustrates typical pipe details where pre-insulated heat-distribution pipes are installed directly in a builder's work constructed trench, complete with pipe supports and removable access cover. A drainage channel should be incorporated and arranged to fall to a collection sump for ease of removal. Provision must be made to allow for thermal expansion of the pipework including allowing the pipes to freely move.

Alternatively, the pre-insulated heat-distribution mains may be obtained factory-made inside a larger steel conduit which has been treated to protect it against corrosion, the heat distribution pipes being internally supported on spacing spiders and guides. The heat-distribution pipes protrude out of the ends of the steel conduit to allow the pipe joints to be welded. On completion of the hydraulic testing procedure of

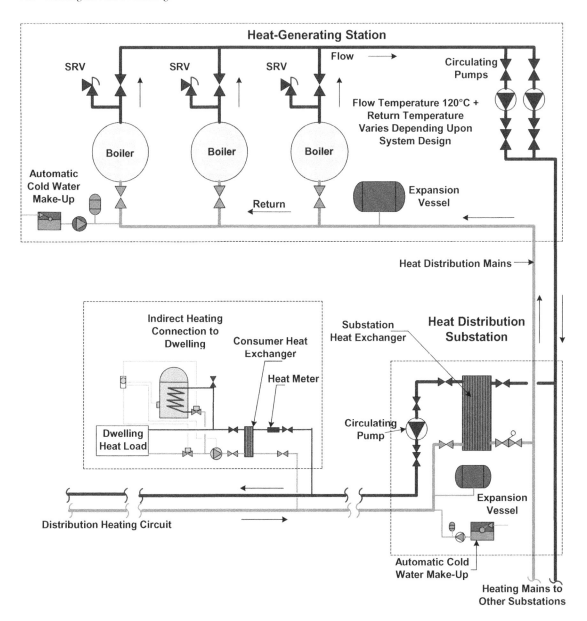

Figure 22.4 Basic high temperature/pressure indirect district heating scheme

the heat-distribution mains, the thermal insulation over the pipe joint may be made using a proprietary insulation pipe sleeve that retains both the waterproof and thermal integrity of the overall heat-distribution mains. The outer steel conduit spacer should be welded together and pressure tested to ensure that it also retains its waterproof integrity.

The pre-insulated heat-distribution mains may also be buried in a trench that is backfilled with a selected material free of sharp stones and similar objects after hydraulic pressure testing. The depth of the trench will be determined by other ground services but should always have a minimum of 500 mm cover. A warning tape should also be laid 100mm above the pipes to give an indication of the presence of the

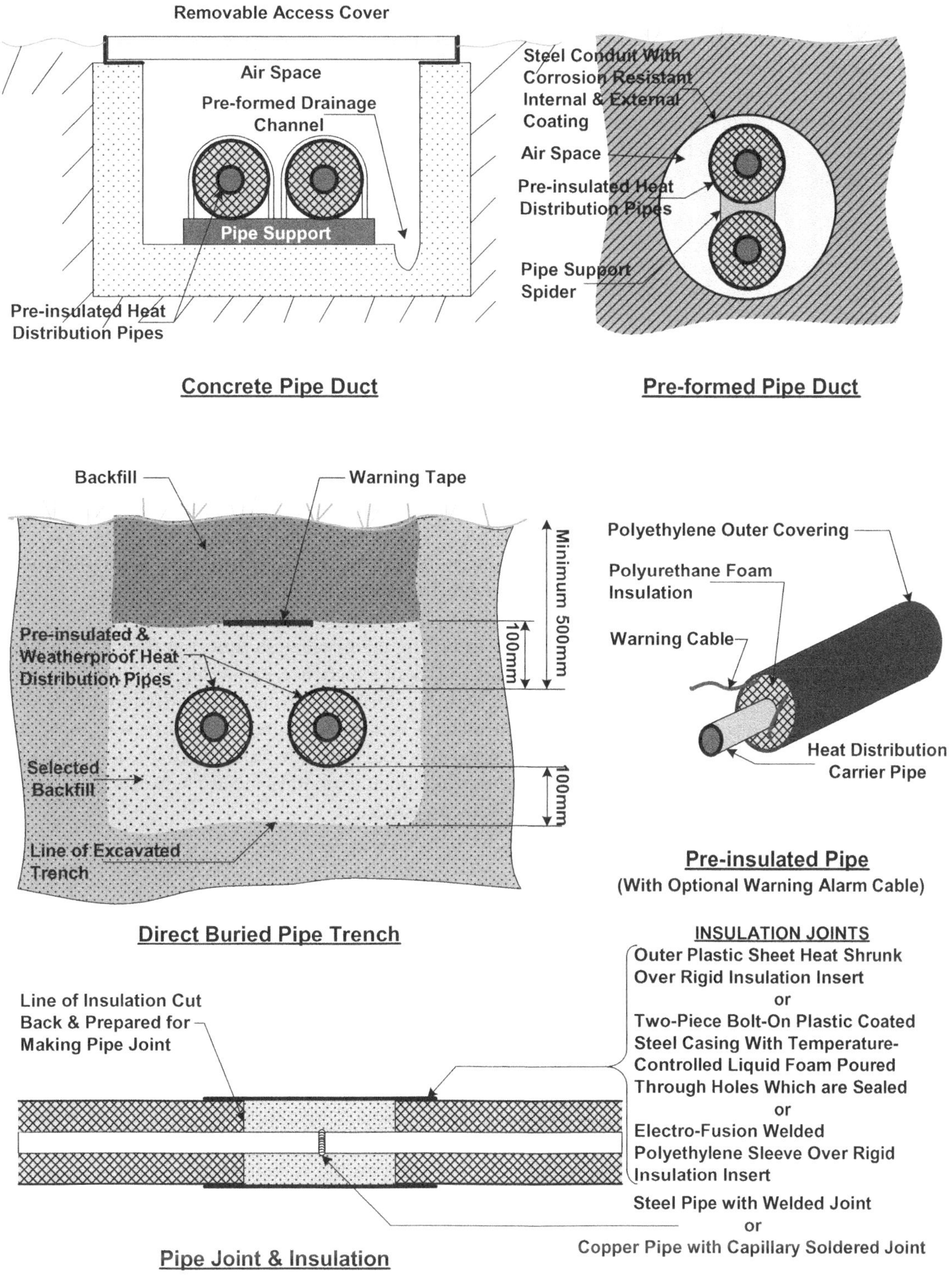
Removable Access Cover
Air Space
Pre-formed Drainage Channel
Pipe Support
Pre-insulated Heat Distribution Pipes
Concrete Pipe Duct
Steel Conduit With Corrosion Resistant Internal & External Coating
Air Space
Pre-insulated Heat Distribution Pipes
Pipe Support Spider
Pre-formed Pipe Duct
Backfill
Warning Tape
Minimum 500mm
100mm
Pre-insulated & Weatherproof Heat Distribution Pipes
Selected Backfill
100mm
Line of Excavated Trench
Direct Buried Pipe Trench
Polyethylene Outer Covering
Polyurethane Foam Insulation
Warning Cable
Heat Distribution Carrier Pipe
Pre-insulated Pipe
(With Optional Warning Alarm Cable)
INSULATION JOINTS
Outer Plastic Sheet Heat Shrunk Over Rigid Insulation Insert
or
Two-Piece Bolt-On Plastic Coated Steel Casing With Temperature-Controlled Liquid Foam Poured Through Holes Which are Sealed
or
Electro-Fusion Welded Polyethylene Sleeve Over Rigid Insulation Insert
Line of Insulation Cut Back & Prepared for Making Pipe Joint
Steel Pipe with Welded Joint
or
Copper Pipe with Capillary Soldered Joint
Pipe Joint & Insulation

Figure 22.5 Heat distribution mains details

heat-distribution mains in the event of any ground excavation, with the wording 'WARNING: DISTRICT HEATING MAINS BELOW' repeated as a continuous statement.

Pre-insulated heat-distribution pipes may be obtained incorporating an electrical alarm cable within the polyurethane foam insulation. This cable is arranged to form a circuit linked to a fault indicator located in the heat generation station or substation, that when cut, or if the outer polyethylene covering is damaged and moisture comes into contact with the alarm cable, will signal an alarm. This will give an early indication that damage has occurred so that a repair may be implemented before corrosion to the carrier pipe has weakened the material.

The pre-insulated pipes may be joined together in a conventional manner such as butt welding steel tube, or capillary soldered joints for copper and stainless steel tube. After the installation has been completed and the piping system has been successfully pressure tested, the space between the insulated ends of the pre-insulated tube must be made good to ensure that complete protection and thermal integrity is maintained. This can be achieved as follows:

- A slip-on plastic sleeve or wrap-around sheet may be heat shrunk onto a sectional rigid foam insulation insert; the ends of the sleeve are usually shrunk over a synthetic rubber ring to seal these ends.
- A two-piece epoxy-coated steel casing may be bolted over the heat-distribution carrier pipe joint that incorporates injection holes that permit liquid polyurethane foam to be injected under pressure into the holes at controlled temperature conditions. This foam sets hard when allowed to cool and the injection holes are plugged off on completion.
- A slip-on polyethylene sleeve may be electro-fusion welded over a rigid polyurethane foam insulation insert, or liquid foam injected as described for the previous method.

The use of pre-insulated pipes can be an advantage in cutting the installation time, but the staff must be fully trained to work on such systems or failures, due to lack of experience, can occur. It should be noted that the exposed ends of the insulated pipes must be kept dry during the installation procedure to prevent moisture from being absorbed into the insulation material and causing premature corrosion to occur. Where these pipes are required to be installed in trenches that could become waterlogged, the joints should be made by raising the pipes above ground to make the joint, then temporarily sealed between the exposed insulated ends before lowering back into the trench when completed.

HEAT METERS AND CONSUMER CONNECTIONS

With the heat being distributed to each individual consumer, some means has to be provided for apportioning the costs of running and maintaining the system. The simplest method is to levy a standing charge to each consumer based on the installed heating load or floor area of the dwelling; however, this does not encourage conservation of energy and is not a fair method as those who use less heat will be subsidising those who use much more.

The easiest method of assessing the heat being used is to install a volumetric type flow meter on the inlet connection after a bypass valve, therefore assuming that the flow temperature is constant. The amount recorded on the meter can be converted to a kilowatt heat load. On this system it is important to arrange the heating controls within the dwelling to stop the flow through the system when no heat is required, otherwise false readings will be recorded.

This arrangement is indicated in Figure 22.6, but is not considered an acceptable or accurate method of metering the amount of heat used and is no longer an approved arrangement.

The most accurate and preferred method of measuring the amount of heat being used is to incorporate one of the many types of heat meters that are commercially available. The manufacturer's specific instructions regarding their location on the heating system should be followed as some are recommended to be installed on the heating return, whilst others are required to be fitted on the heating flow.

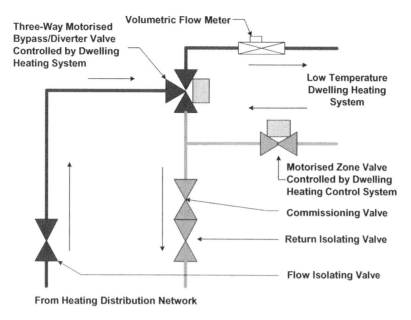

Figure 22.6 Low temperature, direct connection district heating with flow meter

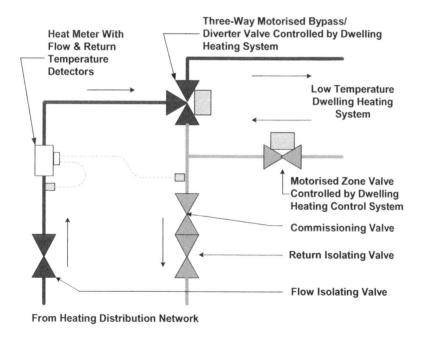

Figure 22.7 Low temperature, direct connection district heating with heat meter

The most common method and arrangement of recording the heat flow passing through the dwelling heating system is indicated in Figures 22.7 and 22.8, where not only the volumetric flow is recorded, but also the temperatures of both the heating flow entering the dwelling heating system together with the temperature of the water being returned back to the heating distribution network. The heat meter measures the temperature difference between the heating flow and return to each dwelling and the heat used is calculated accurately.

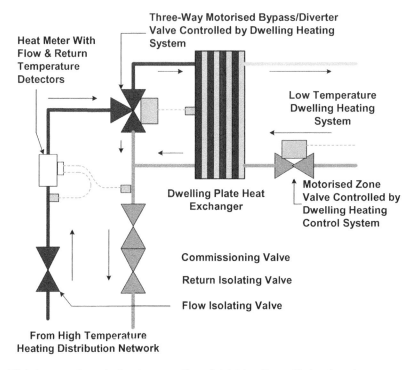

Figure 22.8   High temperature, indirect connection district heating with heat meter

It should be noted that some heat meters are flow sensitive and will only perform correctly when the flow rate is between a maximum and minimum flow, therefore the dwelling heating control system should be capable of controlling the flow between these two restraints to ensure that the meter reading is correct. It should also be appreciated that all heat meters will require periodic checking for accuracy and, if necessary recalibrating, to ensure that they are accurately recording the heat flow passing through.

23

Expansion of Pipework

Any change in temperature will cause a corresponding change in the mass of most substances; this condition has previously been established in Chapter 2 regarding the behaviour of water when heated, which demonstrated there was a change in the volume of the water which had to be accommodated.

Concentrating on the effect that this has on pipework, it will be noticed that any increase or decrease in temperature will have a corresponding increase or contraction in both the length of any pipe and the diameter regardless of the material, the only variance will be the extent of the linear and circumference change that occurs.

The change that occurs in the diameter of the pipe will be minimal and ordinarily does not need any special requirements to protect it unless the pipe passes through a wall unsleeved, where plaster or some other similar finishing material has been applied directly to the pipe: here the increase in the diameter of the pipe could cause the plaster to crack and break off.

The linear increase or decrease may or may not be considerable and will need to be considered as any pipe that has been subjected to a significant temperature change will want to move in length, but if it has been constrained and prevented from moving, a degree of stress will be experienced by the piping material. Over a period of time this could manifest itself by the pipe fracturing or pipe brackets failing, by being pulled from the wall if this stress is considerable.

To enable an investigation into the extent of any linear change, together with the degree of stress that would result, the following information is required:

- The minimum and maximum operating temperatures that will be experienced, together with the initial water fill temperature if different from the minimum operating temperature.
- Maximum fault condition temperature that could occur as a worst-case scenario.
- System piping material or materials if mixed, together with their relevant coefficient of expansion.
- Piping system layout plans that enable the actual pipe lengths to be ascertained and show all changes in direction.

Heating Services in Buildings: Design, Installation, Commissioning & Maintenance, First Edition. David E. Watkins.
© 2011 John Wiley & Sons, Ltd. Published 2011 by John Wiley & Sons, Ltd.
As an aid to lecturers and students, full colour versions of the figures in this chapter may be found at www.wiley.com/go/watkins
These figures are © 2011 by John Wiley & Sons, Ltd.

Table 23.1   Typical coefficient of linear expansion for common piping materials

Material	Coefficient of Linear Expansion m/m per °C
Metallic	
Copper	16.8×10^{-6}
Low Carbon Mild Steel	11.34×10^{-6}
Stainless Steel (Austenitic 304)	17.3×10^{-6}
Lead	29.7×10^{-6}
Aluminium	25.5×10^{-6}
Thermoplastic	
PE Polyethylene	200.0×10^{-6}
PB Polybutylene	130.0×10^{-6}
PVCU Unplasticised Polyvinylchloride	54.4×10^{-6}
PP Polypropylene	90.5×10^{-6}
ABS Acrylonitrile Butadiene Styrene	73.8×10^{-6}

The above coefficients of linear expansion are typical for a temperature range of 0–100°C.

COEFFICIENT OF LINEAR EXPANSION

The coefficient of linear expansion of a material may be described as the change in unit length per degree in temperature difference. It is defined as the proportion of its original length which increases or decreases with every 1°C rise or fall in temperature and is different for each material. Table 23.1 gives the coefficients of linear expansion for some commonly used pipe materials.

The values given in Table 23.1 are typical average thermal coefficients of linear expansion for a selection of common piping materials, and may be used to calculate the extent of any longitudinal expansion that could occur within a temperature range of 0 to 100°C. To be precise, the coefficient of linear expansion will vary for each material depending upon the temperature that it is exposed to. This coefficient may either increase in its rate or decrease, depending upon the temperature range that it is subjected to, but for the purposes of calculating the rate of expansion for low temperature heating systems the values given in the Table 23.1 may be used.

It can also be observed from Table 23.1 that there is a large contrast between the metallic materials and the thermoplastic materials with the coefficients of linear expansion for the latter group being generally much higher.

The coefficients of linear expansion given in Table 23.1 are expressed as the increase in length that will occur for every one degree Celsius change: for example, copper tube is stated as having a coefficient of linear expansion of 16.8×10^{-6}, which means that the decimal point should be moved six places to the left, hence:

$16.8 \times 10^{-6} = 0.0000168$, which is the increase in length in metres for a 1°C temperature rise for copper tube, or for every 1°C temperature rise every metre of copper tube will increase in length by 0.0168 mm.

CALCULATING EXPANSION

Having established the coefficient of linear expansion for various piping materials, the temperature difference that must be applied for the calculation must be obtained.

The temperature variations from cold to hot or vice versa can range from gradual, such as the increase or decrease in room air temperature over a period of time, which will eventually transfer through the pipes, or almost instantaneously, when high temperature water is being circulated through the piping such as in a hydronic heating system. In these system arrangements the pipework can receive a thermal shock, causing the pipework to move suddenly.

The temperature variation normally occurs when a fluid at a higher or lower temperature is passed through the pipe and the expansion or contraction that results must be catered for. This temperature difference, used in any calculation, must allow for both the highest and lowest temperature extremes. For example a low temperature hydronic heating system may be designed to operate at a flow temperature of 82°C and a return temperature of 71°C, giving a design temperature difference of 11°C. However, the fill temperature of the water entering the system could be around 5°C and the temperature safety relief valve set at 95°C, resulting in a possible temperature difference of 90°C, which must be allowed for. For a high-temperature heating system or a steam system the temperature difference would be much higher.

Figure 23.1 illustrates a simple plan view of a copper tube piping system conveying water at an operating temperature of 82°C, but with a fault condition that could reach 95°C. The cold fill temperature is taken as 5°C.

Using the following formula:

$$L_2 = L_1 \times Œ \times \Delta t$$

From the summary of the calculations depicted in Table 23.2, it can be observed that pipe reference A has increased in length by 38 mm, pipe reference C by 31 mm, and pipe reference B has expanded by an insignificant amount and can therefore be ignored.

As both the pipe ends of the example piping system used are anchored and therefore prevented from moving, the pipe will expand towards the middle offset section causing it to deflect as shown in Figure 23.2. This would result in a deflection of 69 mm (38 + 31) over the length of 6 metres. If the piping arrangement permits this thermal movement and increase in length then no special precautions are required. If

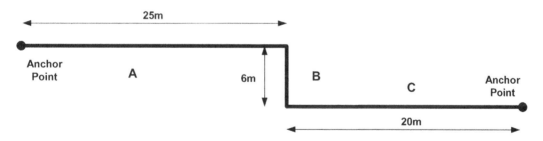

Figure 23.1 Example calculation: plan view of piping arrangement, cold condition

Table 23.2 Summary of pipeline expansion calculations

Pipe Ref	Measured Pipe Length m L_1	Coefficient of Linear Expansion (Œ)		Δt (95 − 5 = 90)	Expanded Total Length m L_2	Expanded Increase mm
		From Table	Calculation			
A	25	16.8×10^{-6}	0.0000168	90	25.0378	38
B	6	16.8×10^{-6}	0.0000168	90	6.009	9
C	20	16.8×10^{-6}	0.0000168	90	20.03	31

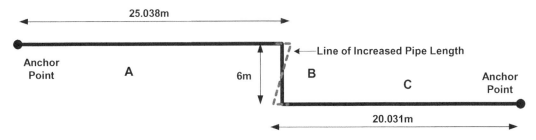

Figure 23.2 Example calculation: plan view of piping arrangement, hot condition

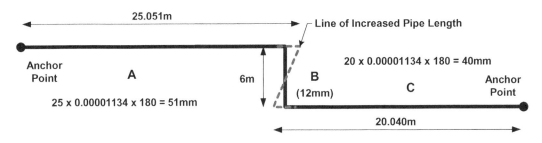

Figure 23.3 Example calculation: plan view of piping arrangement, high-temperature application

this movement and increase in length cannot be accommodated, then provision must be included to permit this movement and reduce the potential stress set-up in the pipe.

Most domestic heating systems, being low-temperature systems, do not involve long straight lengths of pipe and runs are relatively short; therefore expansion is rarely a problem. Even in many commercial buildings which are also low-temperature systems, the piping arrangement is such that there are few long pipeline runs that require expansion considerations as most are catered for in the piping system configuration, with the expansion taken up in changes in direction of the pipes.

However, there are circumstances where, due to the building complexity and constraints preventing the piping arrangement from absorbing the expansion, or where medium- or high-temperature heating systems are employed, then provision for expansion must be incorporated.

Using the same piping layout depicted in Figures 23.1 and 23.2, but for a high-temperature system installed in low carbon steel tube where the maximum temperature that has to be accommodated is 185°C, the effects on the expansion of the system are indicated in Figure 23.3.

In this example it can be seen that although the coefficient of linear expansion is less, due to the higher temperature difference the expansion experienced is higher whereby the central deflection amounts to 91 mm with the 6 m central offset at pipeline reference B growing 12 mm over its length.

PROVISION FOR EXPANSION AND EXPANSION JOINTS

The best approach when planning pipe routes is to allow for natural pipe expansion loops and if possible, to avoid long straight runs of pipework. Natural expansion loops are formed by arranging the piping as shown in Figure 23.4. In this example a branch has been taken off and used as an anchor point to prevent stress being applied to the pipe at this point.

The piping has also been laid out to allow the thermal movement caused by expansion to move towards the natural pipe loop.

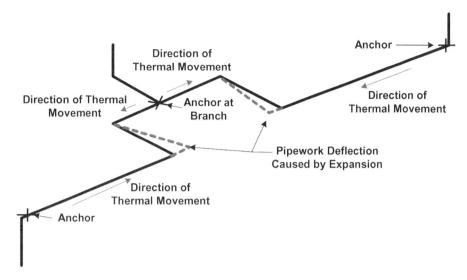

Figure 23.4 Piping arrangement utilising natural expansion loop

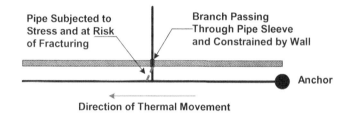

Figure 23.5 Branch subjected to stress by thermal movement

When planning the piping layout the positions of branches and tees should be arranged so that they are not subjected to any significant thermal movement, as stress could cause the pipe to fracture in severe cases. See Figure 23.5 which shows situations to avoid when planning positions of branches and anchor points.

EXPANSION LOOPS

When long straight pipelines are required and it is not possible to arrange the piping to have natural expansion loops, the simplest method of accommodating the thermal movement is to incorporate a fabricated expansion loop – or lyre bend as it is sometimes known, as the shape resembles the Greek harp. These horseshoe shaped expansion loops are fabricated from the same piping material as the piping system but because of the bending constraints of the material, a large amount of space is required to accommodate it, which in many cases makes it unsuitable for most building applications. It is also restricted by the bending stresses of the piping material (see Figure 23.6). However, the life expectancy of these fabricated components is equal to the life expectancy of the piping material, providing that the thermal movement being experienced is not so excessive as to cause fatigue within the metal.

These types of expansion loops should be fabricated from a single length of tube without any intermediate joints. If welds are used they become weak points that are vulnerable to failing, because the continual movement of the loop over a prolonged period of time causes fatigue to occur at the weld.

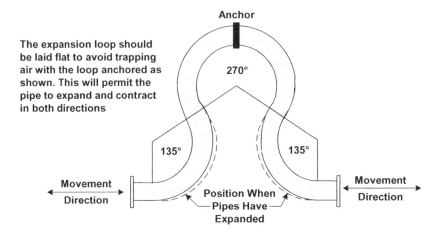

Figure 23.6   Plan detail of fabricated expansion loop (lyre bend)

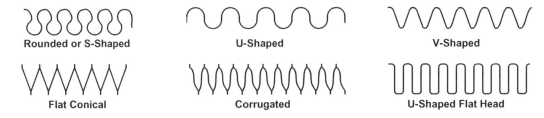

Figure 23.7   Expansion bellows configuration

EXPANSION BELLOWS/COMPENSATORS

The bellows type expansion joint is essentially a flexible membrane that bridges between the adjoining two pipe ends and is free to move without the need for any packed lubricant or sealing gland.

The bellows is generally an all-metal construction with fabrication possible from any commercially available and weldable material. Bellows manufactured from synthetic rubber that are suitable for low-temperature applications involving limited thermal movement are also available, but are normally restricted to being used as a flexible pipe connector as described in Chapter 18, Circulating Pumps.

Traditionally, bellows type expansion joints have been manufactured from an austenitic grade of stainless steel where the thin sheet metal, or multilayer sheet metal, is formed into a corrugated shape which will serve as the bellows to allow axial and angular movement by both compression, to accommodate expansion, and extension to permit contraction.

The design of the bellows shape varies between manufacturers, each proclaiming the virtues of their particular design. Figure 23.7 is representative of a selection of bellows design configurations.

Axial bellows expansion joints are designed to accommodate both compressive and extension movements along the bellows longitudinal axis. Movements available are specified as ± amounts from free length, the free length being the theoretical length before movement. From this free length the expansion bellows unit will provide an equal amount of movement available in either compression or extension, an arrangement that is suitable for piping systems that are used to convey fluids at contrasting temperatures either side of the ambient temperature from cold to hot. For piping systems where it is desirable to utilise the unit's total movement available when it is known that the movement will be in one direction only, the expansion

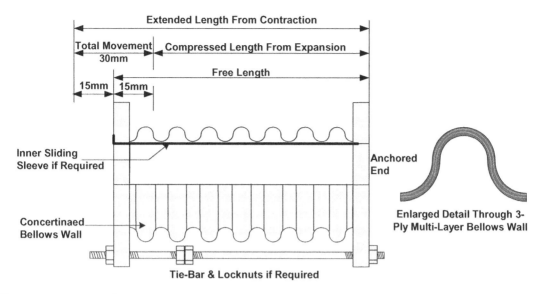

Figure 23.8 Detail of axial expansion bellows

bellows should be installed in either a pre-extension or pre-compressive state. Figure 23.8 illustrates the compressive and extension movement allowance when the total movement permitted is 30 mm; this pre-compression or pre-extension – commonly referred to as cold draw – is achieved by inserting continuous threaded tie-bars through flange holes and adjusting the locknuts evenly, until the required forced movement has been reached. The total movement will be specified by the bellows manufacturer.

The tie-bars may also be used as a safety measure to restrict movement of the bellows beyond its failure point.

Tie-bars as indicated in Figure 23.8 should also be fitted when a hydraulic pressure test is to be applied to the piping system to prevent the expansion bellows from extending longitudinally, or strut instability occurring. Strut instability, or squirm as it is commonly referred to, is a condition where the bellows unit blows out sideways and damages the convolutions of the bellows cylinder.

Figure 23.8 also shows an enlarged section of the bellows wall employing a multi-ply composition, which comprises a series of thin gauge stainless steel sheets un-bonded together rather than a single thicker gauge sheet; this arrangement improves the flexibility of the composition and reduces the stress being applied to the concertinaed bellows unit. It has been shown that this arrangement extends the life expectancy of the expansion bellows due to the reduced stress being applied. The composition may comprise several layers to make up the unit.

Figure 23.8 also indicates an internal rigid sleeve fixed to one end of the expansion bellows if required. An external shroud may also be fitted but is not shown. This addition makes it impossible to install the expansion bellows into a piping system that is initially misaligned. In addition to being practical, the shroud gives this self-guided axial bellows a streamlined appearance. These types of units are supplied in their extended length and held in position at this length by a small set screw. This ensures that they are at all times installed at their correct length, which in turn ensures a trouble-free operation. The set screw is removed as part of the commissioning exercise to release the bellows and allow it to move after the pipe installation has been completed.

The internal sleeve remains static as the bellows expands and contracts and achieves a reduced resistance smooth bore flow for the water through the bellows, reducing the risk of erosion occurring at the internal surface of the convolutions.

The bellows is normally considered the weakest part of any piping system due to the fact that the wall thickness of the convoluted bellows cylinder is less than the thickness of the connecting pipework wall. An external shroud over the bellows convoluted cylinder will provide a degree of protection for personnel against a pressure blow-out when high-temperature liquid is being conveyed, but the shroud cover makes it difficult to inspect the condition of the expansion bellows cylinder.

CYCLIC LIFE

The term cyclic life is applied to the anticipated number of complete expansion and contraction movements that the bellows is expected to accommodate over its working life, before fatigue occurs and causes the convoluted bellows cylinder to fail: this is an important aspect of the bellows design and whole life costing, as discussed in Chapter 26.

Cyclic life is associated with obtaining the correct balance between the pressure containing characteristics and the movement of the bellows. Manufacturers of expansion bellows are not obligated to indicate the number of cycles before failure occurs, but ordinarily the standard design rating used for producing axial expansion bellows is 1000 complete cyclic movements at a service operating temperature not exceeding 300°C.

Any reduction in the total movement length of the axial expansion bellows required will have a corresponding increase in the number of cyclic movements at the reduced length of the total movement, and will therefore increase the expected service life of the units.

The graph indicated in Figure 23.9 depicts a typical cyclic life movement chart whereby, using the axial expansion bellows indicated in Figure 23.8 as an example, a total movement of 30 mm is available at the standard cyclic movements of 1000. If the system operates for 6 months of the year on a programmer requiring the system to turn on twice a day causing two complete piping cycle movements per day, this would equate to 365 movements per year (2×182.5), which means that the bellows would be expected to

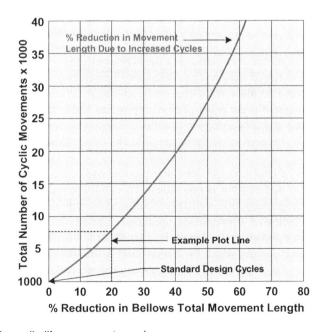

Figure 23.9 Example cyclic life movement graph

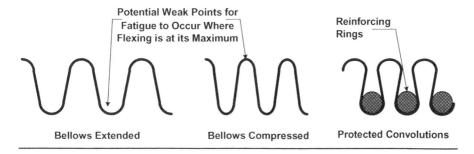

Figure 23.10   Expansion bellows fatigue and protection

fail after approximately 2¾ years. To increase the life expectancy of the bellows unit, it can be seen that by plotting a 20% reduction in the total movement length of the bellows to 24 mm, the number of cyclic movements can be increased to approximately 7500, which equates to a life expectancy of 20 years.

The fatigue manifests itself in the form of cracks and splits at the peaks and troughs of the bellows convolutions, where the maximum flexing of the unit will occur; here it will eventually work-harden and fail. By reducing this amount of flexing by limiting the length of travel of the total movement, the number of movements can be increased before the bellows unit fails.

From this example it can be seen that there is merit in installing an increased number of axial expansion bellows, each with a designed reduced length of total movement than they are capable of, so that the increased installation cost would be offset by the improved life expectancy of the units.

A detailed cost analysis comparing the increased financial expenditure of purchasing and installing additional axial expansion bellows with the increased life expectancy must be undertaken to determine the optimum number of additional bellows that is viable during the design stage. For instance, the previous example showed that the minimum number of axial expansion bellows would have a life expectancy of less than 3 years, which would normally be unacceptable, but by installing additional bellows the life expectancy was increased to 20 years, which in most building developments would probably be an acceptable period of time before having to replace them. Furthermore, if extra expansion bellows were to be installed, the life expectancy may be increased further to approximately 70 years plus, but it would almost certainly be difficult to justify the additional financial cost.

Figure 23.10 indicates the most vulnerable points of the expansion bellows where fatigue or work-hardening is likely to occur due to the continual flexing at these points. It also indicates a method that has been used to protect the rounded profile in the bellows valley of the convolutions, by incorporating a protective ring to reduce the risk of creasing the metal at the tips of the convoluted peaks and troughs.

These convolutions are formed by either rolling the convolutions between external and internal wheels, or by forcing the cylindrical tube radially under hydraulic pressure into the required convolution profile tooling mould.

BELLOWS APPLICATION

The positioning of the axial expansion compensator within the piping system and its relationship with pipe anchors and guides is a critical part of the design exercise to ensure that the complete system functions correctly and safely.

Using the pipeline depicted in Figure 23.1 as an example, the total expansion previously calculated for pipe references A and C equals 69 mm over that total length.

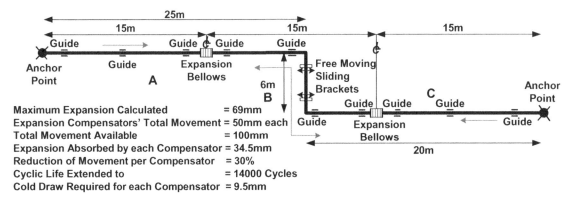

Figure 23.11 Example of compensator application

The expansion bellows selected each have a total movement of 50 mm(±25 mm), this gives a combined total movement of 100 mm available. Each compensator will therefore be required to cope with 34.5 mm of expansion which equates to 70% of their capability.

The expansion compensators will each require a cold draw of 9.5 mm to accommodate the expansion compression requirement.

The expansion previously calculated for pipe reference B is minimal and considered not to require any allowance for expansion; however, this short section of pipe must be free to move laterally along the sliding brackets supporting the pipework.

Compensators are normally produced capable of a total movement of 25–50 mm, although a greater total movement may be manufactured, which will be dictated by the number of convolutions arranged on the bellows – however, the greater the length of the convoluted section of the expansion bellows, the less stable the unit becomes. Figure 23.11 illustrates this example calculation and procedure.

In this example the two bellows have been positioned unrestrained approximately ⅓ and ⅔ of the way along the length of the pipe run; this means that the piping will expand towards the compensators on either side of the bellows, reducing the necessity of the pipe having to expand by the maximum increase in length over the guides, which would be the case if the bellows were located at the pipeline ends by the anchors.

FORCES ON ANCHORS AND GUIDES

Pipeline anchors and intermediate pipeline support guides are just as essential as the axial expansion compensators if the piping system is to operate correctly and safely: without correctly located and designed pipeline anchors and guides the piping system would fail.

Pipeline anchors are a form of pipe support that holds the pipe rigidly, preventing it from moving whilst being anchored into the building structure. At this point it statically braces the piping and compensator by transferring the forces acting upon it by the expanded pipeline to the building structure.

The pipeline anchor, together with its fixings and the structure supporting it, must be capable of withstanding the maximum loads that are likely to be imposed upon it.

Pipeline guides are also a form of pipe support but differ from the pipeline anchors in so far as they should be loose fitting to permit the pipe to move freely through them by guiding the sliding movement of the pipe. The pipeline guides are however still pipe supports, and therefore must be capable of supporting the weight and frictional drag caused by the pipes moving across them.

In order to assess the loadings that will be imposed by the effects of the pipeline expansion on the bellows, anchors and guides, the following must be obtained by either data source or calculation. Each must be examined individually before arriving at the sum of the forces acting upon the anchor point.

- Deflection Load – The deflection load is due to the spring rate of the axial bellows and is obtained from manufacturers' data sheets. A bellows is in fact a spring where a force is required to compress that spring.

$$\text{Deflection force} = \text{spring rate} \times \text{movement}$$

  The movement is considered to be the maximum movement of which the bellows unit is capable within the design parameters of the installation. A stretched bellows having a cold-drawn allowance is trying to pull the anchors together rather than force them apart. In practice this means that for a bellows installed with a 50% cold-draw, half the total expansion should be used for the calculation of the deflection force.

  The deflection force is also referred to as the elasticity force required to stretch and compress the bellows installed through the working movement.
- Frictional Resistance – The frictional force between the pipe moving over the pipe support/guide acting against the direction of movement. This may be calculated using the following formula:

$$\text{Frictional resistance} = \text{coefficient of guide resistance} \times \text{total weight of pipe}$$

  It is normal for the guide manufacturer to advise the coefficient of friction values in their engineering specification, but in the event of this not being available the value of 0.3 may be taken for the majority of metal-to-metal installations. If a polytetrafluoroethylene (PTFE) support lining is used, then a friction coefficient of 0.02 may be used.

  The total weight of the pipe between anchors is the sum of pipe loaded weight including the weight of water plus the weight of thermal insulation.
- Pressure Thrust – This is the force due to internal pressure trying to open out the bellows convolutions.

  Figure 23.12 indicates the effective cross-section of an axial expansion bellows, which is the average diameter of the bellows taking the tip and root of the convolutions as the extremes. Values for the effective area are published in the manufacturer's data sheets.

The resistance encountered by the anchor acts against the direction of the piping movement, as indicated in Figure 23.13. The movement will be in the direction towards the compensator as designed, but the forces will be equal in both directions acting against the anchors at both ends of the pipe. If the pipe run is long and there are intermediate anchors, the forces acting upon them will be the same on both sides and therefore balance out.

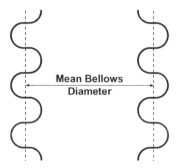

Figure 23.12   Pressure thrusts on bellows

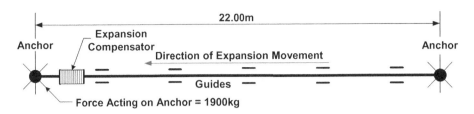

Figure 23.13   Example of expansion forces acting on anchor

For the example depicted in Figure 23.13, it is assumed that the pipe is 150 mm diameter low carbon mild steel medium weight tube 22 metres in length, when at ambient temperature and operating at a maximum temperature 90°C and 5 bar pressure.

The following information has been collated from the stated sources:

Design Information

Pipe weight	19.7 kg m^{-1}	See Appendix 5
Water weight	18.97 kg m^{-1}	See Appendix 5
Insulation weight	6 kg m^{-1}	Manufacturer's data
Total installed pipe weight	44.67 kg m^{-1} (Rounded up to kg m^{-1})	
Coefficient of linear expansion	11.34 × 10^{-6}	See Table 23.1
Effective area of bellows	260 cm^{-2} (0.026 cm^{-2})	See Table 23.3
Axial spring rate	130 N mm^{-1}	See Table 23.3
Coefficient of friction for supports	0.3	See Figure 23.15

Example Load/Force Calculation
Expansion of pipe, 5°C to 90°C

$$0.00001134 \times 85°C \times 22 = 21 \text{ mm increase in length}$$

Thrust due to internal pressure on bellows =
$$500\,000 \text{ N} \times 0.026 \text{ m}^{-2} = 13\,000 \text{ N} \qquad\qquad \textbf{13\,000 N}$$

Deflection load =
$$21 \text{ mm} \times 130 \text{ N mm}^{-1} = 2730 \text{ N} \qquad\qquad \textbf{2730 N}$$

Frictional resistance to pipe movement =
Coefficient of friction × total weight of pipe per metre × length of pipe run
$$0.3 \times 45 \times 22 = 297 \text{ kg} \times 9.807 \text{ N} = \qquad\qquad \textbf{2912 N}$$

Total force = 18 642 N or 1900 kg

From the above example a force of 1900 kg will be acting upon the pipeline anchor, therefore the anchor, together with its fixings and the supporting structure, will have to be capable of absorbing this force.

It is normal to apply a safe working load factor of 100% to this loading making it capable of absorbing a force designed for 3.8 tonnes.

Table 23.3 Extract from typical bellows manufacturer's data sheet

Catalogue Ref	Nominal Dia mm	Supplied Length mm	Movement		Axial Spring Rate		Effective Area	Pressure Rating
			± mm	Total mm	kg/mm	N/mm	sq cm	bars
ABC100	100	160	22	44	10.2	100	130	16
ABC150	150	210	25	50	13.25	130	260	16
ABC200	200	235	26	52	17.33	170	440	16

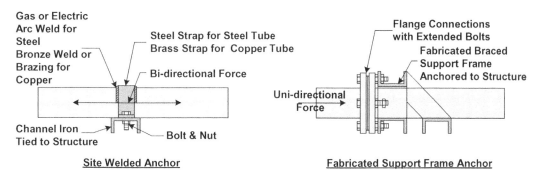

Figure 23.14 Typical pipeline anchors

ANCHORS AND GUIDES

As previously established, the function of an anchor is to provide an unmovable point against which the force being imposed by the pipe expansion against the compression of the bellows can be braced, and to transfer the load to the structure that the anchor is tied into. For this reason the anchor must be tied into a structure that is capable of absorbing this force.

There is no standard or set design for pipeline anchors as they must be made to suit the particular application in question and usually involve a certain amount of site fabrication.

Figure 23.14 details two forms of pipeline anchors of which there are many variations with different forms of construction; however, they all will be required to be tied into the structure and will require a degree of site fabrication to suit the application required.

The welded construction anchor indicated comprises a flat pipe strap bolted to a channel iron support frame, with the top part of the strap that is in contact with the pipe wall welded to the pipe on both sides of the strap. It should be noted that the weakest part of this form of anchor will be the bolts, which must be capable of resisting the shear force that will be exerted on them. To increase the strength of this unit a wider strap or double strap can be used which will double the amount of bolts required. This type of anchor will absorb the force exerted from either direction.

The fabricated support frame arrangement consists of a fabricated channel iron box type frame that permits a pair of mating flanges employing longer securing bolts to be extended through both the horizontal members and vertical members of the channel iron frame and bolted to the frame. This form of anchor will only absorb the force from one direction but generally will be of a stronger construction than the welded strap form; however, the box type support frame must be tied into a structure of equal strength.

If the adjacent structure is not suitable for supporting the load, then the steelwork must be designed to be extended back to a more suitable part of the structure.

The function of a guide is to serve as a pipe support but to be loose enough to allow the pipe to move freely but without snaking, which would cause the bellows to buckle and fail.

As a pipe support, it must be capable of supporting the loaded weight of the pipe and of offering the minimum frictional resistance to the pipe as it moves through the bracket arrangement. It must also be strong enough to resist bending or distorting caused by the drag effect of the pipe outer wall moving over the base of the bracket.

Figure 23.15 illustrates a selection of different forms of pipeline guide designs and as in the case of pipeline anchors, there are many variations of these. The coefficient of friction for each type of guide has also been included for use with the frictional resistance part of the force being exerted calculation.

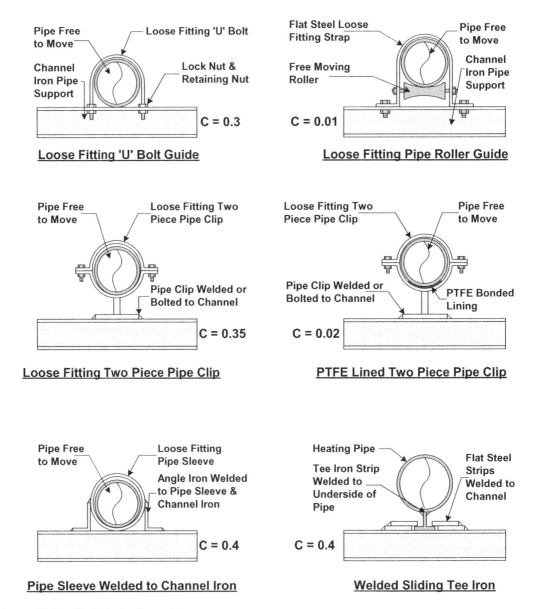

Figure 23.15   Typical pipeline guides

The loose fitting 'U' bolt method is an inexpensive guide in common use for low-temperature heating applications experiencing restricted movement. It comprises a threaded 'U' bolt held in place by two lock nuts to achieve the loose fitting gap between the pipe required.

This arrangement relies on the underside of the pipe sliding or dragging itself across the supporting steelwork, which causes wear and generates noise during thermal movement.

The loose fitting roller method offers the minimum of resistance to the pipeline movement as it assists the pipe in its thermal movement through the guide. It is however a more expensive arrangement that has many variations, such as the roller being fitted on a spindle that spans between the steel strap and permits the roller to move as shown in Figure 23.15, or the roller being arranged on a chair type support separately from the strap type support. A gap between the strap and the pipe must be maintained to permit movements, and the rollers must be lubricated at regular periods.

The two-piece loose fitting standard pipe bracket is also a common arrangement for a pipeline guide, where the bracket is either bolted or welded to the steelwork as shown. If the resistance to the pipe sliding through it is too high, there is a risk that the single vertical bracket support may bend, causing the pipe to bind and drag the bracket until it fails. The inclusion of a PTFE lining pad bonded to the lower part of the bracket reduces its resistance to the pipe sliding through it and improves its properties as a pipeline guide.

The oversized pipe sleeve is used as a common form of site fabricated pipeline guide for larger diameter pipes, which may be just a short pipe off-cut one pipe size diameter larger than the carrying pipe where the pipe slides through it as a metal-to-metal contact. The pipe sleeve is welded to the supporting steelwork as shown.

The final example of pipeline guides illustrated is also site fabricated and normally used for larger diameter pipes; it comprises welding a tee-shaped piece of steel on the underside of the pipe, having a length that will extend beyond the width of the supporting steelwork so that it can slide over the steelwork when subjected to thermal movement without falling off the edge. The tee of the inverted tee shaped steel is contained in the sections of flat steel welded to the supporting steelwork, which will allow the tee shaped steel to slide freely unhindered as shown.

HINGED ANGULAR BELLOWS

Hinged bellows units provide for thermal movement in one plane only and they operate by angulating the bellows. They are non-compressive angular units where the pressure end load is contained by the hinged parts and therefore this type of assembly is ideal where it is not practical to install robust guiding or strong anchors.

The hinges limit the bellows to the angular movement that they are capable of, without damaging the convoluted bellows cylinder where the range of movement is restricted by the protecting integral stops or restraining rods.

Single hinged expansion bellows are usually employed in pairs to provide lateral movement in the plane required, with each bellows being capable of providing an angular deflection of 10° to 15°.

Figure 23.16 illustrates a typical singular hinged angular bellows that appears to be identical in design to a standard axial expansion bellows but with the addition of a hinged brace arranged either side of the bellows; this prevents the bellows from moving in a linear or axial direction but allows it to be deflected in an angular direction. The extent of this angular deflection is limited by the inclusion of the restraining rods that traverse the bellows between the connecting flanges, as shown. Without these rods or integral stops, the angular movement could be in excess of the capabilities of the convoluted bellows cylinder.

Figure 23.17 illustrates the application of hinged angular expansion bellows used in pairs where advantage can be taken of the angle of deflection, by installing them angled towards the expanding pipe in its cold position. This will allow the pipe to expand, which in so doing moves the hinged bellows, causing

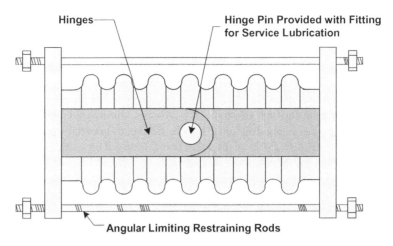

Figure 23.16   Detail of typical hinged angular bellows

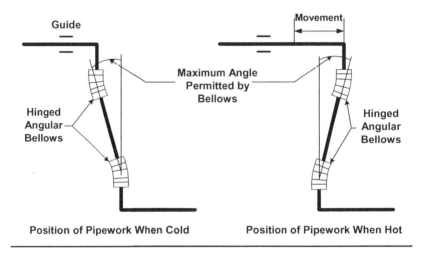

Figure 23.17   Application of hinged bellows used in pairs

them to deflect in the opposite direction. This results in twice the length of movement being available to cater for the increase in pipe length caused by the thermal expansion.

The length of pipe movement available for thermal expansion is determined by the angle of deflection obtainable from the hinged angular expansion bellows, coupled with the length of interconnecting pipe between the pair of bellows. The longer the interconnecting pipe, the greater the pipe expansion movement that can be allowed.

The interconnecting pipe between the pair of hinged bellows must be installed in such a manner that it is allowed to move freely as shown. This means that the pipe must be supported by a length of steel beneath the pipe and at a length that will continue to support the pipe, over its full length of movement.

ARTICULATED BELLOWS

Double-hinged articulated expansion bellows are basically two single hinged expansion bellows combined into a single unit with a common articulated tie bar joining the two hinged convoluted sections together.

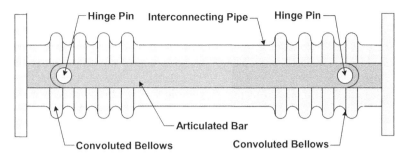

Figure 23.18 Double-hinged articulated bellows

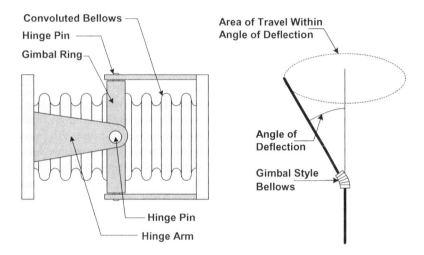

Figure 23.19 Detail of typical gimbal-style angular expansion bellows

Any expansion of the central interconnecting pipe within the limits of the articulated tie bar will simply compress the convoluted bellows and will not exert movements on the adjoining pipework. This type of expansion unit permits lateral movement in one plane only.

The restriction of the angular deflection is provided by integral stops incorporated on the tie bar hinge to prevent the angular deflection being forced past its maximum angle permitted.

The application of this type of expansion bellows unit is exactly the same as indicated in Figure 23.18, the only constraint being that the interconnecting pipe length is fixed and cannot be varied.

GIMBAL EXPANSION BELLOWS

Gimbal style expansion bellows are designed to allow angular rotation in any plane by incorporating two hinges fixed to a common floating gimbal ring centrally located along the length of convoluted bellows. The gimbal ring and hinged components are designed to restrain the end thrust of the expansion joint due to internal pressure and any external forces which are imposed on the joint. The hinges also incorporate an angular deflection limitation to prevent the bellows from being forced past their deflection capabilities.

The gimbal type of expansion joint takes its name from the floating gimbal ring that the two pairs of hinges are fitted to, as shown in Figure 23.19. This enables the bellows to deflect in any direction required, as opposed to the single-hinged angular bellows or the articulated bellows, which can only deflect in one direction. As with the singular-hinged angular bellows, gimbal type bellows must also be installed in pairs.

The angular rotation within the angular deflection governed by the angle deflection limitation of the bellows also indicated in Figure 23.19, and enables the movement of the bellows interconnecting pipe to move in any direction.

The gimbal ring retains its distance from the convolutions of the bellows regardless of the direction in which it deflects, as the curvature of the convoluted cylinder bends within the inner space of the gimbal ring.

PACKED GLAND SLIDING EXPANSION JOINT

Sometimes referred to as a packed type slip joint, this is an older design than the convoluted bellows now in common use for building services applications.

The slip type expansion joint is essentially a pair of telescopic cylinders and is similar to a number of common connecting devices, such as the mechanical gland coupling which has been used for jointing cast iron pipe and compression sleeve couplings, and for jointing plain ended cast iron pipe. The slip type expansion joint comprises a stuffing box and machine finished smooth sliding surfaces, with engineered dimensions and tolerances to enable satisfactory performance in severe service. It is limited to axial movement or rotation about the pipe axis for freedom from binding. A typical packed gland sliding expansion joint is shown in Figure 23.20, which illustrates a single-ended unit. When greater axial movement capacity or traverse is required than a single-ended cylinder unit can provide, then a double-ended joint may be obtained having two stuffing box glands in a common body, usually supplied with brackets for anchoring the unit.

The main features of the slip type expansion joint are the large linear movement that can be accommodated from a single unit when compared to a bellows design expansion joint, and the lower resultant force requiring to be anchored.

The main limitation of any packed gland unit is the difficulty of establishing and maintaining a seal. While ground surfaces and sealing rings are occasionally used, more generally, the seal is dependent upon

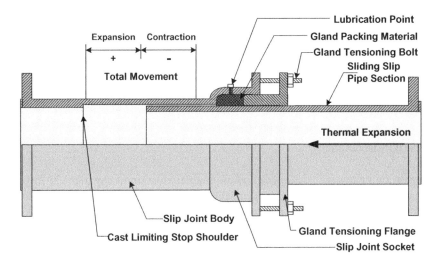

Figure 23.20　Packed gland slip-type expansion joint

packing with limitations on the contacting surfaces of the fluid and its temperature. These packings have in the past been asbestos based but now are generally manufactured from hydrocarbon materials; however, caution should be adopted when replacing old packing material.

A degree of force is required to overcome the friction generated by the packing material, and to reduce this friction. Some designs incorporate a means for lubricating the packing material which requires periodic application when continual pipeline thermal movement is expected.

Maintenance must be anticipated, such as periodically tightening or repacking the stuffing box gland when it shows signs of leaking, or when tightening has reached a point where the gland has the effect of causing excessive friction or binding of the slip pipe section, as the force required to overcome packing friction and operate the joint effectively is a function of the packing material characteristics and gland plate/ flange pressure.

The slip type expansion joint incorporates an integral stop or movement limiter, as shown in Figure 23.20, to prevent the slip pipe section travelling beyond its design capability; it should also be prevented from pulling the slip pipe out of the expansion joint body, by arranging the connecting pipework to be reversed anchored against the back of the connecting mating flange.

The force being exerted on the slip type expansion joint is restricted to the extension of the piping required to move the slip pipe section past the friction of the packing gland material. This force may be considerable on high pressure systems; therefore, the anchor behind the expansion joint will only be required to withstand this force, which is less than could be expected to compress the convolutions of a bellows pattern expansion joint.

Packed gland slip type expansion joints have traditionally been manufactured from cast iron, but are also available made from ductile iron, cast and machined steel, stainless steel and brass, with a domestic heating version being manufactured from copper or brass, with Type 'A' compression coupling type joints and using a synthetic rubber gland packing material. These domestic versions have limited thermal movement available, which is all that is required for a domestic heating system, but they should always be fitted in visible and accessible locations to enable the gland to be inspected on a regular basis and allow the gland to be retightened when necessary.

FLEXIBLE RUBBER EXPANSION JOINTS

As discussed in Chapter 18, Circulating Pumps, flexible bellows manufactured from synthetic rubber materials are primarily used to absorb vibration and noise between rotating machinery, such as circulating pumps and the piping system due to their flexibility.

This high degree of flexibility also makes them suitable for compensating for small installation misalignments and compression caused by expansion. They can be used in a similar way to metal expansion bellows to compensate for axial, lateral and angular movement. As with any other unrestrained expansion joint, flexible rubber bellows will extend when subjected to pressure.

A number of different synthetic rubber materials have been used to manufacture flexible bellows connectors, but they all have limited pressure and temperature properties, normally being restricted to pressures below 16 bar at 20°C and approximately 1 bar at 90°C, making them suitable for only low-pressure and low-temperature heating applications, although reinforcement of the bellows material will raise the pressure/temperature rating by a small percentage as shown in Figure 23.21.

All synthetic rubber bellows have a limited life expectancy as the material will age with time. Life expectancy is also determined by external influences such as ultraviolet light and ozone, but the main influence is temperature: the higher the operating temperature, the shorter the bellows life expectancy. The working life of a synthetic rubber bellows is considerably reduced at temperatures above 70°C.

Figure 23.22 illustrates both the limitations and versatility of synthetic rubber flexible expansion bellows.

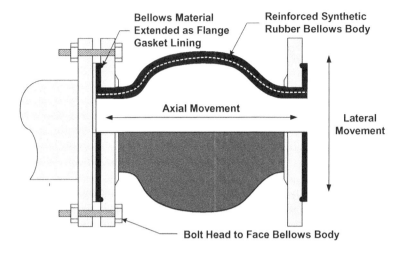

Figure 23.21   Synthetic rubber bellows

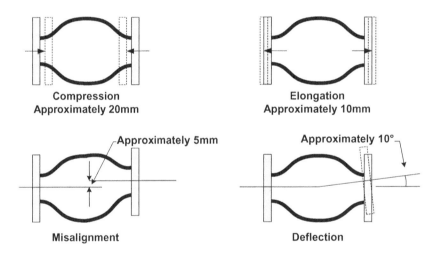

Figure 23.22   Versatility and limitations of flexible synthetic rubber bellows

24

Regulations, Standards, Codes and Guides

There has never been a more regulated time in the building services industry than there is now and the plumbing/heating section of that industry is no exception – in fact it is probably the most regulated area of the building services industry, covering design requirements, material quality, installation competencies and safety procedures. This regulatory situation can make innovation more difficult today than it has ever been in the past, not impossible, just more challenging. This is not meant to imply that this regulatory restriction is a bad thing, but to emphasise that today's engineer – and that means a professional engineer holding an engineering qualification – has a professional obligation to have a working knowledge and understanding of all these regulations, and must undertake to keep abreast of the ever-changing amendments that are made to them.

As the title of this chapter implies, there is a difference in the definition of the terms, regulations, standards, codes and guides, and an understanding of the title of any of these documents will give the reader an indication of their purpose as well as their importance.

Regulations

The term 'regulations' normally applies to a set of mandatory requirements laid down by a recognised authority appertaining to the subject covered. There are numerous regulations covering different aspects of this industry, but the most significant affecting the heating section of the building services industry are:

- The Building Regulations
- The Water Supply (Water Fittings) Regulations
- The Gas Safety (Installation and Use) Regulations

All three of these specific regulations are Statutory Instruments and as such are known as secondary legislation made under the powers conferred by Acts of Parliament. It is this legislation that gives them the authority to be mandatory requirements and therefore they must be complied with without exception.

Regulations such as these made under an Act of Parliament take precedence over any other requirements that may contradict them in any way, which has occurred in the past.

Heating Services in Buildings: Design, Installation, Commissioning & Maintenance, First Edition. David E. Watkins.
© 2011 John Wiley & Sons, Ltd. Published 2011 by John Wiley & Sons, Ltd.
As an aid to lecturers and students, full colour versions of the figures in this chapter may be found at www.wiley.com/go/watkins
These figures are © 2011 by John Wiley & Sons, Ltd.

Other regulations include those that govern health and safety matters, which are equally as important but not included in this text, which is confined to technical aspects of the industry.

Standards

The term 'standards' normally applies to material quality, although in recent times the practice of including system requirements as standards has also been adopted.

Standards specify minimum requirements and dimensional conformity between different manufacturers but in themselves are not mandatory. However, standards are frequently cited in a Statutory Instrument Regulation which does then make them mandatory and means they must be complied with.

The most frequently used standards are:

- British Standards
- British European Harmonised Standards

There are other standards produced by professional bodies and research establishments which make a useful contribution to the advancement of the science and practice of the plumbing and heating industry.

Codes

These are similar to standards but are normally confined to specifying system design requirements, although the custom of publishing British Standard Codes of Practice has now been replaced by publishing them directly as a British Standard.

The North American practice differs from the UK in so far as the definition of the term 'code' is applied to what is called regulations in the UK. For example the USA National Codes include the National Plumbing Code, the National Electrical Code and the National Fire Code.

Guides

Regulations, standards and less so codes, are mainly published by government departments, whereas guides are produced by non-governmental offices such as professional bodies that include:

- Chartered Institution of Building Services Engineers (CIBSE)
- Chartered Institute of Plumbing and Heating Engineering (CIPHE)
- American Society of Heating Refrigeration Air Conditioning Engineers (ASHRAE)

These guides produced by professional bodies are not textbooks but guides containing current engineering data, information and practice with the minimum of explanation. They are designed to be used by qualified engineers as tools and as a source of the current information required to design heating services.

It is not possible to discuss in detail the multitude of documents and publications available under the broad headings of this chapter, because of the amount of data available but the following has been selected as being extra relevant to the subject matter of heating.

BUILDING REGULATIONS

The UK has three separately produced and administered sets of Building Regulations, these are:

- Building Regulations for England and Wales
- Building Regulations Scotland
- Building Regulations Northern Ireland

These regulations are produced by each separate government and although they are very similar to each other they are referenced differently and do vary in places. Any engineer familiar with working in one of these countries would need to acquaint themselves with the Building Regulations of a neighbouring area if they intend to work in that area.

Building Regulations were initially laid before Parliament on 22nd July 1965 and came into operation on 1st February 1966 costing 13/- (65p). They were produced as a Statutory Instrument under the Public Health Acts of 1936 and 1961 and the Clean Air Act of 1956 and were known as The Building Regulations 1965. They have since been amended many times and split into the three divisional UK areas as listed.

The current Building Regulations for England and Wales were made under the powers of the Building Act 1984 and are known as The Building Regulations 2000, although each section has been amended since that date, with some sections amended many times.

The first few editions of the Building Regulations were published as single documents and worded using legal language; they are now published in separate sections with each section being referred to as an Approved Document, which now contains interpretations of the regulations and advice on how to comply with them. At the time of writing, the following documents have been approved and issued by the Secretary of State for the purpose of providing practical guidance with respect to the requirements of The Building Regulations 2000.

- **Approved Document A** – Structure: 1992 Edition, amended 1994 and 2000.
- **Approved Document B** – Fire Safety: 2000 Edition, amended 2000.
- **Approved Document C** – Site Preparation and Resistance to Moisture: 1992 Edition, amended 1992 and 2000.
- **Approved Document D** – Toxic substances: amended 1992 and 2000.
- **Approved Document E** – Resistance to the Passage of Sound: 2000 Edition, amended 2003.
- **Approved Document F** – Ventilation: 1995 Edition, amended 2010.
- **Approved Document G** – Sanitation, Hot Water Safety and Water Efficiency: 2010 Edition.
- **Approved Document H** – Drainage and Waste Disposal: 2002 Edition.
- **Approved Document J** – Combustion Appliances and Fuel Storage Systems: 2002 Edition.
- **Approved Document K** – Protection from Falling, Collision and Impact: 1998 Edition, amended 2000.
- **Approved Document L1, Parts A and B** – Conservation of Fuel and Power in Dwellings: 2002 Edition, amended 2010.
- **Approved Document L2, Parts A and B** – Conservation of Fuel and Power in Buildings other than Dwellings: 2002 Edition, amended 2010.
- **Approved Document M** – Access and Facilities for Disabled People: 1999 Edition, amended 2000.
- **Approved Document N** – Glazing – Safety in Relation to Impact, Opening and Cleaning: 1998 Edition, amended 2000.
- **Approved Document P** – Electrical Safety: 2005 Edition
- **Approved Document to support Regulation 7** – Materials and Workmanship: 1999 Edition, amended 2000.

The Building Regulations have gradually developed since their conception in 1965 to their present format and practically every Approved Document is of interest to those involved with the plumbing and heating services in buildings. They are planned to continue to be developed with further amendments over the next few years.

Approved Document A

This is concerned with structural elements of the building, and charges mechanical services design engineers with the responsibility of ensuring that the structure is capable of supporting any equipment loads that will be imposed on the building, such as boilers, cold water storage cisterns and any fuel storage loads.

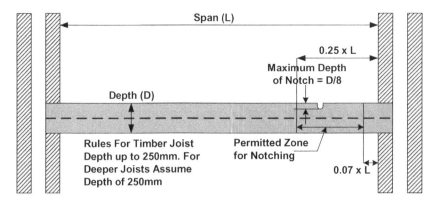

Figure 24.1　Regulations for notching traditional timber joists

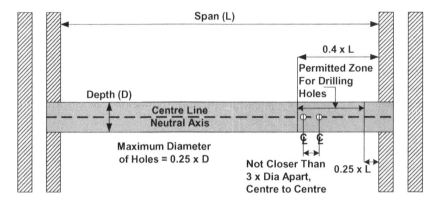

Figure 24.2　Regulations for drilling holes through traditional timber joists

Approved Document A is also concerned with such matters as cutting holes through structural elements of the building, including the drilling and notching of structural timbers, and gives limitational requirements to ensure that the integrity of the structure is not compromised.

The Building Regulations require that any notches and holes cut in simply supported timber floor and roof joists should be within the following limits:

Notches should be cut no deeper than ⅛ of the depth of the joist and no closer to the supporting end than $0.07 \times$ the span of the joist, or no further away than 0.25 (¼) from the supporting end as shown in Figure 24.1.

Notches and holes should be as small as possible whilst being large enough not to restrict pipe movement caused by expansion and contraction. Notches should be formed preferably by drilling a hole at the correct depth and cutting parallel cuts to produce a 'U' shaped notch.

Drilled holes through timber joists should be no greater in diameter than 0.25 (¼) the depth of the joist. They should be drilled on a neutral axis such as the centre line where there is neither compression nor tension, and where more than one hole is required, they should be spaced at a minimum distance not less than three diameters apart, measured from centre to centre.

Drilled holes should be located in an area between 0.25 times the span and 0.4 times the span of the joist from the joist support, as illustrated in Figure 24.2.

More recently many housing developments have adopted the use of composite timber and plywood engineering I beams, or metal web beams, in place of the traditional timber joist, due to construction advantages, but both require a different approach when installing pipework through them.

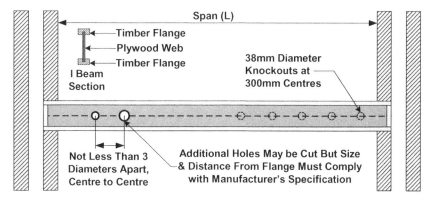

Figure 24.3   Requirements for drilling holes through engineering I beams

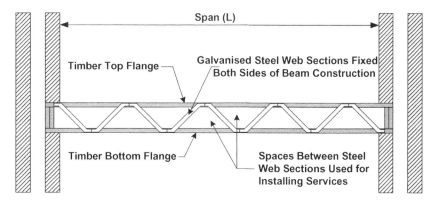

Figure 24.4   Arrangement of metal web engineering beam

Engineering I beams are a composite construction of a plywood web contained between two timber section flanges where the web incorporates 38 mm diameter knockouts spaced at 300 mm centres, but larger holes may be drilled at not less than three diameters apart, measured from centre to centre. The size of the hole is limited to the beam manufacturer's specification (see Figure 24.3).

The same rules apply when drilling through an I beam as for traditional timber joists where horizontal measurements between holes are concerned. However, as the integrity of the I beam is dependent upon the relationship between the timber flanges and the plywood web, they should not be notched as this would seriously affect the beam's strength. Also, any hole that is drilled through the beam should not be any closer than 3 mm from the flange, or as specified by the I beam manufacturer.

The metal web form of engineering beam consists of two timber flange members held apart by galvanised steel profiled sectional web plates fixed to the timber flanges at an angle, as shown in Figure 24.4. The angled steel web plates form a cross-braced effect, providing large triangular clear spaces between them that permit the services to be passed through. These clear spaces cannot be altered in any way, but do provide larger openings that allow bigger diameter pipes such as 100 mm soil and vent pipes to pass through.

The use of engineering beams has had the additional effect of changing the construction techniques that have traditionally been employed on house building and similar projects. Conventionally, pipes have been laid in notches cut into the joists, as the rigidity of the pipes does not easily allow for any other installation method. This dictates that the pipework must be installed before the flooring boards are laid, which prevents other trades working until this is complete.

Flexible piping materials can be threaded through the openings formed in the engineering joists from below, which means this work can be planned for a later date to coincide with what traditionally would have been classified as secondary fix work. This permits continuity of work between first and second fix and reduces the risk of damage being caused by other trades that would have been on-site between these stages.

Approved Document B

This section of the Building Regulations is concerned with fire safety and includes requirements for fire protection services.

Approved Document C

This section of the Building Regulations is concerned with the prevention of moisture entering the building, including damp proof courses and weatherings.

Approved Document D

This section of the Building Regulations is concerned with all toxic substances involved in the construction industry.

Approved Document E

This section of the Building Regulations is concerned with the resistance to the passage of sound that includes sound transferred through pipework.

Approved Document F

This section of the Building Regulations is concerned with ventilation and therefore has an important impact on the design of heating systems.

Approved Document G

This revised section, formerly entitled Hygiene, has now been amended and re-titled Sanitation, Hot Water Safety and Water Efficiency. The requirements appertaining to domestic hot water safety, both unvented and fully open vented systems, are covered where relevant to hydronic heating systems in Chapter 19, Domestic Hot Water.

Approved Document H

This section of the Building Regulations is concerned with the subject of drainage, both foul water and surface water, and waste disposal including the waste from condensing boilers. These requirements for the disposal of condensate waste are explained in Chapter 16, Combustion, Flues and Chimneys.

Approved Document J

This section of the Building Regulations is entitled Combustion Appliances and Fuel Storage Systems which, for those that are involved with heating systems, is of major importance.

Approved Document J covers the requirements for both combustion and ventilation air for all fossil fuel burning appliances, flues and chimneys, details of which are explained in Chapter 16.

Approved Document J also stipulates the requirements for the storage of liquid fuel – such as where the capacity of any oil storage tank exceeds 90 litres but is not greater than 3500 litres, or where the capacity of any liquefied petroleum gas storage is greater than 150 litres. For oil storage tanks that exceed 3500 litres, the local Fire Authority should be advised and any requirements complied with, and the Control of Pollution (Oil Storage) (England) Regulations should also be complied with.

Approved Document J also gives the requirements for protection against pollution and the information that must be available in the form of a durable notice on the action and response to be taken in the event of an oil spillage.

The location of and fire protection for oil storage facilities are discussed in Chapter 12, but the Building Regulations Approved Document Part J also requires the following:

- *Within a Building* – Locate any oil tanks in a place designated as a special fire hazard directly ventilated to outside. The building structure should comply with Approved Document Part B.
- *Less than 1800 mm from any part of a building* – Building walls should be imperforate within 1800 mm of tanks with a minimum 30 minutes' fire resistance to internal fires, and construct eaves within 1800 mm of tanks and extending 300 mm past each side of tanks with a minimum of 30 minutes' fire resistance to external fires and with non-combustible cladding.
- Provide a fire wall between the tank and any part of the building within 1800 mm of the tank and construct eaves as above. The fire wall should extend at least 300 mm higher and wider than the affected parts of the tank.
- *Less than 750 mm from a boundary* – Provide a fire wall between the tank and the site boundary, or boundary wall having a minimum of 30 minutes' fire resistance to the fire either side. The fire wall or boundary wall should extend at least 300 mm higher and wider than the top and sides of the tank.

Although these regulations apply to oil storage tanks having capacities of less than 3500 litres, there is considered to be a significant risk of pollution if the following apply to a storage tank:

- Has a total capacity of more than 2500 litres.
- Is located within 10 m of inland freshwaters or coastal waters.
- Is located where spillage could run into an open drain, or to a loose fitting manhole cover.
- Is located within 50 mm of sources of potable water, such as wells, boreholes or springs.
- Is located where oil spilled from the installation could reach the waters listed above by running across hard ground.
- Is located where the tank vent pipe outlets cannot be seen from the intended filling point.

An oil storage installation should carry a durable label in a prominent position giving advice on what to do if an oil spill occurs, and the telephone number of the Environmental Agency's Emergency Hotline.

Liquefied petroleum gas (LPG) installations are controlled by legislation enforced by the Health and Safety Executive (HSE), or their appointed agents.

Factors which determine the amount of building work necessary for an LPG storage installation to comply with regulations include its capacity, whether or not the tank is installed above or below ground, and the nature of the premises it serves. A storage installation may be shown to comply with the legislation by constructing it in accordance with an appropriate industry Code of Practice, such as that produced by the LP Gas Association prepared in consultation with the HSE.

For an installation of up to 1.1 tonne capacity, where the tank stands in the open air, following the guidance in Approved Document J will normally satisfy the requirements needed to comply with other legislation.

Liquefied petroleum gas tanks should only be installed outside a building and not within an open pit. The tank should be separated from the building's boundary and any source of ignition, to enable the safe dispersal of gas in the event of venting or leaks and – in the event of fire – to reduce the risk of the fire spreading.

Where the LPG storage is to be provided by a bank of free-standing portable cylinders, the installation should conform to the requirements as shown in Figures 24.5 and 24.6, and as discussed in Chapter 14.

The cylinders should stand upright on a firm concrete base or paving slabs 50 mm thick and to be secured against the wall by chains or straps, ensuring ventilation and accessibility. They should be protected against

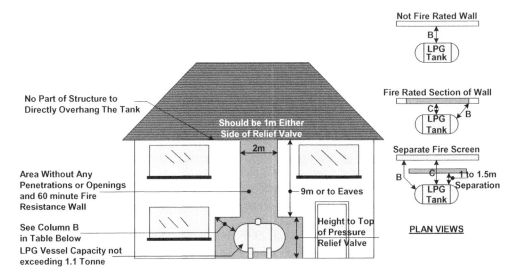

LPG Vessel Capacity Tonnes	Tank With No Fire Wall Dimension B	Tank With Fire Rated Wall Dimension C
0.25	2.5	0.3
1.1	3	1.5

Figure 24.5   Fire separation and shielding LPG storage vessels

possible physical damage, located where they will not cause an obstruction to exit routes from the building, and positioned in non-public areas.

The dimensional requirements given refer to the position of the cylinder isolating valves, which are required to be a minimum of 1m, measured horizontally from any opening, and 300 mm, measured vertically below any opening or heat sources such as flue terminals and tumble-dryer vents.

Because of the density of LPG being heavier than air the storage cylinders should be located a minimum of 2 m measured horizontally from untrapped drain gullies or openings in the ground, such as vents from basements, entrance hatches to cellars, etc.

Approved Document K

Protection from Falling, Collision and Impact applies to all those employed or concerned in any way in the construction industry.

Approved Document L

Approved Document L, Conservation of Fuel and Power in Dwellings, is – as with Approved Document Part J – very important to all those involved with heating systems in buildings.

Approved Document Part L has now evolved into four separate sections, each published separately from each other. These sections are:

- Approved Document Part L1 A – Conservation of Fuel and Power in New Dwellings
- Approved Document Part L1 B – Conservation of Fuel and Power in Existing Dwellings
- Approved Document Part L2 A – Conservation of Fuel and Power in New Buildings other than Dwellings
- Approved Document Part L2 B – Conservation of Fuel and Power in Existing Buildings other than Dwellings

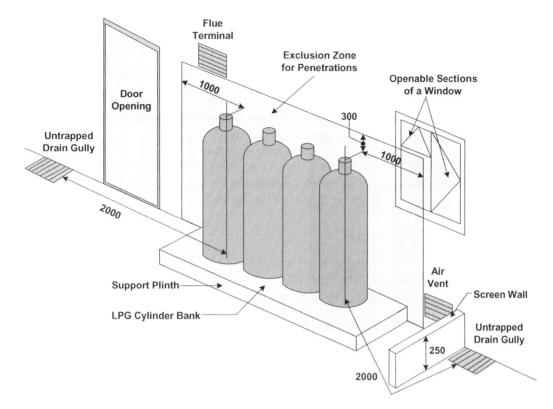

Figure 24.6 LPG cylinder storage requirements

This part of the Building Regulations has been the subject of numerous amendments since the year 2000 and continues to be the focus of further revisions. This is due to it being the vehicle for the UK Government's drive to conserve energy that is consumed within buildings and reduce carbon emissions, in an attempt to meet energy and carbon reduction targets.

To achieve this aim, Approved Document Part L1 requires the following to be incorporated in the design of a building, plus into building refurbishments and any alterations to existing buildings:

Dwellings

Reasonable provision shall be made for the conservation of fuel and power in dwellings by:

a) Limiting the heat loss:
    i. Through the fabric of the building;
    ii. From hot water pipes and hot air ducts used for space heating;
    iii. From domestic hot water storage vessels;
b) Providing space heating and domestic hot water systems that are energy efficient;
c) Providing lighting systems with appropriate lamps and sufficient controls so that energy can be used efficiently;
d) Providing sufficient information with the heating and domestic hot water services so that the building occupiers can operate and maintain the services in such a manner as to use no more energy than is reasonable in the circumstances.

Approved Document Part L2 has the same requirements, but also includes requirements to reduce energy consumption from air conditioning systems.

The loss of heat through the fabric of the building can be reduced by a combination of improving the thermal resistance of the building fabric, plus reducing the air infiltration rate. But this action alone will only satisfy part of the overall objective of reducing the energy consumption from the building, as the heating system will only be responsible for consuming part of the building's energy needs – albeit a large part of the collective energy total.

However, reducing the energy that is being consumed by the heating system will also contribute to reducing the carbon emission, which collectively throughout the country will move us closer to achieving the targets of reducing carbon emissions set by the UK Government.

Approved Document Part L sets out parameters regarding the thermal loss from buildings and lays down requirements of how this can be achieved. In order to meet these requirements, all new buildings must have a certified and have an acceptable SAP rating calculated by the approved method to demonstrate that the energy that will be consumed by the building is within the required limits.

Standard Assessment Procedure (SAP)

The energy performance of a building is determined by the UK Governments Standard Assessment Procedure (SAP) for Energy Rating of Dwellings 2005.

The SAP provides the methodology for the calculation of the Carbon Index which can be used to demonstrate compliance with Approved Document Part L.

Originally there were three approved methods that could be used to demonstrate compliance, namely:

- Elemental method
- Target U-value method
- Carbon Index method

Since publication of the 2006 edition of Approved Document Part L, only the Carbon Index method is considered an acceptable means to demonstrate compliance.

The Carbon Index method calculates the target CO_2 emission rate for the building and the calculated actual CO_2 emission rate for the building as constructed.

The Carbon Index takes into account the fuel being consumed by the heating system's heat raising appliance, plus the carbon being emitted at the electrical generating power plant as a result of the electricity being used by all the building electrical services.

The Carbon Index and Standard Assessment Procedure rating is now required to be calculated using the SAP's downloaded electronic software by a competent person registered by an authorising agency, with both the Target CO_2 Emission Rate (TER) and calculated CO_2 emission for the actual dwelling submitted to the local building control to demonstrate compliance with Approved Document Part L.

Approved Document Part L also sets out minimum standards for heating control systems, endorsing zone controls complete with timing provision and boiler interlocks. It also requires thermostatic radiator valves as a minimum thermostatic control requirement, and fully pumped primary heating circuits for domestic hot water provision.

As the emphasis is now on efficient control systems, the use of weather compensating systems will help to demonstrate compliance.

These requirements are explained in detail in Chapter 11, Controls, Components and Control Sysytems, and Chapter 19, Domestic Hot Water.

Approved Document M

This section, entitled Access and Facilities for Disabled People, is applicable to all those concerned with the construction industry.

Approved Document N

This section, entitled Glazing, is concerned with safety in relation to impact, opening and cleaning of glazed units. This section is not usually of interest to plumbing and heating engineers unless they still undertake glazing work.

Approved Document P

Entitled Electrical Safety, this is the most recent area to be included in the Building Regulations, which for the first time since its conception now incorporates electrical services work.

The implications of the addition of Approved Document Part P are that all form of electrical work in a dwelling must be undertaken by a competent person who has received and successfully completed Part P training, and who is registered with an approved authorising agency. Therefore, anyone installing heating control systems, or wiring up boiler power supplies, must hold a valid Part P competency certificate.

Approved Document Part P is concerned with electrical safety for both the installer and the occupiers of the building, but it should not be construed that anyone holding a valid Part P competency certificate is a qualified electrician, or that Part P training is a shortcut to becoming a qualified electrician – any more than anyone certified as 'gas safe' is a qualified plumbing and heating engineer. See Chapter 10.

Approved Document to support Regulation 7

Covers Materials and Workmanship.

WATER REGULATIONS

The Water Supply (Water Fittings) Regulations, commonly referred to as the Water Regulations, came into force in England and Wales on the 1st July 1999 as a Statutory Instrument made under the provision of the Water Industry Act 1991. As with the Building Regulations, they are mandatory and must be complied with. They replaced the Water Byelaws previously made under the Water Act of 1945.

Scotland has retained the byelaws system by introducing The Water Byelaws 2000, which is almost identical to the requirements contained in the Water Regulations for England and Wales.

The Water Regulations (Byelaws) are an important document and set of requirements that affect the plumbing industry, although their jurisdiction regarding heating systems is confined to the method of supplying water to the system which is explained in detail in Chapter 2, but they do have a greater influence regarding the subject of domestic hot water, which is described in Chapter 19.

The requirements regarding the water being supplied and the method of connection to the heating system are based on preventing a cross-connection between the heating circuit and water supply main, or preventing a backflow of heating circulating water into the water undertaker's water supply main. This is to protect the quality of water being supplied by the water undertaker, which is categorised as Fluid Category 1, and prevent contamination from the heating circulating water, which is categorised as Fluid Category 3, for domestic dwelling, or Fluid Category 4 for heating systems installed in buildings other than dwellings.

The Water Regulations (Byelaws) categorise water into five categories based on their potential hazard to contaminate the water supply.

The examples of the definitions for each fluid category are given as guidance by the Water Regulations Advisory Scheme (WRAS), but also state that the list is not exhaustive. See Table 24.1.

Table 24.1   Fluid categories of water

Fluid Category Designation	Examples of Definitions
Fluid Category 1 Wholesome water supplied by a water undertaker and complying with the requirements of the regulations	Water supplied directly from the water undertaker's mains Water supplied from the mains for drinking water purposes and food preparation
Fluid Category 2 Water in fluid category 1 whose aesthetic quality is impaired owing to a) Change in temperature b) The presence of substances or organisms causing a change in taste, odour or appearance Includes water in a hot water distribution system	**Domestic secondary hot water systems** Mixing of hot and cold water supplies Domestic softening plant with salt regeneration Drink vending machines in which no ingredients or carbon dioxide are injected into the supply or distributing inlet pipe Fire sprinkler system without antifreeze Ice making machines Water cooled air conditioning machines without additives
Fluid Category 3 Fluid which represents a slight health hazard because of the concentration of substances of low toxicity, including any fluid that contains a) Ethylene glycol, copper sulphate solution, or similar chemical additives b) Sodium hypochlorite (chloros and common disinfectants)	**Primary heating circuits and heating systems, with or without additives, for houses (dwellings)** Domestic washbasins, baths and showers Domestic clothes washing and dishwashing machines Home dialysing machines Drink vending machines in which ingredients or carbon dioxide are injected into the supply or distributing inlet pipe Commercial softening plant with salt regeneration Domestic hand-held hoses with flow controlled spray/shut-off Hand-held fertiliser sprays for use in domestic gardens Domestic or commercial irrigation systems, without insecticide or fertiliser additives, and fixed sprinkler heads not less than 150mm above ground

Fluid Category 4

Fluid which represents a significant health hazard due to the concentration of toxic substances, including any fluid which contains

a) Chemical, carcinogenic substances or pesticides including insecticides and herbicides

b) Environmental organisms of potential health significance

Fluid Category 5

Fluid representing a serious health hazard because of the concentration of pathogenic organisms, radioactive or very toxic substances including any fluid which contains

a) Faecal material or other human waste

b) Butchery or other animal waste

c) Pathogens from any other source

Primary heating circuits and heating systems, with or without additives, for buildings other than houses

Fire sprinkler system with antifreeze solutions

Mini-irrigation systems without fertiliser or insecticide application such as pop-up sprinklers or permeable hoses

Food preparation, dairies and bottle-washing apparatus

Commercial dishwashing machines and refrigeration equipment

Dyeing equipment

Industrial disinfection equipment

Printing and photographic equipment

Car washing and degreasing plants

Commercial clothes washing machines

Brewery and distillation plant

Water treatment plant or softeners using other than salt regeneration

Pressurised fire fighting systems

Industrial cisterns

Non-domestic hose union bib-valves

Sinks, urinals, WC pans and bidets

Permeable pipes in other than domestic gardens laid below or at ground level with or without chemical additives

Grey water recycling systems

Any medical or dental equipment with submerged inlets

Laboratories, mortuary and embalming equipment

Bedpan washers

Hospital dialysing machines

Commercial clothes washing plant in health care premises

Non-domestic sinks, baths, washbasins and other appliances

Butchery and meat trades, slaughterhouse equipment and vegetable washing

Dishwashing machines in health care premises

Mobile plant/tankers including gully emptiers

Sewage treatment and sewer cleansing

Drain cleaning plant

Water storage for fire fighting or agricultural purposes

Commercial irrigation outlets below or at ground level and/or permeable pipes with or without chemical additives

Insecticides or fertiliser applications

Commercial hydroponic systems

As previously stated, the Water Regulations have little jurisdiction over heating systems, but it is strongly recommended, or indeed required, that all students of aspects of public health engineering practice become reasonably conversant with them.

THE GAS SAFETY (INSTALLATION AND USE) REGULATIONS

These regulations are also a Statutory Instrument and are concerned with gas safety and the competence of those involved in the installation and maintenance of gas systems and appliances.

BRITISH STANDARDS

The system of producing British Standards has been in existence for many years as a means of establishing a minimum and uniform quality for materials, products and systems, plus ensuring that they are fit for purpose when used within their specified parameters throughout the UK, where they are intended to be used, although some have been adopted by a number of other countries for use in their territories.

British Standards are produced by working committees made up of individuals who are considered qualified in either the manufacture or use of the particular material, product or system concerned. When published they are issued a unique BS number to identify it which will only be used for that particular material, product or system, whereby that unique number will be retired if that specific standard is ever withdrawn.

It is quite common for these standards to be periodically revised, amended or updated to keep pace with changing requirements or conditions. The amendment date is recorded after the unique BS reference number, for example BS 2871-1971. If the particular British Standard has been published in more than one part, the part number will also be recorded for example, BS 2871 Part 1-1971. It is the responsibility of the engineer to satisfy themselves that they are working to the current revision and part of the particular standard concerned.

British Standards have been used nationally throughout the UK. Other countries have equivalent national standards. For some time now, countries within the European Union have been working towards producing a set of agreed harmonised standards that can have a common use across the European Union community, making manufacturing and utilisation easier across this common market.

Harmonised British Standards are denoted by the additional letters EN behind the BS abbreviation, but in front of the unique reference number, with EN standing for European Number. Therefore, BS EN1057-1996, which replaced BS 2871 Part 1, is a harmonised European Standard for copper tube that is in common use throughout the European Union in all the other countries that have adopted it, although they will have issued their own specification reference number for it.

British Standards and harmonised European British Standards now number many thousands and are too numerous to list here, but various standards that are applicable to heating related work are mentioned throughout the text of this volume.

COMPETENT PERSONS SCHEMES

The plumbing and heating industry has evolved over a number of years and various aspects of work now require the operative to hold a valid competency certificate in each one of them and to be registered with the approved authorising agency. These compulsory competency certificates are additional to the main-

Table 24.2 Summary of Competent Persons Schemes for heating systems

Competency	Authorising Agency
Gas systems and appliances. Natural gas and LPG	Gas Safe Register
Oil-fired systems and appliances	Oil Firing Technical Association for the Petroleum Industry (OFTEC)
Solid fuel appliances	Heating Equipment Testing and Approval Scheme (HETAS)
Unvented domestic hot water systems	Approved competency certificate
Energy efficiency	Approved competency certificate
Electrical safety, Part P Certification	Approved competency certificate

stream educational/vocational certificates awarded by City and Guilds and current NVQ/SVQ Level 3, on the successful completion of training.

The first of these Competent Persons Schemes was born out of the gas explosion disaster at Ronan Point, East London in 1968. Following this tragic event a voluntary registration scheme was set up, known as the 'Confederation for the Registration of Gas Installers' (CORGI). Registration with this scheme became compulsory in 1991 with the name changed to 'Council for Registered Gas Installers'.

Today, there is a requirement that any operative involved in the installation or maintenance of gas systems or appliances must be registered with the 'Gas Safe Register'. See Table 24.2.

The authorising agencies' function under the franchise from the Health and Safety Executive is to register operatives that have obtained a recognised valid competency certificate and ensure that the standards are maintained. They are sometimes referred to as 'policing or enforcing agencies' and in certain instances are the subject of controversial debate amongst those who are required to be registered under the scheme.

PROFESSIONAL BODIES

The professional bodies within the plumbing and heating industry should not be confused with the Competent Persons Scheme authorising agencies, nor are they trade associations, trade unions, or any other form of organisation that just requires a membership payment to be made.

They are professional organisations that represent the building services and/or public health engineering industry at a technical level independently of any vested interests, and their views are often sought by government and public media on related matters. They do not promote individuals, manufacturers or companies, but promote the professionalism of the industry. They are made up of qualified people whose membership is graded by academic qualification and experience, all of whom have the common interest of furthering the science and practice of their chosen careers and wish to be part of that development.

The professional bodies are also sponsoring institutions under the control of the Engineering Council UK (ECUK) for the engineering status of its members, where the graded designation of Chartered, Incorporated or Engineering Technician is an engineering status recognised throughout the European Union and beyond European borders.

It is not the intention to appraise each of these professional bodies here, but the author would encourage any student to join the professional body that represents their chosen industry as these have a wealth of information and help available to assist with their chosen career development.

The following is a list of the main professional bodies that represent our industry, but it is not necessarily exhaustive. Their inclusion in the following list does not imply any form of endorsement of any of these professional bodies, but is included to acquaint the entry-level student with their existence and indicate where further information, assistance and a lifelong professional relationship can be obtained.

Chartered Institution of Building Services Engineers (CIBSE)

CIBSE is a professional body that represents all facets of the building services industry, its membership combining on various tasks that include producing a vast amount of technical literature and guidance. The main design tool publications that concern the heating industry include the following:

CIBSE Guide A	Environmental Design
CIBSE Guide B	Heating, Ventilating, Air Conditioning and Refrigeration
CIBSE Guide C	Reference Data
CIBSE Guide D	Transportation Systems in Buildings
CIBSE Guide E	Fire Engineering
CIBSE Guide F	Energy Efficiency in Buildings
CIBSE Guide G	Public Health Engineering
CIBSE Guide H	Building Control Systems
CIBSE Guide J	Weather and Solar Data
CIBSE Guide K	Electricity in Buildings
CIBSE Guide L	Sustainability
CIBSE Guide M	Maintenance Engineering and Management

The list is a small selection of publications available which also forms just one of the numerous benefits and activities of membership.

The Domestic Building Services Panel of CIBSE has also been responsible for publishing – in conjunction with other professional bodies and interested parties – the following:

Domestic Heating Design Guide
Underfloor Heating Design and Installation Guide
Solar Heating Design and Installation Guide

Chartered Institute of Plumbing and Heating Engineering (CIPHE)

CIPHE is also a professional body that represents the public health, plumbing and heating sections of the building services industry at both the design engineering and installer levels.

The membership has combined to produce their main design tool, the Plumbing Engineering Services Design Guide, which has been revised and enlarged over a number of years.

This is just one of a number of publications available, which also form a small part of the numerous benefits and activities of membership.

American Society of Heating, Refrigeration and Air Conditioning Engineers (ASHRAE)

Although a North American organisation, ASHRAE has become a multinational professional body with a very large membership throughout the world.

ASHRAE's research and development are respected worldwide; it has formed relationships with other professional bodies around the world including CIBSE, with a common understanding of sharing technical information and promoting the science and practice of heating, ventilating, refrigeration and air conditioning engineering.

Its membership combines to produce the following design tool publications which are updated on a four-year rolling programme:

ASHRAE Handbook	Fundamentals
ASHRAE Handbook	HVAC Systems and Equipment
ASHRAE Handbook	HVAC Applications
ASHRAE Handbook	Refrigeration

The above forms a very small part of the vast number of publications available.

Institute of Domestic Heating and Environmental Engineers (IDHEE)

The IDHEE is a smaller professional body specialising in domestic heating and environmental engineering matters and has a fewer number of members.

Its aims are as noble as the larger professional bodies, namely promoting the science and practice of the domestic heating market, and it specialises in representing that area of the industry.

Heating and Ventilating Contractors Association (HVCA)

The HVCA is not a professional body, but a trade association made up of and representing heating and ventilating contractors. It has been included in this text as it is important for any student to understand its role within the heating industry, which is to form an agreed set of commercial conditions and contracting issues.

The HVCA has also published a number of publications giving guidance on the installation of various mechanical building services.

25

Testing and System Commissioning

This book has discussed and considered the technical conceptual design criteria considerations through to the detail fundamentals and the installation procedures, which combine to form the bulk of the work required to construct a modern and efficient heating system.

The final obligation before handing the heating installation over to the client is the activity of proving that the system is both fit for purpose and meets the client's design requirements. There is also a commitment to demonstrate the system's operation and provide training for any operatives regarding maintenance and operation.

For many years this has been a standard requirement and contractual obligation for commercial heating installations, but has until recently been neglected in the domestic heating market.

This final testing, commissioning and handover procedure may be summed up as follows:

- Subject the system to a hydraulic pressure test to prove the soundness of the installation, to ensure that there are no fluid leaks and that the mechanical integrity of the system is safe for both the designed operating pressure and possible high malfunction pressures that could result.
- Commission the boiler or heat raising plant together with the associated fuel supply system to ensure that the heat being produced corresponds to the designed heat output for the system. Any combustion appliance should also be checked and if necessary, adjusted to prove that it is operating efficiently and safely.
- Test that all electrically operated equipment and automatic system components are functioning correctly as designed and specified in accordance with the manufacturer's instructions.
- Check that the function of the system control components combines to accomplish the design control philosophy and set any temperature and timed settings.
- Circulate the heating medium fluid around the system and balance each circuit by regulating the flow to correspond to the designed flow rates obtained as a result of the pipe sizing calculation exercise.
- Hand over full operating and maintenance manuals, record drawings, safety file, spares, tools and demonstrate system.

Heating Services in Buildings: Design, Installation, Commissioning & Maintenance, First Edition. David E. Watkins.
© 2011 John Wiley & Sons, Ltd. Published 2011 by John Wiley & Sons, Ltd.
As an aid to lecturers and students, full colour versions of the figures in this chapter may be found at www.wiley.com/go/watkins
These figures are © 2011 by John Wiley & Sons, Ltd.

TESTING FOR SOUNDNESS

As stated earlier, the purpose of a pressure test is to prove the soundness of the installation by ensuring that there are no fluid leaks and that the mechanical integrity of the system is safe and fit for purpose, for both the designed operating pressure and possible high-pressure malfunctions that could result.

This task has always formed part of the contractual obligations of any project regardless of the physical or financial size of the installation. Commercial heating systems require the test to be witnessed by the client's authorised representative and the record sheet of the test recording the test criteria and the result duly signed by the conducting engineer and the witness. Domestic heating systems have not always been pressure tested, relying on just filling the system with water and visually inspecting for any leakage. This poor practice is now becoming less common due to a higher degree of professionalism and awareness of current regulations.

Hydraulic pressure testing will prove that the system will be safe to operate at the system pressure and more importantly, the pressure that could be exerted on the system in the event of a malfunction occurring.

Conducting an approved hydraulic pressure test on the heating system will help expose the problematical micro-leaks that can occur and are discussed briefly in Chapter 18 Circulating Pumps. Micro-leaks is the name given to those minute leaks that will not be large enough to pass water through them when subjected to static pressure from just filling the system up with water, but will allow air to enter the heating system through suspect pipe joints and valve glands installed in pipe sections that will be operating under negative pressure conditions caused by the circulating pump suction effects. Air entering the circulation part of the heating system through micro-leaks will cause continual air-locking problems, with the result of radiators going cold due to air collecting in them and no visual indication of where the micro-leaks are: this should not be confused with other reasons for air locking of heat emitters – see Chapter 21, Water Treatment.

The hydraulic testing of heating systems in new-build projects forms part of a larger testing exercise that includes all of the mechanical building services. These tests are undertaken in conformance with the requirements of BS 6700, which requires the piping services to be tested to 1½ times the working pressure of the system.

The working pressure can be easily assessed for a fully vented open system of hydronic heating as depicted in Figure 25.1, whereby the height of the building is measured in metres and divided by 10.2 to obtain the working pressure expressed in bar.

Therefore, test pressure will be working pressure × 1.5.

Figure 25.1 demonstrates the procedure and application for assessing the working pressure of an open vented hydronic heating system for a high-rise building, together with the resulting test pressure based on 1.5 times the working pressure.

The application of the test pump is shown in two alternative locations. The application of the testing apparatus at the base of the piping system would require the full test pressure of 8.82 bar to be applied, which would produce a pressure of 2.94 bar at the top. If the testing apparatus were to be applied to the top of the piping system then the opposite would be the result, i.e. a test pressure at the test pump of 2.94 would produce a pressure of 8.82 bar at the base due to the effect of the static pressure being exerted.

The general rule of applying a test pressure equivalent to 1.5 times the working or static pressure is acceptable for most applications, but when applied to heating systems that have an extremely low or excessively high working pressure, a sensible compromise should be considered.

Applying this rule to a domestic two-storey dwelling where the static head is 6 m, this would equate to a pressure of 0.58 bar and when multiplied by 1.5 would equal 0.88 bar as a test pressure. In situations involving low working pressures, a minimum test pressure of 2 bar should be adopted to provide a realistic testing pressure.

The opposite situation where a high working pressure exists, would involve extremely high test pressures to be applied, for example a high-pressure system of 30 bar would require a test pressure of 45 bar if this rule is followed. For applications involving high working pressures above 20 bar, a test pressure of 1.25

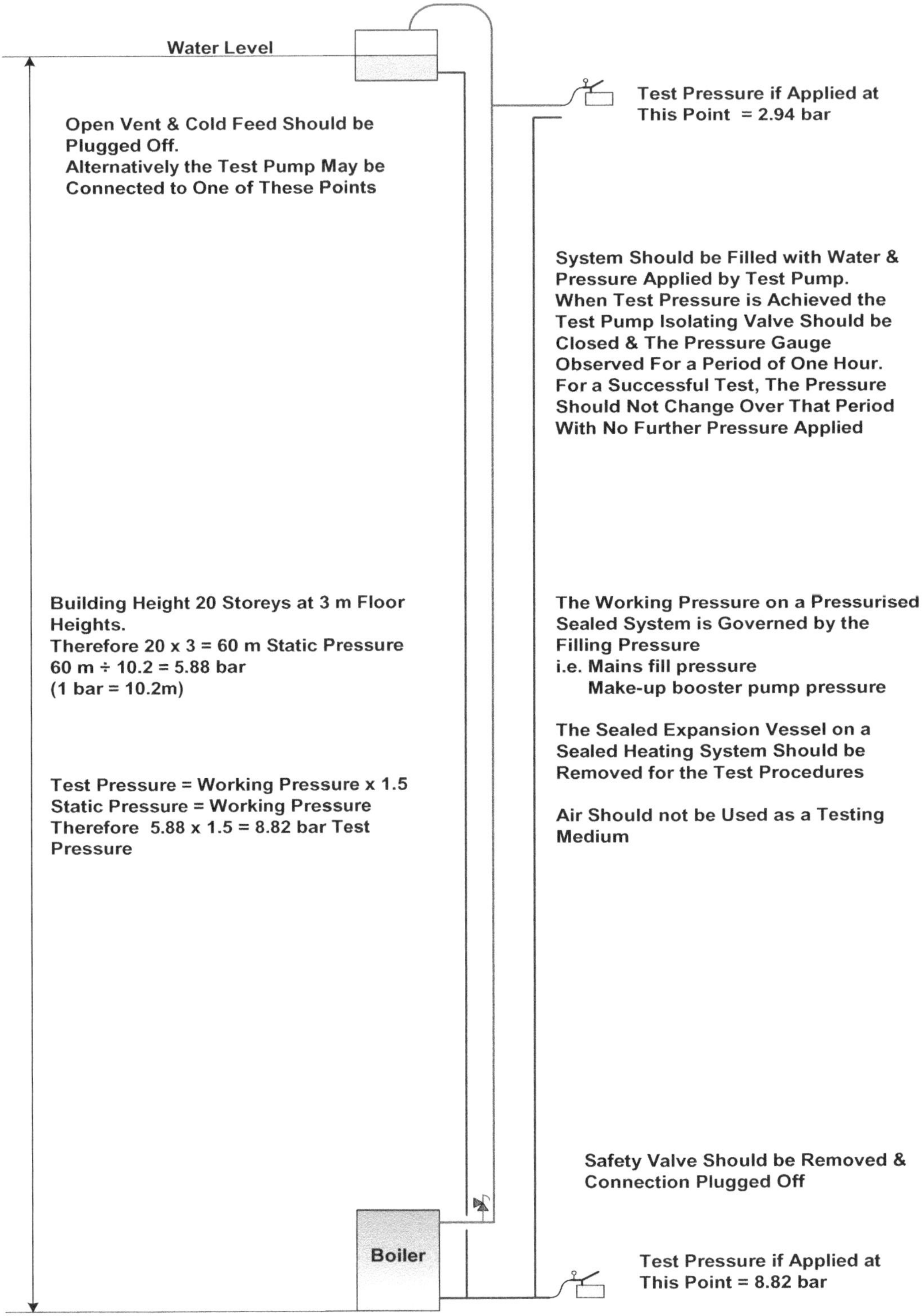

Water Level

Open Vent & Cold Feed Should be Plugged Off.
Alternatively the Test Pump May be Connected to One of These Points

Test Pressure if Applied at This Point = 2.94 bar

System Should be Filled with Water & Pressure Applied by Test Pump. When Test Pressure is Achieved the Test Pump Isolating Valve Should be Closed & The Pressure Gauge Observed For a Period of One Hour. For a Successful Test, The Pressure Should Not Change Over That Period With No Further Pressure Applied

Building Height 20 Storeys at 3 m Floor Heights.
Therefore 20 x 3 = 60 m Static Pressure
60 m ÷ 10.2 = 5.88 bar
(1 bar = 10.2m)

The Working Pressure on a Pressurised Sealed System is Governed by the Filling Pressure
i.e. Mains fill pressure
**   Make-up booster pump pressure**

The Sealed Expansion Vessel on a Sealed Heating System Should be Removed for the Test Procedures

Test Pressure = Working Pressure x 1.5
Static Pressure = Working Pressure
Therefore 5.88 x 1.5 = 8.82 bar Test Pressure

Air Should not be Used as a Testing Medium

Safety Valve Should be Removed & Connection Plugged Off

Boiler

Test Pressure if Applied at This Point = 8.82 bar

Figure 25.1   Working pressure and test pressure assessment

times the working pressure should be considered, providing that the safety or integrity of the system is not compromised.

The test procedure should be applied as follows:

- On completion of the installation, the entire system should be flushed through to remove all foreign matter including surplus jointing materials and flux residuals.
- Before applying any pressure test, check that all components forming the heating system will be suitable for the test pressure to be applied, bearing in mind that standard pressed steel radiators have a maximum working pressure of 8 bar. Any part or equipment that will not withstand the test pressure to be used should be removed from the system and the gap in the pipework produced should be bridged or capped off to allow the test to commence.
- Remove all safety relief valves and seal off any open pipe ends such as open vent pipes and cold feed entry connections. Isolate or remove any sealed expansion vessels.
- The system should be filled with water and observed for obvious leakage under the static pressure of the system, although on larger systems stage testing of smaller piping sections when completed during the installation phase of the project should be considered, to avoid large damaging leaks occurring and to give a degree of confidence when the main test is applied.
- Allow the water in the system to stand for a period of time to permit the temperature of the water to stabilise with the temperature of the surrounding air.
- Apply the test pressure using the test pump and hold the pressure required for a period of one hour without applying further pressure. The pressure on the test pump gauge should be observed and should not fall during that one-hour period whereupon completion of the test may be considered successful. This test should be witnessed by the client's appointed representative who should be asked to sign the record sheet on completion of the test, adding any comments that they may wish to make. The record sheet should include details of the test pressure, time, date, duration, details of test equipment, valid calibration certificates of instruments used and names of the test engineer and client's representative.
- On completion of the test, release the test pressure, drain system down and replace all items of equipment that had been removed for the testing exercise. The system can then be refilled with water and any chemical additives or inhibitors added to the system.

At this stage any fuel lines conveying gas or oil should also be pressure tested. Low pressure gas is tested using air and recorded using a manometer calibrated in millibar. Low pressure oil lines are also tested using air at the pressure equivalent to 1.5 times the working pressure.

Figure 25.2 shows a typical manually operated test pump that comprises a small reservoir to contain water and incorporates an integral pump with hand operated lever, test isolating valve, interconnecting pressure hose and pressure gauge that can be obtained calibrated up to 100 bar. Providing that the heating system can be filled with water by mains pressure or some form of rotating machinery means, then a manual test pump is suitable as very little pumping is required to raise the pressure up to its test pressure level, as water is virtually non-compressible.

If the heating system cannot be filled as described above, then an electrically operated test pump should be used to both fill the system and raise the pressure to its test pressure level.

TESTING MEDIUM

Water is an ideal medium for pressure testing piping systems as it is readily available and virtually non-compressible, requiring very little addition to the system to raise the pressure to the test pressure. It is also a safe medium as in the event of a pipe rupture, such as a pipe joint pulling out, the release of pressure is

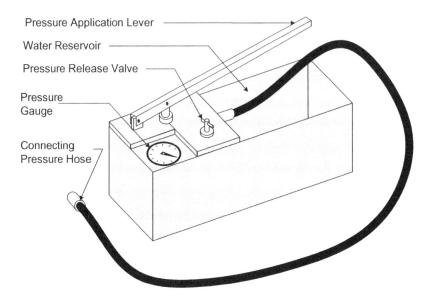

Figure 25.2 Typical manually operated test pressure pump

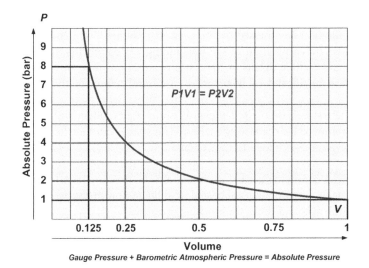

Figure 25.3 Isotherm demonstrating compressibility of air ($P_1V_1 = P_2V_2$)

immediately dispersed without causing a danger to anybody. However, this could cause damage to building finishes and furnishings, etc, and for this reason stage testing should be considered to eliminate this risk occurring on final completion of the building.

For this reason it is easy to be tempted to use air as the testing medium as it is also freely and readily available and will not cause water damage to the building fabric, but unlike water it is compressible and therefore requires the original volume to fill the system to be supplied many times over.

Figure 25.3 illustrates the compressibility of air based on Boyle's Law, $P_1V_1 = P_2V_2$, which states that if the pressure of the gas is increased the volume of the gas will decrease in direct proportion to the increase in pressure. On completion of the piping system it will already be full of air at atmospheric pressure; to double the system pressure would require an equal amount of air to be added to the system. To raise the pressure to 8.82 bar, which is the test pressure required in the example shown in Figure 25.1, would require almost nine times the original volume of air to be added to the system to achieve this pressure. Therefore if the same rupture were to occur, the blast of air escaping from the open fracture would be both prolonged and violent causing a high risk to health to anybody in the near vicinity of the fracture. The volume of water used for the same test would however remain the same as the original filled volume.

For these safety reasons, air should **not** be used as a testing medium on piping systems requiring a test pressure in excess of 0.5 bar, but in exceptional or extenuating circumstances when there is no alternative, the risk assessment and method statement should include the complete exclusion of every unauthorised person from the immediate work area until the test is complete.

TESTING METALLIC PIPING SYSTEMS

The testing of metallic piping systems is quite straightforward and is covered by BS 6700 under the requirements for rigid pipes.

This requires the piping system to be slowly filled with drinking water and all air being vented off and the test pressure applied. It also requires that where there is a >10°C temperature difference between the ambient temperature of the surrounding air and the testing water temperature, a stabilising period of 30 minutes should be allowed before commencing the test procedure. The test pressure of 1.5 times the working pressure should be held for a period of one hour without any drop in pressure occurring.

TESTING THERMOPLASTIC PIPING SYSTEMS

Generally referred to as elastomeric piping for reasons that will become clear later in this text, the properties of these systems are contributory factors that makes the testing of thermoplastic piping systems more complex than the straight-forward process explained for metallic/rigid piping materials.

Metallic/rigid piping materials remain constant in their physical dimensions when subjected to higher pressures until the pressure exerted becomes high enough to fracture or burst the pipe. Thermoplastic pipes behave differently as they are affected by both pressure and temperature change. Manufacturers of thermoplastic pipes normally state the working pressure of their piping materials at 20°C, but any increase in the operating temperature of the circulating medium will result in a dramatic decrease in the working pressure as demonstrated in the typical thermoplastic deration curve in Figure 25.4. This curve accentuates

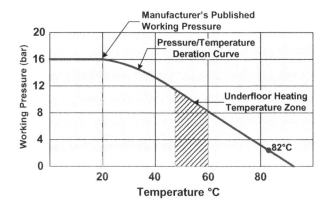

Figure 25.4 Typical pressure deration curve for thermoplastic pipes

the importance of selecting the correct piping material for the service conditions that will exist, as discussed in Chapter 3, Materials. It also emphasises the need to ensure that the test pressure does not exceed the maximum stated working pressure of the thermoplastic material being used.

Thermoplastic piping materials will also exhibit their elastomeric properties when subjected to an augmentation in pressure. They will expand in both diameter and length fractionally and return back to their original configuration when the pressure is released, although some thermoplastic materials will not return fully to their original dimensions, leaving the pipes permanently weakened by the resultant thinner wall and requiring the pressure rating of the pipe to be derated.

This elastomeric effect means that when the piping system has been pressurised to its required test pressure, the pipe will slowly expand fractionally, resulting in a slight drop of system pressure being indicated on the test equipment pressure gauge, giving the impression that there is a minor leak somewhere.

It should also be appreciated that if the test pressure is close to the pressure rating of the piping material employed, then the ultimate pressure rating of the pipe will have to be permanently derated to the lower figure.

For this reason the test procedure in common use for metallic piping systems cannot be used with any confidence when applied to thermoplastic piping systems. This has been recognised in BS 6700, which describes two acceptable methods of testing systems constructed using elastomeric piping materials.

Test procedure A

The preliminary activities prior to the hydraulic pressure test, the calculation of the required test pressure at 1.5 times the working pressure, plus the slow filling, air release and visual inspection of the piping system, are common to all hydraulic pressure tests and remain the same. On successful completion of these preliminary activities the pressure test procedure can commence as depicted in Figure 25.5.

Commence applying the pressure until the required test pressure is achieved and visually inspect the piping system for leakage during this activity. If the pressure falls as a result of the elastomeric properties of the material expanding, recommence pumping until the test pressure is restored. This initial application of pressure should continue for a period of 30 minutes, after which the test pressure should be released to lower the gauge pressure to equal 0.5 times the working pressure, if this is possible, then allow the system to remain charged at this reduced pressure for a period of 90 minutes. The pressure should remain constant or higher than this figure without further addition of pressure during this time whereupon it can then be considered as satisfactory in meeting the test criteria.

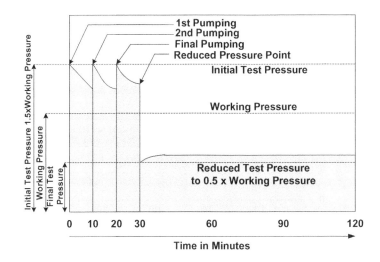

Figure 25.5   Test Procedure A for thermoplastic piping systems

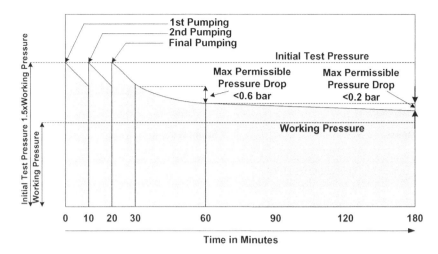

Figure 25.6 Test Procedure B for thermoplastic piping systems

Test procedure A can usually be applied successfully to pressurised sealed hydronic heating systems providing that the reduced test pressure of 0.5 times working pressure is not below the static pressure. However, this test procedure cannot be employed on fully open vented hydronic heating systems, as the working pressure is the same as the static pressure and can therefore not be reduced by 0.5 times the working pressure without part emptying the system. For systems incorporating thermoplastic piping materials, test procedure B should be applied.

Test procedure B

As with test procedure A, the preliminary test activities and preparation work remain common with all hydraulic pressure test procedures, but test procedure B departs from test procedure A by the following process as depicted in Figure 25.6.

Commence applying the pressure until the required test pressure is achieved and visually inspect the piping system for leakage during this activity. If the pressure falls as a result of the elastomeric properties of the material expanding, recommence pumping until the test pressure is restored. Continue with this initial process for a period of 30 minutes after which no further pumping should be applied and the pressure should be noted from the test pressure gauge. Allow the piping system to stagnate for a further period of 30 minutes and note the pressure reading on the test pressure gauge. If the pressure drop is less than 0.6 bar (60 kPa), the system can be considered to have no obvious leakage. Visually inspect the piping system under test for seepage at all points and continue to monitor for a period of 120 minutes. The test criteria are met if a further pressure drop in the piping system does not exceed 0.2 bar (20 kPa).

TESTING SYSTEMS COMPRISING MIXED RIGID AND ELASTOMERIC PIPING MATERIALS

BS 6700 permits test procedures A and B to be used to hydraulically pressure test hydronic heating systems constructed with a mixture of rigid metallic and elastomeric thermoplastic piping materials. However, a more reliable procedure is to disconnect the thermoplastic piping sections from the metallic pipework and test them separately, using the procedures described for each material type.

Figure 25.7 reproduces a typical hydraulic pressure test record sheet that can be used for recording the results of a pressure test.

Medium - Hydraulic

Project/Contract:	
Location and Service:	

Attendance

CRE Engineer:		Date:	
Assisted by:		Time:	
Witness:		Representing:	

Details of Test, and Test Requirements:

Test Pressure (bar): at Start		Start Time	
Test Pressure (bar): at Finish		Finish Time	
Start Temperature (°C):			
Finish Temperature (°C):			

Results:

Reference Documents:

Test Instruments and Number	Calibration Date
Hydraulic hand test pump	

This is to certify that I have witnessed the test carried out on the system recorded and found it to be satisfactory.

Name	Signature	Date
CRE Engineer		
Comments		
Witness		
Comments		

Figure 25.7 Mechanical systems Pressure Test Record Certificate

PRE-COMMISSIONING ACTIVITIES

Following the completion of a successful hydraulic pressure test – including the pre-test activity of flushing the system to clean it, refilling it with water and the post-testing procedure of completing the record sheet of the test – the system can then be protected by adding a corrosion inhibitor to the circulatory water as described in Chapter 21, Water Treatment.

At this stage the various pre-commissioning activities can be undertaken that are required to prove that individual items will perform their function when called upon as part of the larger control system.

Therefore the following functional checks should be done:

- Check the continuity of all control cabling including that the correct connection from point to point has been made.
- Check the manual and automatic operation of all control valves together with any manual override facilities.
- Check the operation of any thermostatic control equipment including the manual operation of any temperature adjustment provision.
- Check that the correct fuel rate is being supplied to the heat raising plant and adjust all associated control components.
- Check the operation of all circulating pumps and adjust the speed operation to suit the designed discharge output setting.

COMMISSIONING THE HEAT RAISING PLANT

The boiler or heat raising plant should be commissioned in accordance with the manufacturer's instructions and as generally described in the individual chapters on the various fuels.

The commissioning should include adjusting the fuel supply to the burner and setting the burner to provide the correct heat output to match the heat requirements calculated at the design stage. The combustion efficiency and discharge should also be checked to ensure that the boiler and associated equipment are functioning correctly as well as safely and efficiently.

COMMISSIONING THE CIRCULATORY SYSTEM

This volume has purposely worked through the design of hydronic heating systems, and included exercises on calculating heat losses from buildings to arrive at the correct heat output of the heat generating plant, plus the selection and accurate sizing of heat emitters, together with the location of the heat emitters. It has also discussed the process of calculating suitable pipe sizes required to distribute the heat load from the centrally located heat generating plant at the correct proportional amount to each heat emitter. The pipe sizing exercise also produced the requisite duty of the circulating pump or pumps needed to force the medium through the piping system and deliver the proportioned calculated heat load. It has also described and explained the various control schemes to suit the heating systems' design criteria and discussed the multitude of components that go to make the system complete.

Having installed all of these components and items of equipment to form the completed hydronic heating system, the scheme has to be set up and made to perform as the design intended.

This chapter has discussed the importance of commissioning the boiler plant and control scheme equipment, but the distribution of heat must be proportionally regulated or the system will not function correctly or efficiently and the heating installation will be considered a failure.

This proportional regulation of heat distribution is termed 'hydronically balancing', which is the important task of arranging the flow of heat to be delivered correctly proportioned for the system to achieve its

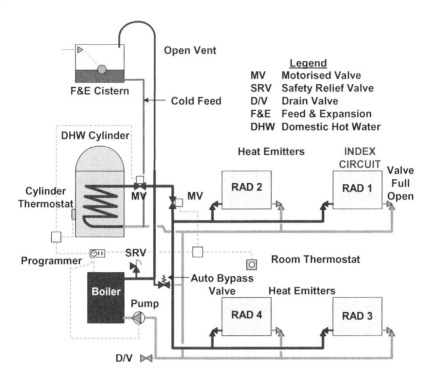

Figure 25.8   Example of hydronically balancing a simple heating system

design performance criteria, and is required to be applied on all heating installations, from small simple domestic systems to large complex commercial schemes without exception.

This task is a straightforward activity on small simple domestic hydronic heating schemes without any additional regulatory equipment and which relies only on the balancing properties of the heat emitters' return valves.

Figure 25.8 illustrates a small domestic two-pipe direct return heating scheme with heat emitter RAD 1 identified as the index circuit; i.e. the circuit with the greatest amount of frictional resistance acting on the flow. To balance this system the heat emitter return valve should be operated to be fully open – offering the minimum resistance to the flow of water through it – with the return valves on the remaining heat emitters proportionally closed.

On a system as simple as the one illustrated in Figure 25.8, this balancing activity of regulating the correct proportion of flow to each heat emitter can be done by trial and error, or by attaching a clip-on thermometer or the probe from an electronic thermometer to regulate the return water temperature from each heat emitter to match the system design return temperature. If the correct design return water temperature is achieved from each heat emitter, then the correct proportion of flow is being circulated through it and the exact amount of heat will be emitted.

Larger heating systems require individual circuits, including large heat load requirements such as the domestic hot water storage calorifier, to be regulated for the resistance to the flow to each circuit to be equal to the index circuit, as indicated in Figure 25.9. This is normally achieved by the use of double regulating valves – see Chapter 3 for details of these valves – where the valve can be regulated to create the resistance equal to that calculated at the pipe sizing design stage so that the object of pressure drop equalisation of each circuit is accomplished.

During the pipe sizing exercise the information in the form of data collated should be recorded, preferably on a schematic drawing of the scheme as shown in Figure 25.10.

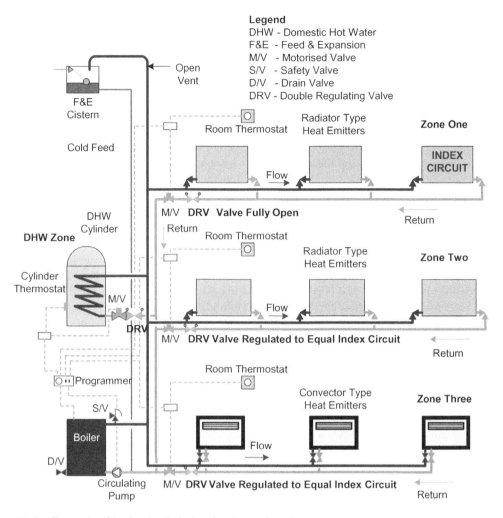

Legend
DHW - Domestic Hot Water
F&E - Feed & Expansion
M/V - Motorised Valve
S/V - Safety Valve
D/V - Drain Valve
DRV - Double Regulating Valve

Figure 25.9   Example of hydronically balancing larger heating system

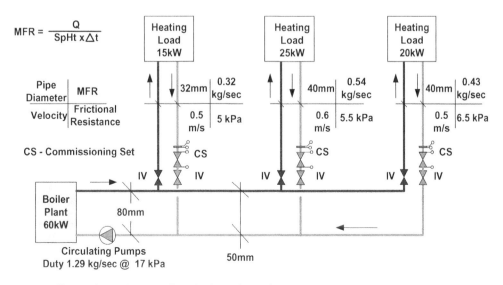

Figure 25.10   Part schematic recording design criteria for commissioning

The schematic drawing demonstrates how the design criteria resulting from the pipe sizing calculations are recorded, where the heating load of each circuit is converted from watts into a mass flow rate expressed in kilograms per second, based on a flow temperature of 82° and a return temperature of 71°. This information now permits the task of pipe sizing to commence and the pipe diameter, mass flow rate, velocity and pressure drop can be recorded for each section or pipe circuit.

The part schematic drawing reproduced in Figure 25.10 indicates the required provision of commissioning sets located on the return at the base of each riser together with isolating valves. These commissioning valve sets are required to commission the heating system by regulating the flow through each piping circuit to match the design figures recorded on the schematic drawing.

The commissioning valve sets indicated on the schematic drawing comprise an oblique pattern double regulating type valve coupled with a metering plate upstream of the DRV. This combination is sometimes referred to as an orifice valve.

This commissioning valve arrangement has been detailed in Figure 25.11, where the combination of these two components enables the commissioning engineer to establish and regulate both the flow rate passing through the valves and the pressure drop encountered. This is achieved by recording the pressures and pressure drop across the two test ports on the metering plate, where the results can then be referenced against the valve manufacturer's flow chart and the corresponding flow rate read off. The flow rate together with the pressure drop across the double regulating valve can then be adjusted to suit the designed flow rate and resistance for each piping circuit. This is also achieved by cross-referencing the pressure drop recordings obtained from the manometer plugged into the test points with the manufacturer's flow charts.

The metering station orifice plate should be installed upstream of the double regulating valve, clamped between two mating flanges and arranged to have a length of straight pipe equivalent to a minimum of five times the pipeline diameter upstream of it.

The commissioning valve settings should be recorded and incorporated in the project's operating and maintenance manual to enable future servicing to be undertaken using the design criteria and original commissioning data. Any future alterations to these original valve settings should be recorded as an amendment to the operating and maintenance manual data.

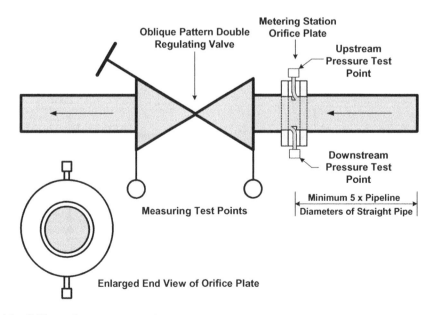

Figure 25.11   Orifice valve arrangement

OPERATING AND MAINTENANCE REQUIREMENTS

The final obligation required to be completed before handover of the system is to compile the operating and maintenance manual and provide any training to the client's representatives or, in the case of domestic heating systems, demonstrate the system to the building's occupier.

The provision of operating and maintenance manuals has always been a requirement for heating systems installed in commercial properties, but this has not always been complied with for domestic heating installations. It is now a mandatory requirement with Part L1 of the Building Regulations that all heating systems including domestic installations are provided with applicable operating and maintenance instructions.

The size of these operating and maintenance instructions will be dependent upon the complexity of the heating system, as a heating system installed in a large complex commercial type building or development will probably require a manual comprising several volumes, whereas the operating and maintenance manual provided for a domestic heating installation will only amount to a few pages of relevant information.

A properly compiled operating and maintenance manual should include the following information and data:

- A fully detailed description of the system including the fundamental operating principles.
- The operating design criteria regarding flow rates and pressure drops recorded during the commissioning stage. The settings for safety valves, thermostats, programmers and any other equipment that has been set up during the commissioning.
- Duties of all equipment including boiler ratings and circulating pump settings together with a schedule of all heat emitter heat emissions. A schedule of capacities for expansion vessels, storage cylinders and feed and expansion cisterns.
- Full details of the fuel being used including the class and characteristics if applicable. The details should include a full description of the fuel storage and/or supply system.
- Specification information on all materials used to construct the heating installation. This should include details of pipes, fittings, valves and all other materials.
- The operating and maintenance manual should also include the contents of the safety file detailing how the scheme was constructed so that during decommissioning the system may be dismantled safely.
- Comprehensive list of all manufacturers of the equipment used in the installation together with their addresses, contact names and order numbers or references used to enable the equipment supplied to be identified.
- Detailed full copies of all relevant manufacturers' installation, operating, maintenance and servicing instructions that include details of obtaining spare parts and any consumable items.
- List of recommended spares and full details to enable the client to be able to obtain them.
- Detailed schedule of all servicing requirements together with recommended inspection periods for all equipment and materials used. It should be appreciated that the complete heating system needs to be serviced, not just the boiler and fuel supply system.
- Complete set of all record, or as-built, drawings to enable the system to be understood by service personnel or by engineers involved in any refurbishment or alteration works.

The above list of operating and maintenance manual requirements might appear to be daunting when applied to a domestic heating system, but a scaled-down version to provide all this information with a realistic and practical set of instructions should not take more than a few pages.

26

Operating Costs and Whole Life Costing

The cost of installing a heating system – regardless of whether it is a large scheme for a commercial development, or replacing an existing system in a small domestic dwelling – is a proportionally large long-term capital investment, one which should take into account the expected life of the building, or projected life of the building before it requires a major refurbishment. Therefore, to enable the scheme to be appropriately evaluated, an accurate assessment of the annual operating costs should be calculated and included as part of the estimated costs submitted with the quotation for consideration.

The cost of operation of any system providing space heating will depend upon a number of variables that include the following:

- The calorific value of the proposed fuel to be used together with the current unit rate costs being charged for that fuel, or fuels if dual fuels are being proposed. Comparison costs for different fuels may be used in the long-term calculations and compared to the installation costs as part of the evaluation process of the quotations received.
- The overall efficiency of the heat raising plant together with the energy consuming requirement of the proposed heating system.
- The designed heating load for the heating system proposed that has been arrived at from the heat loss calculations.
- The expected annual period of the heating season expressed in days for the geographical location of the property to be heated. This can be obtained from a number of published sources based on normal or expected external ambient temperature conditions for the location concerned.
- The normal severity of the heating season based on local experience and records for previous years, noting average external air temperatures experienced.
- The proposed operation of the heating system including the number of hours per day together with the number of days per week if the building concerned is a commercial building.
- The effectiveness of the heating control system including the proposed operation of different heating zones.

Heating Services in Buildings: Design, Installation, Commissioning & Maintenance, First Edition. David E. Watkins.
© 2011 John Wiley & Sons, Ltd. Published 2011 by John Wiley & Sons, Ltd.
As an aid to lecturers and students, full colour versions of the figures in this chapter may be found at www.wiley.com/go/watkins
These figures are © 2011 by John Wiley & Sons, Ltd.

Table 26.1 Comparison of net calorific values of common fuels

Fuel	CV in MJ/kg	CV in MJ/m^3
Butane	45.8	112.9
Propane	46.3	86.1
Natural Gas	–	34.9
Kerosene	43.6	–
Gas Oil	42.7	–

CALORIFIC VALUE

The calorific values for various fuels have been established in the chapters devoted to the particular fuels and have been reproduced as a summary in Table 26.1 for comparison.

The calorific value for individual fuels will have an important influence on the choice of fuel to be selected when designing the heating system, together with the choice of fuels available, both in the short term as well as the long-term availability, within the life expectancy of the building. The calorific value of each fuel is one of the essential factors in determining the running costs of heating systems as it permits the monetary costs for each fuel to be compared.

SYSTEM EFFICIENCY

The efficiency of the heating system is not just confined to the combustion burning heat raising plant, but also includes the arrangement of the control philosophy and its effectiveness in preventing overheating of the building and wastage of heat, which will also result in excess fuel being consumed.

The operating efficiency of fossil fuel burning appliances has improved considerably over recent years since the introduction of the SEDBUK rating system for domestic heating boilers. The UK's largest consumption of fossil fuel is now the combined consumption from domestic residential dwellings, and the UK Government's conservation strategy seeks, through the SEDBUK rating system, to reduce this consumption figure, together with reducing the carbon emission from the combustion of this fuel into the atmosphere.

However, it should be appreciated that although some boiler manufacturers claim efficiencies in the high 90 percents and some even claim efficiencies in excess of 100%, as discussed in Chapter 16 under the subject heading of condensing boilers, the boiler will only operate at these efficiencies in the SEDBUK 'A' rating range if the heating system has been designed to permit condensing of the flue gases if the return water temperature back to the boiler is low enough to allow condensation to occur. Therefore, when applying an appliance efficiency figure to any operating cost calculation, the true boiler operating efficiency must be used. This can be assessed by calculating the percentage of operating time when the boiler is condensing and the percentage of time when it is operating in a non-condensing mode.

Auxiliary operating efficiencies

In addition to the boiler fuel consumption, an allowance must be made for operating any electrically powered equipment such as circulating pumps, pressurisation equipment, domestic hot water heat maintenance tape and control system.

The effectiveness of the control system should also be assessed and improved if necessary to ensure that the heated space is not overheated during milder weather conditions, and that all combustion burning appliances and auxiliary equipment are interlocked to turn off when not required. If this provision has not been incorporated then the heating system's efficiency will be reduced accordingly.

HEATING LOAD

This will have been arrived at through the heat loss calculations conducted for the building as demonstrated in Chapter 7, Heat Requirements of Buildings, together with any heating load requirement for the provision of domestic hot water as established in Chapter 19.

HEATING SEASON/OPERATING PERIOD

To enable the operating costs of a heating system to be assessed, it will be necessary to establish what the proposed operating period for each day will be and the number of days per week that it will be operating for the building concerned, plus the expected normal heating season for the locality of the building where the heating system will be installed.

For a commercial building such as an office block it will be a simple task of establishing the period of time each day that the building will be occupied, together with the number of days per week. For many office buildings where the normal working hours are from 09:00 to 17:00, Monday to Friday, the operating hours would be:

$$17:00 - 09:00 = 8 \text{ hours/day} \times 5 \text{ days per week} = 40 \text{ hours per week operating time.}$$

For other commercial buildings and establishments such as hotels, hospitals, etc, the building will be occupied 24 hours per day, 7 days per week.

For many residential dwellings the method of assessing the expected operating times and heating periods is slightly more complex. Some dwellings will have a normal occupancy of 24 hours per day, requiring heating most of the day with either the heating system programmed to be turned off overnight, or arranged to have a temperature set-back during the early hours of the morning. In other residential dwellings the heating will be set to turned on for two hours in the morning, and programmed to come back on for a longer period in the evening. If the expected use of the dwelling is not known, it is normal to allow for a total combined period of operating time of 8 hours per day for 7 days per week.

The length of the heating season will be dependent upon the locality of the building being heated. It is reasonable to expect that the heating season in the north of the UK is longer than the heating season in the south; also, dwellings constructed in locations of higher altitudes or exposed geographical situations would be expected to have a slightly longer heating season than those constructed in sheltered exposures in towns and cities having lower altitudes.

The actual length of the heating system could vary from one year to the next, but for design purposes and estimating the operating costs, the periods listed in Table 26.2 may be used for calculating the running costs for various regions in the UK.

Table 26.2 Heating season and durations for normal locations

Area	Season	Weeks	Days
Southern England	October to April	30	210
Wales and Midlands	October to mid April	32	224
Northern England	October to May	34	238
Northern Ireland	Mid September to May	36	252
Southern Scotland	Mid September to May	36	252
Northern Scotland	September to May	38 to 40	266/280
Eire	October to mid April	32	224

Certain high-altitude or exposed locations will have an extended heating season.

The normal heating seasons recorded in Table 26.2 for various geographical areas or districts throughout the UK run through the seasons from autumn through to winter and on to spring, with the coldest part of the heating season being in the middle of the term with less severe periods occurring at the start and finish of the heating season.

For this reason it would not be correct to expect the full load of the system's heating capacity to be required every day and every hour of each day over the full period of the heating system, as the external ambient temperature will vary considerably. Also, the external ambient temperature will vary from the night-time temperatures compared to the day-time hours temperatures and if the full heating load were to be used for calculating the operating costs, an unrealistically high figure would result.

A method of compensating for these diverse weather conditions which will enable the heating system's operating costs to be calculated more accurately, is the use of the degree-day method.

DEGREE-DAYS

Degree-days have been used for many years as a means of comparing external ambient temperatures over different periods of the year. These are either British cooling degree-days, used to compare external temperatures during the summer months and employed for air conditioning cooling requirements and cost calculations, or more relevantly for this work, British heating degree-days which have proved to be a useful means for monitoring heating energy consumption for buildings during the heating season/winter months.

The British heating degree-day unit is not an accurate method of determining seasonal operating costs, but of all the means that have been devised for assessing energy costs this method has proved to be the most useful for comparative purposes for winter periods based on the average temperatures experienced over the last 20 years.

The basis of the heating degree-day method assumes that when a building is maintained at an internal temperature of 18°C and the external ambient temperature is 15.5°C or above, no heating to the building will be required due to the 2.5°C natural temperature difference provided by the insulating properties of the building structure and fabric. The difference between the daily mean temperature and 15.5°C, provided that the daily mean temperature is below 15.5°C, is then taken as the number of degree-days for the day in question. For example, if the average external daily temperature over a 24-hour period amounts to 5.5°C, taking into account the daily temperature gradient from the highest temperature during the day and the lowest temperature occurring overnight, there would be 10 degree-days totalled for that day (15.5 − 5.5 = 10); if this average temperature were to remain constant for 7 days then 70 degree-days would be accumulated over this period. These daily degree-day figures would be totalled for the year and recorded as the annual degree-days for the location recorded.

Degree-day data is recorded and compiled monthly by the Meteorological Office for various locations throughout the UK and is published in various industry publications as well as being available on a variety of websites. CIBSE Guide A provides comprehensive month-by-month data on degree-days.

Table 26.3 gives the annual mean totals that have been averaged over the last 15 years for various UK locations. But for applications other than domestic heating systems where a definite specified heating season exists that is less than a year, refer to CIBSE Guide A for individual mean monthly degree-days that can be totalled for the months concerned, which would result in a lower degree-day total for the purpose of the annual energy consumed calculation.

ANNUAL ENERGY CONSUMED

Before any monetary value can be assigned to the annual running cost, the energy consumed must be calculated based on the calorific value of the fuel to be used, plus the efficiency of the heat raising appliance

Table 26.3　Annual degree-days for base external temperature of 15.5°C

UK Region or Weather Station	Annual Degree-Days
Thames Valley (Heathrow)	2034
South Eastern (Gatwick)	2255
Southern (Hurn)	2224
South Western (Plymouth)	1860
Severn Valley (Filton)	1836
Midlands (Elmdon)	2426
West Pennines (Ringway)	2230
North Western (Carlisle)	2390
Borders (Boulmer)	2482
North Eastern (Leeming)	2370
East Pennines (Finningley)	2308
East Anglia (Honington)	2254
West Scotland (Abbotsinch)	2492
East Scotland (Leuchars)	2576
North East Scotland (Dyce)	2668
Wales (Aberporth)	2160
Northern Ireland (Aldergrove)	2360
North West Scotland (Stornoway)	2670

and the heating system, the calculated heat load taken from the heat loss calculations exercise that has to be achieved to satisfy the designed comfort conditions, the planned duration of the heating season if it is a non-domestic heating application, and the total number of degree-days for the geographical location concerned. Once these factors have been established and the energy consumed arrived at, they should remain constant throughout the life of the heating system, providing that no alterations or additions have been made. However, fiscal cost of the fossil fuel is likely to change on numerous occasions, being dependent upon unstable world political conditions affecting the availability and the price of the fuel and hence the annual operating cost. The cost of non-fossil sustainable fuels will be dependent upon their continued development and the ability to provide sufficient quantities to satisfy world demand.

The annual energy consumption may be established by the following equation:

Annual energy consumption

$$Qe = \frac{D \times H \times 3600}{1\,000\,000}$$

Where

Qe = Energy consumed per annum expressed in GJ
D　= Degree-days for specific geographic location
H　= Calculated heat loss for building expressed in kW

For example, consider a building located in the South Eastern geographic area of the UK having a calculated heat loss of 25 kW.

Referring to Table 26.3, the annual degree-days recorded for the South Eastern geographic area of the UK is 2255. If the building concerned is a domestic residential dwelling operating under intermittent control then 2255 degree-days may be used without any correction in the equation. Conversely, if the building in question is a commercial property operating with a specific heating season, the degree-days may be reduced to suit the exact heating season which may be obtained from CIBSE Guide A for the particular months of the heating season. For example, the heating season for the South Eastern geographic area of the UK would be from October to April where the degree-day total would amount to about 1900, which would result in a lower annual energy consumption. This does not apply to domestic residential dwellings where the propensity is to operate the heating system on an as-and-when-required basis, depending on the external ambient temperatures occurring.

For example:

$$\frac{2255 \times 25 \times 3600}{1\ 000\ 000} = 202.95 \text{ GJ/annum}$$

Using a degree-day total of 1900 in the equation for a commercial type building application, the annual energy consumption would be reduced to 171 GJ per annum.

If the reduced degree-day figure is being used for a specific heating season, but the heat raising plant is still being employed to provide the primary energy source for raising the domestic hot water temperature, then the energy required for this period of use must be calculated and added to the annual energy consumption. The same equation can be used for this purpose, but the degree-day total is replaced by the hours per day that the boiler will be required to provide the domestic hot water and totalled for the number of weeks for this summer period between the heating seasons. The heat load will be the amount of energy that will be required only to raise the domestic hot water temperature.

ANNUAL FUEL UTILISATION

Having calculated the annual energy consumption required to meet the heating and domestic hot water demands for the building, then the quantity of fuel that equates to this annual energy consumption may be assessed by the following formula:

Example for Natural Gas

$$Qf = \frac{E \times 1\ 000\ 000}{Cv \times 1000 \times Se}$$

Where

Qf = Annual fuel consumption expressed in GJ
Cv = Calorific value of fuel expressed in MJ m^{-3} from Table 26.1
Se = System efficiency expressed as a percentage

The efficiency of the appliance and the overall efficiency of the heating system must be assessed based on how the system is to be operated together with the efficiency of the appliance corrected to suit its operational use.

Continuing with the previous example with the annual energy consumption rounded up to 203 GJ:

$$\frac{203 \times 1\ 000\ 000}{34.9 \times 1000 \times 0.80\%} = 7270 \text{ m}^3/\text{year}$$

Applying the fuel cost tariff per cubic metre to the total annual fuel consumption:

$$7270 \text{ m}^3 \times £0.33 = £2399.00 \text{ annual fuel operating cost.}$$

The fuel cost tariff per cubic metre of natural gas has been chosen for example purposes only; the actual fuel cost tariff per cubic metre being charged by the energy provider concerned at the time of calculation should be used for the purpose of calculating the annual running costs.

Natural gas is charged in kWh as a unit of energy. To convert from cubic metres the following equation may be applied:

$$kWh = \frac{m^3 \times Cv\ MJ/m^3}{3.6\ kWh\ MJ^{-1}}$$

Example for Class C2 Kerosene fuel oil

$$Qf = \frac{E \times 1\ 000\ 000}{Cv \times 1000 \times SG \times Se}$$

Where

Qf = Annual fuel consumption expressed in GJ
Cv = Calorific value of fuel expressed in MJ kg^{-1} from Table 26.1
SG = Specific gravity of fuel taken from Table 12.1
Se = System efficiency expressed as a percentage

$$\frac{203 \times 1\ 000\ 000}{43.6 \times 1000 \times 0.79 \times 0.80\%} = 7367\ litres/year$$

As with the example of natural gas fuel consumption, the actual fuel cost tariff per litre of kerosene fuel oil being charged by the fuel delivery company at the time of calculation should be used for the purpose of calculating the annual running costs.

The actual cost of all fuels or energy can be unstable for the reasons briefly discussed earlier and will vary considerably; they should be checked each time before use.

ANNUAL AUXILIARIES RUNNING COSTS

The bulk of the annual operating cost will be that associated with the fuel consumption as demonstrated earlier, but all heating systems will require some degree of electrical energy to drive circulating pumps and operate motorised valves as well as general control equipment.

The electrical energy being consumed by these auxiliaries should also be calculated and added to the fuel consumption running costs; this is undertaken by listing all the electrically powered auxiliary equipment together with the electrical power being consumed in watts or kilowatts by each item, which is then totalled and multiplied by the estimated number of hours operating per year. The total will then be a true estimated running cost for the heating system.

WHOLE LIFE COSTING

For many years the necessity of conducting an operating cost exercise for a proposed new heating system has been an essential part of the tender submission to enable a true evaluation to be undertaken and comparisons between schemes offered.

This requirement remains an important aspect of any design and estimate for a proposed heating system as it is not unreasonable for the client or prospective owner to want to know how much it will cost to run the system each year. This is just as applicable to heating schemes for small domestic residential dwellings as it is for heating systems proposed for large commercial buildings.

More recently in today's more environmentally aware world, the practice of contractor submitting a compliant tender that just includes a price for the supply and installation of a proposed heating scheme, with or without a breakdown of costs, together with a detailed running cost analysis, is no longer considered sufficient to evaluate the true cost of any design that is being offered.

It is becoming increasingly common for clients or building owners who are providing the development expenditure to ask for detailed cost of ownership over the life of the building, or asset. This is known as 'whole life costing' which is sometimes referred to as the costs associated with the building from the 'cradle to the grave'.

Whole life costing is applied to the entire building from the structure and its fabric to the services installed within the building, including the heating system.

The building may have been designed to have a normal life expectancy of anything from 40 years to in excess of 100 years, where major refurbishments would have been planned at set periods of the building's life. As part of the overall building design it would be a requirement to identify these renovation periods and, so far as the heating system is concerned, this exercise would be performed as part of the whole life costing application.

The whole life costing procedure would include the following key points:

- *Design cost* – The cost associated with incorporating whole life principles and analysis would be included in the design phase of the project.
- *Installation cost* – The tendered cost for supplying all materials and installing them in compliance with the design requirements, which includes details and cost for the following items.
- *Annual running cost* – The heating system annual energy consumption calculation based upon the set geographic parameters previously described. The calculation data used should be disclosed with the cost submitted.
- *Servicing costs* – A detailed list of annual and periodic servicing requirements for all plant and equipment. The list should incorporate servicing recommendations and the time periods between servicing and inspections. Costs for servicing the system should accompany this information. This information should also be incorporated in the operating and maintenance manuals.
- *Life expectancy* – The normal life expectancy based on the design usage for all equipment and materials should be declared to enable major refurbishment periods to be planned. The replacement cost based on current prices for any equipment and any other materials that are expected to fail within the planned lifetime of the building should be divulged. The life expectancy of particular equipment may have an influence when selecting the equipment at the design phase; for example, a boiler that has a life expectancy of 20 years may be more attractive than a lower cost boiler that has a life expectancy of only 12 years.
- *Decommissioning cost* – The final part of the whole life costing exercise is to establish the cost associated with decommissioning the heating system in a safe and environmentally friendly manner. The method of decommissioning and dismantling the system should be recorded in the health and safety file handed over to the building operator on completion of the project. The requirements for the disposal of the system materials should also be included as part of this exercise, as this aspect could have an influence on the selection of the materials at the design and construction phase. Some materials may be competitively priced when purchased, but their environmental impact in the time required to degrade, or cost associated in reclaiming these materials may be excessively high if this cost has to be borne by the building owner.

The practice of whole life costing is normally applied in full only to major projects at present, but a simplified version is becoming increasingly common for lesser developments, with a variation of this procedure being used to evaluate tender submissions at the estimate stage of the project. The information given by each contractor in their compliant quotation is scheduled for comparison and totalled over a specified design life period and the resulting cost can be a deciding factor in selecting the successful bidder. This method can often show that the lowest installed cost bid is not the cheapest when costs are considered over the life expectancy of the building and the heating system.

Appendices

Heating Services in Buildings: Design, Installation, Commissioning & Maintenance, First Edition. David E. Watkins.
© 2011 John Wiley & Sons, Ltd. Published 2011 by John Wiley & Sons, Ltd.
As an aid to lecturers and students, full colour versions of the figures in this chapter may be found at www.wiley.com/go/watkins
These figures are © 2011 by John Wiley & Sons, Ltd.

APPENDIX 1 COMPARATIVE TABLE OF SHEET METAL GAUGES

(Thickness in mm)

GAUGE No	IMPERIAL STANDARD WIRE GAUGE S.W.G.	BIRMINGHAM WIRE GAUGE or STUBS GAUGE B.W.G.	BIRMINGHAM METAL GAUGE B.M.G.	BIRMINGHAM GAUGE (SHEET) B.G.	FRENCH or PARIS GAUGE P.G.	BROWN & SHARP GAUGE B&S	U.S. STEEL WIRE GAUGE U.S.W.G.	ZINC GAUGE Z.G.
7/0	12.7	–	–	16.9	–	–	12.45	–
6/0	11.8	–	–	15.9	5P 0.3	14.73	11.72	–
5/0	11	12.7	–	14.9	4P 0.34	13.1	10.93	–
4/0	10.2	11.53	–	13.8	3P 0.38	11.68	10	–
3/0	9.5	10.8	0.13	12.7	2P 0.42	10.4	9.2	–
2/0	8.9	9.65	0.15	11.3	P 0.5	9.27	8.4	–
1/0	8.23	8.64	0.18	10	0.6	8.25	7.78	0.1
1	7.6	7.62	0.2	9	0.7	7.35	7.19	0.15
2	7	7.21	0.23	8	0.8	6.54	6.67	0.178
3	6.4	6.58	0.25	7.1	0.9	5.82	6.19	0.2
4	5.9	6	0.3	6.4	1	5.19	5.72	0.25
5	5.4	5.59	0.35	5.7	1.1	4.62	5.26	0.28
6	4.87	5.16	0.4	5	1.2	4.11	4.88	0.33
7	4.5	4.57	0.48	4.5	1.3	3.67	4.5	0.38
8	4	4.19	0.53	4	1.4	3.26	4.11	0.46
9	3.65	3.76	0.58	3.55	1.5	2.9	3.77	0.5
10	3.25	3.4	0.69	3.2	1.6	2.59	3.43	0.58
11	2.95	3	0.79	2.83	1.8	2.3	3.06	0.58
12	2.64	2.77	0.89	2.52	2	2.05	2.68	0.66
13	2.33	2.41	0.97	2.24	2.2	1.83	2.32	0.74
14	2.03	2.1	1.07	2	2.4	1.63	2.03	0.81
15	1.83	1.83	1.19	1.78	2.7	1.45	1.83	0.96
16	1.62	1.65	1.3	1.59	3	1.29	1.59	1.09
17	1.42	1.47	1.4	1.41		1.15	1.37	1.22

18	1.22	1.24	1.52	1.26	3.4	1.02	1.2	1.35
19	1	1.07	1.6	1.12	3.9	0.91	1.04	1.47
20	0.92	0.89	1.65	1	4.4	0.81	0.88	1.6
21	0.81	0.81	1.73	0.89	4.9	0.72	0.8	1.78
22	0.71	0.76	1.83	0.79	5.4	0.64	0.73	1.95
23	0.61	0.64	1.96	0.71	5.9	0.57	0.66	2.13
24	0.56	0.56	2.08	0.63	6.4	0.51	0.58	2.31
25	0.5	0.5	2.29	0.56	7	0.45	0.52	2.49
26	0.46	0.46	2.54	0.5	7.6	0.4	0.46	2.67
27	0.42	0.4	2.84	0.44	8.2	0.36	0.44	–
28	0.38	0.36	3.15	0.4	8.8	0.32	0.41	–
29	0.35	0.33	3.45	0.35	9.4	0.29	0.38	–
30	0.32	0.3	3.81	0.31	10	0.25	0.36	–
31	0.3	0.25	4.22	0.28	–	0.23	0.34	–
32	0.27	0.23	4.62	0.25	–	0.2	0.33	–
33	0.25	0.2	5.08	0.22	–	0.18	0.3	–
34	0.23	0.18	5.49	0.2	–	0.16	0.26	–
35	0.21	0.13	6.05	0.18	–	0.14	0.24	–
36	0.19	0.1	6.35	0.15	–	0.13	0.23	–
37	0.17	–	6.86	0.14	–	0.11	0.22	–
38	0.15	–	7.06	0.12	–	0.1	0.2	–
39	0.13	–	7.34	0.11	–	0.089	0.19	–
40	0.12	–	7.62	0.1	–	0.076	0.18	–
41	0.11	–	–	0.09	–	–	–	–
42	0.1	–	–	0.08	–	–	–	–
43	0.09	–	–	0.07	–	–	–	–
44	0.08	–	–	0.06	–	–	–	–
45	0.07	–	–	0.055	–	–	–	–
46	0.06	–	–	0.05	–	–	–	–
47	0.05	–	–	0.045	–	–	–	–
48	0.04	–	–	0.04	–	–	–	–
49	0.03	–	–	0.035	–	–	–	–
50	0.025	–	–	0.03	–	–	–	–

APPENDIX 2   TEMPERATURE COMPARISON AT ATMOSPHERIC PRESSURE

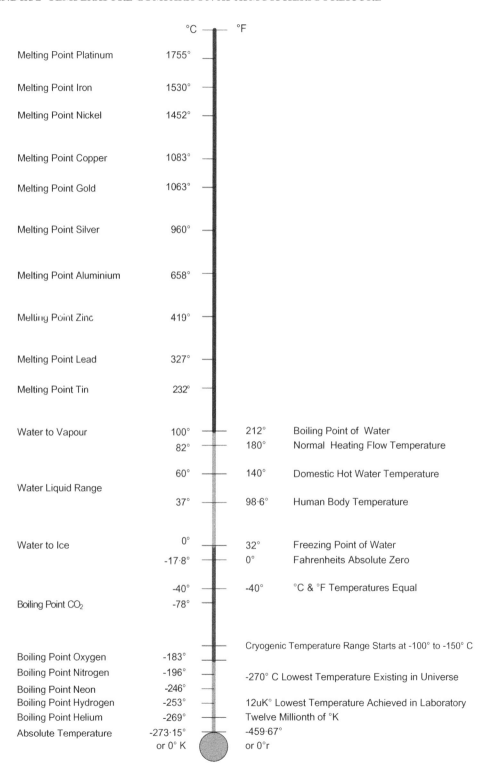

	°C		°F	
Melting Point Platinum	1755°			
Melting Point Iron	1530°			
Melting Point Nickel	1452°			
Melting Point Copper	1083°			
Melting Point Gold	1063°			
Melting Point Silver	960°			
Melting Point Aluminium	658°			
Melting Point Zinc	419°			
Melting Point Lead	327°			
Melting Point Tin	232°			
Water to Vapour	100°		212°	Boiling Point of Water
	82°		180°	Normal Heating Flow Temperature
	60°		140°	Domestic Hot Water Temperature
Water Liquid Range				
	37°		98·6°	Human Body Temperature
Water to Ice	0°		32°	Freezing Point of Water
	-17·8°		0°	Fahrenheits Absolute Zero
	-40°		-40°	°C & °F Temperatures Equal
Boiling Point CO_2	-78°			
				Cryogenic Temperature Range Starts at -100° to -150° C
Boiling Point Oxygen	-183°			
Boiling Point Nitrogen	-196°			-270° C Lowest Temperature Existing in Universe
Boiling Point Neon	-246°			
Boiling Point Hydrogen	-253°			12uK° Lowest Temperature Achieved in Laboratory
Boiling Point Helium	-269°			Twelve Millionth of °K
Absolute Temperature	-273·15° or 0° K		-459·67° or 0°r	

APPENDIX 3 MESH/MICRON RATING

Mesh	Holes Per Sq ins	Holes Per Sq cm	Aperture		Micron Rating	Open Area %	Wire Diameter	
			Inch	mm			mm	Inch
2	4	0.6	0.39	10	10000	64	2.5	0.1
2½	7	1	0.31	8	8000	68	2	0.08
3	9	1.4	0.27	7.1	7100	70	1.4	0.055
3½	12	1.8	0.24	6.3	6300	72	1.25	0.05
4	16	2.5	0.2	5.1	5100	65	1.219	0.048
5	25	4	0.15	4	4000	64	1.106	0.04
6	36	6	0.13	3.3	3300	61	0.914	0.036
8	64	10	0.097	2.5	2500	60	0.71	0.028
10	100	15	0.078	2	2000	58	0.56	0.022
12	144	22	0.058	1.5	1557	54	0.56	0.022
16	256	40	0.044	1.13	1131	51	0.46	0.018
18	324	50	0.041	1.07	1075	50	0.38	0.015
20	400	62	0.034	0.89	894	48	0.34	0.014
24	576	89	0.027	0.71	713	45	0.34	0.014
28	784	121	0.024	0.63	633	44	0.27	0.011
30	900	139	0.022	0.57	572	43	0.27	0.011
36	1296	200	0.017	0.45	451	41	0.25	0.01
40	1600	248	0.015	0.4	401	40	0.23	0.0092
50	2500	387	0.012	0.31	315	38	0.19	0.0076
60	3600	560	0.009	0.25	251	35	0.17	0.0068
70	4900	759	0.008	0.21	210	34	0.15	0.006
80	6400	995	0.007	0.18	185	34	0.13	0.0052
100	10000	1550	0.0058	0.15	152	34	0.1	0.004
120	14400	2240	0.0046	0.12	120	34	0.09	0.0036
130	16900	2620	0.0043	0.11	112	33	0.08	0.0032
140	19600	3038	0.004	0.105	110	33	0.07	0.0028
150	22500	3481	0.0039	0.1	100	33	0.07	0.0028
165	27225	4220	0.0035	0.09	90	32	0.06	0.0024
180	32400	5020	0.0031	0.08	80	32	0.06	0.0024
200	40000	6200	0.0029	0.076	76	32	0.051	0.002
230	52900	8200	0.0027	0.075	75	31	0.04	0.0016
250	62500	9680	0.0023	0.061	61	31	0.04	0.0016
270	72900	11300	0.0021	0.056	56	30	0.04	0.0016
300	90000	13924	0.0019	0.05	50	29	0.036	0.0014
325	105625	16375	0.0018	0.046	46	28	0.036	0.0014
350	122500	18990	0.0016	0.042	42	27	0.03	0.0012
400	160000	24800	0.0015	0.038	38	26	0.025	0.001
500	250000	38750	0.0009	0.025	25	25	0.025	0.001

1 micron = 0.001 mm (0.0000393 inch).

25 microns = 0.001 inch approximately.

The term 'mesh' which is commonly used has been derived from the old imperial unit of measurement of mesh per inch, which is the square root of the number of perforations per square inch; hence a mesh rating of 10 would be 10 × 10 = 100 holes per square inch of filtration area.

The term 'micron' is used in the metric system of measurement to denote one millionth of the main unit, hence micrometre (μm) is used to denote one millionth of a metre. In filtration terms the micron rating relates directly to the aperture in millimetres, which is 1000 times larger than the hole size, hence an aperture of 0.1mm will have a micron rating of 100.

APPENDIX 4 COPPER TUBE BS EN1057 (INTRODUCED 15/8/96, FORMERLY BS2871)

Nominal Outside Diameter	NOMINAL WALL THICKNESS mm											
	0·5	0·6	0·7	0·8	0·9	1·0	1·1	1·2	1·5	2·0	2·5	3·0
6	o Z	*RXW*				R*						
8	o Z	*RXW*		*R Y*		R*						
10	o Z	*R X*	*R W*	*R Y*		R*						
12	o Z	*R X*				R*						
15	o Z		*R X*	*R Y*								
18*						*R Y*						
22		o Z		*R Y*		R*			R*			
28		o Z		R*	*R X*	R*		*R Y*	R*			
35				R*		R*						
42					*R X*			*R Y*	*R Y*			
54								*R X*				
64*									*R Y*	*R Y*		
66·7				o Z	o Z			*R X*	R*	R*		
76·1						o Z		*R X*		o Y		
88·9*										*R Y*		
108									*R X*	R*		
133*								*R X*			*R Y*	R*
159								o Z				R*
219*									*R X*			R*
267*								o Z	R* / o Z	*R X*		R*

*Contained in BS EN1057, but generally not available in UK except as specials.
R Recommended in BS EN1057 in R220, R250 and R290 tempers.
o Not recommended in BS EN1057, but may still be obtained to order.
X Available as half hard (bendable) temper R250.
Y Available as annealed (soft) temper R220 and half hard temper R250.
Z Available as hard (thin wall) temper R290.
W Available as annealed (soft) temper R220.

RXY Same as BS 2871

APPENDIX 5 DIMENSIONAL TOLERANCES OF LOW CARBON MILD STEEL TUBE CONFORMING
TO BS EN10255, 2004. FORMERLY BS1387

Lightweight Tube (L Green and L1 White)

Nominal in	Dia mm	EN10255 Specification mm	L Wall Thickness mm	L1 Wall Thickness mm
¼	8	13.5	2.0	2.0
⅜	10	17.2	2.0	2.0
½	15	21.3	2.3	2.3
¾	20	26.9	2.3	2.3
1	25	33.7	2.9	2.9
1¼	32	42.4	2.9	2.9
1½	40	48.3	2.9	2.9
2	50	60.3	3.2	3.2
2½	65	76.1	3.2	3.2
3	80	88.9	3.2	3.6
4	100	114.3	3.6	4.0
5	125	139.7	4.5	–
6	150	165.1	4.5	–

Grades L and L1 tube are contained in BS EN10255, but not normally used in the UK.

Lightweight Tube (L2 Brown)

Nominal in	Dia mm	EN10255 Specification mm	Outside Max mm	Dia Min mm	Wall Thickness mm	Weight Kg/m
¼	8	13.5	13.6	13.2	1.8	0.515
⅜	10	17.2	17.1	16.7	1.8	0.670
½	15	21.3	21.4	21.0	2.0	0.947
¾	20	26.9	26.9	26.4	2.3	1.38
1	25	33.7	33.8	33.2	2.6	1.98
1¼	32	42.4	42.5	41.9	2.6	2.54
1½	40	48.3	48.4	47.8	2.9	3.23
2	50	60.3	60.2	59.6	2.9	4.08
2½	65	76.1	76.0	75.2	3.3	5.71
3	80	88.9	88.7	87.9	3.2	6.72
4	100	114.3	113.9	113.0	3.6	9.75

Medium Weight Tube (M Blue)

Nominal in	Dia mm	EN10255 Specification mm	Outside Max mm	Dia Min mm	Wall Thickness mm	Weight Kg/m	Capacity L/m
¼	8	13.5	13.9	13.3	2.3	0.641	0.0622
⅜	10	17.2	17.4	16.8	2.3	0.839	0.1208
½	15	21.3	21.7	21.1	2.6	1.21	0.2035
¾	20	26.9	27.2	26.6	2.6	1.56	0.3665
1	25	33.7	34.2	33.4	3.2	2.41	0.5855
1¼	32	42.4	42.9	42.1	3.2	3.10	1.018
1½	40	48.3	48.8	48.0	3.2	3.57	1.38
2	50	60.3	60.8	59.8	3.6	5.03	2.225
2½	65	76.1	76.6	75.4	3.6	6.43	3.70
3	80	88.9	89.5	88.1	4.0	8.37	5.12
4	100	114.3	114.9	113.3	4.5	12.2	8.68
5	125	139.7	140.6	138.7	5.0	16.6	13.26
6	150	165.1	166.1	164.1	5.0	19.7	18.97

Heavyweight Tube (H Red)

Nominal in	Dia mm	EN10255 Specification mm	Outside Max mm	Dia Min mm	Wall Thickness mm	Weight Kg/m	Capacity L/m
¼	8	13.5	13.9	13.3	2.9	0.765	0.048
⅜	10	17.2	17.4	16.8	2.9	1.02	0.10
½	15	21.3	21.7	21.1	3.2	1.44	0.175
¾	20	26.9	27.2	26.6	3.2	1.87	0.327
1	25	33.7	34.2	33.4	4.0	2.94	0.52
1¼	32	42.4	42.9	42.1	4.0	3.80	0.93
1½	40	48.3	48.8	48.0	4.0	4.38	1.276
2	50	60.3	60.8	59.8	4.5	6.19	2.07
2½	65	76.1	76.6	75.4	4.5	7.93	3.53
3	80	88.9	89.5	88.1	5.0	10.3	4.92
4	100	114.3	114.9	113.3	5.4	14.5	8.38
5	125	139.7	140.6	138.7	5.4	17.9	13.04
6	150	165.1	166.1	164.1	5.4	21.3	18.7

Tube supplied with plain or screwed ends complete with a steel female threaded socket.
Tube is supplied with plain black steel finish, varnished, epoxy coated or galvanised to protect against corrosion.

APPENDIX 6 HYDROSTATIC DATA

1 litre of water weighs 1 kg at 4.4°C
1 cubic metre = 1000 litres = 1000 kg = 1 tonne at 4.4°C
Head of water in metres $\times$ 9810 = pressure in $N\,m^{-2}$
Head of water in metres $\times$ 9.81 = pressure in $kN\,m^{-2}$
Pressure in $kN\,m^{-2} \times 0.102$ = head of water in metres
1 bar pressure = 1000 mbar = 100 000 $N\,m^{-2}$
1 millibar (mbar) = 100 $N\,m^{-2}$
1 bar pressure = 10.2 head of water in metres
Atmospheric pressure at sea level = 101 $kN\,m^{-2}$ = 10.33 m head of water

APPENDIX 7 COMPOSITION OF COPPER ALLOYS (COMMON)

Classification	Old BS Designation	Common Alloy Name	Copper	Zinc	Tin	Lead	Nickel	Aluminium	Manganese	Phosphorus	Antimony	Iron	Silver	ISO Designation
Bronze or Gilding Metal	CZ125	Cap Copper	95	5										CuZn5
	CA101	Aluminium Bronze	90					10						CA104
		Copper Coinage	95·5	1·5	3									
	PB101	Phosphor Bronze	90		9·5					0·5				CuSn4/8
	CT1	Gunmetal	90·5		9·5									G-CuSn
	LG1/4	Leaded Gunmetal	85	5	5	5								G-CuSnPb
	G1	Admiralty Gunmetal	88	2	10									G-CuSnZn
		Bronze Casting Metal	88	10	2									
		Bell Metal	80		20									
	CZ103 80%	Gilding Metal	80	20										CuZn20
Alpha Brass		English Brass	66·7	33·3										
	CZ107	Common Brass	65	35										CuZn35
		Dutch Brass	75	25										
	CZ106	Cartridge Brass	70	30										CuZn30
	CZ108	Basis Brass	63	37										CuZn37
	CZ111	Admiralty Brass	70	29	1									CuZn28Sn1
	CZ110	Aluminium Brass	76	22				2						CuZn20AL12
	63%	Pot Metal	71·7			28·3								
Duplex Alpha Beta Brass	CZ132	DZR Brass	62	36		2								CuZnPbAg
	CZ123	Yellow or Muntz Metal	60	39·5		0·5								CuZn40
	CZ120	Machine Brass	58	40		2								CuZnPb
		Gedges Metal	60	38								2	0·1	
	CZ112	Navel Brass	60	39	1									CuZn38Sn1
		Manganese Brass	58	40					2					
	NS108 55%	German Silver	55·2	24·1			20·7							CuNiZn
		Monel Metal	30				70							
		White Brass	4		96									Not
		Babbitts Metal	3·7	8·3	88									Copper
		Britannia Metal	2	0·5	89·5						8			Alloys

APPENDIX 8 COMPOSITION OF SOFT SOLDERS

General	BS EN29453	Composition %	Melting	Temp °C	Comments
Group	Alloy No	(Volume)	Solid	Liquid	
Tin-Lead	1 and 1a	Sn63 Pb37	183*	183	Tinmans Solder
Alloys	2 and 2a	Sn60 Pb40	183	190	
	3 and 3a	Pb50 Sn50	183	215	Fine Soldering
	4	Pb55 Sn45	183	226	
	5	Pb60 Sn40	183	235	Fine Soldering
	6	Pb65 Sn35	183	245	
	7	Pb70 Sn30	183	255	Plumbers Wiping Solder
	8	Pb90 Sn10	268	302	
	9	Pb92 Sn8	280	305	
	10	Pb98 Sn2	320	325	
Tin-Lead	11	Sn63 Pb37 Sb	183*	183	}Contains between 0.12 to 0.5 of
Antimony	12	Sn60 Pb40 Sb	183	190	}antimony
Alloys	13	Pb50 Sn50 Sb	183	216	}
	14	Pb58 Sn40 Sb2	185	231	
	15	Pb69 Sn30 Sb1	185	250	
	16	Pb74 Sn25 Sb1	185	263	
	17	Pb78 Sn20 Sb2	185	270	
Tin-Antimony	18	Sn95 Sb5	230	240	
Tin-Lead	19	Sn60 Pb38 Bi2	180	185	
Bismuth	20	Pb49 Sn48 Bi3	178	205	
Alloys	21	Bi57 Sn43	138*	138	
Tin-Lead-Cadmium	22	Sn50 Pb32 Cd18	145*	145	
Tin-Copper	23	Sn99 Cu1	230	240	Capillary Fittings
Alloys	24	Sn97 Cu3	230	250	Capillary Fittings
Tin-Lead	25	Sn60 Pb38 Cu2	183	190	
Copper Alloys	26	Sn50 Pb49 Cu1	183	215	
Tin-Indium	27	Sn50 In50	117	125	
Tin- Silver	28	Sn96 Ag4	221*	221	
Lead Alloys	29	Sn97 Ag3	221	230	
	30	Sn62 Pb36 Ag2	178	190	
	31	Sn60 Pb36 Ag4	178	180	
	32	Pb98 Ag2	304	305	
	33	Pb95 Ag5	304	365	
	34	Pb93 Sn5 Ag2	296	301	

1a, 2a and 3a have less antimony and bismuth impurities than alloy numbers 1, 2 and 3.
*Eutectic solders

Leadless solders suitable for wholesome (potable) water pipe joints
Pb – Lead, Bi – Bismuth, Ag – Silver, As – Arsenic, Sn – Tin, Cd – Cadmium, Zn – Zinc, Fe – Iron, Sb – Antimony, Cu – Copper,
Al – Aluminium, In – Indium.
All solders can contain up to 0.2% of impurities made up of Pb, Sn, Sb, Bi, Cd, Cu, Ag, As, Zn, Al, Fe and In.

APPENDIX 9 SI PREFIXES

Multiplication Factor			Prefix	Symbol
1 000 000 000 000 000 000	=	10^{18}	exa	E
1 000 000 000 000 000	=	10^{15}	peta	P
1 000 000 000 000	=	10^{12}	tera	T
1 000 000 000	=	10^{9}	giga	G
1 000 000	=	10^{6}	mega	M
1 000	=	10^{3}	kilo	k
100	=	10^{2}	hecto*	h
+10	=	10^{1}	deka*	da
UNIT 1				
−0.1	=	10^{-1}	deci*	d
0.01	=	10^{-2}	centi*	c
0.001	=	10^{-3}	milli	m
0.000 001	=	10^{-6}	micro	μ
0.000 000 001	=	10^{-9}	nano	n
0.000 000 000 001	=	10^{-12}	pico	p
0.000 000 000 000 001	=	10^{-15}	femto	f
0.000 000 000 000 000 001	=	10^{-18}	atto	a

*Not SI recognised units and should not be used for engineering purposes.

APPENDIX 10 LIGHT GAUGE STAINLESS STEEL TUBE AUSTENITIC TYPE 304 OR 316 (BS EN10312)

Size mm	Max O D mm	Min O D mm	Wall Thickness mm	Max Working Pressure bar
6	6.045	5.94	0.6	260
8	8.045	7.94	0.6	200
10	10.045	9.94	0.6	160
12	12.045	11.94	0.6	130
15	15.045	14.94	0.6	105
18	18.045	17.94	0.7	102
22	22.055	21.95	0.7	83
28	28.055	27.95	0.8	91
35	35.070	34.965	1.0	75
42	42.070	41.965	1.0	69
54	54.090	53.965	1.0	58

APPENDIX 11 ELEMENTS AND CHEMICAL SYMBOLS

Atomic No	Element	Chemical Symbol	Melting Point °C	Boiling Point °C	Atomic No	Element	Chemical Symbol	Melting Point °C	Boiling Point °C	Atomic No	Element	Chemical Symbol	Melting Point °C	Boiling Point °C
1	Hydrogen	H	−259	−252	41	Niobium	Nb	2477	4744	81	Thallium	Ti	303	1457
2	Helium	He	−272	−268	42	Molybdenum	Mo	2610	4640	82	Lead	Pb	327	1725
3	Lithium	Li	180	1342	43	Technetium	Tc	2140	4265	83	Bismuth	Bi	271	1560
4	Beryllium	Be	1287	2471	44	Ruthenium	Ru	2335	3900	84	Polonium	*Po*	254	962
5	Boron	Be	2075	2550	45	Rhodium	Rh	1964	3695	85	Astatine	*At*	302	337
6	Carbon	C	3642	sub	46	Palladium	Pd	1554	2963	86	Radon	*Rn*	−71	−62
7	Nitrogen	N	−210	−196	47	Silver	Ag	961	2162	87	Francium	*Fr*	27	677
8	Oxygen	O	−219	−183	48	Cadmium	Cd	321	765	88	Radium	*Ra*	700	1140
9	Fluorine	F	−220	−188	49	Indium	In	156	2072	89	Actinium	*Ac*	1050	3198
10	Neon	Ne	−249	−246	50	Tin	Sn	232	2270	90	Thorium	*Th*	1750	3850
11	Sodium	Na	98	892	51	Antimony	Sb	630	1587	91	Protactinium	*Pa*	1230	
12	Magnesium	Mg	650	1107	52	Tellurium	Te	450	990	92	Uranium	*U*	1132	3818
13	Aluminium	Al	660	2450	53	Iodine	I	114	183	93	Neptunium	*Np*	637	3900
14	Silicon	Si	1410	2680	54	Xenon	Xe	−102	−108	94	Plutonium	*Pu*	640	3238
15	Phosphorus	P	44	280	55	Cesium	Cs	29	690	95	Americium	*Am*	994	2011
16	Sulphur	S	119	445	56	Barium	Ba	714	1640	96	Curium	*Cm*	1340	3100
17	Chlorine	Cl	−101	−35	57	Lanthanum	La	920	3464	97	Berkelium	*Bk*	1050	
18	Argon	Ar	−189	−183	58	Cerium	Ce	795	3443	98	Californium	*Cf*	900	
19	Potassium	K	64	760	59	Praseodymium	Pr	931	3127	99	Einsteinium	*Es*	860	
20	Calcium	Ca	840	1440	60	Neodymium	Nd	1021	3027	100	Fermium	*Fm*	1527	
21	Scandium	Sc	1540	2730	61	Promethium	*Pm*	1032	3000	101	Mendelevium	*Md*	827	
22	Titanium	Ti	1668	3260	62	Samarium	Sm	1072	1790	102	Nobelium	*No*	827	
23	Vanadium	V	1900	3450	63	Europium	Eu	828	1439	103	Lawrencium	*Lr*	1627	
24	Chromium	Cr	1873	2200	64	Gadolinium	Gd	1312	3000	104	Rutherfordium	*Rf*		
25	Manganese	Mn	1245	2061	65	Terbium	Tb	1336	2800	105	Dubnium	*Db*		
26	Iron	Fe	1536	2861	66	Dysprosium	Dy	1407	2567	106	Seaborgium	*Sg*		
27	Cobalt	Co	1495	2927	67	Holmium	Ho	1461	2600	107	Bohrium	*Bh*		
28	Nickel	Ni	1453	2730	68	Erbium	Er	1497	2868	108	Hassium	*Hs*		
29	Copper	Cu	1083	2562	69	Thulium	Tm	1545	1727	109	Meitnerium	*Mt*		
30	Zinc	Zn	420	906	70	Ytterbium	Yb	819	1196	110	Darmstadtium	*Ds*		
31	Gallium	Ga	30	2240	71	Lutetium	Lu	1663	3327	111	Roentgenium	*Rg*		
32	Germanium	Ge	937	2830	72	Hafnium	Hf	2222	4603	112		*Uub*		
33	Arsenic	As	817	sub	73	Tantalum	Ta	2996	5425	113		*Uut*		
34	Selenium	Se	217	685	74	Tungsten	W	3410	5555	114		*Uuq*		
35	Bromine	Br	−7	58	75	Rhenium	Re	3180	5596	115		*Uup*		
36	Krypton	Kr	−175	−153	76	Osmium	Os	3033	5012	116		*Uuh*		
37	Rubidium	Rb	39	688	77	Iridium	Ir	2446	4428	117		*Uus*		
38	Strontium	Sr	777	1380	78	Platinum	Pt	1769	3825	118		*Uuo*		
39	Yttrium	Y	1510	2927	79	Gold	Au	1063	2856					
40	Zirconium	Zr	1553	3580	80	Mercury	Hg	−38	357					

Sub – Sublimation, change directly from a solid to a vapour when heated.
Radioactive.
Names not yet proposed for elements beyond 111.

APPENDIX 12   BEAUFORT WIND SCALE

Beaufort Number	Description of Wind	Wind Speed		
		m/s	km/hr	mph
0	Calm	0.5	1.8	1.125
1	Light Air	1.5	5.4	3.37
2	Light Breeze	3	10.8	6.75
3	Gentle Breeze	6	21.6	13.5
4	Moderate Breeze	8	28.8	18
5	Fresh Breeze	11	39.6	24.75
6	Strong Breeze	14	50.4	31.5
7	Moderate Gale	17	61.2	38.25
8	Fresh Gale	21	75.6	47.25
9	Strong Gale	24	86.4	54
10	Whole Gale	28	100.8	63
11	Storm Force	32	115.2	72
12	Hurricane	36+	129.6	81+

Meteorologists report wind speeds at an altitude of 10 metres in open country.

Exact conversion $m/s \times 3.6 = \dfrac{km/hr}{1.6} = mph$

1.0 m/s = 3.6 km/hr = 2.25 mph

APPENDIX 13   COMPARISON OF BSP AND NPT THREADS

Pipe Size Inches	Threads per Inch		Gauge Pipe Diameter mm		
	BSP	NPT	BSP	NPT	
½	14	14	21.7	21.33	Compatible
¾	14	14	26.8	26.7	
1	11	11½	33.8	33.6	
1¼	11	11½	42.5	42.3	
1½	11	11½	48.4	50.2	Use With Caution
2	11	11½	60.2	60.2	
2½	11	8			
3	11	8			Not Compatible
4	11	8			
6	11	8			

Heating equipment imported from North America may sometimes be supplied having pipe threads conforming to the US Standard National Pipe Thread (NPT). These are only compatible on ½ inch and ¾ inch British Standard Pipe (BSP) sizes.

The pitch angles of the threads are also different being 55° on BSP threads and 60° on NPT threads.

APPENDIX 14 PROPERTIES OF WATER

Temperature °C	Absolute Vapour Pressure kPa	Specific Heat Capacity kJ/kg	Density kg/m³	Temperature °C	Absolute Vapour Pressure kPa	Specific Heat Capacity kJ/kg	Density kg/m³
0.01	0.61	4.21	999.8	62	21.84	4.1861	982.1
1	0.66	4.2096	999.8	64	23.91	4.1874	981.1
2	0.71	4.2088	999.9	66	26.15	4.1886	979.9
3	0.76	4.2075	999.9	68	28.56	4.1898	978.9
4.4	0.83	4.2051	1000	70	31.16	4.191	977.7
5	0.87	4.204	999.9	72	33.96	4.1921	976.6
6	0.93	4.2019	999.9	74	36.96	4.1934	975.4
7	1	4.1997	999.8	76	40.19	4.1947	974.3
8	1.07	4.1974	999.8	78	43.65	4.1962	973.1
9	1.15	4.1952	999.7	80	47.36	4.198	971.8
10	1.23	4.193	999.7	82	51.33	4.1999	970.6
11	1.31	4.1913	999.6	84	55.57	4.202	969.3
12	1.4	4.1897	999.4	86	60.11	4.204	968
13	1.5	4.1883	999.3	88	64.95	4.206	966.7
14	1.6	4.1871	999.2	90	70.11	4.208	965.3
15	1.7	4.186	999	92	75.61	4.2099	964
16	1.82	4.1852	998.9	94	81.46	4.212	962.7
17	1.94	4.1845	998.7	96	87.69	4.2141	961.2
18	2.06	4.1839	998.6	98	94.3	4.2165	959.8
19	2.2	4.1834	998.4	100	101.3	4.219	958.3
20	2.34	4.183	998.2	102	108.78	4.2217	956.9
21	2.49	4.1826	997.9	104	116.68	4.2246	955.5
22	2.64	4.1821	997.7	106	125.04	4.2274	954
23	2.81	4.1817	997.5	108	133.9	4.2302	952.6
24	2.98	4.1814	997.2	110	143.26	4.233	951
25	3.17	4.181	997	115	169.13	4.24	948.9
26	3.36	4.1806	996.7	120	198.53	4.248	943.1
27	3.56	4.1801	996.5	125	232.18	4.259	939.1
28	3.78	4.1797	996.2	130	270.12	4.27	934.8
29	4	4.1794	995.9	135	313.16	4.28	930.5
30	4.24	4.179	995.6	140	361.36	4.29	926.1
32	4.75	4.1784	995	145	415.63	4.3	921.5
34	5.01	4.1781	994.6	150	475.97	4.32	916.9
36	5.94	4.1781	993.6	155	543.45	4.33	912.2
38	6.62	4.1784	993	160	618.05	4.35	907.4
40	7.38	4.179	992.2	170	792.03	4.38	879.3
42	8.2	4.1798	991.4	180	1002.7	4.42	886.9
44	9.1	4.1806	990.6	190	1255.2	4.46	876
46	10.09	4.1812	989.8	200	1555.1	4.51	864.7
48	11.16	4.1817	988.9	210	1908	4.56	852.8
50	12.33	4.182	988	220	2320.1	4.63	840.3
52	13.61	4.1823	987.2	230	2797	4.7	827.3
54	15	4.1828	986.2	235	3063.5	4.73	820.6
56	16.51	4.1833	985.2	240	3348	4.78	813.6
58	18.15	4.1841	984.3	245	3652.4	4.82	806.5
60	19.92	4.185	983.2	250	3977.6	4.87	799.2

APPENDIX 15 TEMPERATURE CONVERSIONS

°Celsius	°Fahrenheit	°Celsius	°Fahrenheit	°Celsius	°Fahrenheit
−273.15	−459.67	43	109.4	90	194
−200	−328	44	111.2	91	195.8
−100	−148	45	113	92	197.6
−40	−40	46	114.8	93	199.4
0	32	47	116.6	94	201.2
1	33.8	48	118.4	95	203
2	35.6	49	120.2	96	204.8
3	37.4	50	122	97	206.6
4	39.2	51	123.8	98	208.4
5	41	52	125.6	99	210.2
6	42.6	53	127.4	100	212
7	44.6	54	129.2	120	248
8	46.4	55	131	140	284
9	48.2	56	132.8	160	320
10	50	57	134.6	180	356
11	51.8	58	136.4	200	392
12	53.6	59	138.2	250	482
13	55.4	60	140	260	500
14	57.2	61	141.8	300	572
15	59	62	143.6	350	662
16	60.8	63	145.4	400	752
17	62.6	64	147.2	450	842
18	64.4	65	149	500	932
19	66.2	66	150.8	600	1112
20	68	67	152.6	700	1292
21	69.8	68	154.4	760	1400
22	71.6	69	156.2	800	1472
23	73.4	70	158	900	1652
24	75.2	71	159.8	1000	1832
25	77	72	161.6	1100	2012
26	78.8	73	163.4	1200	2192
27	80.6	74	165.2	1300	2372
28	82.4	75	167	1400	2552
29	84.2	76	168.8	1428	2600
30	86	77	171.6	1500	2732
31	87.8	78	172.4	1600	2912
32	89.6	79	174.2	1700	3092
33	91.4	80	176	1760	3200
34	93.2	81	177.8	1800	3272
35	95	82	179.6	1900	3452
36	96.8	83	181.4	1982	3600
37	98.6	84	183.2	2000	3632
38	100.4	85	185	2500	4532
39	102.2	86	186.8	2760	5000
40	104	87	188.6	3000	5432
41	105.8	88	190.4	4000	7232
42	107.6	89	192.2	5000	9032

Conversions   °F to °C = $\frac{5}{9}$ (°F − 32)   °C to °F = $\frac{9}{5}$ °C + 32

APPENDIX 16 METRIC CONVERSION FACTORS

Unit of length

Metre: symbol m,   Sub-unit millimetre: symbol mm

or

$1000\,mm = 1\,m$, $1000\,m = 1\,km$ (kilometre)

Conversion Factors

Inches to mm	$\times 25.4$	mm to inches	$\times 0.03937$
Feet to metres	$\times 0.3048$	Metres to feet	$\times 3.281$
Yards to metres	$\times 0.9144$	Metres to yards	$\times 1.094$
Miles to kilometres	$\times 1.609$	Kilometres to miles	$\times 0.6214$

Unit of area

Square metre, (sq m): symbol m^2 Sub-unit sq mm: symbol mm^2

or

$1\,000\,000\,mm^2 = 1.0\,m^2$ $1\,000\,000\,m^2 = 1\,km^2$

Conversion Factors

Sq ins to sq mm	$\times 645.2$	Sq mm to sq ins	$\times 0.00155$
Sq ft to sq m	$\times 0.0929$	Sq m to sq ft	$\times 10.76$

Unit of volume

Cubic metre: symbol m^3,   Sub-unit cubic mm: symbol mm^3

or

$1\,000\,000\,000\,mm^3 = 1\,m^3$.

Conversion Factors

Cu inch to mm^3	$\times 16390$	mm^3 to cu inch	$\times 0.000061$
Cu ft to m^3	$\times 0.02832$	m^3 to cu ft	$\times 35.31$

Litre

$1.0\,dm^3 = 1$ litre $= 0.001\,m^3$

Conversion Factors

Imp gallons to litres	$\times 4.546$	Litres to imp gallons	$\times 0.22$
US gallons to litres	$\times 3.785$	Litres to US gallons	$\times 0.265$
US gallons to imp gallons	$\times 0.833$	Imp gallons to US gallons	$\times 1.2$

Flow rate

Litres per second L/s: $1.0\,l\,s^{-1} = 1\,kg\,sec^{-1}$

Conversion Factors

Imp gpm to ls^{-1} × 0.07577   L/s to imp gpm × 13.2
US gpm to ls^{-1} × 0.0631   L/s to US gpm × 15.84

Heat energy

Joule: symbol J

Conversion Factor

1 Btu = 1.055 kJ, or 1055 J

Heat flow

Watt: symbol W
1000 watts = 1 kW

Conversion Factors

Btu/hour to watts   × 0.2931   Watts to Btu/hour   × 3.412
Btu/sq ft/hr to $W\,m^{-2}$ × 3.155 $W\,m^{-2}$ to Btu/sq ft/hr × 0.317

Heat transfer coefficient

Btu/ft2/hr/°F to $W\,m^{-2}\,°C-$ × 5.678 $W\,m^{-2}\,°C-$ to Btu/ft2/hr/°F × 0.176

Pressure

Pascal: symbol Pa
Bar: symbol b
Newton per square metre: symbol $N\,m^{-2}$
1 Pa = $1\,N\,m^{-2}$, 1000 Pa = 1 kPa or $1\,kN\,m^{-2}$, 100 000 Pa, or 100 kPa = 1 bar.

Conversion Factor

Lbs/sq in (psi) to bar × 0.06895   bar to psi × 14.5

Mass

Kilogram: symbol kg
1000 grams = 1 kg,   1000 kg = 1 tonne

Conversion Factor

Pounds (lbs) to kg × 0.4536   kg to pounds × 2.205

APPENDIX 17 PRESSURE CONVERSION

Pressure Head Metres	Pressure bar	Pressure kPa	Pressure psi	Pressure Head Metres	Pressure bar	Pressure kPa	Pressure psi
1	0.098	9.8	1.42	35	3.43	343	49.7
2	0.196	19.8	2.84	40	3.92	392	56.8
3	0.294	29.4	4.26	45	4.41	441	63.9
4	0.392	39.2	5.68	50	4.9	490	71
5	0.49	49	7.1	55	5.39	539	78.1
6	0.588	58.8	8.52	60	5.88	588	85.2
7	0.686	68.6	9.94	65	6.37	637	92.3
8	0.784	78.4	11.36	70	6.86	686	99.4
9	0.882	88.2	12.78	75	7.35	735	106.5
10	0.98	98	14.2	80	7.84	784	113.6
15	1.47	147	21.3	85	8.33	833	120.7
20	1.96	196	28.4	90	8.82	882	127.8
25	2.45	245	35.5	95	9.31	931	134.9
30	2.94	294	42.6	100	9.8	980	142

10.2 m head of water equals 1 bar pressure. 10 bar pressure equals 102 metres head of water.

APPENDIX 18 HEAT FLOW CONVERSION

kW	Btu/hr	kW	Btu/hr	kW	Btu/hr
0.001	3.412	11	37 532	35	119 420
0.25	853	12	40 944	40	136 480
0.3	1 023	13	44 356	45	153 540
0.4	1 365	14	47 768	50	170 600
0.5	1 706	15	51 180	55	187 660
0.6	2 047	16	54 592	60	204 720
0.75	2 559	17	58 004	65	221 780
0.8	2 730	18	61 416	70	238 840
0.9	3 071	19	64 828	75	255 900
1	3 412	20	68 240	80	272 960
1.5	5 188	21	71 652	85	290 020
2	6 824	22	75 064	90	307 080
3	10 236	23	78 476	95	324 140
4	13 648	24	81 888	100	341 200
5	17 069	25	85 300	200	682 400
6	20 472	26	88 712	300	1 023 600
7	23 884	27	92 124	400	1 364 800
8	27 296	28	95 536	500	1 706 000
9	30 708	29	98 948	750	2 559 000
10	34 120	30	102 360	1000	3 412 000

APPENDIX 19 APPROXIMATE VISCOSITY CONVERSION

Kinematic Centistokes	Redwood No1 Seconds	Redwood No2 Seconds	Saybolt Universal Seconds	Saybolt Furol Seconds	Degrees Engler	Example
1.00	29.00		31.00		1.00	Water at 20°C, Kerosene at 38°C
2.00	30.20		32.60		1.10	
3.00	32.90		36.00		1.20	Gas Oil at 38°C
4.00	35.50		39.20		1.30	
5.00	38.60	5.28	42.40		1.37	
6.00	41.80	5.51	45.60		1.43	
7.00	43.10	5.60	46.80		1.48	
8.00	46.00	6.03	52.10		1.64	
9.00	48.60	6.34	55.40		1.74	
10.00	51.30	6.66	58.80		1.83	Class E Oil at 38°C
20.50	85.50	10.11	100.00		3.02	
34.00	136.00	15.39	160.00		4.64	
43.00	170.00	18.90	200.00		5.92	
65.00	255.00	28.00	300.00	32.70	8.79	
108.00	423.00	46.20	500.00	52.30	14.60	
151.00	592.00	64.60	700.00	72.00	20.44	
194.00	763.00	83.00	900.00	92.10	26.28	
216.00	846.00	92.30	1000.00	102.10	29.20	Class F Oil at 38°C
259.00	1016.00	111.00	1200.00	122.00	35.10	
302.00	1185.00	129.00	1400.00	143.00	40.90	
345.00	1354.00	148.00	1600.00	163.00	46.70	
432.00	1693.00	185.00	2000.00	204.00	58.40	
540.00	2115.00	231.00	2500.00	254.00	73.00	
648.00	2538.00	277.00	3000.00	305.00	87.60	
755.00	2961.00	323.00	3500.00	356.00	102.00	Class G Oil at 38°C
863.00	3385.00	369.00	4000.00	408.00	117.00	
971.00	3807.00	415.00	4500.00	458.00	131.00	
1079.00	4230.00	461.00	5000.00	509.00	146.00	
1295.00	5080.00	554.00	6000.00	610.00	175.00	

APPENDIX 20 VISCOSITY–TEMPERATURE RELATIONSHIP

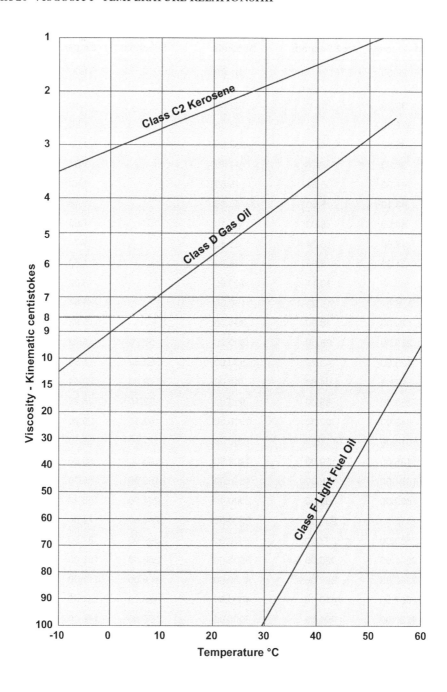

APPENDIX 21 ALTITUDE AND PRESSURE CORRECTIONS

Altitude metres	Pressure kPa	Pressure mm Hg	Average Air Temperature °C	Boiling Point of Water °C
−1500	120.4	903	25	104.8
−1000	114.5	858.7	22	103.5
−250	104.4	783	17	101
0 (Sea Level)	101.3	760	15	100
150	99.4	745.5	14	99.6
250	98.4	738	13	99.4
500	95.5	716.2	12	98.5
750	92.6	694.5	10	97.5
1000	89.9	674.2	8	96.7
1500	84.6	634.5	5	95
1609 (1 Mile)	84.3	632.2	5	94.9
2000	79.5	596.2	2	93.5
3000	70.1	525.7	−4	90
3218 (2 Miles)	68.2	511.5	−6	89.2
4000	61.6	462	−11	86.5
6000	47.2	354	−24	79.9
8000	35.6	267	−37	73.4
10 000*	26.4	198	−50	66.4
20 000	5.5	41.25	−56	34.5
30 000	1.2	9	−60	10

*Stratosphere begins at 8000–15000TS>m above sea level.

Index

Heating Services in Buildings: Design, Installation, Commissioning & Maintenance, First Edition. David E. Watkins.
© 2011 John Wiley & Sons, Ltd. Published 2011 by John Wiley & Sons, Ltd.
As an aid to lecturers and students, full colour versions of the figures in this chapter may be found at www.wiley.com/go/watkins
These figures are © 2011 by John Wiley & Sons, Ltd.

Printed and bound by CPI Group (UK) Ltd, Croydon, CR0 4YY

27/10/2024

14580153-0004